2000 years of a river crc

MoLAS Monograph Series

1 Excavations at the Priory and Hospital of St Mary Spital, London,
Christopher Thomas, Barney Sloane and Christopher Phillpotts
ISBN 1 901992 00 4

2 The National Roman Fabric Reference Collection: a handbook,
Roberta Tomber and John Dore
ISBN 1 901992 01 2

3 The Cross Bones burial ground, Redcross Way, Southwark,
London: archaeological excavations (1991–1998) for the London
Underground Limited Jubilee Line Extension Project,
Megan Brickley and Adrian Miles with Hilary Stainer
ISBN 1 901992 06 3

4 The eastern cemetery of Roman London: excavations 1983–1990,
Bruno Barber and David Bowsher
ISBN 1 901992 09 8

5 The Holocene evolution of the London Thames: archaeological
excavations (1991–1998) for the London Underground Limited,
Jubilee Line Extension Project, Jane Sidell, Keith Wilkinson,
Robert Scaife and Nigel Cameron
ISBN 1 901992 10 1

6 The Limehouse porcelain manufactory: excavations at 108–116
Narrow Street, London, 1990, Kieron Tyler and Roy Stephenson,
with J Victor Owen and Christopher Phillpotts
ISBN 1 901992 16 0

7 Roman defences and medieval industry:
excavations at the Baltic House site 1995–6
Elizabeth Howe
ISBN 1 901992 17 9

8 London bridge: 2000 years of a river crossing,
Bruce Watson, Trevor Brigham and Tony Dyson
ISBN 1 901992 18 7

London bridge

2000 years of a river crossing

Bruce Watson, Trevor Brigham and Tony Dyson

MoLAS Monograph 8

Museum of London Archaeology Service

Published by the Museum of London Archaeology Service

Reprint 2006

A CIP catalogue record for this book is available from the British Library

Production and series design by Tracy Wellman
Typesetting and design by Susan Banks
Reprographics by Andy Chopping
Managing Editor: Sue Hirst/Susan M Wright

Printed by the Lavenham Press,
Lavenham, Suffolk CO10 9RN

Front cover: elm pile foundation for the 15th-century bridge abutment; detail of 1666 painting showing north end of London Bridge during the Great Fire; the modern bridge looking north with the medieval bridge superimposed

Back cover: reconstruction of houses and congestion on medieval London Bridge (P Jackson); collapsed piles from the 11th-century foreshore revetment; the 18th-century modification of the medieval bridge shown during the 'Great Frost' of 1814

CONTRIBUTORS

Principal authors	Bruce Watson, Trevor Brigham, Tony Dyson,
Tree-ring analyses	Ian Tyers
Reconstructions	Peter Jackson
Historical commentaries	Vanessa Harding, Laura Wright
Musicology	Richard Lloyd
Timber studies	Damian Goodburn
Prehistoric pottery	Louise Rayner
Roman pottery	Robin P Symonds, Jo Groves
Medieval and later pottery	Jacqueline Pearce
Accessioned finds	Angela Wardle
Medieval wooden statues	John Cherry
Animal bones	Kevin Rielly
Plant remains	John Giorgi
Human bones	Jan Conheeney
Building materials and geological samples	Ian M Betts
Saga evidence	Jan Ragnar Hagland
Bridge material reused at Stanwell House	Raymond Gill
Graphics	Susan Banks, Hester White, Jeannette van der Post
Photography	Andy Chopping, Maggie Cox
Project managers	Richard Malt, Tracy Wellman
Academic advisers	Nicholas Brooks, David Harrison
Editors	Sue Hirst, Susan M Wright

CONTENTS

FIGURES

TABLES

FOREWORD

Professor Nicholas P Brooks FBA
Department of Medieval History
University of Birmingham

It is a great pleasure to welcome the publication of *London bridge: 2000 years of a river crossing*. The volume marks a fitting and handsome culmination to an outstanding and sustained programme of collaborative research by the Museum of London Archaeology Service (MoLAS). It records in fascinating detail the results of scientific investigations and of archival researches of the highest quality. Most periods of the astonishing history of the bridge have hitherto been obscure, neglected or misunderstood. Here, however, we are presented with a new, wonderfully researched and fully integrated interpretation. It raises our knowledge of every century of the history of London bridge to entirely new levels. The British public have long been dimly aware of the importance and interest of London bridge. We may recall from childhood the nursery rhyme recording its collapse ('London Bridge is broken down') or may have admired one or more of the paintings or engravings of the medieval bridge, with its astonishing superstructure of gates, houses and shops. Such popular images of the great bridge reflect London's role as the capital city of Britain and the significance of its precise location on the River Thames. For many centuries London was both the lowest convenient bridging point on the river and also the highest point to which cargo-laden vessels could easily be brought for unloading. London bridge has, therefore, been the fulcrum of British communications and trade since Roman times. In the development of politics and of the economy, there is no British institution or building of longer-lasting importance than London bridge.

It is an unfortunate fact that bridge studies have, until the last decade, been the Cinderella of archaeological and historical research. Roman bridges were agencies of Romanisation just as much as Roman roads, forts or villas; medieval bridges were as important in the medieval political and economic landscape as castles or churches. Yet, relatively speaking, bridges have attracted much less research. Their neglect is easily understood. River crossings continue in use today. Most bridges have been partially or totally rebuilt in modern forms as traffic has increased. Moreover, the archaeological investigation of earlier bridges normally involves underwater excavation, one of the newest and most expensive forms of archaeology. Bridges have, therefore, been much less accessible in the architectural and archaeological record than castles or churches. In addition the written sources for bridge history are very scattered – save for a very few exceptionally important bridges, like London, which acquired enduring wealth and substantial archives at an early date. Historians more usually have had to assemble the rare notices of particular bridges buried within vast national and local records. Only with difficulty can a coherent story be created from such evidence.

This volume (only the second scholarly history of a major British bridge to have been published) is a pioneering work of the greatest importance. The three principal authors and the team of specialists that they have gathered have not only been able to reconstruct for the first time the fortunes of London bridge over some two millennia, they have also set the bridge within a much wider context, which they have largely had to piece together themselves. Their work is, therefore, of fundamental importance not only for local London history, but also in much wider British and European terms for the development of communications and social power. It also has much to teach us of building techniques in timber and stone, and of the changing roles of states, oligarchies and urban communities in the provision of major public works. This campaign of outstanding research has been superbly conceived and realised. We can justifiably salute the contribution to knowledge that these scholars and the Museum of London Archaeology Service have achieved. An ambitious project has now been splendidly fulfilled.

SUMMARY

London has only developed as a port and city because of the Thames estuary, which offers an excellent navigable routeway stretching from the North Sea westwards far into central England. It was at London, where the channel narrowed and the topography was favourable, that the Romans decided to bridge the tidal estuary. The largely indirect evidence for the existence of the Roman bridge is sufficient to propose a sequence of three bridges, the latest built after *c* AD 90 and before *c* AD 120. The first and second bridges were both probably timber-built, while the third was possibly of composite construction with masonry abutments and piers, and a timber superstructure. The incidence of Roman coins recovered from the River Thames during dredging in 1824–41 suggests that the third phase of bridge might have gone out of use by AD 330.

The date of the replacement of the bridge in the Saxon period is uncertain; it probably did not happen before the late 9th-century reoccupation of London. Documentary evidence indicates that the bridge had been replaced by *c* 1000. The archaeological investigation of the southern bridge abutment site in 1984 revealed a sequence of Saxo-Norman timber bridge structures, the earliest of which was constructed from timber felled *c* AD 987–1032. At least five phases of short-lived timber bridge caissons and abutments appear to have been built between the late 10th or early 11th century and *c* 1160–78, many of them washed away by floods. Throughout the later 11th century the Southwark bridgehead suffered from erosion and repeated attempts were made to stabilise the foreshore by building revetments.

Construction of the first stone bridge by Peter, chaplain of St Mary Colechurch, took place between *c* 1176 and 1209. The southern abutment contained timbers felled during 1187–8, confirming that construction took place during 1189 or perhaps 1190. The stone bridge was 276.09m long, and supported by 19 piers surrounded by starlings. Between the piers were 19 stone arches, one of them spanned by a drawbridge. The original width of the roadway is unclear, but it was probably 6.09m, very likely reduced to 3.66m by the presence of buildings on both sides of the bridge; there is evidence of people living on the bridge by 1281. On the east side of the central pier was the chapel of St Thomas the Martyr. The bridge also formed part of the City's defences: it featured a gatehouse (Stonegate) or barbican on the second pier from the southern end, while the drawbridge spanned the seventh opening from the southern end.

Medieval bridges often fell victim to floods and ice flows but the masonry London Bridge appears to have suffered only two catastrophic collapses during the late medieval period, in 1281–2 and 1437. On both occasions the collapse was preceded by a period of neglect, probably due to financial problems, which may have meant that the routine maintenance programme was reduced. London Bridge had a fundamental design weakness: its piers were all founded on sillbeams and short piles protected from erosion by massive starlings which, to judge from the Bridge House accounts, needed constant repair. The recorded observations of 'cracked' arches before the two collapses occurred strongly suggest that the pier foundations had been undercut by erosion, presumably because of insufficient starling maintenance.

In 1281–2 the accumulation of ice caused five of the bridge's arches to collapse and, in February 1282, the king instructed the mayor and the wealthier citizens to contribute towards the cost of building a temporary wooden bridge. In January 1437, a week after damage to the fabric was recorded, the gatehouse (Stonegate) and its two adjoining arches collapsed. Again, a temporary wooden bridge was constructed across the gap. The 1437 collapse clearly prompted a major rebuilding of the southern end of the bridge, including the enlargement and partial rebuilding of the abutment. Study of the Bridge House records shows that from the early 1460s until well into the 1490s most of the remaining portion of the bridge was rebuilt arch by arch.

By the mid-17th century the expansion of London and Westminster was generating more traffic than London Bridge, with its narrow roadway, could cope with. Thus the roadway was widened during the late 17th century and in 1722 the keep left rule was introduced as an attempt to organise the traffic flow.

Various attempts to build a new bridge at Westminster from 1664 onwards were all opposed by the Corporation of London. However, in 1738 the construction of a new bridge at Westminster began and it was completed in 1750. This new bridge brought a shower of criticism upon London Bridge. There were not sufficient funds available to rebuild London Bridge but it was further modified in 1757–62 – the bridge widened by building across part of the starlings and all the buildings on the bridge demolished to create an even wider roadway and pavements for pedestrians; the removal of one of the central piers created a wider new central arch to improve water flow and navigation.

By the 1820s it was decided to rebuild the bridge, which was in a poor structural condition. Work on a new bridge constructed alongside started in 1824; this included the realignment of the approach roads which resulted in the area of the Southwark Bridge Foot, and the foundations of its adjoining medieval and later buildings, being buried under a new warehouse basement. After the completion of the new bridge in 1831, old London Bridge was demolished. Fragments of the medieval bridge on both the north and south sides of the river escaped destruction and were sealed under the new foundations. The remains of the medieval bridge at the northern end were uncovered during redevelopment in 1921 and 1937, while the remains at the southern or Southwark end were rediscovered in 1983 and archaeologically examined before redevelopment in 1984.

ACKNOWLEDGEMENTS

This publication is the product of a joint venture between English Heritage and the Museum of London Archaeology Service (MoLAS) to publish backlog sites identified in the London post-excavation review (DGLA 1989; see now Hinton and Thomas 1997).

The London Bridge project was funded by English Heritage. The excavations at Fennings and Toppings Wharves in 1984 were supervised for the Department of Greater London Archaeology (DGLA) by George Dennis; the site staff who laboured in difficult and wet conditions were: James Drummond-Murray, Gary Harding, Martin Leyland, Ken MacGowan, Gina Porter, Geoff Potter, Mark Steel, Angus Stephenson (surveyor), Steve Trow and Ian Tyers (dendrochronologist). Thanks are due to Brian Kerr of English Heritage and Gordon Malcolm, Peter Rowsome and Barney Sloane for management of the post-excavation programme. The summary was translated into French by Dominique de Moulins and into German by Friederike Hammer. The index was compiled by Susanne Atkin.

Thanks are due to Alex Bayliss, English Heritage Scientific Dating Co-ordinator, for calibrating the Fennings Wharf and Essex Holocene radiocarbon dates; to Anthea Lycett, Lady Ashcombe's private secretary, for arranging access to the Sudeley Castle statue in 1998 for the purpose of photography and study, and to Jean Bray, Sudeley Castle archivist, for providing the documentary material concerning the acquisition of the statue; to Brian Spencer for commenting on the draft text of Chapter 14.7 and to Alan Pipe (MoLSS) for identifying the animal bones referred to in Chapter 14.10. Tony Dyson wishes to acknowledge once again the assistance of the staffs of the Corporation of London Record Office and Guildhall Library.

The graphics for this publication (except Fig 7) were produced by the MoLAS graphics section; the CAD work on which many of the plans are based was undertaken by Josephine Brown. Fig 7 is reproduced by kind permission of English Heritage and was drawn by Judith Dobie.

The photographs are the work of MoLAS with the following exceptions and acknowedgements. Kind permission to reproduce the view of the Trier bridge (Fig 19) was given by the Rheinisches Landesmuseum, Trier; the view of the Apollodorus bridge (Fig 20) by the British School at Rome; the reconstructions of the medieval bridge on Figs 62 and 68 by the artist, Peter Jackson; the Orleans view of medieval London Bridge (Fig 64) by the British Library; the two portions of the Wyngaerde panorama (Figs 65 and 111) by the Ashmolean Museum of Oxford; the three views of the London Bridge model from the 1996 Royal Academy *Living Bridges* exhibition (Figs 69, 82 and 109) by the Royal Academy of Arts, London, and the model makers, Andrew Ingham and Associates; the photograph of the statue of St Benedict (Fig 80) by courtesy of the Trustees of the British Museum; the photograph of the Sudeley Castle statue (Fig 81) by Lady Ashcombe; the view of the formwork of the central arch (Fig 121) by courtesy of the Trustees of Sir John Soane's Museum; Cooke's views of the demolition of medieval London Bridge (Figs 125, 127 and 129) by the Corporation of London Guildhall Library; the view of Herne Bay pier (Fig 134) by Harold Gough of the Herne Bay Record Society; and the view of the Stanwell House balustrade (Fig 135) by Raymond Gill of the Barnes and Mortlake Historical Society. Figs 75, 120, 131, 132 and 133 are the work of Ryszard Bartkowiak and are reproduced by kind permission of the varous owners. Fig 146 is by Ian Tyers. The following images were supplied by the Museum of London Picture Library (MoLPL): Figs 53, 67, 70, 71, 72, 79, 116, 118, 119, 122, 123 and 128.

'The Knight in the triumph of his heart made several reflections on the greatness of the *British* Nation; as, that one *Englishman* could beat three Frenchmen; that we cou'd never be in danger of Popery so long as we took care of our fleet; that the *Thames* was the noblest river in Europe; that *London Bridge* was a greater piece of work than any of the Seven Wonders of the World; with many other honest prejudices that naturally cleave to the heart of true *Englishman*.'

Joseph Addison (1712)

1

Introduction

Bruce Watson with Ian Tyers

London Bridge is probably the capital's most famous landmark. Countless people have crossed the successive bridges that have spanned the River Thames here throughout the last two millennia. The most famous of these bridges was the great medieval stone structure which stood for over 600 years and has been immortalised in nursery rhymes, songs, paintings and engravings. The theme of the present publication is the bridging of the Thames in London from the Romano-British period until the present day, taking account of both the archaeology and the history of the many London bridges. President Franklin D Roosevelt said in 1931 (Steinman and Watson 1941, epigraph): 'There can be little doubt that in many ways the story of bridge building is the story of civilisation. By it we can readily measure an important part of a people's progress.'

1.1 The archaeological background to London Bridge

Almost all the archaeological data discussed in this report derive from the archaeological investigation of one area of the London borough of Southwark adjoining the east side of the modern bridge abutment (TQ 3284 8036), on a site formerly occupied by the warehouses of Fennings and Toppings Wharves (Fig 1). The area is now occupied by an office block known appropriately as 'Number One London Bridge'. The first archaeological work here was carried out at Toppings Wharf (TQ 3286 8034) in 1970–2 (Sheldon 1974). During this period the southern portion of the wharf was excavated by members of the Southwark Archaeological Excavation Committee (TW70). In 1984, while ground reduction was proceeding on the northern portion of Toppings Wharf, a watching brief was carried out by members of the Museum of London's Department of Greater London Archaeology (DGLA) to record the archaeological deposits and structures within this area of the wharf (TW84) (Fig 2).

Archaeological work started on Fennings Wharf in 1983 when an evaluation trench within the standing building located part of the medieval bridge abutment (FW83) (Young et al 1984, 231–2). From May to September 1984, following the demolition of the warehouse, a programme of archaeological work was carried out on the site (FW84) by members of the DGLA during groundworks (Fig 2). The aim of this archaeological programme was to investigate and record the remains of the medieval bridge and its associated structures before they were destroyed by redevelopment. The southern part of the site (the former warehouse basement) was dug as a standard open-area investigation (areas 1 and 2). Owing to truncation caused by the construction of the 19th-century warehouse basement, the archaeology was largely confined to features dug into the natural geology. At the northern or riverside portion of the site the nature of the archaeological investigation was largely determined by civil engineering

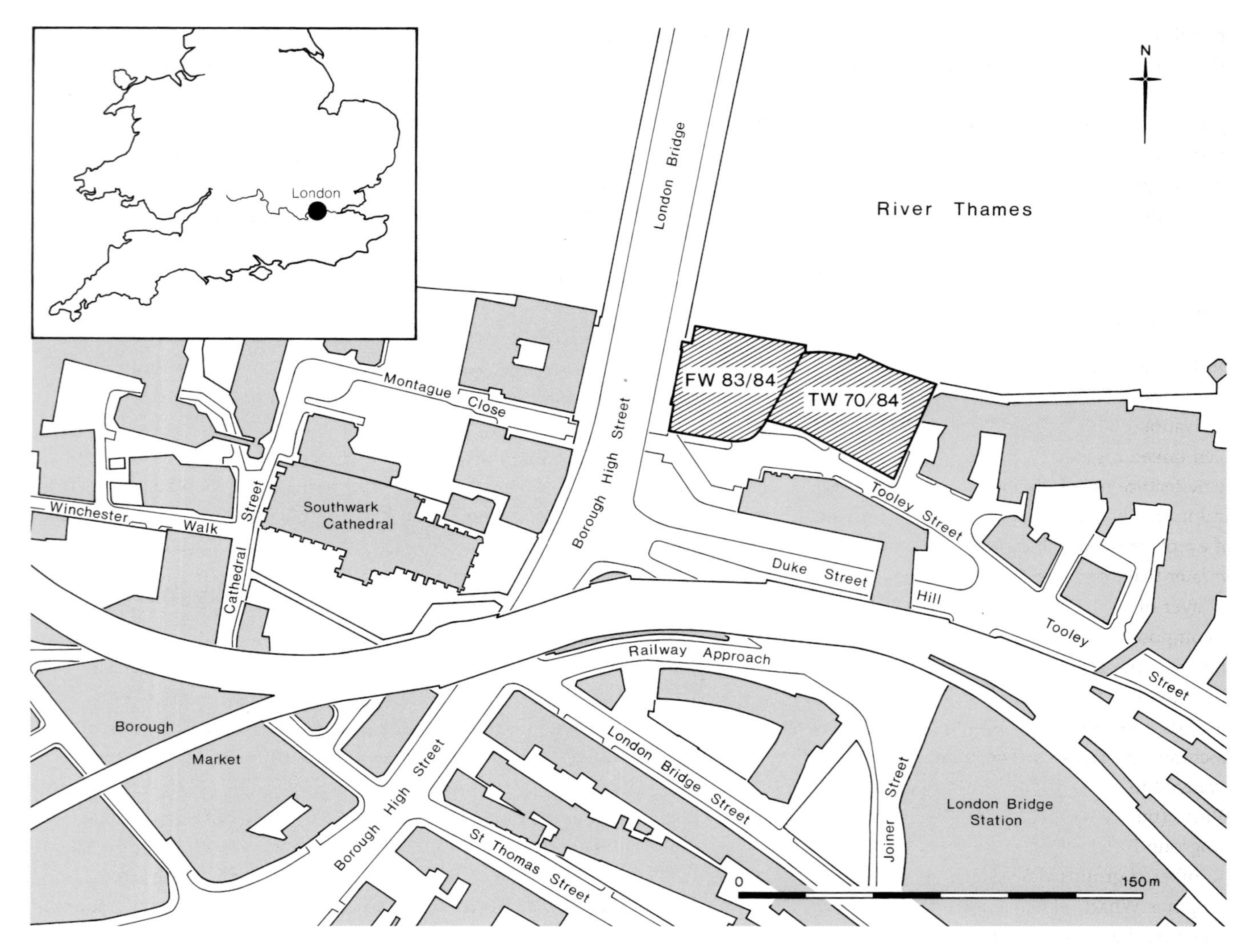

Fig 1 Site location map showing areas of FW83, FW84, TW70 and TW84 within the modern street plan; insert shows location of London within south-east England

constraints. Groundworks were carried out within a series of 5–7m square, closewall-shored 'bays' or trenches (area 3) on the internal side of the modern riverwall (Fig 3). These trenches were machined down by the contractors in spits of several metres, the trench re-shored and then one or two days allowed for archaeological recording before the next spit was removed (Fig 4). Thus it was possible to detect and record all major structures, although not in the detail that controlled excavation would have provided. The working conditions within these trenches were always difficult and frequently noisy and wet. To ensure that the riverwall was properly supported during ground reduction the trenches were dug alternately. This method of working was rather like being given most of the pieces of several large jigsaw puzzles in random order and trying to determine from these the importance, or even the exact number and subject, of the various puzzles. This was in fact impossible until we had all the pieces at the end of the operation. An account of how the post-excavation work was structured has already been published (Watson and Tyers 1995, 63). The excavation on this site was one of the first in London where an attempt was made to sample comprehensively a very large assemblage of timbers, of a variety of species, for subsequent tree-ring dendrochronological and other analysis (see Chapter 14.2). Later, during December 1984, a final excavation (area 4) was undertaken to investigate the area of the Saxo-Norman bridge caissons (Fig 2; Fig 5). An interim account of the sequence of bridge structures was published during the post-excavation project (Watson 1997c).

All finds and site records from the 1970–84 investigations are stored in the Museum of London Archaeological Archive. The research archive is also lodged with the Museum of London and can be consulted by prior arrangement. The tree-ring dendrochronological research archive is currently held at the University of Sheffield dendrochronology laboratory.

1.2 The structure of the publication

Our aim has been to present the archaeological and historical evidence as a single chronological narrative, with reference to specialist data which are presented as a series of appendices (Chapter 14). The chronological information is derived from four sources: stratigraphic analysis, finds analysis, tree-ring dendrochronological analysis and documentary study. Each source has its own strengths and weaknesses, so that the results are not always strictly compatible. The principles behind each are outlined here. The graphical conventions used in this report are shown on Fig 6.

Stratigraphic analysis

During the post-excavation process a relative sequence of events was determined from the site records for each area of excavation or investigation and these strings of events were then linked together. Normally this is done by identifying the same feature or soil deposit in adjoining areas of excavation, and in this way it is possible to build up a site-wide sequence of events from which can be established what events are earlier or later than, or contemporary with, any pit, wall foundation or layer of soil (Watson and Tyers 1995, 64). Such a sequence is composed of stratigraphic 'events' referred to as groups. At this stage the dating of the groups is only relative, pending the provision of absolute or calendar dates from the analysis of finds, tree-ring or dendrochronological samples, and from documentary research. The archaeological data have been arranged in two separate sequences, one for Fennings Wharf (numbered groups with an 'F' prefix, eg group F12, subgroup F12.1 etc) and the other for Toppings Wharf (numbered groups and subgroups with a 'T' prefix). The Toppings Wharf sequence includes all the 1970–2 material, which has been integrated with the rest of the site. In the light of new evidence the opportunity has been taken to reassess and in some cases reinterpret certain aspects of the original publication (Sheldon 1974) (see Chapter 14.5).

Finds analysis

The largest class of dating evidence comprises those objects such as pottery, coins and building materials found within the archaeological deposits as they are being dug away. From the study of their typology such items can be dated with varying degrees of precision, so that dates can be assigned to the deposits which produced the finds. Normally a date range such as 'AD 70–120' can be assigned to the backfilling of a pit, although problems can be caused by the presence of residual or occasionally intrusive material. Almost all the finds dates given in this publication are derived from pottery. Context numbers of finds are denoted in the text as [101], small find numbers as <101> and sample numbers as {101}.

Tree-ring analysis

Ian Tyers

This is the technique of tree-ring dating, which employs the biological process of tree-growth to determine calendar dates for the period during which the sampled trees were alive (English Heritage 1998). The technique is applicable in temperate parts of the world where trees have a growth season, marked as an annual growth ring (see Chapter 14.2). The amount of wood laid down in any one year by most trees is determined by the climate and other environmental factors. Trees over relatively large geographical areas exhibit similar patterns of growth. This enables tree-ring dates chronologists to assign dates to samples by matching the growth pattern with other ring sequences previously linked together to form reference chronologies (Fig 7).

The successful matching of a particular ring sequence with a specific reference chronology, while providing absolute calendar dates for the ring sequence, does not in itself always provide a precise date for the felling of that tree. This is because the full complement of annual rings right to the bark edge is required for the felling date to be established with absolute precision, and these are often missing. There are several reasons why many of the tree-ring dendrochronological samples considered in this report had an incomplete sequence and, as a consequence, cannot be used to provide a precise date for the construction of the structure with which they were associated (see Chapter 14.2). Firstly, the outermost rings of an oak tree, the sapwood, are significantly less resistant to physical damage and to insect and fungal attack than the heartwood and as a result often do not survive. Secondly, sapwood is often removed during the physical conversion of the whole, round log into squared timbers. In such cases, the dendrochronologist can only assign a date after which the timber must have been felled. However, a date range within which the felling took place can be estimated if some sapwood rings are present, based on the assumption that the tree would originally have had a minimum of ten and a maximum of 55 sapwood rings (Hillam et al 1987). Where the full sapwood complement survived on the samples analysed then the date of felling can be determined with absolute historical precision.

However, these dates cannot be taken without qualification as the date at which the associated structure was built. Timber need not have been used immediately after felling, although during the medieval period there is little evidence for its long-term storage and it was normal practice to use 'green' unseasoned timber (Charles and Charles 1995, 46). Full confidence in a particular felling date as an indicator of the date of construction is only possible where it is replicated in other samples from the same structure. The reuse of timbers is a further problem in the dating of waterfront revetments as the impression is that timbers from the previous structure were often salvaged for reuse in its successor. Of course reused timbers are normally identified during recording on site, by features such as the presence of relict joints (Brigham 1992b), but simple beams are harder to identify when reused.

Radiocarbon determinations

Many of the prehistoric deposits and structures discussed in Chapter 2.2 are dated by radiocarbon dates. To allow these

Fig 2 Plan of Fennings and Toppings Wharves showing areas of archaeological investigation and extent of the original and enlarged masonry bridge abutment

dates to be linked with other sources of dating evidence they have to be converted or calibrated from radiocarbon years into calendar years. All the radiocarbon determinations cited in this publication have been calibrated using CALIB v2.1 (Stuiver and Reimer 1986). Samples of atmospheric origin have been calibrated using the data sets published by Pearson et al (1986). Samples of marine origin have been calibrated using the data set of Stuiver et al (1986, fig 10B). The date range has been calculated according to the maximum intercept method (Stuiver and Reimer 1986), with calibrated date ranges cited in the text at two sigma (95% confidence) as cal BC. Dates are quoted in the format recommended

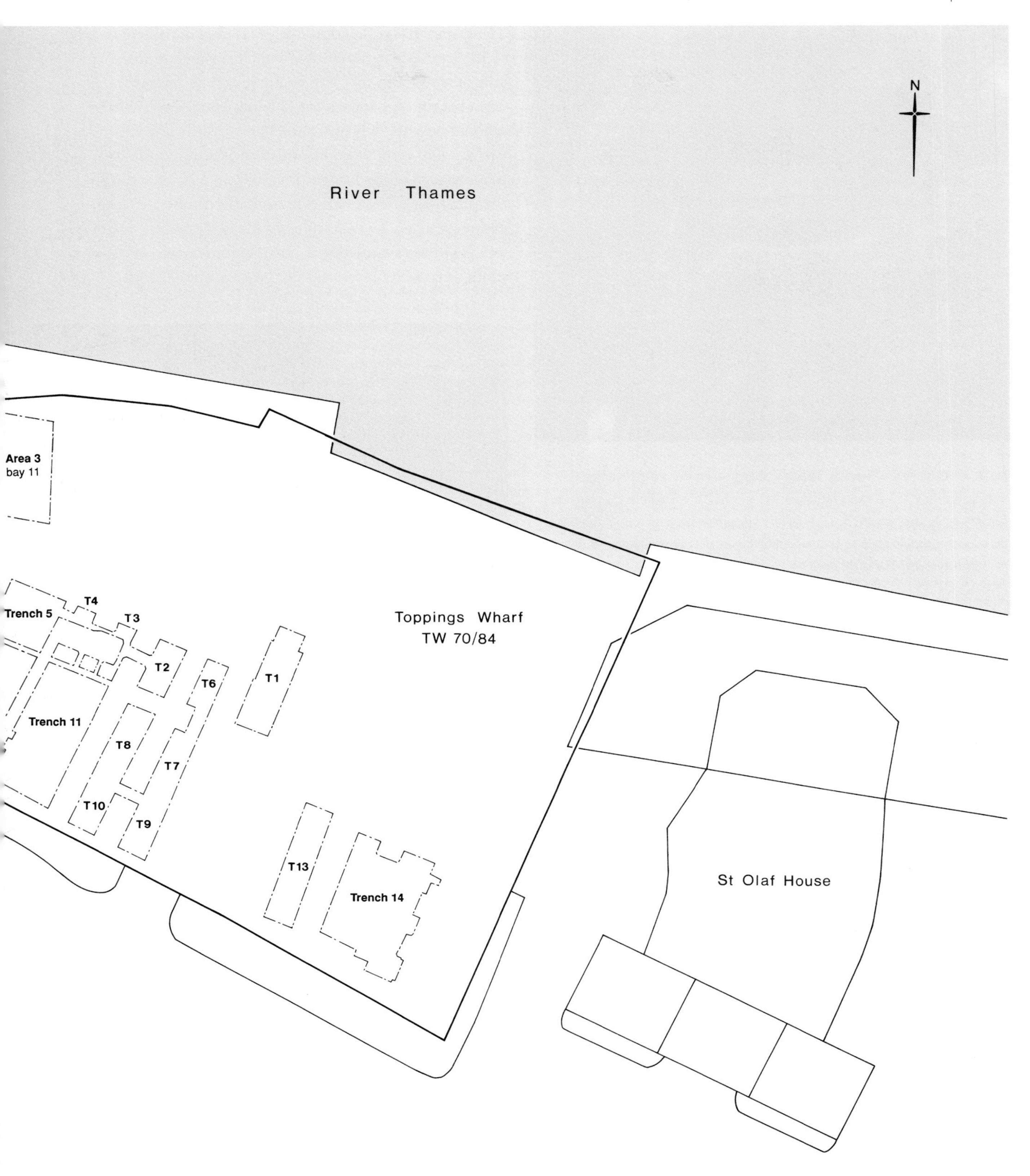

by Mook (1986), with end points rounded outwards to ten years. Interpolated determinations from environmental sequences have been calibrated and are quoted in the form *c* 3770 cal BC (*c* 5000 BP). For further details see the notes to Chapter 2, where the date range in years before present (BP) is also cited.[1]

Documentary analysis

There is a large volume of historical documentation for London, which enables some of the structures and rebuilds located on site to be linked with recorded events, thus helping to confirm or refine the dating evidence from other sources.

The proper name 'London Bridge' is only used here to refer to the medieval stone bridge, which was begun in *c* 1176 and survived into the 19th century, and its successors. The main source of evidence for London Bridge is the Bridge House accounts, which from 1381 onwards provide a vast amount of detailed information relating to the daily maintenance of the bridge (see Chapter 10), including the purchase of materials, the sources of stone and timber, and wage rates. Analysis of early maps, drawings, paintings and engravings has made it possible to reconstruct the rest of the medieval bridge and the post-medieval properties of

Fig 3 Working view of Fennings Wharf during groundworks and archaeological investigation in 1984, looking east along line of 'bays'; to the right is the upstanding masonry of the west side of the 15th-century bridge abutment; the 'bays' each side of the masonry have already been finished demonstrating the piecemeal nature of the investigation of this area of the site (MoL)

Fig 4 Working view of Fennings Wharf looking north, showing the west wall of the 15th-century bridge abutment being removed during ground reduction (MoL)

Fig 5 Working view of Fennings Wharf, looking west across area 4 during excavation in December 1984 (MoL)

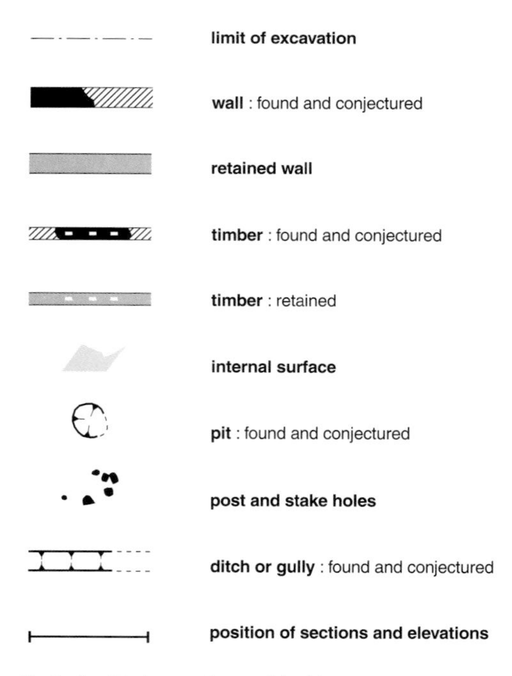

Fig 6 Graphical conventions used in this report

the Southwark bridgehead in considerable detail.

There is documentary evidence for many of the medieval and post-medieval events discussed in the text, such as the construction of the Colechurch bridge or 13th-century erosion of the Southwark foreshore. Where possible these historical events are linked with the relevant archaeological evidence. However, in some instances a precise correlation is not possible, often due to the date range assigned to the relevant archaeological events.

In the text, dates are simply assigned to events and phases of activity, and the rationale and evidence for these dates is discussed in Chapter 14.1.

Before 1971 the UK monetary system consisted of pounds, which were sub-divided into 20 shillings, each of which was further subdivided into twelve *denarii* (popularly known as pence). Since decimalisation in 1971 pounds have been subdivided into 100 pence. All sums of money are quoted in the text as cited in £, s and d, since modern equivalents would be misleading. Dyer (1989, xv) provides the following reminder on which to base an approximation to current values: 'a skilled building worker earned 2d per day in 1250, 4d per day in 1400 and 6d in 1500'. Similarly for weights and measures readers should note that liquids may be cited by the gallon (4.5 litres), which contains eight pints (0.57 litre); dry goods may be cited by the pound (abbreviated to lb) (1lb equals 0.45kg), which comprises 16 ounces (1oz equals 28g). Dimensions of structures may be quoted in feet (1ft equals 0.3m), each of which comprises 12 inches (1in equals 25.4mm).

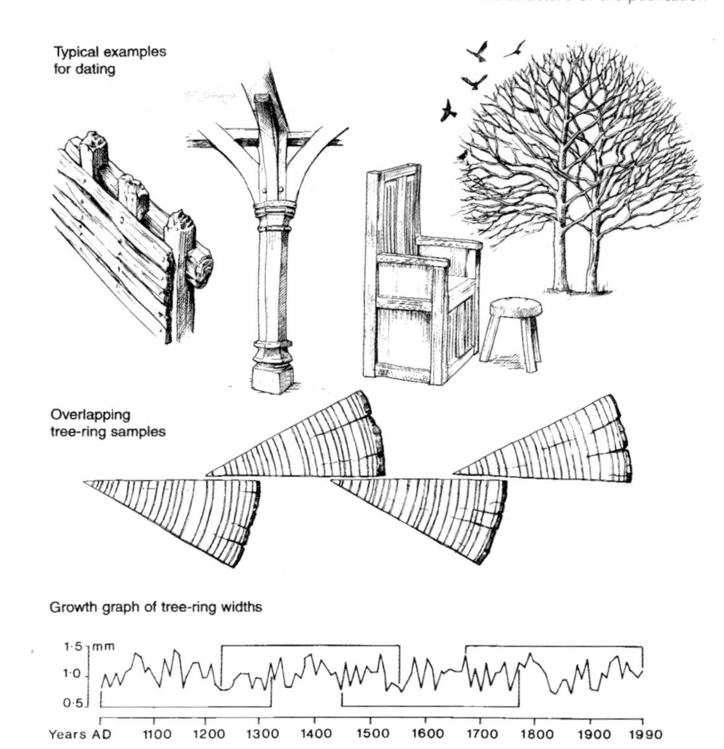

Fig 7 Diagrams showing construction of tree-ring chronology: overlapping tree-ring samples show (to right) the rings from a modern oak tree which was felled at a known date, with (to left) a chronology constructed by overlapping the growth ring patterns from successive older samples; the source of these older timbers can be items of furniture, medieval roofs or archaeologically excavated structures such as waterfront revetments (top left) (© English Heritage 1998, fig 2; illustrator Judith Dobie)

2

The Thames and the Southwark waterfront in the pre-Roman period

Trevor Brigham

2.1 Geology and natural topography of Southwark

The study area of the Roman and medieval bridges is located on the south bank of the River Thames, immediately downstream of the present London Bridge (Fig 1). The area is naturally low-lying, within the flood plain of the river. The underlying Tertiary geology is blue-grey Palaeocene London Clay, with a surface at between -5.5 and -6.0m OD, sealed by a series of islands formed from an eroded 7.0m thick layer of Pleistocene flood plain gravels and sands.

The island of which the study area forms part consists of gravels generally around 1.2m OD, rising to 1.3m OD near the bridge approach (District Heating Scheme Areas B–C, Graham 1988). A study of the topography around the bridgehead allows the line of the pre-Roman riverbank to be traced to the east at Toppings Wharf and Fennings Wharf, to the west at New Hibernia Wharf (Bird and Graham 1978) and further to the south-west in the Winchester Palace (Yule in prep; Yule and Rankov 1998) and Courage's Brewery study areas (Cowan in prep). On these sites, the gravels are generally slightly lower, at 1.0–1.2m OD in the south, falling away towards the north. At Fennings Wharf (F1.1) there is a rapid drop to below -1.5m OD, a trend which continues around the eastern fringes of the island to the so-called 'Guy's Channel', which passed close to the Toppings Wharf site. Here (T1.1), the gravels fall from about 0.8m OD in the west to Ordnance Datum on the east side of the site, continuing below -1.2m OD near the north end of the channel (District Heating Scheme Area G, Graham 1988).

2.2 The study area: prehistoric activity

At Toppings Wharf and Fennings Wharf (groups T1.2, F1.2: not illustrated) sandy prehistoric subsoils formed on the gravels. At Toppings Wharf this horizon was *c* 0.2m deep, with a surface at up to 1.0m OD in a band across the south end of the site, falling away to below 0.54m OD further north, where it was probably eroded by the river. The deposit contained abraded Late Bronze Age or Early Iron Age pottery (*c* 1500–100 BC) and some undatable flint blades and blade fragments, which confirm a prehistoric formation date. At Fennings Wharf a layer of redeposited natural was overlaid by subsoil with a surface at 0.85–1.18m OD in the south-west (designated as OA1 on Fig 8). This deposit contained no artefacts, but it is assumed to have formed at the same date as the similar horizon at Toppings Wharf.

This period has been dated independently from a number of sites by radiocarbon dating of peats and organic silts which formed in channels and on areas of mudflat beyond the settlement area (Tyers 1988; Sidell et al in prep). Of these, the nearest observation to the west at New Hibernia Wharf

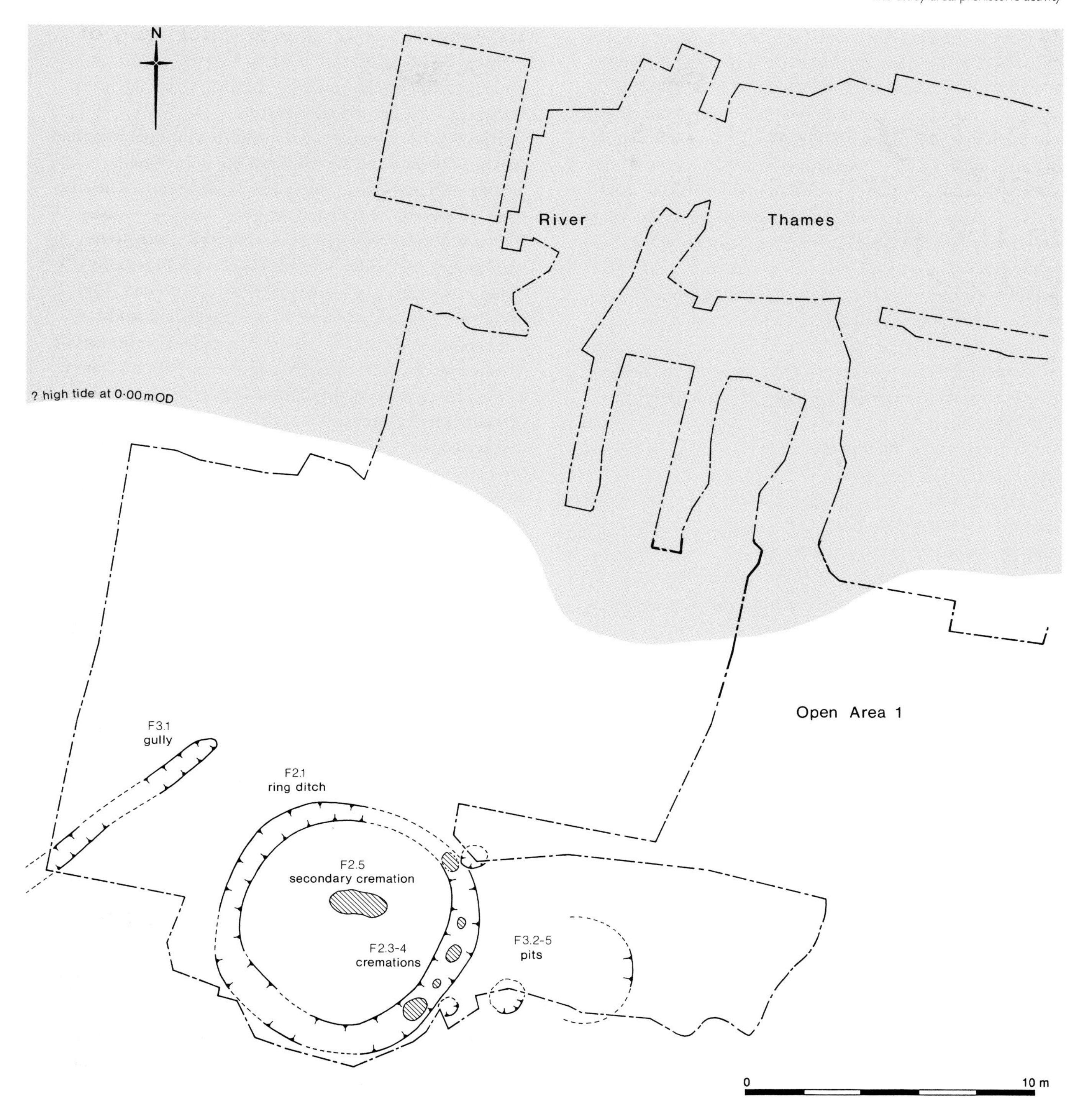

Fig 8 *The prehistoric foreshore and prehistoric features on Fennings Wharf*

included organic deposits dated to 930–540 cal BC.[2] Peats at the Willsons Wharf site to the east of Toppings Wharf were dated to 1430–1030 cal BC, 1155–800 cal BC and 900–410 cal BC.[3] Some of the most recently dated deposits were peats representing fen and alder carr at Bramcote Green, Bermondsey, which formed *c* 1400–1000 cal BC (Thomas and Rackham 1996, 251),[4] and others at Hays, Dagenham, dated to 1380–990 cal BC (Meddens 1996, 326).[5]

At Fennings Wharf, a series of important structural features were uncovered over an exposed area of natural gravels in the south-west (Fig 8). These included a circular ring-ditch at least 1.0m deep, with an internal diameter of 7.0–7.4m (F2.1). The ditch had a narrow V- or U-shaped section, eroded at the top. At least one possible cremation was located within the primary sand and gravel infills, with evidence of *in situ* burning (F2.2; not shown). A further series of spreads of charcoal and cremated bone were located in the secondary silty sand infills (F2.3). The eight radiocarbon dates from these charcoal spreads confirm that the ring-ditch was in use as a cremation cemetery *c* 1900–1600 cal BC.[6] The eastern side of the ditch was subsequently recut and the infill of this area contained another three spreads of charcoal and cremated bone (F2.4).

All these spreads of charcoal and cremated bone are probably derived from the dumping or scattering of material from nearby pyres, as there is little evidence of *in situ* burning. The exact number of individuals represented by the cremations is uncertain, but it appears that there were at least eight children or juveniles and only one adult present (White in prep). The pottery from the ring-ditch consisted of a few small, abraded sherds of fairly undiagnostic pottery, attributed to the second millennium BC. The close range of the available radiocarbon dates indicates that the ring-ditch was only in use for a short period of time, during the early second millennium BC, but the associated pottery indicates a longer period of use or perhaps reuse of the central feature.[7] The flint work recovered from the ring-ditch consisted of 155 fragments, including four blades and one complete scraper. About a tenth of the flints were burnt.

In the centre of the ring-ditch was a 2.04m by 0.8m oval feature containing silty gravel, cremated bone and pottery (F2.5), which may be interpreted as a secondary not a primary cremation as the pottery from its fills dates to the Late Bronze Age/Early Iron Age transition period (*c* 1000–600 cal BC). An area of 'compact silty loam' may represent the last remains of a mound covering this central feature. The angle of the tip lines within the ring-ditch fills suggested that natural processes had probably largely levelled the mound even before the area was inundated by the rising river.

In the vicinity of the ring-ditch, but not stratigraphically related, were several other prehistoric features, dated to the Late Bronze Age (*c* 1000–700 cal BC) by a few flint-tempered pot sherds and three fragments of perforated clay slab. Immediately to the west was a narrow 7.5m long gully aligned north-east to south-west, with no discrete fill (F3.1). To the east was a group of four small pits (groups F3.2–F3.5) infilled with gravels, sands and sandy silts, as well as an area of later subsoil (F3.6: not shown).

This concentration of prehistoric features is important as it demonstrates human activity on the edge of the river in an area that was subsequently inundated. It can be assumed that this area was not considered marginal by the builders of the ring-ditch, although it is unclear how far away the riverbank lay because of aggressive erosion in both the pre- and post-Roman periods. A conservative estimate would locate the Late Bronze Age bank somewhere near the north edge of the site, at least 20m away, but it could have lain considerably further out into the modern river (Sidell et al in prep).

2.3 The study area: pre-Roman fluvial deposit

The existing topography was modified in the pre-Roman period by the deposition of fluvial silts and clays across the study area (not illustrated). This deposition reached 1.4m OD in the south of Fennings Wharf (group F4), and remained consistently at *c* 1.0m OD across the lower-lying area of Toppings Wharf to the east (T2), sealing the prehistoric features and subsoil horizons on these sites and, therefore, marking a retreat of the settled area.

The flood deposits are all thought to have been laid down during a period when the sea level rose relative to the land broadly identifiable as Devoy's Thames IV transgression (Devoy 1979; see also Chapter 3.4 and Glossary). Marginal areas inhabited during the Late Bronze Age and Early Iron Age were inundated, many of them only occupied because an earlier phase of falling sea level (Devoy's Tilbury IV regression) had allowed expansion on to former mudflats, salt marshes and beaches. At Bramcote Green, the onset of the formation of estuarine clay silts has been suggested as *c* 1000 cal BC, continuing on that site until it was reclaimed relatively recently (Thomas and Rackham 1996, 251). It remains unclear whether the clay silts were laid down fairly rapidly over a short period, or slowly over a protracted timescale. They do not show the visible laminations that would imply incremental seasonal deposition, although at Bramcote Green it has been suggested that they represented seasonal flooding of wet pasture (Thomas and Rackham 1996, 249). Tidal and biological disturbance (such as worm action) may in any case remove evidence of internal structure. In Southwark there may have been periods where the estuarine deposits were disturbed, eroded and subsequently relaid in the long Iron Age hiatus between the creation of the latest prehistoric features and the next phase of structural activity in the early Roman period.

Flood clay deposits have been recorded across much of the Southwark settlement, sealing the earlier peats and subsoils, with a surface level in the same range as that noted in the study area: at 1.0–1.4m OD (Yule 1988). Identical deposits have, for example, been recorded at Montague Close (Dawson 1976, 41) and in the Winchester Palace study area (Yule in prep) immediately upstream of the modern bridge, and these can be seen as part of the same horizon. Areas lying above this level seem to have largely escaped the effects of flooding. At 15–23 Southwark Street in the south-western corner of the island, for example, only one small area was affected in the vicinity of a Late Bronze Age or Early Iron Age ditch. It was inundated, leaving flood clays to a level of 1.06m OD (Cowan 1992, 10).

A similar level (1.04m OD) was attained in the Courage's Brewery study area (Cowan in prep) where the clays sealed surfaces and foreshore peats laid down *c* 1000 cal BC (Tyers 1988). Roman pottery has been found in the upper part of the deposits, normally of mid-1st-century date (eg Toppings Wharf, Winchester Palace and Courage's Brewery Area M), but slightly later in other areas (eg Courage's Brewery Area D). It is likely that the earliest pottery was restricted to those areas protected by the first embankments and revetments where settlement began almost immediately. Conversely, the later pottery may come from areas initially beyond the fringes of the settlement, where the clays were not sealed until the late 1st or early 2nd century and may actually have continued to form in the early Roman period.

The name of the river itself may throw some light on its early character. *Tamesis* (Thames) was first documented by Julius Caesar in 51 BC in his account (*De bello Gallico*) of the second invasion of Britain three years previously (Caesar, V). The origin of this Celtic river-name is uncertain, but it is possibly derived from the same linguistic root as a number of other English river-names such as the Tame, Team and Thame, meaning 'dark river' (Ekwall 1928, 402–5). This could be a reference to discoloration by sediment, which would certainly have been characteristic of the Iron Age river, judging from the depth and extent of the pre-Roman flood clays. However, more recent research suggests that Thames is one of a number of river-names based on the root '*ta*-' or '*te*-', meaning 'to flow' (Rivet and Smith 1979, 466).

3

The Thames and the Southwark waterfront in the Roman period

Trevor Brigham

3.1 The eastern area: Toppings Wharf

River embankment: mid-1st century

The resumption of occupation in the study area following the transgressive phase (see Glossary) is represented at Toppings Wharf by an east–west ditch at least 18.0m in length, up to 2.7m wide and 1.2m deep (T3.1), with a base at between -0.1 and -0.2m OD (Fig 9; Fig 13). In the east, this feature was truncated by medieval erosion (see T6/F11), but appears to have survived to its original dimensions for the westernmost 7.0–8.0m. Recorded in section on the northern (riverward) side of the ditch was a truncated gravel bank, which was at least 1.0m wide and 0.2m high with its surviving top at 1.3m OD. In two out of the three sections drawn through the sequence in 1974 a thick layer of clay seems to have been applied to the surviving back of the bank, probably as waterproofing. The bank and ditch are interpreted as a flood embankment, protecting the area to the south from the prevailing high tidal levels suggested by the extensive earlier fluvial deposition (Chapter 2.3). It is assumed that the bank was at least partly composed of upcast from the ditch, and it is, therefore, likely that it contained material of a similar volume, suggesting an original width of *c* 2.5m and a height above the contemporary foreshore of up to *c* 1.0m. This would place the top of the bank at *c* 2.0m OD, a level comparable to that of the contemporary waterfronts on the north bank (Milne 1985; Brigham 1990b; Brigham et al 1996b; Brigham 1998).

Near the western end of Toppings Wharf, a row of posts (T3.2: Fig 9) was driven into the gravels along the southern edge of the bank, and it seems likely that these had originally supported revetting planks or wattlework to support the edge of the feature. The T6/F11 erosion horizon had, however, destroyed any evidence of a similar row of posts on the riverward (northern) side of the embankment, which would have been necessary to protect it from the river (Fig 9). The reconstructed plan of the eastern end of the bank is shown curving towards the south, a trend suggested both by the natural contours and by the shifting alignment of later buildings (see below).

Although the ditch was partly, if not primarily, a linear quarry to provide materials for the embankment, it may also have acted as an intervention drain. In other words, it collected water from the occupied area and any seeping through the bank, and then allowed it to be voided into the river at low tide through a sluice or one-way 'flap valve'. The ditch was, however, infilled with a variety of sands, silts, clays and gravels in preparation for the construction of the first buildings probably between *c* AD 60–80. The date of the embankment and ditch themselves is not known, although the excavator considered that the ditch was not open for very long, since there was little evidence for primary silting. The three available cross sections (Sheldon 1974, figs 4, 6 and 7) (Fig 13; Fig 112) do, however, show considerable irregularity suggestive of erosion, and the lack of primary silting may simply be the result of regular cleaning. It is possible, therefore, that the embankment dates from the initial stages of occupation *c* AD 50–60.[8]

It is notable that no westward continuation of the ditch and bank was observed crossing the eastern part of Fennings Wharf, although the ditch seemed to be narrowing where it reached the slightly higher ground predominant in the area. However, a low embankment could have been constructed on the existing ground surface in the western part of the study area by importing relatively small amounts of gravel from nearby without the need for quarrying.

Occupation of the protected area: mid- to late 1st century

The infilling of the ditch enabled the area south of the embankment to be used for building purposes (Fig 10). A group of six buildings (B1–B6) and their associated alleyways was recorded (T4), aligned with the embankment and of a similar date to the ditch infill (*c* AD 60–80). There is only scattered evidence for early activity antedating these structures, perhaps contemporary with the ditch and bank. This includes several random stakeholes and a small pit below Building 1, a gully between Buildings 1 and 2, into which the later alley gravels subsided and 'pre-Flavian' (ie AD 50–70) pottery below the alleyway and Building 2, and in a pit below Building 5.

The walls of the buildings, particularly Buildings 2 and 3, seem to show a curvature in the general south-easterly alignment trending towards the mouth of Guy's Channel. Although there is no archaeological evidence, it is assumed that the buildings were also aligned with an east–west street some distance further south, and it is interesting to note that the proposed curvature is matched by the modern line of Tooley Street and the 19th-century warehouses which formerly occupied the site. Two of the buildings (B5 and B6) were partly constructed over the backfilled ditch; all of these structures have already been described at length in the original published report (Sheldon 1974, 9–23), and are only summarised here, beginning in the east. The original numbering has been retained, but arabic numerals have been substituted for Roman in line with recent usage.

Building 1

Building 1 (T4.1) appears to have been circular, with an estimated diameter of at least *c* 10.0m (not the 8.5m suggested by Sheldon: see Fig 10), although only the curving west wall was found. The wall was *c* 0.15m thick with stakeholes at intervals, which implies a wattle and daub construction. There was an eavesdrip gully around the perimeter, and within were three successive floors, the earliest of which contained charcoal and burnt materials, and was littered with metalworking debris, mainly iron slag with some bronze. The second and third clay floors had hearths, possibly associated with the metalworking, and there was further slag associated with each, totalling *c* 8.4kg. The latest floors contained pottery of the early 2nd century, but since most of the building was truncated it is not clear how long it remained in use.

It is possible to speculate that the occupant was a metalworker of native origin, attracted to a settlement which was still not fully Romanised, although there is no other evidence to support a connection between building form and ethnicity.[9]

Between Buildings 1 and 2 was a gravelled area with at least five metalled surfaces, forming an alleyway around 3m wide. The lowest surface contained a sherd described as 'pre-Flavian', and, therefore, presumably of *c* AD 50–70.

Building 2

Only around 3.0m of the east wall of Building 2 survived, with the remains of an east–west return at the north end of the main section (T4.2). This building appears to have been supported by substantial posts set *c* 0.6m apart (ie two Roman feet, *pedes monetales*) with wattle and daub infill (Fig 10). The earliest floor was gravel over clay make-up containing more 'pre-Flavian' pottery; later surfaces were truncated by a post-Roman wall, although an area of clay may have been part of an internal wall added later.

Although there was no evidence from either building for activity later than the early 2nd century, brown silts (T5.3) and dark earth deposits (T5.4) containing Roman pottery and rubble were dumped in the intervening alleyway above 1.7m OD. This demonstrates that the buildings – or their successors – probably remained in use through the 2nd century at least.

Between Buildings 2 and 3, an area approximately 14.0m wide was badly damaged or destroyed by post-medieval intrusions, but on the basis of the widths of Buildings 5 and 6 (*c* 5.0–6.0m) it is probable that there was a further property for which there was no other evidence.

Building 3

Two walls of Building 3 were recorded (T4.3). Both were thought to be internal partitions, joined together in the north-west to form an L-shape, and seem to have been narrow wattle and daub structures with five postholes more or less on the wall lines (Fig 10). One was faced with white plaster. The floors were of fire-reddened clay occupying an area at least 4.9m north–south and 2.7m east–west. There was no sign of a north wall, although a horseshoe-shaped hearth was recorded in the open area between the surviving edge of the clay floor and the infilled T3.1 ditch, suggesting that the original wall lay close to the edge of the floor. There was a fragment of clay in the south-west corner of the building which may have formed the only surviving section of the west wall. It may, however, have been shared with the east wall of Building 4, in which case it was probably removed by later alterations to that structure. The building was apparently still occupied in the early 2nd century and later clay layers suggest that it was rebuilt subsequently, without replacing the internal partitions, although these later layers may have been derived from final

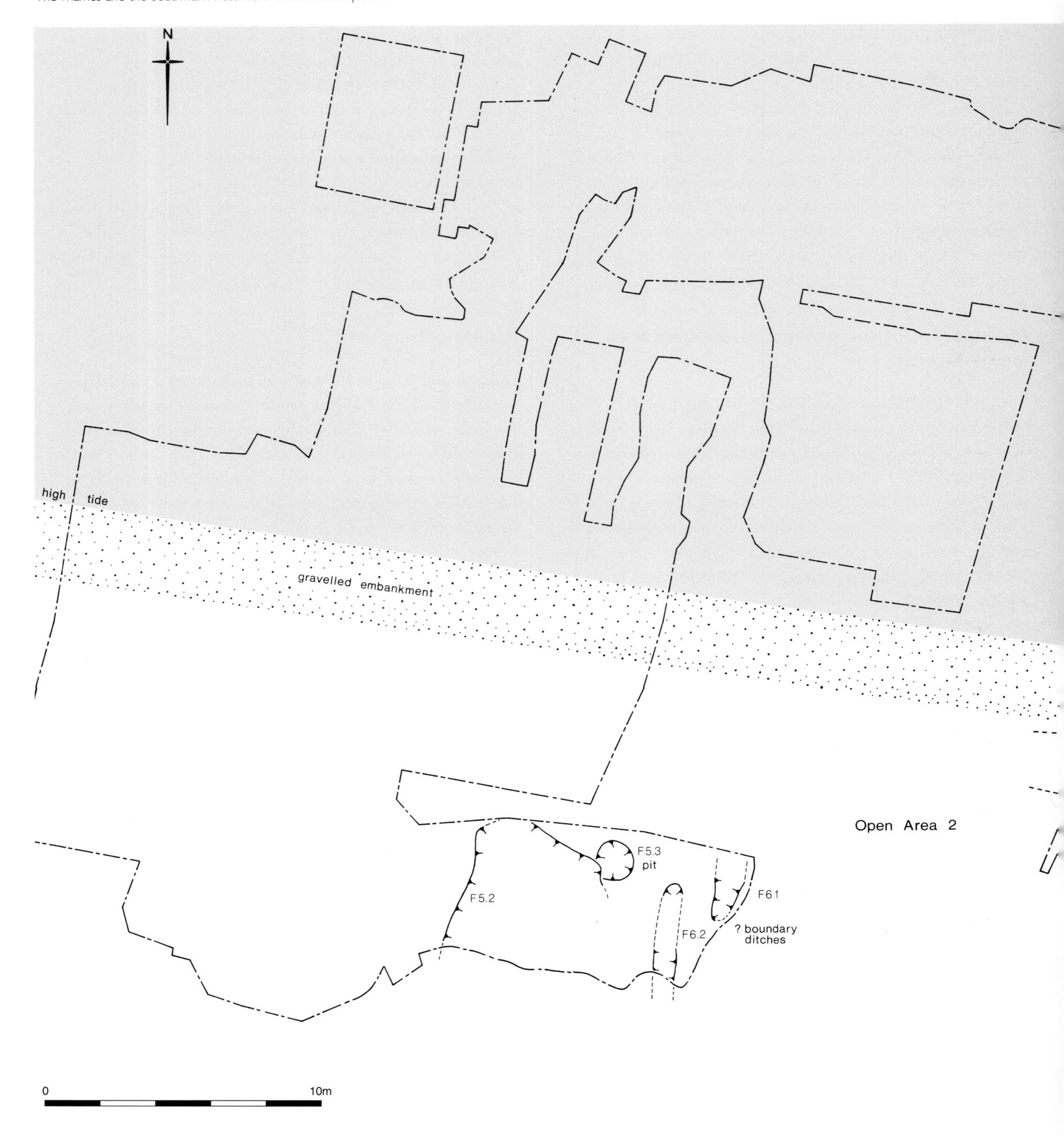

Fig 9 *Plan of primary Roman activity and the construction of the gravel embankment*

demolition since they were sealed by overlying rubbish dumping (T5.3).

Building 4

The neighbouring Building 4 (T4.4) was immediately adjacent to Building 3, but differed from this and the other structures on site in its construction. Three walls survived enclosing an internal area *c* 3.7m wide, 4.5m long. All were massive, up to 0.8m thick, and built of mud-brick, cob, or rammed earth rather than wattle and daub, with some traces of internal plastering (Fig 10). The west wall contained three postholes near the outer edge which may have formed part of an internal supporting framework. The first floor was of gravel, with a hearth beyond the north end at the same level, followed by a clay floor.

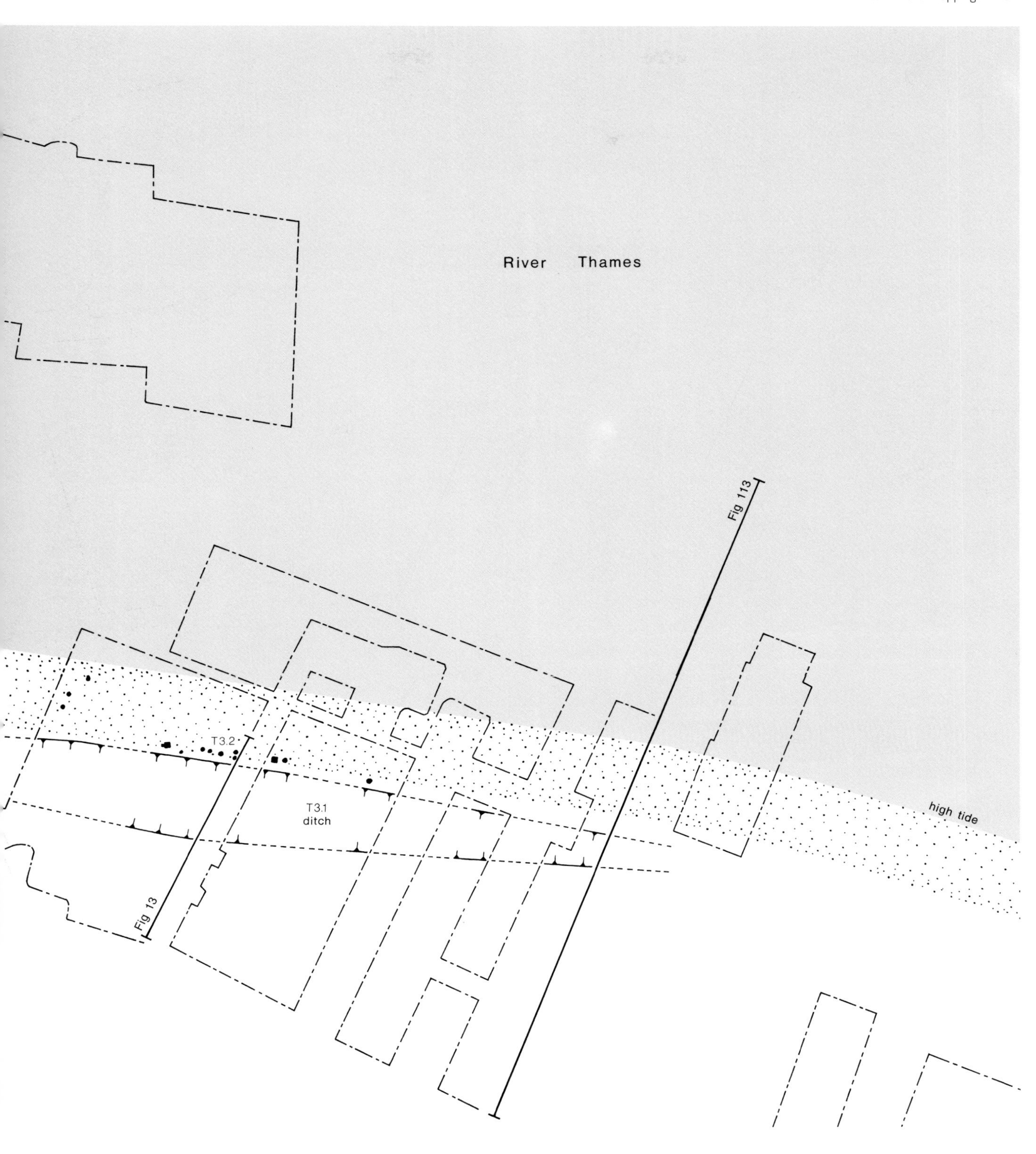

The south wall appears to have been demolished subsequently, and a third floor, of gravel, was extended over it (Fig 11). A second hearth was built on the floor in an alcove in the west wall; it is not clear whether this recess had been a doorway, or the space for a chimney associated with the hearth. The original east wall was replaced by a ragstone and tile foundation (T4.7) about 0.8m wide and supporting a burnt baseplate, which implies that the superstructure was still of mud-brick, cob, pisé or an allied material. A deep post pit 2.1m beyond the recorded north end of this wall suggests that there was a change of construction at the rear of the building where it approached the infilled T3.1 ditch. The 0.4m post may have supported an outshot roof or wall structure. It is possible that these changes were allied to the extension of the later gravel floor across the demolished southern internal wall.

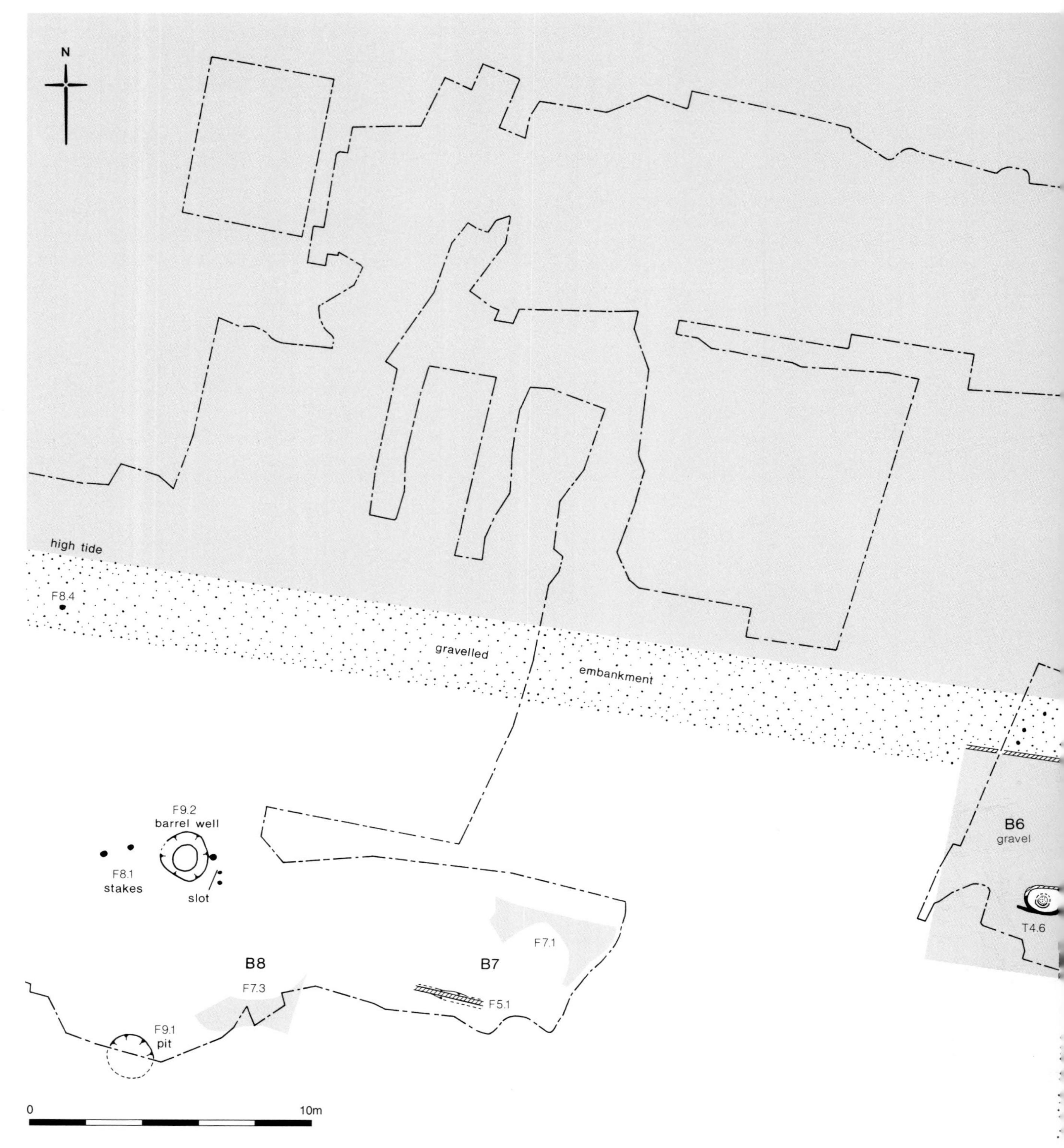

Fig 10 Plan of early Roman buildings, showing all primary structural activity and other contemporary features

Building 5

Building 5 (T4.5) consisted of several rooms in the area recorded, subdivided by wattle and daub partitions; the northernmost room was built over the infilled T3.1 ditch (Fig 10). This and the neighbouring room had clay floors, but the dividing wall was later removed and a new floor laid across the extended area. In the new room was a hearth or furnace of clay, ragstone and tile, with a rake-out area to the south, bordered by tiles and containing metalworking debris (bronze and iron slags). The main north–south external wall forming the west side of the rooms faced on to an adjoining alley, and the north wall (Fig 13), facing the revetted bank, contained postholes for the supporting frame of the building, with an eavesdrip gully alongside the west wall.

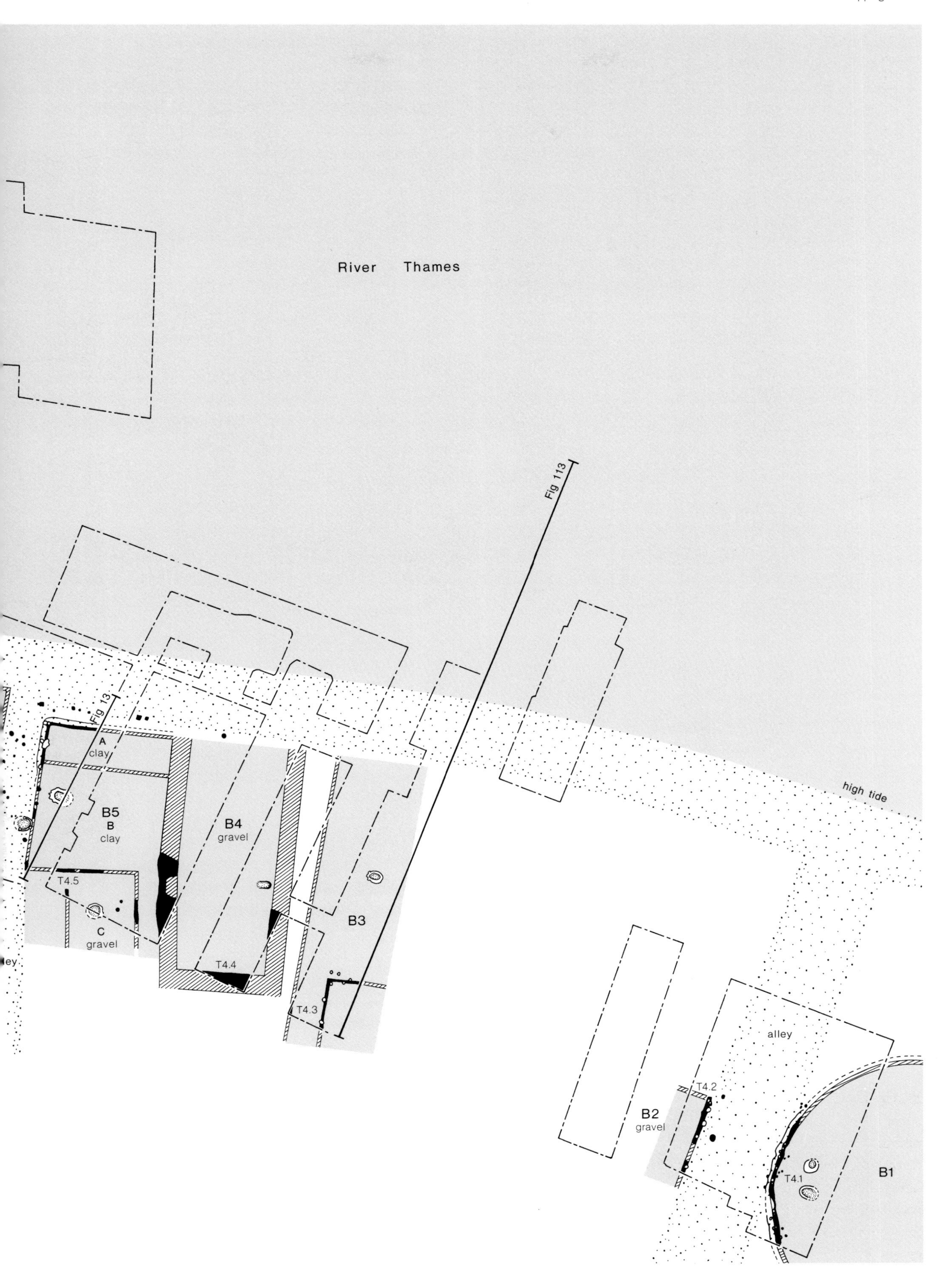
River Thames
Fig 113
Fig 13
high tide
A
clay
B5
B
clay
B4
gravel
T4.5
C
gravel
T4.4
B3
T4.3
alley
T4.2
B2
gravel
T4.1
B1

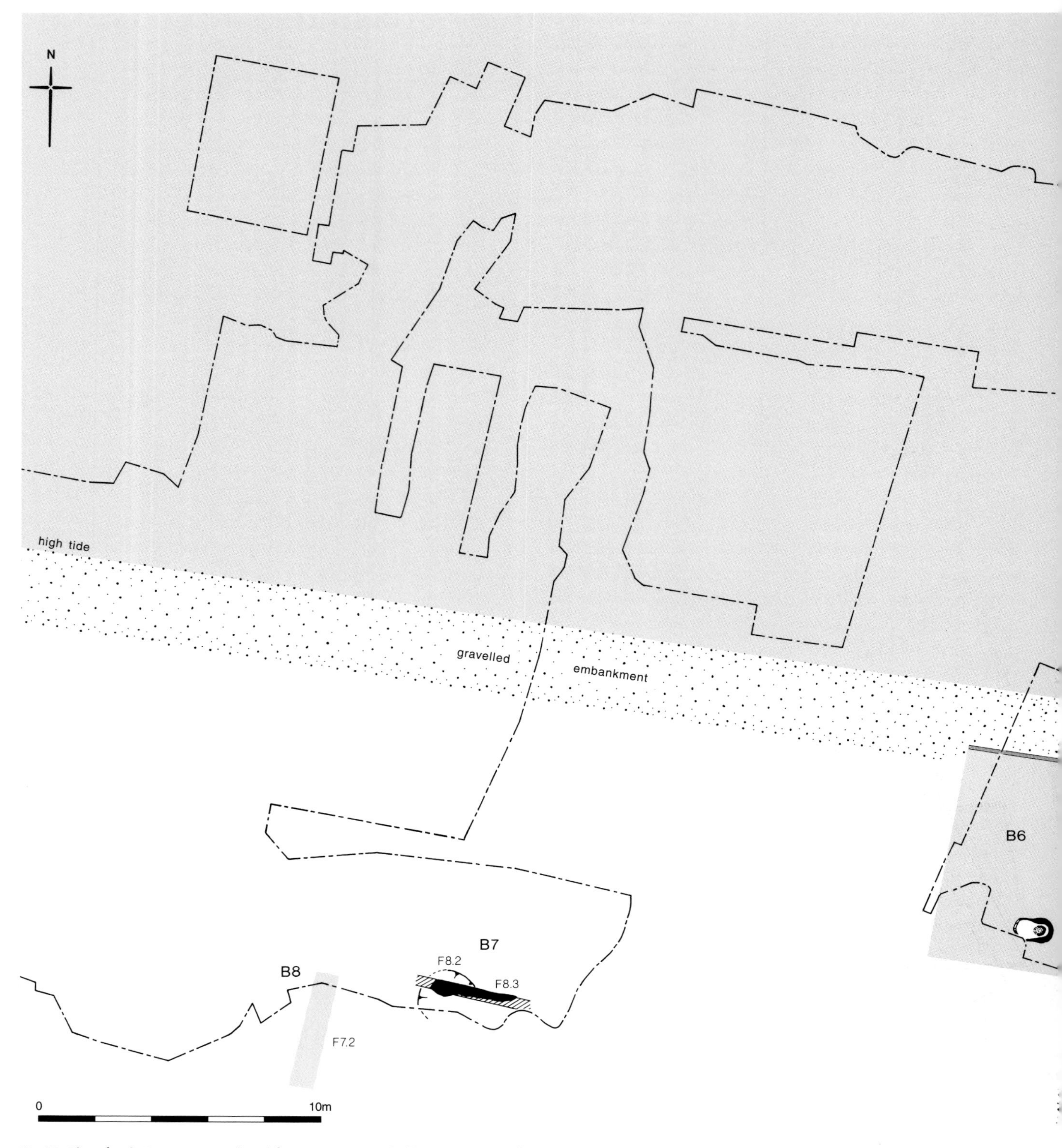

Fig 11 Plan of early Roman structural modifications to existing buildings and retained features

The floor in the southern room was initially clay, with a burnt hearth structure placed centrally, followed by two gravel surfaces. This was bordered by narrow walls to east and west. The western wall may have partitioned off a separate room facing the alley, or possibly a verandah, since the area had a gravel surface, while a 0.6m wide strip of unsurfaced gravel between the eastern wall and the west wall of Building 4 suggested an internal corridor.

The latest evidence for occupation of the building appears to have been a reconstruction of the east wall (T4.8), consisting of a clay core with a painted plaster facing on the east side (Fig 11).

Between Buildings 5 and 6 was a 1.8m wide metalled alleyway, with a hearth on the west side, probably abutting the eastern wall of Building 5. Domestic rubbish build-ups on the surface preceded two phases of remetalling.

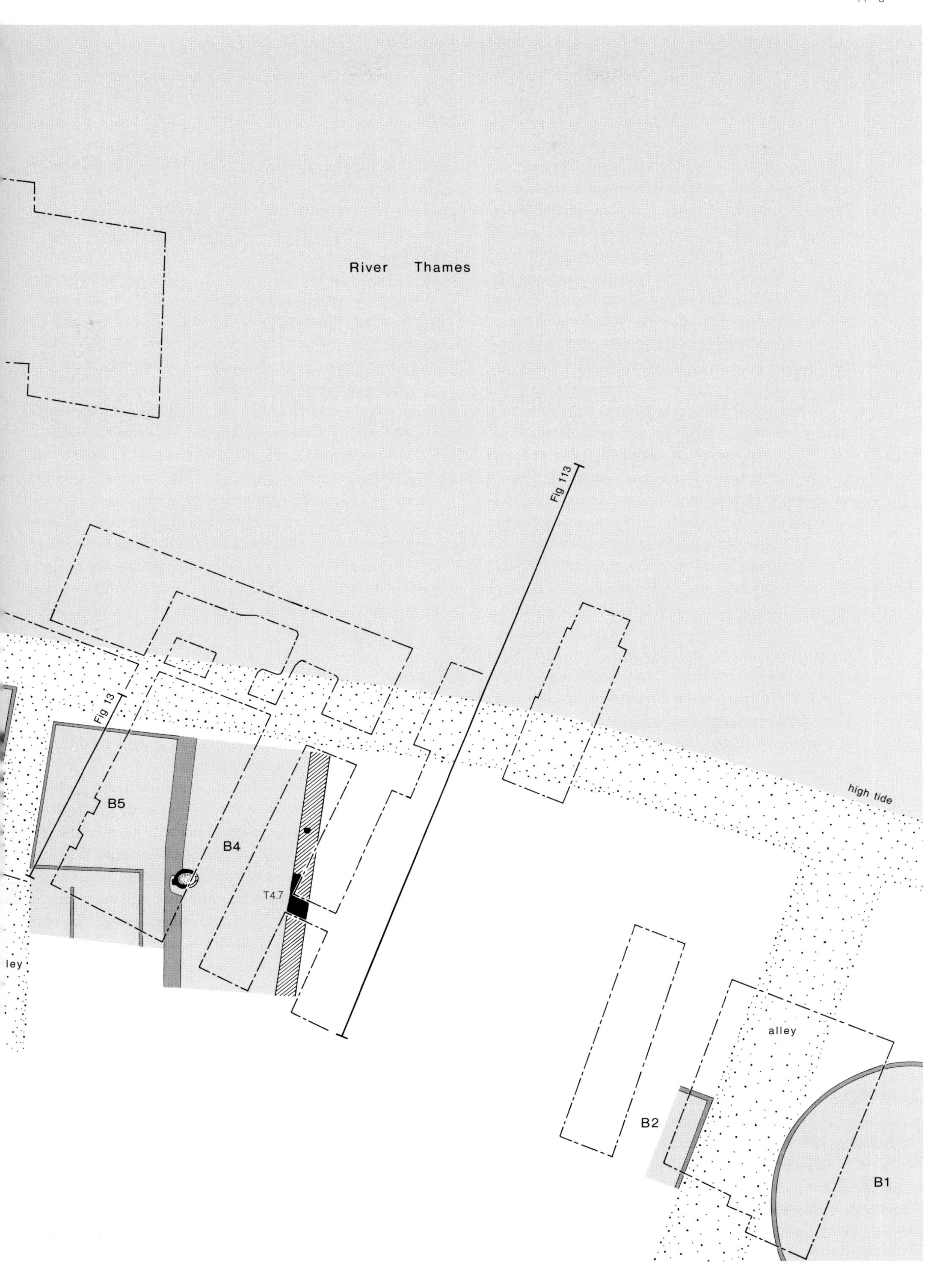
River Thames
Fig 113
Fig 13
B5
B4
T4.7
high tide
ley
alley
B2
B1

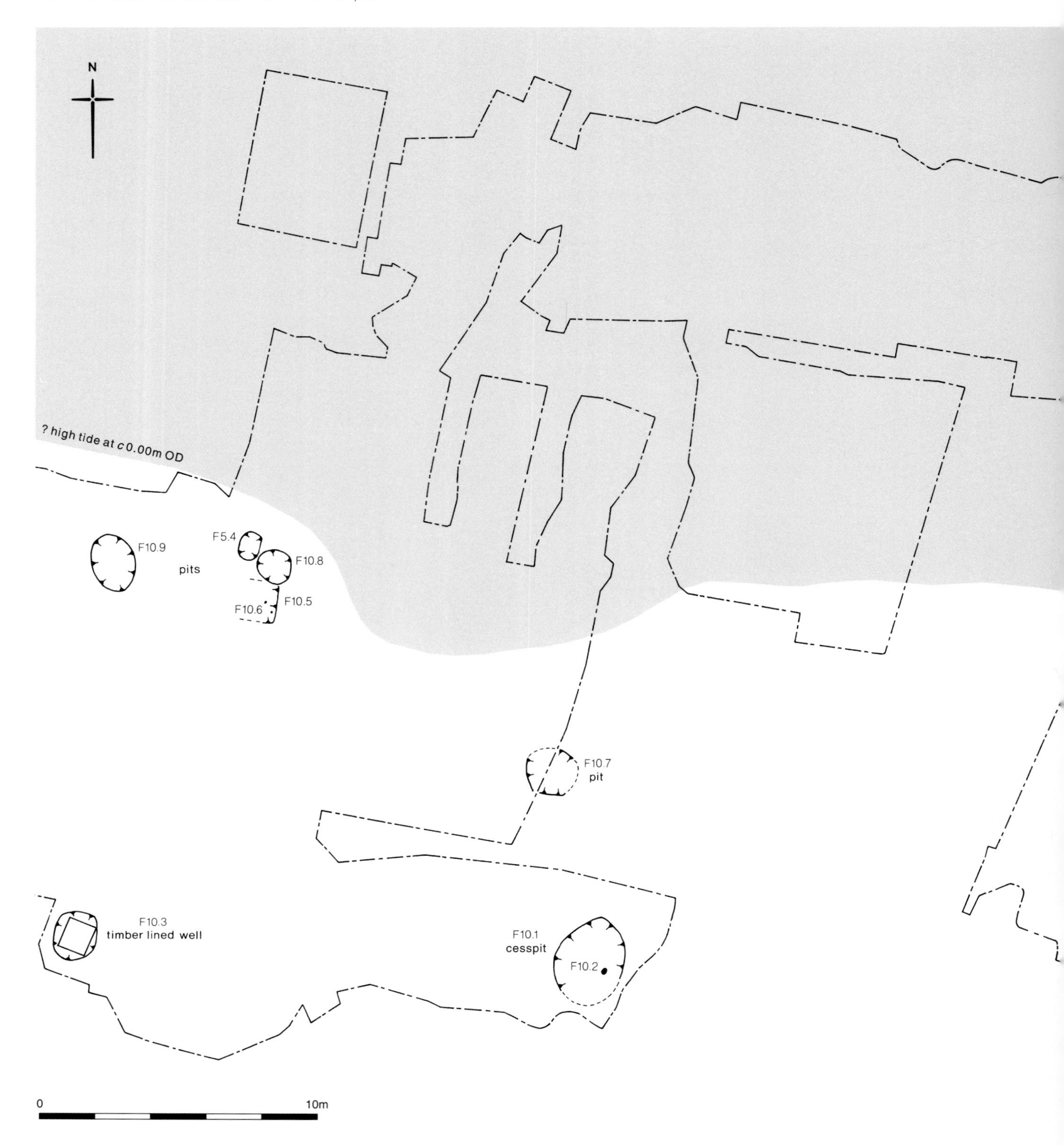

Fig 12 Plan of all later Roman features

Building 6

Only the east wall of Building 6 survived (T4.6). Of wattle and daub, it was bounded externally by an eavesdrip gully running along the alley edge (Fig 10). There were two successive floors, the first of clay and the second of gravel, each with a large horseshoe oven built against the wall and opening to the west. The walls of the ovens were constructed largely of tiles, clay and plaster slabs, the second having a floor of yellow tiles (Fig 11).

Buildings 5 and 6 and the intervening alley were sealed by dumps of sandy gravel and green silts, containing pottery dated to the late 1st or early 2nd century (T5.3), which suggest abandonment. It is not clear whether these structures were rebuilt subsequently, but a well was cut through the latest deposits (see T5.1 below), demonstrating continued occupation in some form (Fig 12).

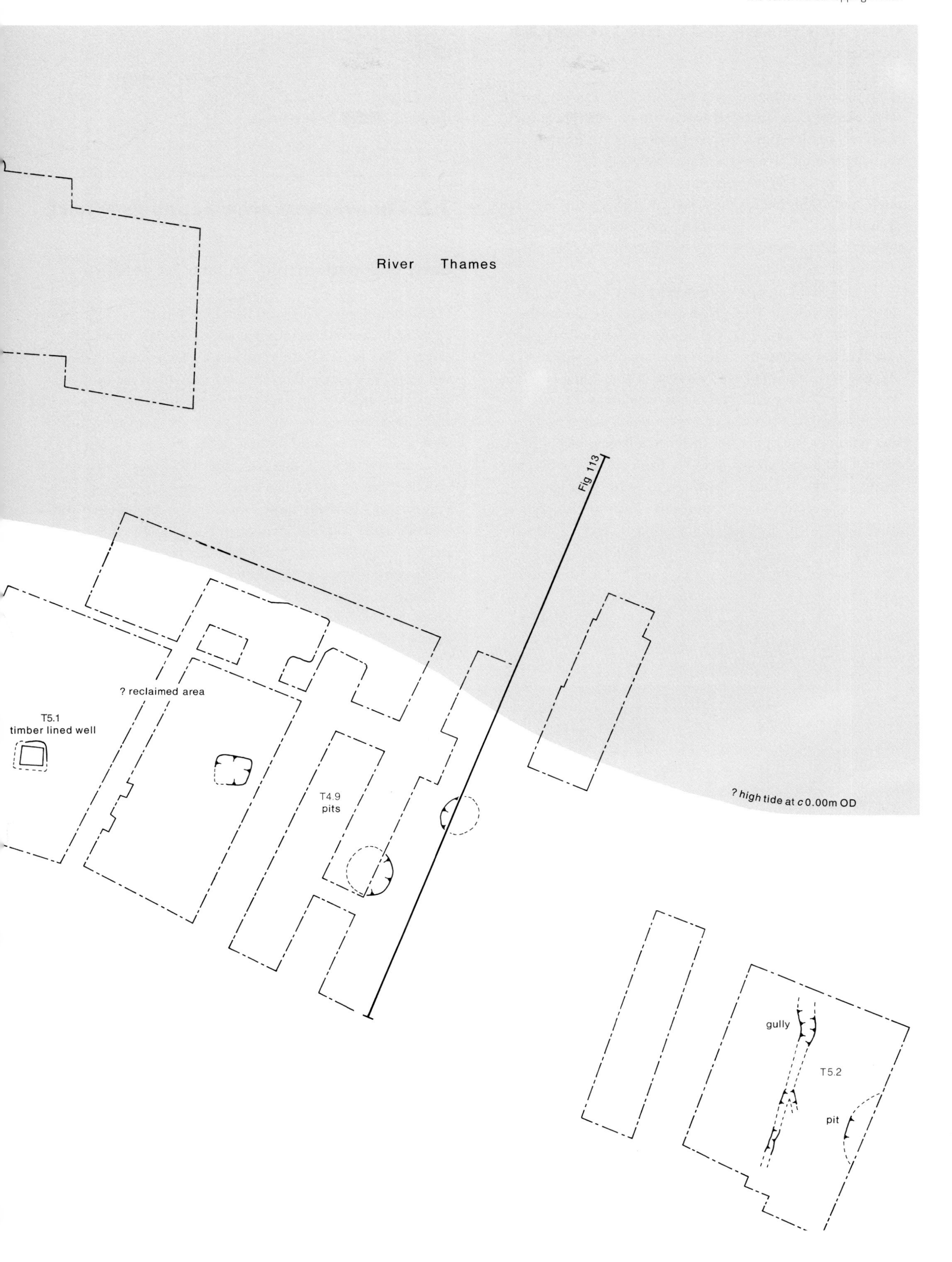
River Thames
Fig 113
? reclaimed area
T5.1
timber lined well
T4.9
pits
? high tide at c 0.00m OD
gully
T5.2
pit

Later activity: early 2nd to late 3rd/early 4th century

From the early 2nd century to the end of the Roman period there was no clear structural activity on the site. There was, however, a sparse but fairly even pattern of cut features showing that the area was still occupied (Fig 12).

Two pits (T4.9) were cut through the backfilled T3.1 ditch, and to the south a third was cut through the latest floor of Building 3. The fills contained early 2nd-century pottery, which conforms with the suggested date for the later phases of some of the buildings.

Within Building 6, a 2.5m deep timber-lined well (T5.1) was cut through the T5.3 gravel dumps on the line of the former east wall (Fig 13). The lining consisted of oak planks lapped at the corners to create a self-supporting structure. At the base was a dump of sand and gravel containing two lead weights and, above this, brown clay containing a barrel base complete with the remains of staves, which may have been a later relining. The upper fills included a large quantity of plain and red-lined wall plaster, possibly from one of the buildings nearby, as well as pottery dated to *c* AD 120–140/160.

Other later Roman cut features included the corner of a pit (T5.2) in deposits overlying Building 1, and a gully cut through the T4.1 alley gravels between Buildings 1 and 2. This contained 2nd-century pottery, and also a 4th-century coin. There was also an intrusion at the west end of the site of which only part was recovered, containing 4th-century pottery (Sheldon 1974, 24), probably to be identified as the small unnumbered pit shown on the site plan as cutting the infilled T3.1 gully.

Although this scattered evidence points to some occupation in the area continuing to the end of the Roman period, the density of these features over a period of over 250 years is not substantial. It is possible that the focus of occupation moved southwards towards the suggested street, or northwards over the later reclaimed foreshore; evidence for the latter would have been removed by post-Roman erosion. The key to this could be the T5.4 dark earth overlying Building 2 and its adjoining alleyway, which probably represents a deposit that was originally ubiquitous but lost to later erosion and truncation. If this were the case, even fairly deep late Roman foundations could also have been removed. The presence of ceramic building materials, including box-flue tiles and tesserae at Fennings Wharf, seems to point to the existence of a substantial building with a hypocaust system in the area (see Chapter 14.11), although this may have been located to the south nearer Tooley Street.

3.2 The western area: Fennings Wharf

Early occupation: mid- to late 1st century

The earliest Roman activity at Fennings Wharf was represented by several scattered features that cut the earlier fluvial clays (Fig 9). The earliest features included a large hollow (F5.2) and a pit (F5.3) located on the high ground in the southern part of the site. Although there is no datable material associated with these features, they are assumed to be Roman because of their proximity to dated features of the following phase. The pit contained two box-flue tiles, which together with a residual wall tile from a post-Roman context (see Chapter 14.11) suggest that a centrally-heated masonry-walled building was located nearby, very different from the buildings at Fennings and Toppings Wharves. Such a building was found to the south-east of Toppings Wharf in District Heating Scheme Area F, immediately south of St Olaf's church (Graham 1988, 49), dated after AD 150. This may also have been the source of a number of tesserae found mainly in the fills of later pits (see F10.1, F10.3 on Fig 12). An unstratified roofing tile impressed with the procuratorial stamp PPBRILON may have come from the same building, since they were apparently used in the construction of public buildings (Betts 1995, 222), implying a degree of procuratorial control. However, similar tiles have been found on sites with no obvious local government association.

A little to the south-east, two parallel north–south gullies (groups F6.1, F6.2) were cut to the east of the F5.2 and F5.3 features. The gullies were *c* 1.0m apart, and it is not clear whether they were contemporary or successive. They have been interpreted as early markers for land allotment, antedating the construction of the first buildings in the area but on the same alignment. The fills of both were dated by ceramic

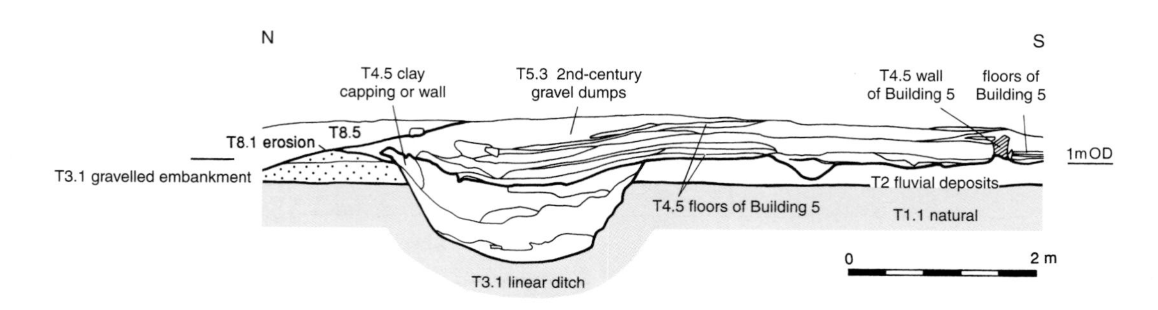

Fig 13 Redrawn section through the early Roman ditch and building sequence on Toppings Wharf (excluding post-Roman deposits except 13th-century erosion line T8.1)

evidence to AD 40–100.

It is believed that during the 1st century the western portion of the site was an unoccupied, external area, designated Open Area 2 on Fig 9, where rubbish pits could be dug. Later this situation changed and Buildings 7 and 8 were constructed here.

Building 7

To the west of the gullies were traces of a building (B7). These consisted of a truncated east–west cut (F5.1: Fig 10) interpreted as the beam slot of a wall line, while to the north over the F6 gullies was an area of gravel make-up (F7.1), followed by three successive floors, the lowest at 1.22m OD in the east, rising to 1.60m OD centrally. These were mainly of sandy clay with an area of charcoal on the earliest floor in the east possibly suggesting the presence of a hearth. Between the second and third floors was a gravel make-up, raising the final floor to 1.76m OD. The initial floor contained pottery dated AD 150–250, implying a construction date no earlier than the mid-2nd century, and was cut by a feature of similar date range (see group F10.1: Fig 12) which marked its eventual demise. The floor area was at least 4.5m east–west, and the building must, therefore, have been at least as wide. It is possible that the F5.1 slot was part of a wall line at the south end of the room containing the F7.1 floors.

A circular pit (F8.2: Fig 11) possibly 2.0m in diameter cut through the west end of the F5.1 slot; it contained some pottery of AD 40–80, substantially earlier than the date of the floors and presumably residual if the wall and floors were contemporary. The silty clay fill of the pit was cut by a replacement slot (F8.3) which contained hard-packed chalk rubble blocks which may have formed a foundation at least 1.07m in depth (Fig 11). The trench was packed with clay backfill containing pottery dated AD 40–160. If this was constructional packing, it suggests that the foundation may have been contemporary with the F7.1 floor sequence further north. It was clearly a replacement of the earlier F5.1 foundation which lay directly beneath, suggesting a mid-2nd-century reconstruction of the building. One possibility is that the F5.1 wall was the north wall of an early version of Building 7, subsequently extended with the addition of the F7.1 floors and the new F8.3 wall.

Building 8

To the west, a clay surface was laid over the F4 fluvial deposits at 1.53m OD, followed by a second surface (F7.2) which probably belonged to a structure contemporary with and adjacent to Building 7 and of a minimum width of *c* 3.5m (B8: Fig 10). These were sealed by dump or demolition layers, possibly for later floors, reaching 1.75m OD. Two successive floors and their associated make-ups (F7.3: Fig 10) were recorded a little further west at the same general level (1.51m and 1.59m OD respectively) and may have been extensions of the F7.2 floors (Fig 11). The first surface was an extensively-burnt thin brickearth slab covered with charcoal. The upper floor was sealed by compacted dumping which may have been make-up for further surfaces. Neither F7.2 nor F7.3 was closely datable because of the small size of the associated finds assemblages.

Isolated activity in the west

To the north-west (Fig 10), in a heavily truncated area in the centre of the F2.1 ring-ditch, were two clusters of five large undated postholes, each up to 0.2m in diameter and *c* 0.06–0.21m deep. There was no evidence that the posts had decayed *in situ*, and they had almost certainly been removed. Possibly associated with the posts was a 0.7m+ long north–south slot with a shallow stakehole at the north end (F8.1: Fig 10). These features may have been related to a building to the west of the F7.3 floors, but lay to either side of a well lined with a reused, silver fir barrel (F9.2) and cut into the natural gravel. Packing around the barrel contained pottery of AD 40–100, which may place the construction of the well in the same period as the F8.2 pit. The well was, however, backfilled with mainly dark grey silty clays containing residual material including a coin of Nero (AD 64–8), but dated by pottery of AD 180–300 (F9.3), which suggests that it remained in use for a considerable time, possibly until the late 3rd century.

Analysis of the botanical remains from the initial infill and final backfill showed marked similarity between the two assemblages. They contained few plant remains but some charred cereals (spelt wheat, wheat, barley) and hazel shell fragments, as well as seeds of wetland and disturbed habitats, almost certainly the area around the well (see Chapter 14.9).

About 6.0m to the south was a truncated rubbish pit (F9.1) *c* 1.5m in diameter and at least 0.66m deep (Fig 10). The backfill contained a good group of late 1st-century pottery (AD 70–100) including samian vessels (five of them stamped), which suggests that it was possibly contemporary with the similar F8.2 pit and F9.2 well.

The pit contained a sizeable assemblage of bones, much larger than that obtained from the F9.2–.3 well fills. This showed a preponderance of pig with a lesser quantity of cattle and sheep/goat, although domestic and wild birds (chicken, duck, teal) were also present. The high proportion of pig bones is relatively rare on Roman sites of early date, and was possibly evidence of a high-status diet since a preference for pig was also noted at Winchester Palace and at the Fishbourne villa site (see Chapter 14.8).

To the north, on the projected line of the F1.1 gravel riverbank, was an isolated posthole (F8.4) of similar dimensions to the F8.1 features and containing pottery dated AD 180–300 (Fig 10). This places it at the same period as the F7.1 building (B7). It is not clear whether it was associated with the embankment and removed at a much later date or its position was simply coincidental, since the bank was redundant and had probably been levelled by this time.

Later activity across the area: mid-2nd to 4th century

In the later Roman period, a group of pits and wells of a date broadly comparable to the infilling of the F9.2 well was cut across the area (Fig 12). This included a large 1.73m deep unlined cesspit (F10.1) in Building 7, containing a large group of pottery dated AD 180–250. The pit may have been in use during the lifetime of the building, although this is unlikely. Botanical analysis of one of the fills recovered a large quantity of charred cereal grains (bread wheat, barley, rye, oat) as well as seeds characteristic of stored grains, such as corncockle and brome (see Chapter 14.9). Rye is uncommon in the Roman period, and oat may also not have been exploited in the south, but appeared as a weed among other grains. Two other fills contained very different assemblages, including a large number of fruit seeds (grape, fig, apple/pear, elder, blackberry/raspberry, and some *Prunus* species) and some seeds of the opium poppy, which were used for flavouring. The pit would, therefore, appear to have been used for the disposal of both cess (fly puparia and insects were present) and household waste, some of it related to cereal processing.

The fill was cut by a single posthole (F10.2). Five metres to the north was a sub-circular cut (F10.7), up to 1.6m across, 0.6m deep, with a highly structured series of organic and sandy silt fills containing pottery dated AD 180–300 (Fig 12).

At the south-west corner of the F2 ring-ditch was a square timber-lined well at least 1.34m deep (F10.3). The remains of the lining consisted of two levels of planking lap jointed at the corners to create a self-supporting frame. The planks used were of what have now been identified as two standard widths: *c* 420–430mm and 300–310mm, broadly equivalent to 1.5 and 1.0 Roman feet (*pedes monetales*) respectively (cf Brigham et al 1996a, 25). The well appears to have been relined or braced at the bottom with four planks of varying size (two of 1.5 Roman feet), which may represent an attempt to prevent the underlying natural sand from washing into the well and undermining the structure. These have been tree-ring dated to after AD 132, although they were almost certainly reused and the date is, therefore, a *terminus post quem* for the relining. Certainly the well was infilled at a much later date (F10.4), and included pottery from the uppermost fill dated AD 200–300 (probably later 3rd century). The fill contained a number of tesserae, which together with several from the fill of F10.1 suggest the presence of a building with a tessellated floor in the area (see Chapter 14.11). This may be the same structure which was the source of two box-flue tiles and a wall tile; one possibility is that they originated in a building recorded in area F of the District Heating Scheme (Graham 1988, 49) a little to the south-east, although similar buildings may have been constructed along the line followed by modern Tooley Street. The fill included roe deer bones and seems to suggest a high standard of diet, possibly reflecting the relatively common occurrence of pig and other game animals and birds.

Some distance to the north beyond the projected line of the Toppings Wharf (T3) river embankment was a cluster of features including a square pit (F10.5) cut into the exposed natural gravels 13m north of the F10.3 well. Only the eastern half was found, but the pit had probably been 1.5m square, truncated to 0.5m deep. The infill contained pottery dated AD 150–250, and after backfilling was subsequently cut by two 0.1m square-section stakes (F10.6).

A circular pit (F10.8) lay immediately to the north. About 1.3m in diameter, the pit was at least 0.64m deep but had also been truncated by erosion occurring before the 13th century (Chapter 6.1, groups F11.1, T6.2). The fills contained pottery dated AD 250–400, suggesting that the pit replaced the F10.5 pit (Fig 12). An ovoid pit 1.8m by 2.0m (F10.9) was cut into the underlying gravels 4.5m further west; 0.65m deep, it had like the others in the area been truncated by the much later erosion. The primary fill, which contained the remains of a small stake, included pottery dated AD 300–400, which makes this pit potentially the latest Roman feature surviving on site (see Chapter 14.1 and 14.4). It contained the remains of cereal bran fragments and fly puparia, as well as corncockle, implying cess disposal, although there was also a diverse range of local wild aquatic and wetland plant seeds (see Chapter 14.9). A large number of blackberry/raspberry seeds imply that they were a popular item for consumption as food or drink. As in the earlier period a high proportion of pig bones was present, although cattle and sheep/goat were more common (see Chapter 14.8). Amphibian bones – possibly frog or toad – indicated that the pit was open. Notably, the pit contained a large number of fish bones, particularly eel (201 examples), but including smaller numbers of herring, smelt, sprat and other species. These imply the seasonal exploitation of the estuarine fishing grounds, either to produce fish for eating or for conversion into fish paste (*garum*).

An undated pit (F5.4: Fig 12) together with an area of redeposited sand (F5.5) were located in the same area; originally all thought to be early Roman, their position in relation to the projected line of the river embankment suggests that they were late features.

Although the later Roman horizons had been substantially truncated by post-medieval cellaring, it is clear that sporadic activity continued across the study area until the end of the Roman period. It is also clear that, while 1st-century features were confined to the area south of the line of the T3.1 ditch and embankment, later features spread beyond these confines: the bank must have ceased to function as a boundary to the settlement. However, further evidence for this expansion was destroyed by a return to strongly transgressive conditions similar to those which predominated in the period between the construction of the F2.1 Late Bronze Age/Early Iron Age ring-ditch and the construction of the T3.1 Roman ditch and bank. These led to extensive erosion of the northern half of the site (see Chapter 6.1).

3.3 The Southwark waterfront in the Roman period

The position of the study area on the Roman waterfront and its identification as a possible site for the southern bridgehead combine to make Fennings and Toppings Wharves a site of considerable importance. Until recently there was some uncertainty regarding the location of Southwark's Thames frontage at this period and its development from the 1st century onwards. The extent of Southwark has been shown to be fairly constant throughout the Roman period, although an attempt was made to show the differing extent of the northern islands at high and low tide in the 1st century (Milne 1985, fig 49). Changing river levels, although their relevance was recognised, were not taken into account, and it was not, therefore, possible to map the pattern of expansion beyond the original boundaries of the settlement. Work on several major publication projects has, however, gone some way towards rectifying this deficiency. These projects include Winchester Palace (Yule in prep) and Courage's Brewery (Brigham et al 1996a; Cowan in prep), although the body of evidence is still not comparable to the extensive corpus of structural material from the northern bank (Milne 1985; Brigham 1990b). This discrepancy between the two sides of the river can now be seen to be due to the relatively slight character of such early waterfront structures as survive in Southwark, the subsequent erosion of the later Roman riverfront along the Thames frontage in the 13th century and the greater complexity of the topography of the south bank.

As a result of recent work, archaeologists are now able to map the outline of Roman Southwark with some confidence, using a combination of depositional and topographical evidence together with records of the fragmentary remains of timber revetments (most recently Yule 1988, fig 3, and in prep; Cowan in prep) and reliable tree-ring dating to build on foundations laid down earlier (Graham 1978). A better understanding of changing river levels, based originally on north bank evidence (Brigham 1990b), enables the model to be altered for different periods (see below).

At 64–70 Borough High Street, a channel was revetted in the mid-1st century and again in the early 2nd (Graham 1988), with sufficient surviving for a reconstruction of its original height to be attempted. A contemporary Thames waterfront at the Winchester Palace site, immediately to the north-east of the study area (Yule 1989, 31–9, and in prep) was also extended early in the 2nd century.

In 1989–90 extensive well-preserved 2nd-century revetments and associated surfaces were found lining the 'Guy's channel' watercourse at Guy's Hospital (Taylor-Wilson 1990), in this case modified in the 3rd century. Once again, the height of these structures can be reconstructed.[10] More recently, two phases of revetment, one dated AD 231–2, have been found just upstream of Tower Bridge near Tooley Street (D Seeley, pers comm).

3.4 The Roman river level

It has been estimated that since the last Ice Age mean sea level (MSL) in northern Europe has risen by some 25m in relation to the land surface, itself altered as a result of isostatic changes caused by the removal of the kilometres-thick mass of ice. There is good evidence, however, that this strong trend towards a rising sea level took the form of a series of transgressive phases punctuated by much shorter regressive interludes (see Glossary). These phases were identified and classified for the Thames estuary by Devoy (1979), basing his work on radiocarbon dated organic samples obtained from deposits formed during regressive phases at Tilbury. Devoy's work has not been accepted universally, partly because of problems in dating samples and providing accurate levels that take adequate account of the compaction of organic deposits since their formation, and partly because conditions clearly varied considerably from place to place, just as they do today. However, the closely dated sequence of structures in central London from the Bronze Age to the present day provides a uniquely accurate record of tidal oscillations.

Following on from work on the 14th-century riverfront at Trig Lane, City of London (Milne and Milne 1982, 60–2), the excavation of Roman waterfront sites at Pudding Lane, Peninsular House and Miles Lane enabled evidence for Roman river levels to be collated (Milne 1985, 79–86). A graph of the combined results was published (ibid, fig 50) using the upper and lower levels of 1st-century quays, 14th-century revetments, and recent figures as fixed points to produce a 2000-year curve. These results were plotted against the projected mean high water spring level of the Thames at Tilbury from 3500 BP onwards (after Devoy 1979) and mean sea level at Foulness (after Greensmith and Tucker 1973).

In 1990 Milne's graph was substantially modified and extended for the Roman period (Brigham 1990b, fig 13) to include the evidence for 2nd- and 3rd-century structures from New Fresh Wharf, Billingsgate Lorry Park, Swan Lane, Seal House and elsewhere. More recent sites at Regis House, Cannon Street Station, Thames Exchange, Vintners' Place and Bull Wharf have confirmed the validity of the graph (Fig 14). Taken together, the evidence collected suggested that mean high water spring (MHWS) and mean low water spring (MLWS) tidal levels in the mid-1st century lay between *c* 1.25–1.5m OD and -0.5m OD respectively, falling by the mid-3rd century to *c* 0.0m OD and -2.0m OD. A substantial and progressive decline in levels through the Roman period was indicated, which, it was argued, coincided with the 'Tilbury V' regression recorded both in the inner Thames estuary (Devoy 1979) and at Foulness (Greensmith and Tucker 1973).

At the time, there was no further evidence to extend the downward curve beyond *c* AD 250, but recent work on a section of the riverside defensive wall at Three Quays House near the Tower of London (Grainger 1996) suggests river

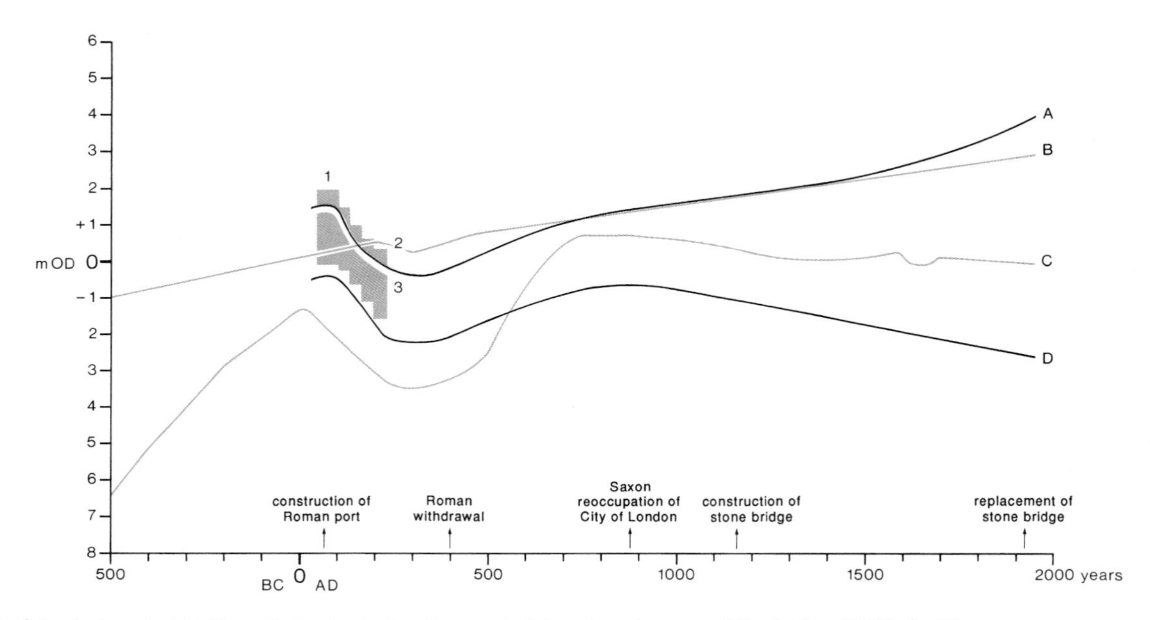

Fig 14 Graph of river levels in the tidal Thames levels, showing how the water levels have changed over time (after Brigham 1990b, fig 12)
Key: 1 – top height of Roman Port of London quay advances showing the falling river level; 2 – top of the foundation raft of the riverside wall at Three Quays House; 3 – ground level at Summerton Way, Thamesmead; A – mean high water, City of London; B – mean high water, spring, inner Thames estuary; C – mean sea level, outer Thames estuary; D – mean low water, City of London

levels were still in decline at the time of its construction in the AD 270s. MHWS appears to have fallen below Ordnance Datum (Fig 14). This has been confirmed more recently by excavations at Summerton Way, Thamesmead, where very low-lying 3rd- and 4th-century Roman occupation and a possible field system were encountered on an area of former foreshore, presumably protected by embankments (Lakin 1997 and 1999).

The presence in Southwark of occupation surfaces below the high tidal levels seemingly contradicted the evidence of the relatively reliably dated structural sequence from the north bank. At Toppings Wharf and elsewhere these surfaces were normally laid directly on the pre-Roman flood clays at c 1.2–3.0m OD, suggesting that the early Roman high water level lay somewhat below c 1.0m OD. This was apparently confirmed by the level of surviving revetments, although there was a recognition that the tops of these structures may have been lost through decay and later truncation (Graham 1978, 510–13). If the river had risen higher than 1.0m OD, therefore, Southwark would have been exposed not only to irregular storm surges and astronomical high tides but would have been at risk even during normal periods of high spring tides, which could occur some 228 times a year (Milne and Milne 1982, 61). Clearly, this apparent low river level was problematic, since it would have left the north bank quays, with a foreshore level at Ordnance Datum, unusable for most of the day, and then only approachable by small river craft.

However, the widespread evidence of inundation to c 1.1–1.3m OD, marked on so many sites by the presence of flood clays, demonstrates that the river did indeed regularly rise above the suggested level until the mid-1st century, but then stopped immediately the area was settled. We can be fairly certain of this, because where details are available the clays are almost invariably described as showing no sign of weathering or plant growth, so that it is likely that the silts were still being deposited until shortly before they were sealed beneath the earliest Roman occupation levels. To achieve the necessary protection for the settlement, it would clearly be necessary for Southwark to be protected by embankments, allowing the area behind them to be empoldered, as proposed by Yule (1988, 17) and Brigham (1990b, 134), although such embankments had not been recognised in the archaeological record. Toppings Wharf is, therefore, an important site in that it provides the first clear evidence for a bank potentially reaching a height of c 2.0m OD. A mid-1st-century structure in the Winchester Palace study area consisting of five lines of timber piles can also now be seen as an eroded embankment (Yule in prep). Additionally, a 1st-century revetment at 64–70 Borough High Street may have attained at least 1.75m OD (Graham 1988, 17), and it seems likely that other revetments of the early period could be restored to similar levels.

Now that these reinterpreted remains allow the main obstacle to establishing the 1st-century river level to be removed, the continuing fall in river levels throughout the Roman period can be used to explain the expansion of the Southwark settlement over former mudflats and drainage ditches, and buildings being constructed at levels which would have been untenable in the 1st century. These include a recently-published sunken-floored timber warehouse in the Courage's Brewery study area which was constructed in AD 152 at the west end of the settlement in a previous intertidal zone (Brigham et al 1996a) and the masonry building complex in the Winchester Palace study area (Yule in prep).

3.5 Post-Roman river levels

The apparent fall in river levels during the Roman period was reversed subsequently with the resumption of the rise in MSL which has prevailed since the last Ice Age. In London this led inevitably to inundation and burial of the latest Roman waterfronts and other marginal areas, initially by bands of highly erosive gravels and subsequently by finer river silts and clays. While the erosion was slowed and, in places, halted on the north bank by the substantial wharves and the late 3rd-century riverside defensive wall, the situation in Southwark was more serious as the area was reduced to a size similar to that of the 1st-century settlement. This process as it affected the study area is discussed in Chapter 6.1.

There is no date in London for the onset of the current transgressive phase, although it does not appear to have begun during the Roman period. At Fennings Wharf the earliest post-Roman foreshore feature was riverine erosion, which is undated (see Chapter 6.1). A later phase of riverine erosion is dated to the early to mid-11th century.

The first structural evidence of attempts to regulate the river in the post-Roman period comes from near the Alfredian harbour at Queenhithe (pers comm Julian Ayre). Revetments built *c* AD 980–1014 at the Bull Wharf site were recorded in 1995 at up to 1.45m OD, with a base at *c* 0.5m OD. These may have been eroded at or just above mean high water neaps (MHWN, the lowest regular high tides), but it must be assumed that the contemporary occupation surface lay above this level. A slightly later embankment of AD 1014–35 survived to 1.15m OD; this may have been severely eroded at MHWN or, alternatively, represent some kind of revetted hardstanding on the foreshore rather than a waterfront. Clearly, however, these structures demonstrate a substantial rise in river levels since the end of the Roman period (Fig 14).

A contemporary jetty at New Fresh Wharf (Steedman et al 1992, 120) survived to between 1.2m and 1.6m OD, which may give a better indication of the level of 'dry land' at the turn of the first millennium. This higher level is corroborated by the evidence of subsequent embankments at Billingsgate and New Fresh Wharf which were surfaced above 1.9m OD, a height maintained until at least the 14th century at Trig Lane. The decay of timbers at *c* 1.2–3.0m OD implies a level for MHWN at this period (Milne and Milne 1982, 61) (Fig 14). Taken together, the evidence suggests that tidal levels, after a substantial rise during the early medieval period, levelled out for a considerable time, with a MHWN level at *c* 1.2m OD, a MHWS of perhaps *c* 1.7m OD and occasional highest astronomical tides (HAT) approaching 1.9m OD. This would explain the level of the 13th-century erosion horizon at Fennings and Toppings Wharves (Chapter 12.2), which reached 1.8m OD.

In the subsequent period, river levels have risen dramatically (Fig 14), with MHWS at London Bridge currently at 3.9m OD, and HAT at 4.7m OD (Admiralty 1977), rising between 1791 and the 1970s by at least 0.73m (Everard 1980).

4

Roman London bridge

Trevor Brigham

4.1 Introduction

There has been considerable past discussion regarding the existence and position of a Roman river crossing between the main Roman nucleus on the north bank and the smaller, but important, outlying settlement on the south bank (eg Dawson 1969, 1970, 1977; Merrifield 1970; Merrifield and Sheldon 1974; Milne 1985, 44–54; Rhodes 1991). An incidental reference to a bridge across the Thames was made by the 3rd-century writer Dio Cassius who referred to Celtic auxiliaries crossing the river in pursuit of British tribesmen during the Claudian invasion (Cassius, LX.20). Dio Cassius, writing about events which had taken place some 150 years earlier, may have confused the existence of a ford with a bridge, or with a later Roman crossing by the main army across a temporary structure (Merrifield 1965, 34). There is no mention of a bridge over the Thames in Julius Caesar's account of his second invasion of Britain, where he states that the Thames was only fordable in one place (Caesar, V). Merrifield (1983, 15–16) suggests that an Iron Age ford existed at Westminster, or possibly at Brentford, although the 1994 *Time Team* investigation in the grounds of Lambeth Palace failed to locate the presumed approach road to the Westminster ford. The investigators concluded that 'in the final analysis, the case for any river crossing at Westminster appears to be a slight one' (Sloane et al 1995, 370).

There is also no convincing evidence that the technology existed to construct a bridge of sufficient length to cross the Thames prior to the Roman period. On the foreshore at Vauxhall, a Bronze Age timber structure was discovered in 1998, radiocarbon dated to *c* 1870–1310 cal BC.[11] It consisted of two irregular lines of piles, *c* 4.0m wide and *c* 18.0m long, leading straight out into the river (Haughey 1999, 18–19). It has been interpreted as a bridge across the main channel of the river, but it is more likely to have been either a deepwater jetty or a causeway bridge to an island. Certainly the design of the Vauxhall bridge is very similar to that of the Eton rowing lake Bronze Age causeway bridges found within an infilled Thames meander (Allen and Welsh 1996, 125).[12] Other structures undoubtedly await discovery or reinterpretation; closer to the study area, a series of undated pre-Roman piles forming two roughly parallel lines found at Courage's Brewery in 1988 may also have formed part of a causeway or platform (Cowan in prep).

Roman London only developed as a port and a major urban centre because of the Thames estuary, which offers an excellent navigable routeway stretching from the North Sea westwards far into central England. It was at London, where the channel narrowed, that the Romans decided to bridge the tidal estuary (Fig 15). The site was carefully chosen, as recent topographical studies show that the approach roads were sited on natural promontories of

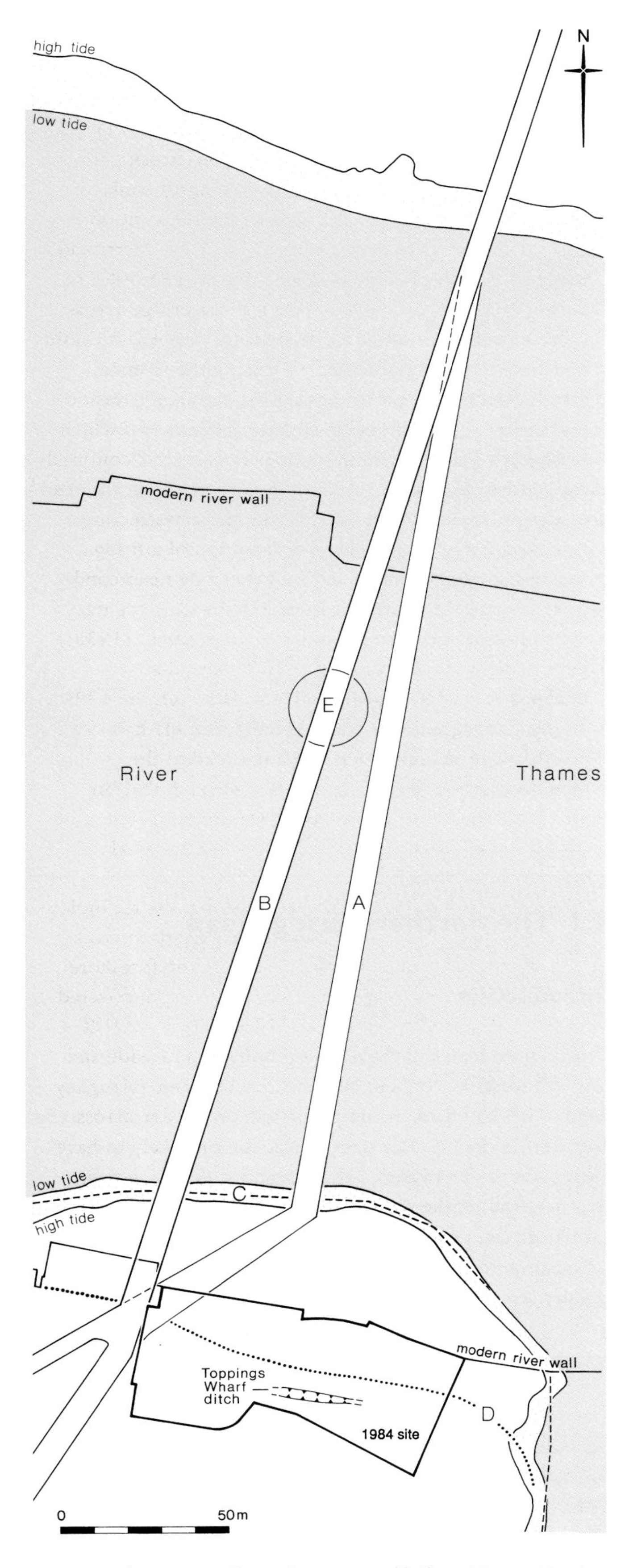

Fig 15 *Map of Roman River Thames showing proposed bridge positions: position A is the previously plotted position of the Roman bridge, which is in the same position as the medieval stone bridge; B is the revised plot based on a reappraisal of the evidence; zone C is the previously plotted inter-tidal zone; D marks the revised position of the high tide mark; E is the findspot of the coin concentration*

high ground on each bank to make the task of bridging easier. Roman London was an important road junction: immediately south of the Thames, Watling Street, running westward from the Kent ports, converged with Stane Street running north from Chichester (Fig 16). Once across the river north of this intersection, this road divided again, with separate routes leading off to Colchester, St Albans and Silchester.

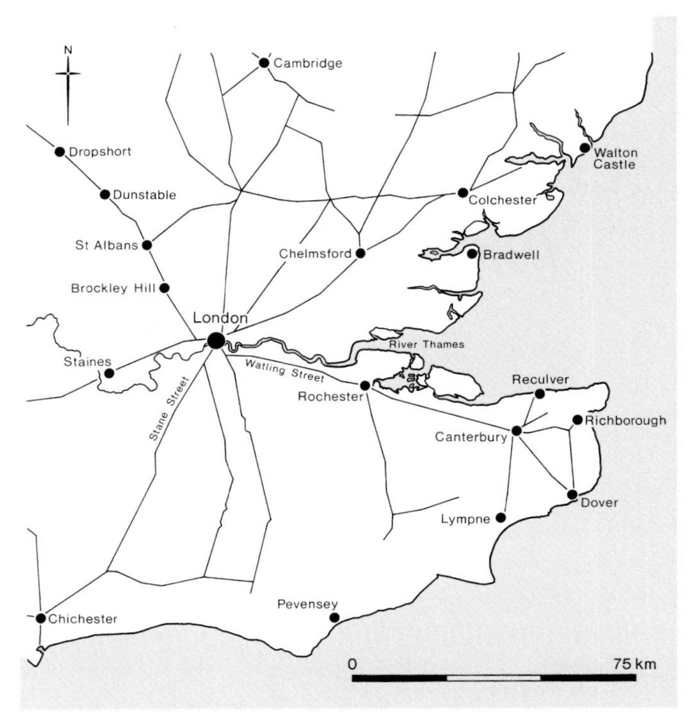

Fig 16 *Map of Roman road network in south-east England, showing location of London*

Archaeological evidence for the bridge

Archaeological evidence for the bridge can be said to fall into three main categories:

1) actual structural remains, including the structure identified as a pier base at Pudding Lane;
2) indirect archaeological evidence, including the presence and position of approach roads and negative evidence (ie where the bridge was not);
3) artefactual evidence, including coins, bronzes and other finds recovered from the river during the construction of the new London Bridge and the demolition of the old bridge in 1824–32.

This evidence will be reviewed in this chapter; first the northern bridgehead will be described and then the southern, followed by a summary of the artefactual evidence, discussion of which has been dealt with more fully elsewhere (Parsloe 1928; Merrifield 1987; Rhodes 1991). Aspects of the bridge's history will also be considered and likely reconstructions discussed.

4.2 The construction of the Roman bridge: a chronological framework

Bearing in mind the fragile and circumstantial nature of much of the evidence outlined below, the following sequence of Roman bridges is proposed as a model (Fig 17).

Bridge 1 A timber bridge built by the army after AD 43, possibly *c* AD 52, on the line from Fish Street Hill to the southern abutment of the modern bridge (Fig 17, a). This may have required restoration or repair after the Boudican revolt: even if not actually damaged or destroyed, the maintenance of the bridge must have suffered.
Bridge 2 A temporary timber bridge built *c* AD 85–90 from the Pudding Lane site to a point on the south bank near the modern bridge abutment (Fig 17, b), possibly on or just beyond the western boundary of Fennings Wharf. No southern abutment was found, and the bridge may simply have consisted of a pier constructed in the river at either end, a row of anchored pontoons or piled caissons, and removable ramps and decking.
Bridge 3 A new permanent bridge built on the line of the original structure after *c* AD 90 and before *c* AD 120, possibly with masonry piers and a timber decking capable of maintenance and replacement (Fig 17, c–d; Fig 18).

4.3 The northern bridgehead

Introduction

Structural evidence for the northern bridgehead (see Glossary) (Fig 17) is more complete than that for the southern, largely because the riverbank in Southwark was severely eroded in the post-Roman period prior to the 13th century. The erosive process on the north bank was arrested by the late 3rd-century riverside wall, the remains of massive timber quays of the earlier 3rd century, and by the construction of effective flood defences in the late Saxon period, leaving the 1st- and 2nd-century sequences intact.

The Pudding Lane pier

The main direct evidence for the bridge has come from excavations at Pudding Lane in 1981–2, where timber structures dating to the 1st century were recorded in a fine

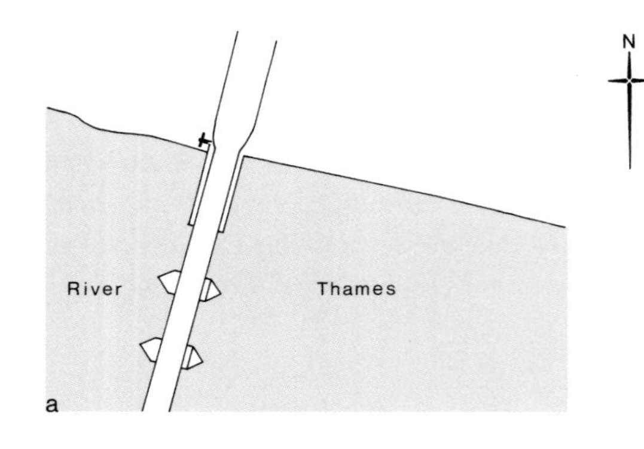

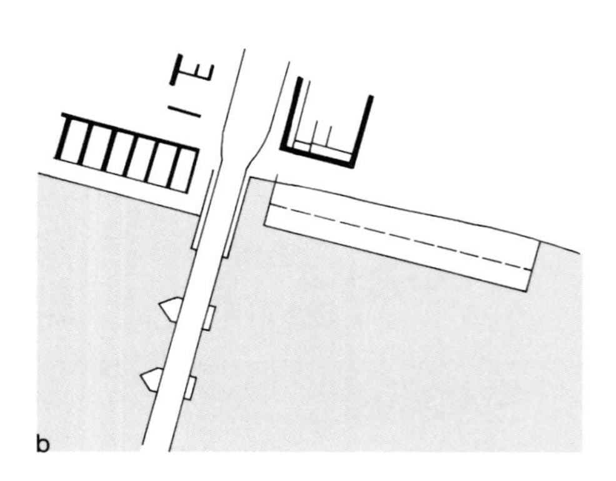

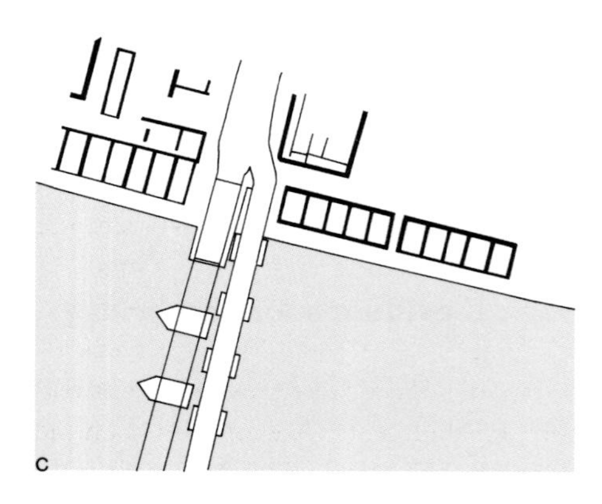

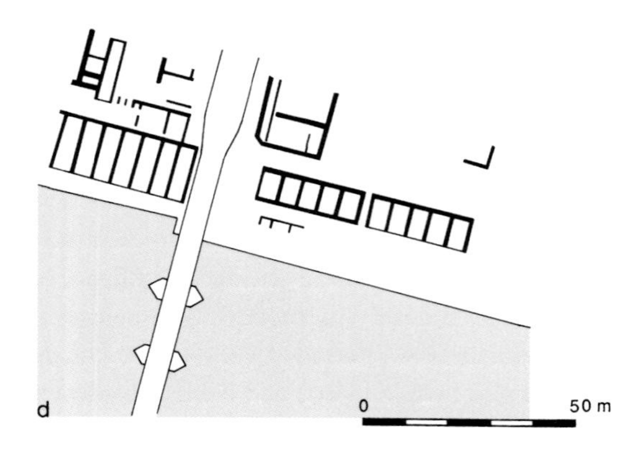

Fig 17 Plan of northern Roman bridge abutment showing the Pudding Lane and Regis House waterfronts, the 'bridge pier' and approach road: a – mid-1st-century (initial waterfronts and bridge 1); b – late 1st-century (extended waterfronts, buildings and temporary bridge 2); c – early 2nd-century extended waterfronts, buildings and bridge 3; d – late 3rd-century (final line of waterfront and riverside wall)

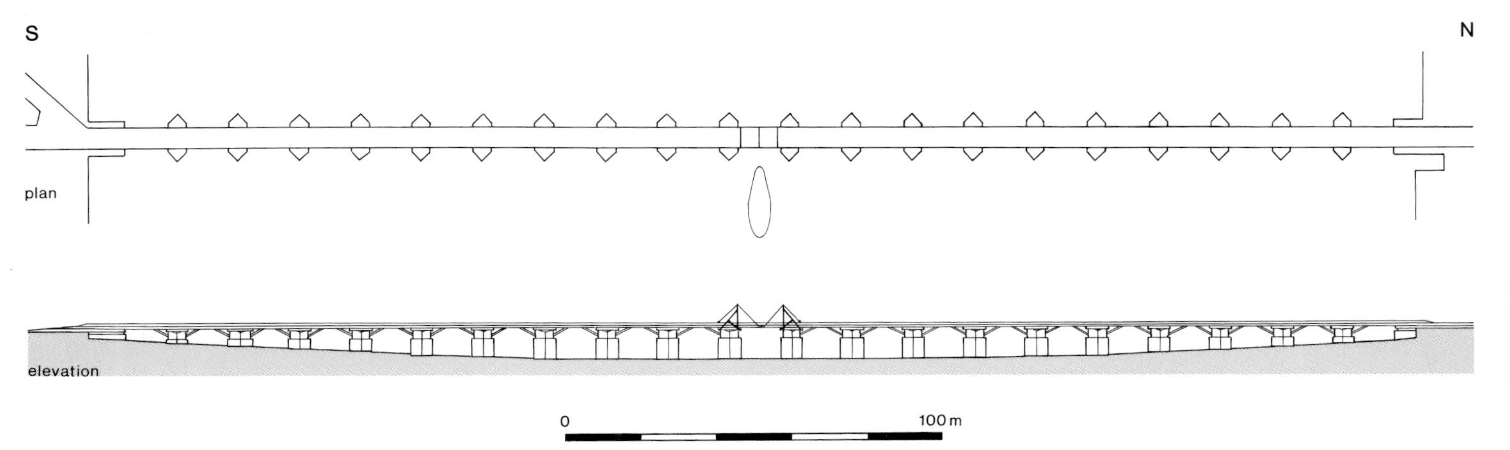

Fig 18 Plan and elevation of final phase of Roman bridge

state of preservation (Milne 1982; Milne 1985, 44–54). These structures included an early embankment and a revetment dated between AD 59–74 (possibly *c* AD 60–5), a landing stage dated AD 69–91 (possibly *c* AD 80), a box structure dated AD 78–118 (suggested date *c* AD 85–90), and a quay which replaced the landing stage, dated AD 86–106 (suggested date *c* AD 95–100).

Of these, the box structure was an anomaly, being a free-standing structure positioned across the front of an inlet at the west end of the earlier landing stage (Fig 17, b–c). It was not possible to recover the entire plan: the southern and western limits lay beyond the permitted limits of excavation. Consequently the structure has been reconstructed as a rectangular feature *c* 5.0m north–south and *c* 7.0m east–west, the west wall coinciding with the line of the modern street frontage of Fish Street Hill. The structure consisted of at least seven tiers of squared oak beams laid contiguously, attached at the corners by lap joints, and braced internally by east–west tiebeams. The top of the east wall of the structure included a shallow morticed beam, the mortice passing through the timber below and suggesting a superstructure of different form which had been removed, leaving the surviving top at 2.2m OD, some 0.2m above the contemporary wharf. The general details of construction were similar to those of the Regis House quay (AD 63–4), and to the Pudding Lane landing stage (*c* AD 80); like the landing stage, the box structure was not infilled until the superstructure was dismantled and the waterfront had moved forward to a new line in the early 2nd century.

The fact that the structure had originally survived to a higher level suggested that it was not simply an adjunct to the landing stage. Milne interpreted it as a pier base for the Roman bridge, encouraged by its proximity to the supposed position of the bridge and the lack of any apparent alternative function. There were, however, two main problems with this interpretation, which were recognised at the time. The most serious of these was that a construction date of *c* AD 85–90 suggested either that there was no bridge previously or that it had been built elsewhere. Secondly, the short life of the pier suggested a complete rebuild of the bridge within a very short period (probably less than 30 years), again on an alternative site. If Milne's interpretation was correct, three successive bridge structures were necessarily constructed between the mid-1st and early 2nd centuries, two of them apparently sited elsewhere. This was a solution suggested by Milne, but one which could not be developed further at the time owing to the lack of supporting evidence (Milne 1985, 54). Some years prior to the discovery of the Pudding Lane pier, Ralph Merrifield had already noted with some prescience that 'it is quite possible that the first bridge was a temporary military structure, such as a pontoon bridge, later replaced by a permanent bridge built on wooden piles – in a different position, since the first bridge must have continued in use until the completion of the second' (Merrifield 1965, 152 fn 37).

There were other subsidiary problems that tended to emphasise the impermanent nature of the pier and imply that it was intended to be a temporary installation. Firstly, the lack of contemporary infill within the pier was unsuitable for a large bridge which was intended to be durable. Such a structure would have required a solid foundation not dependent solely on the strength of its timbers and joints – a particularly important factor in damp conditions conducive to rot. Secondly, no abutment was encountered on the solid ground to the north of the inlet when the northern fringes of the area were examined in a later excavation at 37–40 Fish Street Hill (Bateman 1986). Thirdly, the archaeological evidence suggested that the line of the pre-existing approach road lay to the west below the present Fish Street Hill, so that the pier would seem to have been constructed off-line for no good reason. Fourthly, building the bridge in front of an inlet meant an additional span to bridge the gap, when it would have both been simpler and made better structural sense to fill the inlet and construct a solid abutment there (Fig 17, a–c).

These problems combine to suggest that the pier was not intended as part of a permanent bridge. It may instead have been the eastern end of a longer pier extending below Fish Street Hill to the true line of the bridge, but that raises the same problems of dating as those noted above, since it suggests that there was no bridge before *c* AD 85–90, and that it was modified to such an extent that the Pudding Lane end of the pier became redundant within a few decades. While not impossible, this interpretation implies that London's most important topographical feature was a victim of poor planning. It may be suggested that the pier was the eastern end of a landing stage, but in that case it should have been integrated with the existing wharves at Pudding Lane and Regis House. The inlet behind also remained unfilled until the early 2nd century, which seems a strange omission since it would have enlarged the available wharf area at a time when the port was beginning to expand rapidly.

Three bridges

The most likely interpretation of the evidence, and the one adopted here, is a modified version of Milne's original hypothesis of three bridges, using further evidence not available until recently. In the light of this, it is now possible to hypothesise that the original invasion-period bridge of *c* AD 50 (bridge 1) was probably not intended to be permanent and may even have been damaged during the Boudican revolt (AD 60–1). At any rate, the Pudding Lane pier probably represents part of a second temporary bridge (bridge 2) built slightly off-line to allow the main bridge to be replaced in a more permanent form (bridge 3) in its original position as it came up for major rebuilding work. Far from being poorly planned, therefore, a methodical approach to constructing the river crossing was adopted. The essentially temporary nature of bridge 2 was reflected in some of the points noted earlier: its lack of infill; the absence of evidence for a substantial abutment; and the awkward position off the line of the main road, sandwiched between the line of the road, some slightly later warehouses, and in front of an inlet. The inlet itself had to be traversed by an extra bridge span, and remained unfilled until the extension of the waterfront south of the pier.

The new evidence which has come to light since the Pudding Lane excavation was obtained from two sites.

The evidence from 37–40 Fish Street Hill, 1985

In 1985 excavations took place at 37–40 Fish Street Hill immediately to the north – the last quarter to be investigated of a block bounded by Lower Thames Street, Pudding Lane, Monument Street and Fish Street Hill (Bateman 1986). This site provided evidence for early activity on the natural hillside, followed *c* AD 60–70 by the construction of a massive masonry structure overlooking the riverbank, with several layers of gravel metalling on the west side. Crucially, the outer wall of the building lay just to the west of the edge of the inlet defined shortly afterwards by the construction of the landing stage, and subsequently confirmed by the later quay. The line of the Pudding Lane bridge pier lay astride the projected line of the wall, requiring an awkward bend in the line of the approach road. In addition, the gravels seen at Fish Street Hill were not cambered, which suggests that the actual edge of the street was a little to the west, making the obligatory bend still more acute. It seems far more likely, therefore, that the main bridge was always to the west and reached by a straight approach road, since the evidence from Fish Street Hill showed that the road was already in existence by the time the Pudding Lane box structure was built. Introducing an unnecessary bend would have been illogical.

The evidence from Regis House, 1929–31 and 1994–6

The second relevant site was Regis House (Brigham et al 1996b) on the west side of Fish Street Hill, recorded during major excavations in 1994–6 and observations by Gerald Dunning of the Guildhall Museum in 1929–31 (GM 248). This evidence includes two crossed beams recorded by Dunning in front of the natural riverbank on the eastern site boundary, and resting directly on the London Clay which formed a promontory at this point (Fig 17, a). The upper beam, aligned east–west, continued beyond the perimeter under the bridge approach, implying that these timbers were the earliest feature in the area. These may well represent the dismantled abutment of the earliest phase of the first bridge. In line with the rear of the structure further west was a revetment constructed in AD 52, forming the earliest riverbank.

Dunning also recorded a north–south revetment on the site boundary to the south of the crossed beams, intended to protect a tongue of land on the line of the bridge approach and projecting at least 15m south of the AD 52 waterfront. One possible explanation is that this structure replaced the initial abutment. The fact that there was no similar structure at Pudding Lane, where the inlet occupied a comparable position, must demonstrate that the first bridge lay beneath Fish Street Hill, probably towards the western side of the road. The revetment was replaced in AD 63–4 by a massive timber quay whose timbers stopped at the site boundary on Fish Street Hill, presumably against the revetted abutment. A line of contemporary warehouses behind the quay also stopped short of the street, presumably to allow access down from the approach road to the new wharf.

A later east–west revetment constructed at Regis House in AD 102 still took account of the abutment, if Dunning's plan and a contemporary section have been correctly interpreted.

Two sections recorded in 1995 on the eastern site boundary provide evidence for the approach road, dumped gravels making up an embankment behind the line of the revetment and sloping southwards from 3.4m OD to 2.8m OD to carry the road on to the abutment. At the southern end of the site, Dunning's east–west section across the street edge showed that gravels were dumped up to the edge of the revetted abutment, probably after the AD 102 revetment was built and very probably after the Hadrianic Fire of *c* AD 125. By that time the waterfront had moved southwards at least once at both the Pudding Lane and Regis House sites, incorporating the abutment firmly within the quayside. The approach road was necessarily extended at a higher level to serve new buildings laid out on the west side. Dunning's section showed gravels continuing above 3.0m OD, and a section to the north proved that the road was regularly resurfaced, reaching an eventual level at its margins of 4.4m OD, opposite the internal floors of a series of new post-Fire buildings. To this, a further 0.5m should be added to give an approximate level for the crown of the road.

Summary

In summary, the evidence shows beyond reasonable doubt that there was an early bridge abutment immediately beneath Fish Street Hill, antedating AD 63 and possibly constructed *c* AD 52. The abutment was defined on the west side by a revetment projecting well beyond the contemporary riverbank, while the east side lay somewhere below Fish Street Hill, bounded by an inlet. Two earlier crossed beams may represent an initial Conquest-period structure, possibly the landward abutment of a pontoon bridge. Because the Regis House site was built up after AD 63 the inlet on the eastern side at Pudding Lane was the only suitable site for a replacement bridge which may have been constructed *c* AD 85–90 as a temporary measure.

4.4 The southern bridgehead

The study area

A major part of this evidence is summarised in Chapter 3 as part of the presentation of the Fennings and Toppings Wharves site sequences. Early features only a little later than the establishment of bridge 1 survived in truncated form in the area behind the original river embankment. Any elements of a bridge approach and abutment in that area ought, therefore, to have survived as well, although admittedly the temporary bridge tentatively identified at Pudding Lane (bridge 2) would have left few traces, particularly as the north bank evidence suggests that it was largely dismantled. The complete absence of masonry or timber remains identifiable as part of bridges 1 and 3, however, strongly suggests that no permanent southern abutment appears to have been located in the study area. Indeed, the fact that later erosion was so complete in the north demonstrates the absence of any substantial Roman installations capable of resisting the river. It is in any case clear from the density and distribution of Roman features of all dates (Fig 9; Fig 12) that there was no suitable space for the bridge and its approach road on any part of the study area. In the absence of any direct evidence for the bridge, we must, therefore, turn to the indirect evidence.

The approach roads

The general line of the two bridge approach roads where they pass through the main part of the Southwark suburb has been fairly well understood for at least 25 years (Fig 15), and recent excavations have not substantially modified this. The line of the easternmost route, Road 1, has recently been found to run immediately to the west of the Borough High Street Ticket Hall excavation for the Jubilee Line extension at London Bridge Station. The arguments for the location of the bridge abutment have been rehearsed in a number of papers and articles (eg Dawson 1969, 1970, 1977; Merrifield and Sheldon 1974, 1983; Sheldon 1978), and it has previously been plotted at the point where the roads were all thought to meet the presumed line of the Roman waterfront, some distance to the north of the present riverwall. This extended position (Fig 15, position A) was intended to take into account the erosion of Roman deposits in the early medieval period, but led to the road junction being extended considerably as the plotted lines actually met some distance to the south of the supposed line of the waterfront. On the basis of finds evidence the eastern road, Road 1, has always been assumed to be the earlier of the two, being part of Watling Street (Sheldon 1978, 24–5; Graham 1988, 24). Sheldon would put the construction of both in the consolidating governorship of Didius Gallus (*c* AD 52–7), which would place the first Regis House waterfront, the bridge and the roads into the same phase of operations. Whatever the date, the primacy of Road 1 makes its alignment crucial to the position of the bridge abutment.

A reappraisal of the location of the southern abutment

The reanalysis of Toppings Wharf now shows that although the Roman waterfront moved northwards as the river level fell, and by the 3rd century may indeed have been located somewhere at the edge of the present river, the original 1st-century frontage contemporary with the presumed construction date of the bridge was actually some distance to the south. This immediately shifts the likely position of the bridgehead to the south and west along the axis of Road 1 (Fig 15, position B). Assuming that there was a revetted abutment similar to the structure described on the north bank, it is likely that the north end of Road 1 changed course slightly to align itself with the approach ramp. This would also have aligned the road with the buildings and waterfront in the study area, instead of at an angle. Road 2 may have turned eastward for a short distance, parallel to the waterfront. Such a realignment was suggested by Dawson, who excavated this road at Montague Close just to the west of the bridgehead (Dawson 1976, 42). The advantage of this relocation of the southern bridge abutment is that the line of the bridge now becomes a direct continuation of the approach road on the north bank (the present Fish Street Hill and Gracechurch Street), which led up to the front gate of the forum. The bend necessitated by the postulated alignment disappears, and the junction with the south bank roads is far less awkward. An impressive view of the forum would have met travellers from the south as they crossed the bridge.

4.5 Artefactual evidence for the bridge

19th-century evidence from the river

In addition to the relatively recent structural and topographical evidence, there is evidence for the bridge from the river itself, in the form of coins and artefacts dredged up and excavated during the construction of Rennie's bridge and the demolition of its medieval predecessor. This included finds made during work to construct the new abutments and approach roads, which are not, strictly speaking, related to the Roman bridge. The material found included coins dated from Augustus to Honorius and a series of bronze figurines and statues, or parts of statues, including the well-known head of Hadrian found in 1831 or 1832. These extensive finds, many of which are now in the British Museum, were reported in a wide variety of journals, often by the antiquary Charles Roach Smith (eg Smith 1840, 1841, 1846) and others (Knight 1834). Many of the coins and some of the bronzes found were in dredged material which had been removed and used for resurfacing work on the banks of the Surrey Canal at Deptford, on the towpath between Hammersmith and Barnes, and at Putney (Smith 1846), where some artefacts were still being recovered as late as 1845–6. The reports of the finds have already been drawn together and summarised (Parsloe 1928) and the coins recently discussed (Rhodes 1991); they will, therefore, only be discussed briefly here.

Amongst the earliest finds (in 1825) was a silver figure of Harpocrates, possibly located during excavations for the southern abutment, where the lead figure of a horse was also found, although Smith believed it came from the Thames, presumably from the one of the new piers under construction (Parsloe 1928, 194). The main concentration of finds came in the next decade during the removal of the medieval piers opposite the second and third arches of the new bridge immediately to the north of the chapel pier. Although this point was only a third of the way across the medieval bridge, it can now be seen to coincide almost exactly with the centre of the Roman bridge (Fig 15, E), extensive medieval reclamation having taken place on the north bank while the south bank has remained relatively static. The survival of the cluster of finds may owe much to the shifting of the centre of the channel southwards as a result of the uneven reclamation, since the most extensive dredging in the medieval and post-medieval periods can be expected to have occurred where the channel was deepest.

The concentration of artefacts below the piers of the Colechurch bridge clearly pointed to the existence of a Roman bridge in the vicinity. The volume of material recovered during excavations for the abutments and piers of Rennie's new bridge was much smaller, and this itself narrowed down the probable location to the area downstream of the new bridge.

Votive offerings

The position of the objects in relation to the bridge centre makes sense if they were votive offerings, and their number, nature and variety in any case preclude the likelihood of their being casual losses. Some, including weights, the balance beam of a set of scales, and a model of a ship's prow, reflect the trading nature of the town. There were also objects in other materials, including the head of a marble statuette, pottery, and weapons including a knife and a javelin head.

The coins found included medallions of Aurelius, Faustina and Commodus, which came from the central piers of the old bridge. Smith considered that these objects were probably thrown into the river as commemorations of particular events, presumably such as the building and repair of the bridge, the accession or death of an emperor, and possibly visitations, such as those made by Hadrian and Septimius Severus (Smith 1846, 113). More recently, however, Michael Rhodes has suggested that the periodicity of the coin deposition in most cases simply reflected fluctuations in the production of Roman coinage, something which was not appreciated in the 19th century, and further that the coins were votive offerings to the divinity of the bridge or the river, rather than commemoratives (Rhodes 1991, 183).

The true volume of finds of all categories cannot be established, since although Smith recovered many coins and objects as the dredging progressed, and removed others from their secondary deposition in gravel riverbank consolidation, many were undoubtedly never recovered, while further objects were sold by workmen before Smith was able to retrieve them. The smaller coins of the 4th century may have evaded collection in the first place. The number of artefacts known, however, is sufficiently large to suggest that there was a long-standing tradition of making offerings.

By contrast, the presence of the images of pagan deities, some mutilated, such as representations of Apollo, Atys and Mercury, has been attributed to a fit of Christian zeal (Smith 1846, 113). Certainly, there are examples of temples being demolished and sacred images mutilated and ritually buried, as was the case at Uley in Gloucestershire (Merrifield 1987, 96). However, Rhodes has suggested more convincingly that the London bridge offerings in some cases represented the standard practice of ritually depositing representations of afflicted body parts, so that the images could be pagan offerings. The mutilation of the majority of these objects – for example a group of five figurines – by the removal of limbs, and in some cases by scarring, may represent ritual 'killing' before being offered up (Rhodes 1991, 183–4). The figurines were of a type kept either in family shrines or as votive offerings in temples, and their use in this way is, therefore, not surprising. Some coins may also have been deliberately bent and broken.

On the Continent, annual rites seem to have been held which may throw some light on the origins of this practice (Steinman and Watson 1941, 43). In Rome itself a pageant was held on the ides of May, involving a parade of members of the *collegium pontifices* (the priestly 'college' of bridge-builders), *praetores* and Vestal Virgins. After collecting a series of rush images at 24 shrines passed *en route*, the Virgins threw the images into the River Tiber from the Pons Sublicius. Similar practices were observed in Germany, including the town of Halle and parts of Bavaria. All of these rites may hark back to actual human sacrifices in much earlier times, represented by the celebrated 'bog bodies' found in northern Europe, particularly Denmark (Glob 1971).

A middle view of the reasons for the disposal of these objects was discussed in some detail by Ralph Merrifield (1987, 96–106). As he suggested, precisely the same methods seem to have been used to dispose of unwanted pagan images as had been used to make offerings to the deities of earth and water that they represented: '... it may even be doubtful whether a deposit of this kind was made by pagans as an offering or by Christians as a means of despatching a dangerous instrument of their spiritual enemy to the place where it belonged.'

One particular antiquarian reference (Anon 1833b) seems to point to the discovery of a group of 200 coins within one of the medieval bridge starlings which, while it may have been a votive offering or the contents of a lost purse or bag, could equally have been an *in situ* foundation offering. If this were the case, then at least one of the medieval starlings coincided with the position of a Roman bridge pier. Many of the other coins seem to have come from a single location opposite the second arch of the new bridge. This concentration makes it extremely unlikely that the coins (and other finds) were simply deposited by the river itself, which has been suggested in the past (Parsloe 1928, 192–4). As Rhodes pointed out, the good condition of the majority of the Roman coins demonstrates that they had been recovered from undisturbed silting in anaerobic conditions (Rhodes 1991, 182).

The concentration of offerings has led Rhodes to suggest the presence of one or more shrines on the bridge. He cited the discovery of three (of four) surviving altars on the bridge at Cendere Çay in Turkey and of two matching altars to Oceanus and Neptune during reconstruction work on the medieval bridge at Newcastle-upon-Tyne following a flood in 1771 (Rhodes 1991, 184). The presence of such altars seems very likely on a bridge crossing a river of such great significance as the Thames, and a maritime theme including statues of sea gods would have been highly appropriate to London with its reliance on commerce. At the same time as the altars, votive coin deposits similar to those found in London, although not so rich, were found at Newcastle (Bruce 1885, 1–11). These dated from Trajan's reign in the early 2nd century to an unidentifiable 3rd-century radiate, and included a coin of Faustina the Elder, struck to commemorate her deification, and perhaps ritually deposited to invoke her favour.

A substantial Carausian-Allectan peak observed among the London bridge coin deposits may, therefore, be seen as tokens to enlist divine support for the imperial pretenders but, perhaps more likely, as donatives to welcome the fleet of Constantius Chlorus who rescued London from Allectus after the latter's men had ransacked it in AD 296. A gold medallion known from a single example from Arras was struck to commemorate the event, depicting a grateful personification of London kneeling to the hero (Merrifield 1983, 200 and fig 49, 201). The medal shows a war galley on the Thames in front of a representation of the gates of London, but as the medallion was struck at Trier the depiction of gate towers flanking the bridge cannot be conclusive, although the riverside defensive wall had stood for at least 20 years by this time and such an arrangement must be highly probable.

The striking of the medallion in itself suggests that the event was seen to be of particular significance. As Merrifield suggested: 'To London, the creation of the Romans and always more Roman than British, this restoration and the ending of a long period of isolation (AD 286–296) must have been particularly welcome' (Merrifield 1983, 200). What better way for a crowd gathered on the bridge to give thanks than to shower the imperial fleet with coins of the usurpers? It is also possible that the radiate coins of the preceding period (AD 217–73), including many copies, were deposited at the same time, since some hoards of the period – for example two from Verulamium (discovered 1930–3), one from Peterborough (found 1904) and one from Coleford, Gloucestershire (1852) – also contained coins of Carausius (and Allectus in the latter instance). It is clear that these 'barbarous' radiates remained in circulation and production as small change to the end of the 3rd century (Hill 1967; Rhodes 1991, 187). The volume of coins deposited at this time may, therefore, have been even greater than is apparent from the date of the coins themselves.

Finally, the famous bronze head of the Emperor Hadrian (discovered near the Southwark side of the bridge) may have been deposited during this period, since it does not fit easily into the normal pattern of votive deposition. He was not an unpopular emperor, but it is possible that his statue was seen as an imperial symbol and was dismembered during one of the several insurrections which affected Britain, of which the Carausius/Allectus episode was the most notable. Parts of several other life-size bronze statues were found elsewhere, including two hands found in Lower Thames and Gracechurch Streets, the latter probably from a statue in the forum (RCHM 1928, 44). In the case of the London Bridge find, it is even possible that only the head was removed, ritually deposited and replaced by the portrait of an imperial pretender, such as Carausius or Allectus, although this is not paralleled elsewhere.

4.6 Structural evidence for the bridge

In association with many thousands of Roman coins, Charles Roach Smith noted an 'abundance of broken Roman tiles' (Smith 1846, 113), which may constitute the only evidence for the construction of a masonry bridge since there can be no other reasonable explanation for their presence. Unfortunately, no further details were given. In this respect, the 'conglomerate' referred to in connection with some of the coins and metal finds (Smith 1841, 154–8) may in fact have been the remains of Roman concrete which when set under water can appear as a natural concretion, similar to Hertfordshire pudding stone. Such conglomerate was also encountered 'among stout oaken piles with iron shoes' found in the 19th century 'slightly to the east of the site of Peter of Colechurch's erection, and extending from the Surrey to the Middlesex shore; the conglomerates were huge and composed of ferruginous matter. Among them were numerous Roman coins, chiefly first and second brass of the higher empire' (Cuming 1887, 166–7). It was supposed that the conglomerates were the result of corrosion around the iron-shod piles, although this explanation alone would hardly account for their 'huge' size. It is possible that the corrosion had actually impregnated Roman concrete or gravel infill at the base of the piers where the piles were located to form a mass resistant to erosion and dissolution.

Cuming was certainly of the opinion that the iron of the shod piles was of a quality produced by the Romans, and not 'the porous, ropy iron characteristic of the Teutonic forge', an observation supported by the presence of Roman coins. The presence of the conglomerates immediately downstream of the medieval bridge may have been due to movement by the strong river current, later dredging and, perhaps, demolition or clearance of the Roman bridge footings to make way for Colechurch's bridge. Such clearance must in any case be postulated to account for the complete absence of substantial remains of the Roman structure during the removal of the medieval bridge. Observers such as Stow, writing at the very end of the 16th century, fail to mention any finds of Roman masonry or timberwork in the river.

4.7 The coin evidence: a date for the bridge?

In his recent review of the coin evidence, Michael Rhodes (1991) suggested a date for the establishment of the bridge by studying the rate of loss, although this could admittedly only take into account minting dates, not the actual date of deposition, since there was no context for the finds. On this basis, Rhodes suggested a construction date of AD 68–81, when there was an apparent peak in deposition, matching contemporary loss rates in the city. This hypothesis was based partly on the presence of an earlier peak of coin loss in Southwark (AD 43–68), whose initial development would, therefore, seem to have outstripped the city until some time after AD 68. Southwark could, therefore, have been an early focus of settlement, with the north bank only beginning to develop more rapidly with the construction of the bridge.

However, as already discussed in some detail above, work at Regis House in 1994–6 (Brigham et al 1996b) supports a construction date for the first bridge of *c* AD 50. In addition, new evidence for pre-Boudican activity in the city now shows that Southwark and the north bank both developed at the same early period: good evidence for buildings of this date has been recovered from the south bank at London Bridge Station, and from the north bank at Regis House and No 1 Poultry, for example, with very early post-Boudican restoration work at Regis House (AD 63–4) and 74–75 Cheapside (AD 61–2) dated by tree-ring dating (Tyers 1995, 6). The low coin loss rate along the line of the bridge for the earliest period cannot easily be explained, therefore; perhaps the practice of using the bridge habitually as a place for depositing votives did not develop immediately. Certainly the number and density of coins found within one small area of the riverbed suggest the existence of one or more pagan shrines on the third phase of the bridge (Rhodes 1991, 184).

The number of Roman coins being thrown into the river declined sharply during the period AD 330–48. Rhodes (1991, 190) has interpreted this decline as implying that the 'bridge had been swept away' by this date and replaced by a ferry. Certainly this decline in coinage is not connected with the abandonment of the northern bridgehead, as this area was occupied until *c* AD 400 or possibly a little later (Watson 1998a, 100). The number of mid- to late 4th-century coin finds in Southwark suggests that a high level of economic activity was sustained until after AD 388 (Hammerson 1988, 425). One possibility is that the sample of 4th-century coinage found during dredging has been biased by the relative ease with which the river current scattered some of the small, lightweight coins issued during this period. Another possibility is that following Constantine's edict of toleration in AD 313 and the documented presence of a bishop in London by AD 314 (Ireland 1996, 204), pagan shrines such as the one on the bridge quickly went out fashion. However, it should be noted the practice of throwing money into the river did not entirely cease during the early 4th century, as coinage of Honorius (AD 395–423) has been recovered from the site of the bridge (Parsloe 1928, 193).

Rhodes has suggested that the cost of maintaining such a massive structure may have been too much to endure, particularly in a period when the town forum basilica had itself been demolished. This is certainly a possibility: the forum was demolished in the late 3rd or early 4th century (Brigham 1990a, 77), and the port facility had also been largely abandoned and dismantled in the mid-3rd century prior to the construction of the riverside defensive wall *c* AD 270 (Brigham 1990b, 140). Other public buildings demolished in the late 3rd or early 4th century include the structure identified as the Governor's Palace near Cannon Street Station, and the amphitheatre. Southwark also suffered, moreover: parts of the large and possibly public building at Winchester Palace were demolished for example (Yule in prep).

On balance, Rhodes's hypothesis depends upon the importance the Romans placed on maintaining the road link at a time when they were demonstrably not interested in maintaining either a large-scale port or an expansive public works programme. It is important to note that the river was still an important resource, bringing in a high proportion of regional Romano-British ceramics, although imported wares were increasingly rare from the late 3rd century onwards (Symonds and Tomber 1991). This may suggest that there was still a port facility in the area, although possibly not in London itself. If road and river traffic had declined to a sufficiently low level, it is possible that the bridge was no longer maintained in the 4th century, particularly if the maintenance depended on the receipt of tolls and customs. However, during the last quarter of the 4th century the legal responsibility for maintaining major bridges lay with the leading landowners of the *civitas* territory. According to the Theodosian code (issued AD 370, 382 and 390) bridge building was one of a number of public duties citizens were expected to undertake. Bridge building was considered menial work, one of the *sordida munera*, from which all men of high status were exempt (Mommsen and Meyer 1905, xi.16.18, p 390). To what degree these codes were followed in distant provinces like *Britannia* is uncertain. In particular it depends upon how London was governed in the late 4th century; the available evidence is both limited and contradictory, but it is possible that the bridge might have been maintained for its strategic significance, even if its commercial importance had declined.

4.8 Roman bridge construction

Continental parallels

The available Continental evidence for timber bridges, summarised in O'Connor (1993, 132–49), includes depictions on the, respectively, early and late 2nd-century AD triumphal columns of Trajan and Marcus Aurelius. As O'Connor pointed out, the use of scaffolding and timber centring for the construction of masonry bridges in effect meant building a timber bridge first. Familiarity with the very different technology for constructing timber bridges was, therefore, a prerequisite. Building in timber in effect used beams as lintels, supported by cantilevers or braces; building in stone used the properties of shaped stone voussoirs or rubble set in a concrete mass to construct self-supporting arches.

Details of many timber bridges can be examined in a recent republication of 19th-century plates of Trajan's column (Lepper and Frere 1988). These include a bridge of boats with a braced pile timber abutment and handrails (pls VII–VIII, scenes iv–v), several examples with piles supported by arched braces or cross braces, and with plank decking and handrails (pls XII, scene xii; XIII, scene xiv; XIV, scene xvii; XVI, scene xxi; LXXIV, scene ci), and at least one with unbraced piles and handrail (pl XLI, scene lviii). There is also a depiction of the Roman army at work building two small bridges across the ditch of a fort under construction (pl XV, scene xix). In addition to this evidence, there is a well-known description of a temporary bridge in Caesar's *De bello Gallico* (Caesar, IV). It would be dangerous, however, to regard this as representative of a temporary bridge crossing the Thames in the Claudian period, since Roman carpentry techniques had undoubtedly changed in the intervening century. A temporary bridge with braced piles but no handrails shown on Trajan's column is much closer to the period (Lepper and Frere 1988, pl XCVII, scene cxxxi).

A bridge recorded crossing the Rhine at Mainz (Cüppers 1969, 198) was based on rectangular caissons *c* 12m by 7.5m, whose construction closely resembled the 2nd-century quay constructed in London at Swan Lane (Brigham 1990b, 150–6). Dovetailed lap joints were used to attach three levels of tiebeams to the caisson wall, and the four tiers of beams which formed each wall were joined with free tenons set in small mortices. The caisson wall and the tiebeams were lined on both sides by rows of piles that may have helped to guide the timbers into position during construction. There was also a regular series of supporting piles within, and a triangular grid forming a cutwater at one end, the cutwater itself being formed of massive ashlar blocks. A bridge at Laupen (Cüppers 1969, 185) used a different arrangement with rectangular caissons *c* 6m by 5m of several levels of paired beams clasping posts set at the corners and midway along the longer side, with triangular cutwaters *c* 3m long built integrally at one end. This structure used lap joints more reminiscent of the 1st-century waterfronts in London and the Pudding Lane pier.

Mortices set into the top tier of beams of both bridges were presumably cut to retain the base of timbers supporting the bridge deck.

At Trier on the Mosel (Fig 19) a bridge was constructed *c* AD 44–70 with ashlar piers and a timber deck supported by understruts springing from the piers (Cüppers 1969, 65–70). This was replaced by a similar bridge *c* AD 140, and in various versions survived in this form to the late 14th or early 15th century. The piers were constructed within double-walled cofferdams, 27m by 14m, which in four instances were constructed of horizontal baulks, lap jointed at the corners, a method which changed in the three piers situated in the deeper

Fig 19 *Reconstruction drawing of a pier of the Roman bridge at Trier, showing the cofferdam (Landesmuseum, Trier)*

Fig 20 *The Apollodorus bridge on Trajan's column (from Richmond 1982, pl 13, © British School at Rome)*

central channel. Here, grooved piles were set in pairs to retain a double skin wall of planking, which was slotted into position. Integral cutwaters were incorporated at both ends. Within the cofferdams, the alluvial silts covering the riverbed were excavated to more solid gravel, and 0.5m diameter iron shod piles were driven through to support timber base frames on which the piers could be built.

A final Continental parallel is a bridge constructed across the Danube between Turnu-Severin and Drobeta on the Romanian-Yugoslav border *c* AD 104–5 by the engineer Apollodorus (Fig 20). For this bridge there are both written details (Cassius, LVIII.13) and an illustration on Trajan's column (Lepper and Frere 1988, pl LXXII, scenes xcviii–xcix; Richmond 1982, 113). The column shows massive masonry piers which Dio Cassius quoted as being 20 in number and 170 feet (*pedes monetales*) apart, although this great distance (50.32m) was more likely to refer to the centres (O'Connor 1993, 142). The piers were 60 *pedes monetales* wide (17.76m), giving a true span of 110 *pedes monetales* (32.56m). This was still a considerable distance to bridge with timber girders, and could not have been achieved without an elaborate supporting framework.

This bridge, at almost 1.1km long, was three times the length of the London crossing and far higher. Apollodorus left a substantial vertical distance between the top of the piers and the level of the bridge deck to allow the use of complex trusses. Not only was this more economical financially and in the use of masonry, the potential instability of very tall, heavy piers was much reduced. The trusses also allowed the piers to be set much further apart than would otherwise be possible because the long timber spans could still be supported adequately. Where the height of the bridge deck above the river was constrained by that of the banks – as was the case in London – such complex solutions could not be used with the available materials, and the spacing of the piers was, therefore, determined by the available length of timbers.

Trajan's column shows that the piers of the Danube bridge supported triangulated supports for a series of understrutted trusses beneath a timber deck with handrails. The understruts formed angular, segmented arches, each one with three levels connected by radial timbers attached to the underside of the deck girders. Each arch consisted of four segments presumably with cross braces running across the width of the bridge to connect at least two rows of trusses, one at either side of the bridge, and quite probably four or more, given the width (Fig 20). The arches would have supported the longitudinal girders, which in turn must have supported lateral joists to carry the planked deck. The deck presumably carried a metalled roadway of some kind, possibly laid on a clay or sand bed. In building the bridge, Apollodorus may have used a modified form of roof truss developed to span buildings whose width exceeded the limitations of the available timber resource; the sectional arches would have been ideal for buildings with barrel-vaulted ceilings such as bathhouses.

British parallels

As on the Continent, evidence for timber bridges in Britain is uncommon compared to the more durable remains of masonry structures, such as those recorded on Hadrian's Wall (Bidwell and Holbrook 1989) and in its hinterland (Dymond 1961). In those areas timber bridges have been identified from lack of masonry evidence rather than positive remains, for example at Hunwick Gill, Stockley Gill and Willington South Dene on Dere Street. These were represented solely by earthen embankments carrying the roadway on to the bridge (Dymond 1961, 139–42).

There are, however, a few examples where sufficient details can be obtained to build up a picture of timber bridge building in the province. Of these, the collapsed remains of a bridge crossing the Nene at Aldwincle, Northamptonshire, found in 1968–9 provide the clearest evidence (Jackson and Ambrose 1976). The superimposed remains of at least three successive bridges were found (Fig 21), and it is likely that many of the elements of the latest bridge were salvaged from its predecessors. The width of the Nene in the Roman period was estimated to have been *c* 24.4m, which was considerably less than the Thames at *c* 350–360m, and yet among the recorded timbers of the final bridge (bridge 3) were several of similar dimensions to those employed in the 1st-century London quays. It can confidently be asserted from this that timbers used in the Thames crossing were likely to be at least as large.

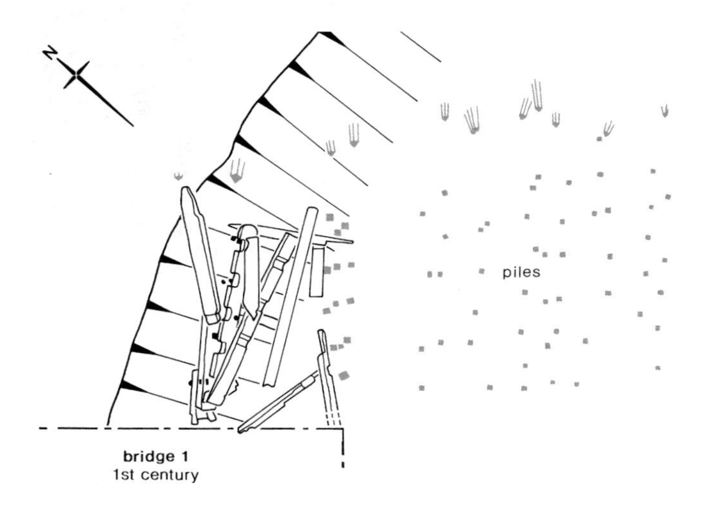

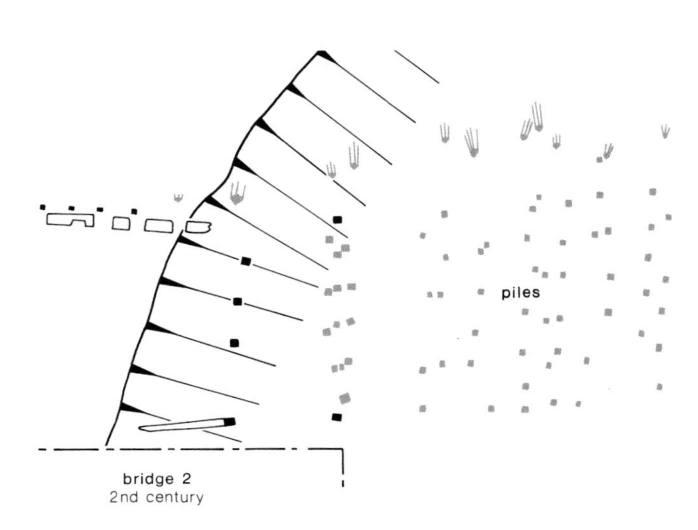

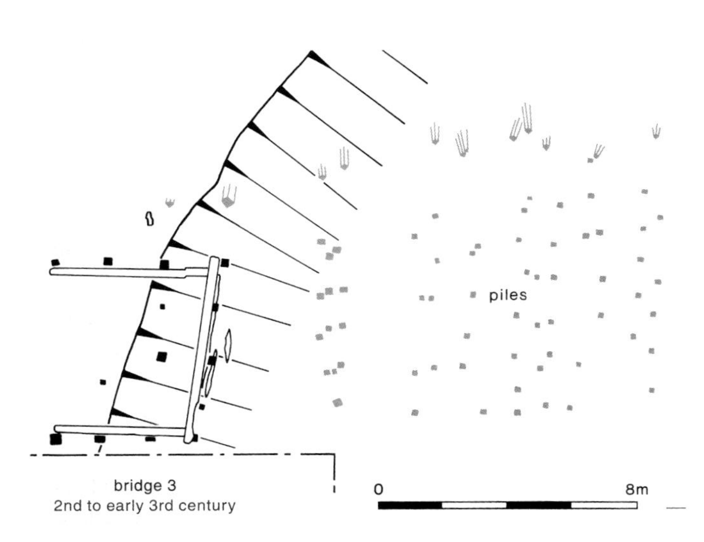

Fig 21 Three plans showing the sequence of Roman bridges at Aldwincle (redrawn from Jackson and Ambrose 1976, fig 3)

At Aldwincle the largest timbers were 6.05m long and one example was 0.51m by 0.41m in section, while two further examples were 0.43m square. The first of these beams was cut at 0.74–0.86m intervals with 0.18–0.23m deep half-lap housings wide enough to take cross beams of secondary scantling, 0.36–0.43m wide. One of the other two timbers had shallow housings 0.36m wide and only 40–80mm deep, set *c* 0.9–1.0m apart, with a longer slot 0.61m by 0.28m deep on the opposite face. There was a set of smaller timbers up to 6.0m long and 0.33–0.36m square, with similar slots. A single plank fragment 2.29m long, 0.45m wide and 51mm thick was also noted (Jackson and Ambrose 1976, 45).

Not all of the timbers can be identified as having a particular function, but those that can throw considerable light on the possible construction of the bridge. It can be suggested that the deck was formed in the same way that a large timber-framed floor is still built today. Substantial horizontal girders formed the edges of the deck, with binders laid across the width, retained by half-lap joints. The size of the joints suggests that some of the binders were of similar dimensions to the girders, but most were slightly smaller; it is not clear whether there was an alternating pattern of large and small, or whether this depended on the availability of timber. Parallel to the girders, lighter joists were lapped across the binders, forming the base for crossplanking. This retained the metalled road, which was possibly laid on sand, matting, or clay rather than directly on the planks. The tops of the girders may have acted as kerbs. Since the single recorded example of a girder was 6.05m long, four spans would be required to cross the river, and the deck was supported on lines of piles or trestles set around half that distance apart. It is assumed that tiebeams connected the heads of the piles or trestles, and that the girders were either jointed to these or simply rested on them. The width of the binders was similar, and allowing for the lap joints, the road deck would appear to have been *c* 5.0–5.5m wide. The abutments may have been reused from the previous bridge, being horizontal beams retained by piles to form a three-sided box.

A second example of a timber bridge was discovered at Wallasey Pool, Birkenhead, Cheshire, in about 1850 (Massie 1849–55; Watkin 1886). This structure (Fig 22) was *c* 30.48m long, and consisted of four rows of squared oak beams laid *c* 1.83m apart. Each row consisted of three contiguous tiers of timber in three separate lengths 10.16m long. The timbers forming the rows were 0.45m wide and 0.23m deep (corresponding to 1.5 *pedes monetales* – one *cubit* – and one *span* respectively). There were traces of joists crossing the span, although these were too decayed to ascertain any further details. The timbers crossed a deep channel filled with silt, but beneath this were the remains of two stone piers, which appear to have been *c* 8.4m long, compared with a width for the bridge deck itself of *c* 7.0m, and *c* 3.1m wide, although they were badly damaged. Curiously, the available section (Massie 1849–55, 57) leaves a gap between the tops of the piers and the timbers, which are shown continuing straight through the river silt with no visible means of support. There

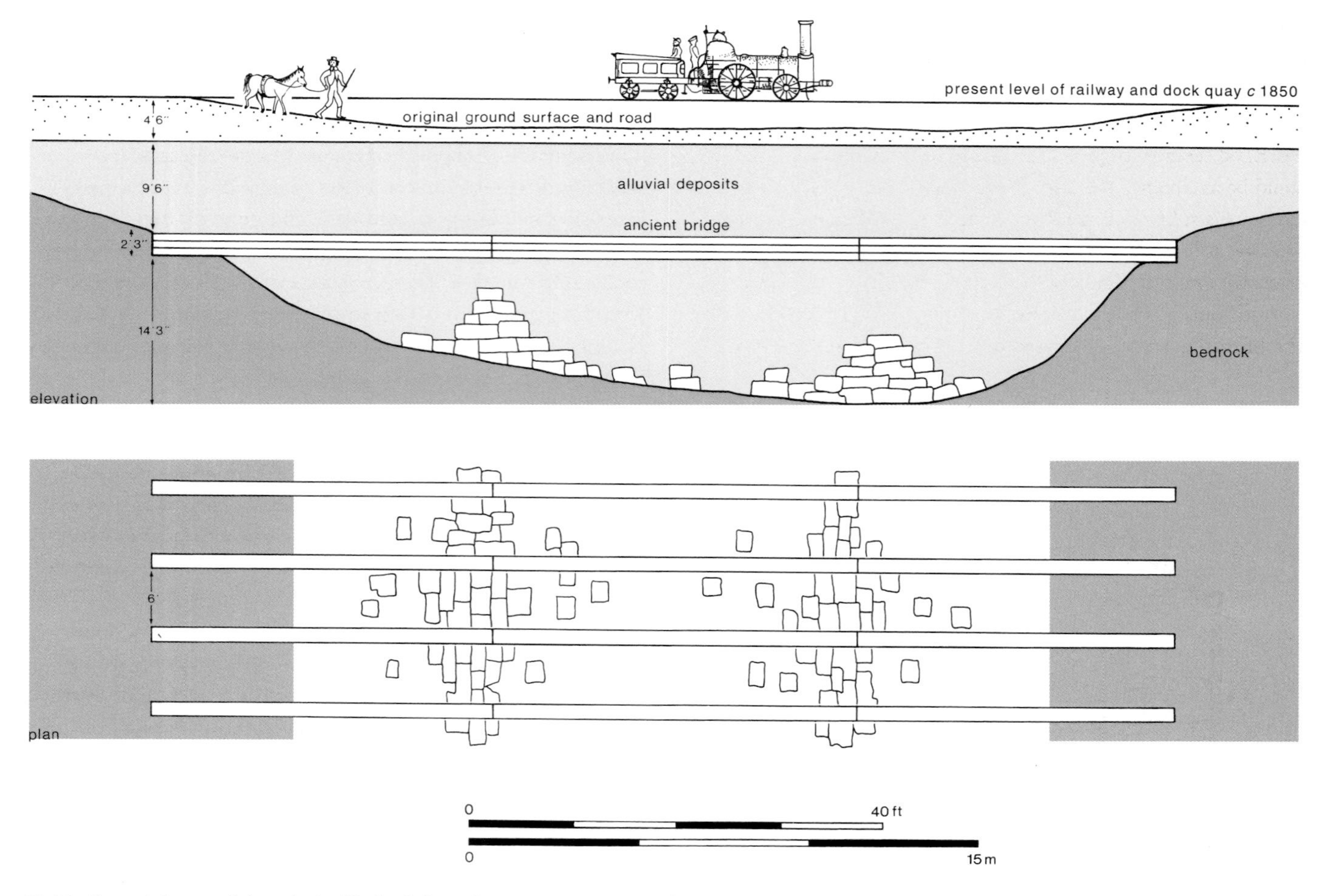

Fig 22 *Plan and elevation of the Birkenhead bridge (redrawn from Massie 1849–55, 57)*

is, therefore, no way of determining how the timbers were attached to the piers, if at all, although the piers occurred directly below the joints in the spans.

The beams had mortices in the upper surfaces and, immediately adjoining, angled housings along one edge, clearly for the attachment of a trussed handrail, identical to those seen on many of the Trajan's column illustrations, including Apollodorus' bridge across the Danube. As Massie suggested, the tenons of the handrail posts and the feet of the crossed rails served to bind together the various tiers of the compound girder, although it is not clear from the report whether the joints passed through all three levels.

It is assumed that the Romans adopted composite girders because large timbers were not available, although their dimensions suggest that the individual beams were halved from baulks *c* 0.45m square. This would suggest that 18 trees were used for the main beams, not 36 as Massie suggested (Massie 1849–55, 56). It seems likely that the beams were used to create a compound girder, but that each layer may also have had different functions: the upper level may have acted as a kerb for the roadway, as well as supporting the handrail; the middle tier could have supported the joists of the road decking; and the lower tier was the subframe, resting on the piers and stiffening the structure.

A consideration of the bridge at Rochester needs to be included in any discussion of the Roman London bridge (Brooks 1994, 3–11). Rochester lay on the line of the main road between Canterbury and eastern Kent (including the harbours of Dover and Richborough), and London (Fig 16). In terms of the land route, a crossing on the Medway was almost as important as one on the Thames and the first bridge at Rochester can, therefore, be assumed to have been built at about the same time. Lying in the same geological region, the bridges are also likely to have been built in a similar way.

The only direct evidence for a Roman bridge at Rochester was revealed during the construction of a cast-iron structure in 1851. It consisted of unmortared ragstone rubble containing substantial timbers, including iron-shod piles driven into the river gravel, and possibly framing. The timber was predominantly oak but included some elm and beech, although this was almost certainly part of a medieval rebuild (Brooks 1994, 9 fn 12), since neither timber was much used in the Roman period. The surface of the rubble was found just under the 19th-century riverbed to a depth of between *c* 3.96m (13ft) and 7.62m (25ft). It is not clear whether the timberwork had formed part of a caisson or cofferdam, but the unmortared rubble suggests that it filled a timber structure. For reasons that have been discussed previously, ragstone would not form a suitable facing for an underwater structure, at least in the way that the Romans habitually used it in the south-east.[13] It is notable that there is literary evidence for the construction of the bridge deck in the 11th century using three beams per span to support the plank deck itself, and it has been suggested that the Saxon bridge may have reused Roman piers. The lengths described seem to suggest that the longest timbers employed could be as much as 15.24m (50ft), a little less if the width of the piers is subtracted (Brooks 1994, 16–25). Timbers of such prodigious length have not been found in Roman London despite extensive excavations along the waterfront. Their existence cannot be ruled out, given that oak trees can grow to up to c 45m high. The limitations are not dictated by the length of timbers available

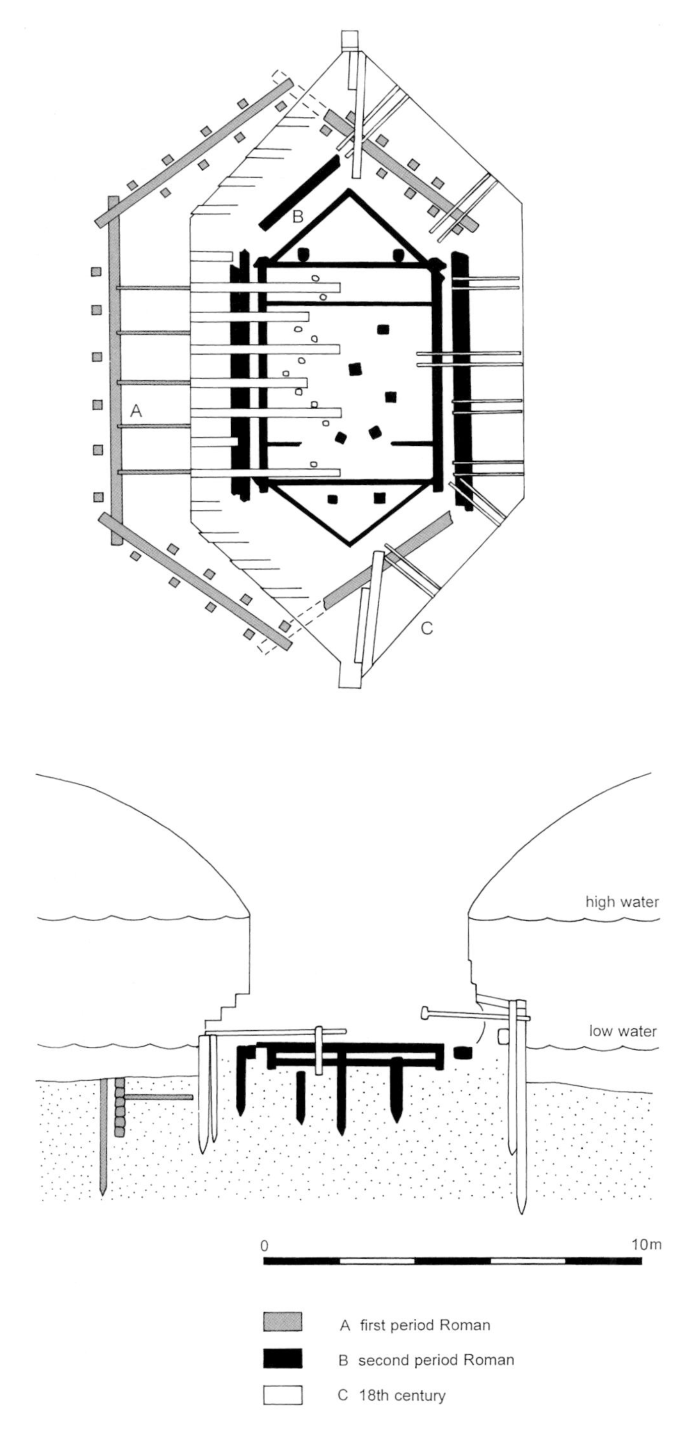

Fig 23 Plan and elevation of the Tyne bridge pier (redrawn from Bruce 1885, 6)

but by the strength and carrying capacity of unsupported spans: hence the complex trusses of Apollodorus' bridge and the simpler understrutting used in the Trier bridges and many of those illustrated on Trajan's column (Fig 19; Fig 20).

A type of caisson structure similar to the Mainz and Laupen examples can now be recognised from the 19th-century excavation of a bridge pier (Fig 23) during rebuilding work at Newcastle-upon-Tyne (Bruce 1885). There has been some confusion over the original identification of the various periods recorded by Bruce, who suggested that there were remains of 1) a Roman, 2) a medieval and 3) a more modern pier (belonging to a bridge rebuilt after flood damage in 1771). A recent work (Bidwell and Holbrook 1989, 99–102) has suggested that Bruce confused timbers of the medieval starling of the pier in question (the fourth from the south end of the medieval bridge) with the medieval pier resting upon it, and the remains of the medieval pier with a Roman pier. However, a plan of the damaged medieval bridge shows that the medieval pier was central to the starling, and not offset as Bruce showed, and that the carpentry of both the 'medieval starling' and the 'medieval pier' identified by Bidwell and Holbrook was more typically Roman than medieval. Although admittedly tentative, if this is the case, then both structures 1) and 2) are of considerable importance and will, therefore, be described, as far as the details are available from illustrations.

1) Bidwell and Holbrook's 'medieval pier', Bruce's Roman pier

This was constructed with a rectangular frame of two surviving levels of contiguous horizontal squared beams lap jointed at the corners, with integral cutwaters and traces of at least two tiebeams. Mortice and tenon joints are described and, although their location is not mentioned, they may well have served to attach the tiebeams to the side walls of the pier. To either side were double rows of beams, possibly supported on piles, and a single beam parallel to one end of the pier. These beams may represent the remains of a caisson. Squared piles within the structure probably supported the decking, and at least one had an iron-shod point. The timbers are described as black and friable oak, the tops of the piles being angular, and it is likely that all the timbers were desiccated as a result of encasement within the medieval structure.

2) Bidwell and Holbrook's 'medieval starling', Bruce's medieval pier

The main part of this structure lay outside the 1771 pier, but parts of the south wall were encased within. The structure was located below the riverbed, at a considerably lower level than 1), and consisted of six surviving tiers of contiguous squared timbers forming a box structure with angled ends acting as cutwaters, the beams lap jointed at the corners. The cutwater timbers had been let down into the water between staggered rows of squared iron-shod piles, while the timbers of the long sides apparently had a single row of external piles and were braced internally by a series of equidistant tiebacks. The joints between the tiebacks and the wall are shown schematically, but are unclear – they may have been dovetailed. The oak used was described as having a greenish decay on the surface, but solid brown within, and this would reflect its position in waterlogged ground.

The line of the east walls of both structures coincided, and structure 1) appears to have cut through the wall of 2), which was at a lower level. From this evidence it appears that 2) was the earlier.

The carpentry used in both structures, as far as these details can be determined, was quintessentially Roman. There are no known English examples of medieval cofferdams using contiguous squared baulks, and both closely resembled Roman quay structures found in London, as well as the bridge piers at Trier and Laupen in Germany. Another factor in the identification of these structures as Roman was the discovery of a considerable number of coins, and also stone blocks, which Bruce considered had Roman tooling, with the marks of cramps (Bruce 1885, 10), although the piers of the medieval bridge were paved with large slabs as recorded in the damaged structure in 1772 (Bidwell and Holbrook 1989, 100). The episodic incidence of coin loss suggests that there may have been two phases of construction, in the early 2nd century (Hadrianic) and a rebuild in the late 2nd or early 3rd century (Severan), which would match the two main phases of construction or reconstruction on Hadrian's Wall itself.

The medieval piers were apparently based on timber platforms resting on piles, with rubble infill, but it seems perfectly possible that these were simply built on existing Roman caissons, in some cases at least. Bruce described another pier (the third from the north end) which consisted of loose 'freestone' supporting 5.6m of concrete encasing a frame of horizontal timbers supported by piles. The corresponding medieval pier rested on the concrete (Bruce 1885, 10). Bidwell and Holbrook suggested that the substructure was also medieval, concluding that there was no definitive evidence for a Roman bridge (Bidwell and Holbrook 1988, 100), although the medieval use of concrete in such circumstances is unknown, whereas Roman concrete is well attested, as for example in the harbour mole foundation at Caesarea where it surrounded a similar timber frame and supported a masonry superstructure (Oleson 1988).

A recent re-evaluation of the two bridges at Piercebridge (Fitzpatrick and Scott 1999) summarises the evidence for an initial timber phase with piled piers, although these may have supported stonework. First exposed in the same 1771 floods as the Newcastle structures, in subsequent work (1933) the southern abutment is stated as having had what are described as 'cross-members' set well down in the river gravel. The corresponding abutment of the later stone bridge, which was sited *c* 200m away, included a row of five angled sockets set 1.0m apart similar to those recorded at Trier. These are assumed to have retained the bases of understruts supporting a timber deck, the sockets themselves being 0.25–0.32m across and 0.12–0.14m deep reflecting the scantlings of the missing beams. The angle of the sockets was remarkably consistent, between 57° and 59° above horizontal.

This potential evidence and that summarised in the preceding pages enable us to make some attempt to discuss possible reconstructions of the three phases of London bridge identified above.

4.9 Bridge 1: AD 50–85/90

It has been suggested above that a bridge was constructed in the years following the invasion, probably *c* AD 50, and using army labour and engineering skills. Necessity would probably have determined that this was built with local material which, in the absence of suitable stone, was almost certainly timber (Fig 17, a). The later development of the port demonstrates that the army was well-versed in using oak as a building material and, while a bridge built in AD 43 or shortly thereafter may only have been of a temporary nature, there is no reason why one built *c* AD 50, at a comfortable remove from the invasion, should not have been a substantial, semi-permanent structure. We can see from excavations on both banks of the Thames that there was considerable investment and effort involved in the construction of the approach roads, including the laying of substantial timber foundations beneath Road 1 in Southwark (Sheldon 1978, 22). It would be incongruous to take such pains to build a well-planned and executed infrastructure merely to serve a hastily built makeshift.

If we accept that bridge 1 was constructed of timber, there are several ways this could have been achieved. One method would be to sink long piles into the riverbed and use these to construct trestles. In this respect it would have resembled some of the structures depicted on Trajan's column and mentioned by Caesar. An alternative solution would be to construct caissons, probably on the foreshore, which could be floated into the required position and sunk, perhaps by filling them with clay or rubble. The caissons may have been double-walled, as Vitruvius suggested (V.12), with the gap filled with clay to create a watertight structure which could then be drained. This system was apparently used in Trier (Cüppers 1969) and also to construct the concrete base of the harbour mole at Caesarea (Oleson 1988). The use of a watertight cofferdam would allow excavation of the riverbed to a solid stratum for the driving of piles. The bridge structure would then be constructed on the caissons.

Otherwise, a single-walled caisson could simply be lowered into position and filled by emptying baskets of clay or rubble into it, unless it was constructed with contiguous horizontal beams to produce a watertight structure similar to those recorded at Mainz, Laupen and Newcastle. In these examples, we see possible parallels to the quay structures which characterised the Roman port in the 1st to 3rd centuries, and this type of construction may have been used for the London bridge: that is the use of heavy timber caissons built with massive squared timbers and infilled to provide the base for the superstructure, with the caissons acting as starlings. It should be noted, however, that in the early period at which the first bridge is all thought to have been built, before the Boudican revolt, there is no evidence of the use of this type of construction method in London, and known examples of bridge piers elsewhere date from the late 1st or early 2nd century. The contemporary waterfronts in London, including the bridge abutment identified at Regis House, were simple pile-and-plank revetments or piled embankments. The form of the earliest bridge must, therefore, remain conjectural. A pontoon bridge is one possibility, although this is unlikely to have lasted some 30 years, and it seems more likely that the piers were constructed using contiguous piles driven in at low water, then filled with clay or rubble to form the base for the bridge itself.

This bridge would still have been in service at the time of the Boudican revolt and, as discussed below (4.13), recent evidence from near London Bridge Station suggests that the Iceni used it to cross into Southwark and extend their destructive operations to the suburb (Drummond-Murray and Thompson 1998). There may have been attempts to damage or destroy the bridge, but this cannot be determined and we must in any case assume that it was restored.

Some evidence has recently been found which, if correctly interpreted, suggests who may have been involved in the rebuilding work. In AD 63 a quay was built at the present Regis House up against the bridge abutment, and a new reading of a stamp at the end of one of the massive horizontal quay-front timbers has been seen to refer to the presence of a Thracian cohort: the inscription *IRAECAUG* can be restored as [T]hraec[orum] Aug[usta], 'The [numeral] Augustan Cohort [or Ala] of Thracians' (Hassall and Tomlin 1996, 449). If true, the involvement of the army in the restoration of the harbour is demonstrated by the initial provision of raw materials for the construction of the quay, and a well-preserved fragment of scale armour (*lorica squamata*) from the quay infill is also suggestive of this link. The use of timbers on this monumental scale strongly suggests the presence of an expert army unit which, if there had been substantial damage to the bridge in AD 60–1, might very possibly have been employed in rebuilding work and employing the same techniques that they were to use in the construction of the first substantial port facilities.

4.10 Bridge 2: AD 85/90–100/20

Timber structures have a relatively limited life before they require major reconstruction work; a replacement cycle of around 30 years has been indicated in waterfront structures of the Roman and medieval periods, and a similar figure can be suggested for timber buildings. A bridge combining attributes of both types of structure may, therefore, have required substantial rebuilding work by the AD 80s, and the Pudding Lane pier, built *c* AD 85–90, can, therefore, be interpreted as a temporary replacement constructed alongside the line of the bridge proper as part of the rebuilding programme (Fig 17, c). At the time that bridge 2 was constructed there were other major public building works in the town which underlined the permanence and success of Londinium, and the need for a suitable, reliable and long-lasting bridge should be seen against this background. The great investment in terms of capital expense, labour and materials would only have been worthwhile if the new bridge was properly built, even if this meant delaying its construction while funds were raised and materials sourced. The temporary bridge would, therefore, have needed to be substantial enough to survive indefinitely; the Pudding Lane pier remained extant for between 20 and 30 years before being buried beneath the infill of a new quay or revetment *c* AD 120, and it may have been in service for any length of time within this period.

In summary, the rebuilding programme would encompass four clearly defined stages:

1) construction of the temporary bridge, involving building temporary abutments and pier bases, with road diversions; on the north bank, this was achieved by bridging over an existing inlet next to the old bridge;
2) removal of the old bridge, including piles, trestles or piers, and identifying any reusable elements;
3) construction of the new bridge;
4) restoration of the approach roads and removal of the temporary bridge.

The discovery of the pier at Pudding Lane in 1981 (Milne 1982, 271–6) led to an attempted reconstruction of the bridge (Milne 1985, fig 31) with a cantilevered superstructure narrower than the pier itself supporting the spans of the road decking. While not impossible, there is no archaeological, documentary or graphical evidence that this type of construction was used by the Romans. In this case, however, a deep mortice on the top of the surviving pier suggests a change of superstructure which was not accounted for in the reconstruction offered, since the mortice, which penetrated two levels of beams, was clearly designed to support a substantial vertical post. Similar mortices were visible on the Mainz and Laupen structures, where they probably supported the road decking.

As an alternative reconstruction, the pier should be seen as a rigid caisson with a relatively light decking, compared with the massive and timber-intensive cantilevered structure suggested. This would support the hypothesis that bridge 2 was temporary. Instead of an abutment, for which there was no evidence, a simple ramp consisting of a similar decked frame could be used to cross the inlet. For the main span, one possibility was a pontoon bridge with simple frames of girders supporting a plank decking; another that the deck was supported on pile clusters, rather like Caesar's temporary bridge described in *De bello Gallico* (Caesar, IV). It is important to note here the 19th-century record of iron-shod piles surrounded by conglomerate found immediately downstream of the medieval bridge, which may have formed part of this structure, although they could equally have been disturbed from the main bridge as suggested (Cuming 1887, 166–7).

4.11 Bridge 3: AD 90/120–400

While it is possible that a new more permanent structure was built as early as c AD 90, the major public building programme continued into the early 2nd century. This included the construction of an enormous new forum basilica and a new revetment built immediately upstream of the bridge at Regis House in AD 102. It is possible that it was in this context that a new bridge was built as a suitably imposing approach to the new forum (Fig 17, d).

The more permanent structure required implies the use of masonry, and it should be borne in mind that an 'abundance of broken Roman tiles' was discovered during the 19th-century excavation of the piles of the old London Bridge (Smith 1846, 113). While this may have been deposited within timber caissons as consolidation, this would have been wasteful when ragstone rubble was generally more cheaply available as a by-product of masonry construction elsewhere in London, and appears to have been used in this way in Rochester (Brooks 1994, 9). It is suggested, therefore, that the tiles were part of the bridge structure, and in this respect the Roman use of tiles in London (or more properly in this context, bricks) should be considered.

In the 1st century brick was rarely used in solid walling, and its use was confined largely to bonding courses, quoins, jambs, arched door and window openings, and features such as culverts. There was no building stone in the immediate vicinity, and the nearest durable material was ragstone from the Kentish Greensand Beds which was probably imported by river via the Medway. This stone is a hard limestone which cleaves into fairly small roughly rectangular blocks but is not capable of producing a fine ashlar finish. It was not much used for quoining, possibly because of its tendency towards irregularity and its susceptibility to spalling (flaking through frost damage), which is why brick was adopted for this purpose. In the 1st century, masonry construction was largely confined to public building projects. These included the waterfront buildings at Regis House (Brigham et al 1996b), Pudding Lane/Peninsular House (Milne 1985), Miles Lane (Miller 1982), Cannon Street Station 'Governor's Palace' (Marsden 1975), Huggin Hill public baths (Rowsome 1999) and, further afield, the first forum (Marsden 1987). Where ceramic building material was used in mass walling, it was restricted to small areas, such as a pier in the south wall of a masonry building at 37–40 Fish Street Hill (Bateman 1986). Otherwise, some early 'brick' structures consisted of thinner reused roof tile, with the flanges laid outwards to resemble brick. Examples include a temple next to the first forum and a small building at Suffolk House, next to the Governor's Palace (Brigham and Watson 1995).

In the late 1st and early 2nd centuries, however, brick-faced concrete and brick mass walling became fashionable throughout the Empire. The houses and tenement blocks of Ostia epitomised the use of this relatively new material, and its use became more widespread in London, although ragstone continued to predominate. The load-bearing capacity of Roman brick used on its own seems to have been appreciated, and narrower walls could be used to carry the same load as a more substantial 'conventional' ragstone structure. This made it particularly suitable for piers, where compactness allowed less floor space to be taken up. Its use in this way was seen, for example, in the second forum, where brick piers replaced conventional ragstone and brick string course walling in an early rebuild of the basilican nave (Brigham 1990a, 67–8). It was also used extensively in the south wing of the forum, which was completed in the new era, whereas the slightly earlier east and west wings of the forum were of conventional materials (Brigham 1992a, 90–1). The south wall of the possible public building at 37–40 Fish Street Hill was also partially rebuilt in brick in the early 2nd century (Bateman 1986, 238).

It can be suggested, therefore, that brick was used to construct the piers of bridge 3. Such a bridge may have had masonry arches, although ragstone is not a suitable material for wide arches, since it cannot be formed into voussoirs, so an arched decking would either have to be in brickwork, or suitable stone brought in from elsewhere. This would, however, have required a high structure to accommodate the arches, and evidence from northern Europe suggests that the bridge was more likely to have been provided with a timber decking like its predecessors. Parallels include the mid-1st- and mid-2nd-century bridges at Trier, and the early 2nd-century bridge constructed by Apollodorus. Apart from decorated structures such as the monumental arch and screen of the gods found in fragments in the late 3rd-century riverside wall (Hill et al 1980), ashlar was not used – or at least has not survived – in London, and it is unlikely that bridge 3 featured stone starlings. The small, roughly rectangular ragstone blocks used for facings in land structures had wide joints which would have eroded quickly underwater. Timber caissons were, therefore, the most likely option. There was no shortage of suitable building timber in the region. In London, large oak baulks had been used since at least as early as the construction of the Regis House quay in AD 63–4, and could still be obtained when the last quays were built at Billingsgate and New Fresh Wharf in the second quarter of the 3rd century.

4.12 A suggested reconstruction of bridge 3

The previous section has already discussed possible construction details of the bridge and some parallels. This section will discuss more detailed aspects of the available and comparative evidence, including Roman structural carpentry.

Roman carpentry

Recent work on timber structures of the period, particularly in London, has provided good evidence for the range of techniques available to the Roman carpenter, including buildings (eg Brigham et al 1996a; Goodburn 1991), and quay structures (Brigham 1990b; Brigham et al 1996b; Taylor Wilson 1990; Parry 1994). These confirm the existence of a Roman carpentry tradition for which parallels can be found elsewhere: for example at Carlisle (Caruana 1983), Valkenburg, Netherlands (Glasbergen and Groenman-van Waateringe 1974) and Xanten, Germany (von Petrikovitz 1952).

The basis of the Roman tradition was the use of squared timbers for almost all structural purposes and the employment of a limited range of joints. The most commonly used joints in what we might term 'monumental carpentry' included the following.

Mortice and tenon – for attachment of posts to horizontal beams, and from at least as early as the mid-2nd century in the derivative form of free tenons to attach two contiguous tiers of timbers.
Notch-and-tenon – common scarf joint from at least the mid-1st century, the tenon sometimes slightly dovetailed to resist withdrawal.
Half-lap or saddle joint – for attachment of subsidiary element to main element, such as revetment or quay tieback, joist or tiebeam, allowing subsidiary element to cross and regain its full depth – widely used in the 1st and early 2nd centuries.
Simple or straight lap – as above, but smaller element does not cross main timber, being housed instead.
Lap dovetail – lap joint for use in same circumstances as the simple lap from the early or mid-2nd century, dovetail replaced largely by weaker **half dovetail** from late 2nd century.
Edge-halved scarf with dovetailed abutment – common scarf joint used from at least as early as mid-1st century to mid-2nd century.
Angled lap – used for the attachment of angle braces.
Angled recess – used for inserting the horizontal elements of a wattle frame between the uprights of a timber-framed building.

The universality of these joints is conclusive evidence for the techniques available to the builder of a Roman timber bridge. In contrast with the medieval period, innovation was virtually non-existent, and this inflexibility meant that the joints used were not always appropriate for their task. This at least enables us to predict which joints would have been used in particular circumstances.

The caissons

The proposed reconstruction of bridge 3 (Fig 24; Fig 25; Fig 26) is based on timber caissons (see Glossary), constructed using contiguous squared timbers lap jointed at the corners and with internal braces; dovetail joints were not commonly adopted in London for this purpose until the mid-2nd century. The resulting structure resembles the Pudding Lane pier, but is also based on parallels from Newcastle, Laupen and Mainz and, therefore, has integral cutwaters, which in this case lie on both the upstream and downstream sides, since the Thames was (and still is) a tidal river. As discussed earlier, this solution has been adopted in view of the lack of suitable building stone which would otherwise have allowed the construction of masonry starlings within cofferdams similar to those used to build the seven piers of the second Trier bridge, and also to those specified in Vitruvius (V.12: pile-and-plank double-skinned structures with clay packing).

The timbers forming the caissons are shown as being of the same dimensions as those employed in the mid- to late 1st-century quay structures, *c* 0.5m square, with internal braces *c* 0.40–0.45m square. In permanently wet conditions, oak of this size could be expected to last indefinitely. A roadway of *c* 6.0m would require caissons of at least that length and, (given the evidence from Newcastle) to accommodate the pier above, as much as *c* 8.0–10.0m. The cutwaters would have added a further *c* 7.0–8.0m to the total length. The width of the caissons varied in known examples from slightly wider than the length of the long side (Newcastle 1) or the same (Laupen) to around one-third (Mainz). In practice, the width was governed by the available length of the timbers acting as tiebeams, which in London was generally not much more than *c* 5.5m (Regis House), some of which was taken up by the joints. A width of not much more than *c* 6.0m for the caissons is, therefore, likely.

It is assumed that the top of the caissons lay roughly midway between high and low water (*c* 0.3m OD in AD 100) as was the case with the medieval bridge (Fig 49; Fig 95). A study of the depth of the riverbed in the centre of the channel suggests that on that basis the caissons would need to be *c* 6.0m deep, graduating to perhaps 1.0m deep near the banks, where the river was much shallower. Using timbers 0.5m square to form the outer walls would require 12 tiers. A calculation of the weight of the deepest caissons as illustrated, including tiebeams and cutwaters, and taking a weight for green oak of 1073kg/cu m, suggests a total of *c* 156 tonnes. The piers nearest the riverbank, of perhaps two tiers (ie 1m deep) would have weighed only a sixth of the amount, *c* 26 tonnes. This is still a substantial load for the cranes of the day, and it is clear that the caissons could not have been prefabricated and lowered into position in one piece. The presence of guide piles in the Newcastle and Mainz bridges suggests that one tier could be lowered at a time, presumably using two or more boats or rafts positioned either side: even using this method, a single tier would have weighed *c* 13 tonnes. This would explain the complex nature of some of the caisson structures recorded elsewhere, since the frames would need to have been rigidly constructed although, once submerged, the river would have had a supporting effect.

The infill of the caissons, bearing in mind that they had to support the piers, must have been solid. There is no evidence from London for true hydraulic cement like the *pozzolana* used in the Mediterranean, although crushed tile was used as an additive to increase the speed of setting and improve water-repellent properties, producing a pink mortar. The references to the discovery of 'conglomerate' on the 19th-century riverbed (above, 4.6) suggests that concrete may have been

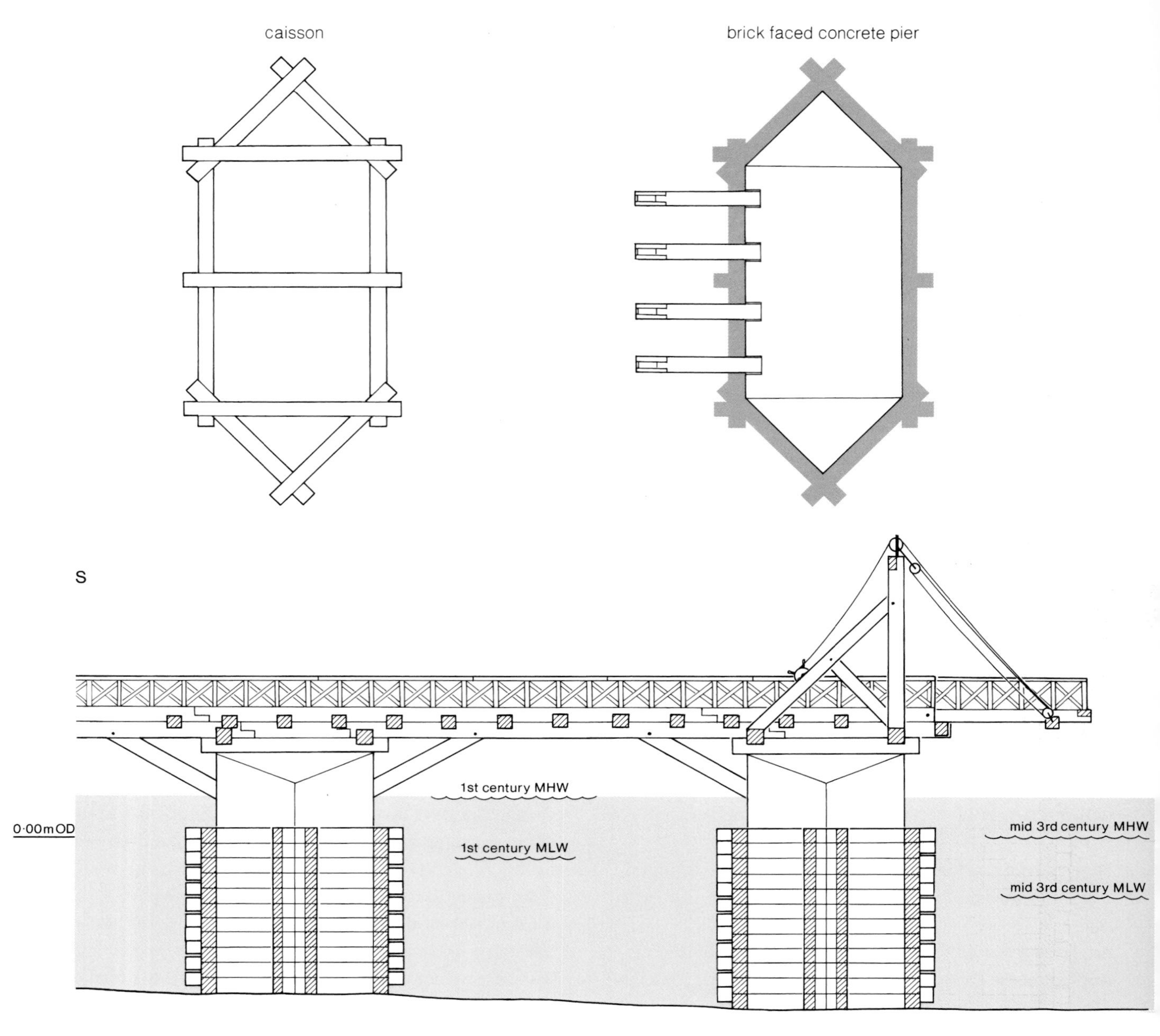

Fig 24 Reconstruction of elements of the final late 3rd-century bridge from caisson to road deck in both plan and elevation, including the central drawbridge

used, as was the case in one of the bridge piers recorded at Newcastle, although this may have been simply a by-product of iron corrosion. Clay is an alternative possibility, since when puddled it makes a very firm foundation and has the added attraction of being impervious. A mixture of clay and ragstone rubble is the most likely infill, rubble being used, for example, at Rochester.

The spacing of the caissons would to some extent be constrained by the available timbers. In a bridge where there was headroom to develop complex trusses, as was the case with Apollodorus' bridge on the Danube, this was not so much of a problem. British examples include the Hadrianic bridge at Chesters (bridge 1), which had piers 10.8m apart, the Severan bridge at Willowford (bridge 3) with piers 7.31–7.77m apart, and a broadly contemporary bridge at Piercebridge with piers 10.5–11.6m apart (Bidwell and Holbrook 1989, 1–14, 103–7, 110–12). The piers of the Trier bridge were apparently *c* 11m apart (Cüppers 1969) and a figure of *c* 11–12m seems likely for the London bridge, requiring a total of some 20 caissons *c* 5.0–6.0m wide.

The piers

The caissons are shown supporting brick piers (Fig 24; Fig 26), which would need to be long enough to carry the width of the roadway (*c* 6.0m), with cutwaters adding a further *c* 5.0m. The piers would have rested on timber piles driven into the infill of the caisson, and possibly a frame of horizontal beams, similar to Newcastle 2 and Trier. The piers would have had to be wide enough to be sufficiently stable, possibly being *c* 4.0m, but because they were smaller than the caissons this would increase the spans between piers by at least *c* 1.0m. This could have been mitigated by the design of the timber frame supporting the roadway (see next section).

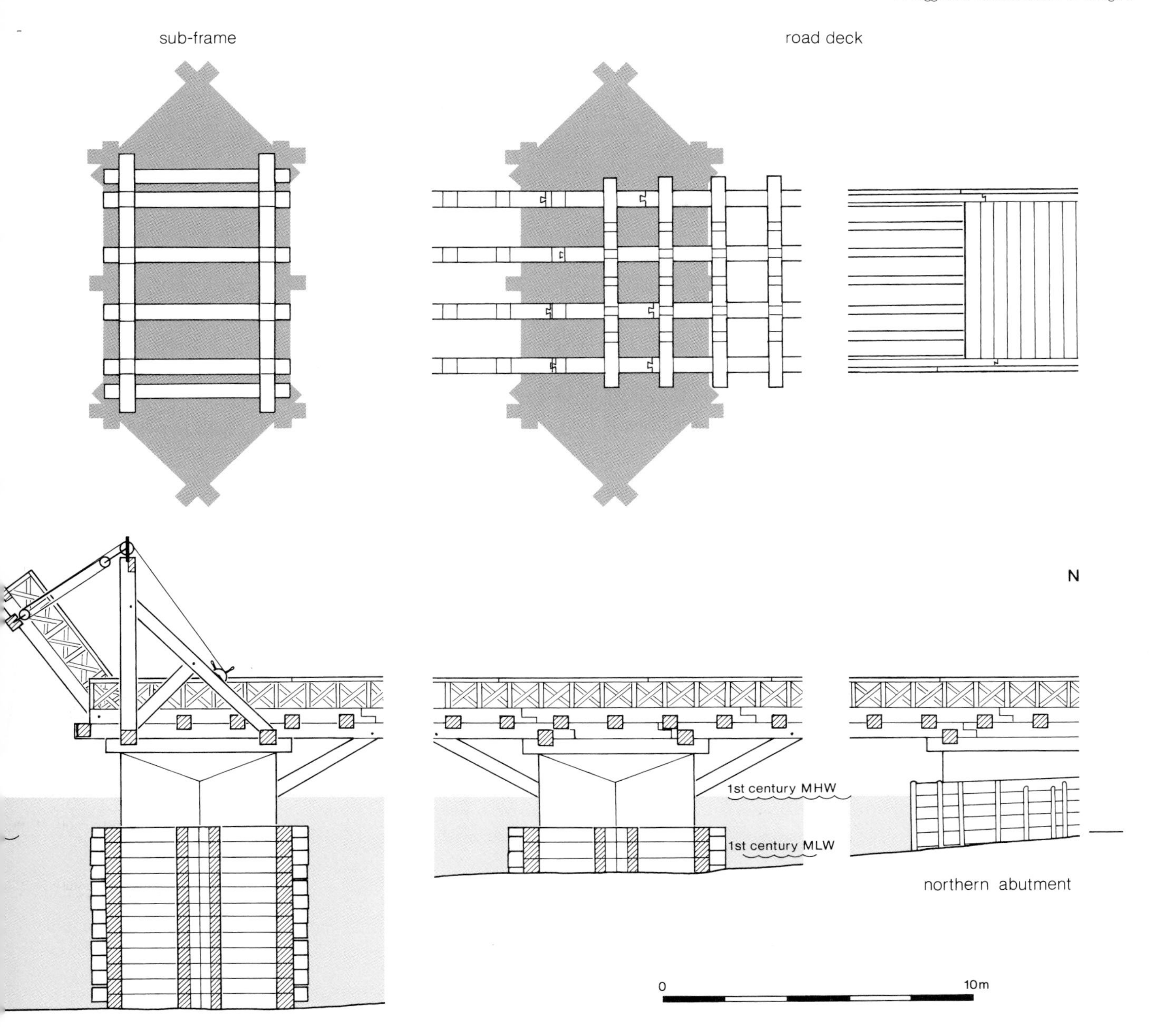

The minimum height of the piers would have been defined by the level of the top of the caissons and the required level of the bridge deck. The evidence currently suggests that the contemporary approach road on the north bank was at *c* 3.0m OD. On the south bank the earliest surface of the western road, Road 2, was recorded at Montague Close about 50m south-west of the bridge abutment at *c* 1.8m OD, continuing to rise towards the north. The final surface was over 3.0m OD (Dawson 1976, 41). The same road was found in District Heating Scheme trench A a few metres nearer to the bridge, but only the cambered edges were recorded at *c* 1.5m OD. Road 1 was observed in District Heating Scheme trench D at a similar level, about 25m south of the abutment (Graham 1988, 20–22). The late 1st- and early 2nd- century high tide level is all thought to have begun to decline from its previous level while still attaining *c* 1.2–1.3m OD (Fig 14). The bridge deck must have been at the same level or slightly higher than the northern approach road to allow sufficient clearance beneath it during times of flood. A height greater than *c* 3.0–3.5m OD would have required a substantial embankment on the south bank to raise the road to bridge level, unless the end spans of the bridge deck were canted. The starlings of the medieval bridge were constructed about halfway between high and low water, and if this practice had been adopted for the Roman bridge a level of *c* 0.2–0.3m OD can be suggested. Allowing for the thickness of the bridge deck, the piers were, therefore, likely to have been a minimum of about 2.5m high.

The bridge deck

The bridge deck, judging from the Aldwincle and Birkenhead bridges, Trajan's column and evidence of Roman carpentry in London, probably rested on frames constructed on the heads of the piers (Fig 24; Fig 26). These frames are likely to have

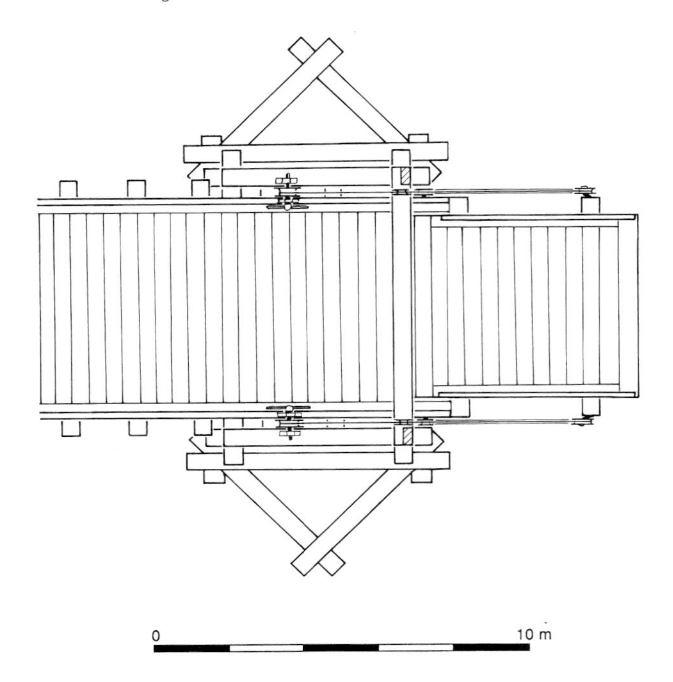

Fig 25 Plan of drawbridge section of final late 3rd-century bridge

been constructed in a similar manner to the caissons, with horizontal beams lapped at the corners, and with cross beams lapped at intervals to make the frame rigid. On this the initial girders could be laid, one line down each side and, on the Birkenhead model, probably a further two underneath the deck at around 1.5m centres. These timbers can be envisaged as being of the largest available size, *c* 0.50–0.55m, and may have been laid with two tiers for added support. Certainly, the Birkenhead bridge had three levels of smaller timbers totalling 0.68m in depth, and a reconstruction of the second bridge at Trier shows two levels of squared baulks (Fig 19).

From the point of view of the feasibility of spanning the gaps between the piers, as has already been seen, timbers in excess of 15.0m were apparently used in the Saxon bridge at Rochester with no reference to understrutting. In London the longest timbers found to date include a possible *c* 10.0m quay-front beam of *c* AD 200 found at Vintry House (D Goodburn, pers comm), an 8.85m example of *c* AD 70 at Miles Lane (Miller 1982), an 8.2m beam from Regis House, and a 7.95m beam of *c* AD 225 at New Fresh Wharf (Miller et al 1986, 96–8). It must be stressed, however, that the baulks used in the quays were simply immobile building blocks where mass and not length was the main consideration. The quay timbers are, therefore, only an indication of those available, and much longer baulks could almost certainly be found for structures specifically requiring large spaces to be spanned. A good example of this is the nave of the great basilica built *c* AD 100–20 (Brigham 1990a; Milne 1992a). This was *c* 14.0m wide, and may well have been spanned by tiebeams consisting of single lengths of timber, unless composite girders were used.

On the safe lengths for unsupported timbers, one set of figures quoted (Yates and Gibson 1994, 25 fn 58) suggests that baulks 0.45m square with bridge decking 50mm thick could span just over 15.0m safely. The suggestion here, of possibly two tiers of girders 0.5m square supporting a rigid deck structure over 12.0m and supported by understruts, would clearly be more than adequate in providing a safe, firm and durable bridge.

The Aldwincle bridge timbers show that lap joints would be cut on the upper faces of the girders to allow slightly smaller binders, probably of the same gauge as contemporary quay tiebacks (0.40–0.45m), to be laid across the width of the bridge at intervals of perhaps *c* 1.5m centres. These would have the effect of linking the four rows of girders together in the form of a row of rigid boxes. These were then lap jointed to lay smaller joists – the next most common gauge timber was 0.25–0.30m – at closer intervals of perhaps around 0.50–0.75m to support the planking which ran across the width of the bridge. Roman planking was generally supplied in either 0.3m or 0.45m widths (Brigham et al 1996a, 25), reflecting the size of the baulks from which they were cut (1 and 1.5 *pedes monetales*). The planks would need to be thick to support the weight of traffic, probably at least 50mm. The width of the roadway was probably *c* 6.0m, in common with many other parallels.

The combined weight of the timber decking and the loading would have ensured that the girders required further support in the form of angled braces or understruts. The understruts would have the effect of supporting up to between half and two-thirds of each span, reducing the unsupported section to 5 or 6m. The existence of angled braces is known in London from reused timbers which had originally formed part of framed buildings from Cannon Street Station (Goodburn 1991) and Guildhall Yard (D Goodburn, pers

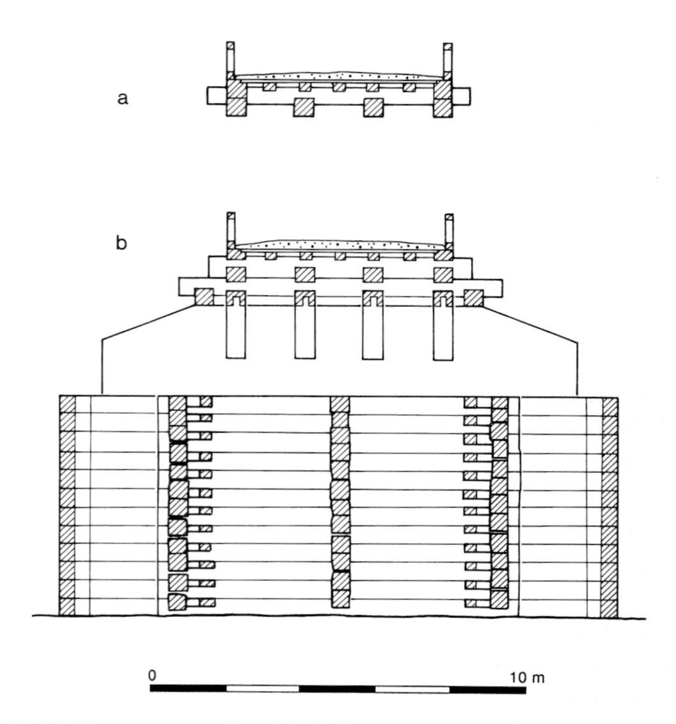

Fig 26 *Transverse sections through the bridge at: a – mid-span of the road deck; b – pier position*

comm). Examples are also shown on many of the bridges depicted on Trajan's column, including Apollodorus' bridge, and they have also been used on the reconstruction of the 2nd-century bridge at Trier. As in that case, the lower ends of the struts may have rested in a recess within the sides of the pier, possibly set on timber pads or a baseplate. The upper ends of the braces would have been attached to the underside or side of the girders, using either a simple lap joint or mortice and tenon. Two 3rd-century quays at Billingsgate Lorry Park (Brigham 1990b) and New Fresh Wharf (Miller et al 1986, 96–8) show that lap joints in monumental carpentry were commonly secured by spikes or large nails set in countersunk holes; the countersinking was then filled with cement to protect the iron from the elements.

A handrail would have been provided, probably in the cross-braced form commonly illustrated on Trajan's column, and as probably used on the Birkenhead bridge. Mortices would have been cut in the upper girders for the posts of the rail, and the cross braces either prefabricated and attached to the posts, or, as in the Birkenhead example, attached by angled joints to housings cut in the outer edge of the girders.

Finally, the discovery of two wrecked sailing boats upstream of the bridge raises the question of how they passed through (Marsden 1994). This and, in particular, the question whether there was a drawbridge on the bridge will be discussed next.

Navigation

It is suggested below (4.13) that the bridge may have had an additional function as a tariff barrier, a place where tolls could be collected from vessels passing beneath. This raises the topic of shipping and river navigation, taking account of the depth of the river, the headroom beneath the bridge, the spacing of the caissons and the size of vessels using the river in the Roman period.

The depth of the Roman river has not previously been calculated, and is somewhat problematic because of the apparent 1.5m fall in tidal levels between the later 1st and mid-3rd centuries. Soundings taken above London Bridge in the early 19th century, away from the effects of natural scouring and deliberate dredging, showed a difference in depth of *c* 3.66m (12ft) between the sides and the central channel. The Roman river of *c* AD 100 was, however, 50% wider than the river of *c* 1800, and the same profile would produce a difference in depth of *c* 5.5m. Since the 1st-century foreshore lay at Ordnance Datum, this gives an absolute level of -5.5m OD, a figure which can be supported by calculating the depth of the mid-3rd-century river, which was only *c* 10% wider than in *c* 1800 on account of reclamation on both banks. The difference in depth between the sides and central channel would, therefore, be *c* 4.0m which, from the contemporary foreshore level of -1.5m OD, gives the same result: -5.5m OD. The depth of water at high and low tide respectively would have been *c* 7.0m and 5.0m respectively in the 1st century, reduced to *c* 5.5m and 3.5m in the 3rd century. The latter figure was the minimum depth necessary if the river was to remain navigable for fully-loaded seagoing vessels in the later Roman period.

Some supporting evidence for the level of the riverbed can be obtained from the discovery in 1962 of a mid-2nd-century boat at Blackfriars, 1.2km upstream of the bridge. This wrecked vessel lay 120–130m south of the Roman riverbank with the base at -3.59m OD (Marsden 1994, 91), the depth of the Thames in this area; the river was wider at Blackfriars than at the bridge, but the boat lay about a third of the way across the channel. The base of the late 3rd-century County Hall ship, found near the southern riverbank in 1910, lay below *c* -2.0m OD (Marsden 1994, 111), suggesting that it was abandoned at the contemporary low tide level.

The suggested depth of the Roman riverbed agrees with that of the early 19th-century level, where the foreshore on the south bank lay at *c* -2.0m OD (Fig 14), which with the 3.66m differential gives a figure for the centre of the channel close to that suggested for the Roman river. Engineering drawings produced by the Corporation of London in 1967 (GM, 106, drawing no. 043/B/1051) show a level of *c* -6.1m OD for what appears to be the top of the London Clay below the central arch of the 1832 bridge. Allowing for some river gravels, this would give a very similar figure to the Roman level. The riverbed rose towards the sides to *c* -3.0 to -3.5m OD.

The conclusion must be that the base of the river has remained at much the same level since the Roman period, with dredging and scouring constantly removing intervening build-up. It is not surprising, therefore, that Roman artefacts were found so close to the surface during the demolition of the medieval bridge in the 1830s.

The suggested reconstruction of the bridge allows sufficient room for the three known Roman vessels in London to navigate through. The largest of the three, the Blackfriars boat (Blackfriars boat 1, Marsden 1994, 33–95), had a beam of *c* 6.12m, with a working draught of between 1.0m and 1.5m fully loaded. At the time of its loss in the mid-2nd century, this would have allowed a considerable depth of water below the keel at low tide, probably *c* 3.0m, with a comfortable clearance either side of *c* 2.5m. The second largest, the County Hall boat (Marsden 1994, 109–29), had a beam of *c* 5.06m, with a working draught of perhaps a little over 1.0m. At the time of its abandonment at the end of the 3rd century, this would have allowed a clearance under the keel at low tide of *c* 2.5m, and between the caissons of *c* 3.0m to either side. The third vessel, found at New Guy's House in 1958 (Marsden 1994, 97–108), was a much smaller, flat-bottomed river boat, probably a lighter. It had a beam of 4.25m, but a draught of perhaps only 0.5m fully loaded. It was abandoned at the end of the 2nd century, at which time it would have cleared 3.5m below the keel, and *c* 3.5m to either side.

All the known vessels could, therefore, comfortably navigate the river and the bridge, even at low tide and after the river levels had fallen away in the later period. The main problem remaining, however, is the question of headroom.

The Blackfriars boat was masted, and the County Hall vessel was almost certainly a sailing vessel as well. The mast of the Blackfriars boat could not be unstepped and, unless it had either been demasted before being abandoned above the river, or the mast was hinged at a point above the base, it must have been able to pass through the bridge (Marsden 1994, 67–70). It is not inconceivable that boats had pivoting mast sections, since these are easy to operate with efficient tackle, but the bridge was certainly not raised sufficiently for any boat to pass through with its mast in place. The mast of the Blackfriars boat, for example, would have been at least 10.8m high and probably nearer 13.0m; the bridge must, therefore, have featured a movable section (Fig 25). The simplest form that this could take would be a lifting bridge with counterbalance beams. Given the span between piers of *c* 11.0m, and the need to clear the mast and yards of passing vessels, it is likely that there would be two opposing bridges, probably replacing the central span of the bridge, although the drawbridge constructed for the medieval bridge for the same reasons was off-centre (Fig 25). A properly balanced bridge of this type could easily be operated by hand, although a system of pulleys and windlasses would have made the task both easier for the operators and safer from the point of view of vessels passing beneath.

If the drawbridge section was central, it may explain why the concentration of votive finds was immediately adjacent but on the city side of the span. Inhabitants of the main settlement donating votive offerings probably crossed the bridge to a position over the deepest point of the river but avoided standing on the lifting section itself. There may even have been bridge shrines located to either side.

4.13 Who built Roman London bridge?

In the Roman Empire roads and bridges which were considered necessary could be commissioned by anyone with sufficient authority, including local government officials such as *aediles* or *curatores* through to governors and emperors at the other end of the scale (O'Connor 1993, 36). The actual design work was probably the province of army engineers specialising in their field and coordinating legionary or auxiliary work forces led by specialists, although the army unit as a whole had joint responsibility for all construction work, since there was no specifically-tasked corps of engineers (Webster 1979, 118). Subsequent day-to-day maintenance generally remained a local responsibility.

It has already been suggested that the first river crossing was built by the army to shorten the route between the south coast ports at Dover and Richborough and the expanding frontier to the north. This was almost certainly carried out by direct order of the military governor, and there may initially have been no consideration of the commercial advantages of such a bridge. Its importance as the lowest crossing point on the river would have been established very rapidly, however, as Londinium became the hub of a new communications network covering south-eastern Britain, with the roads and the river as the spokes, and the bridge as the linchpin (Fig 16). The bridge may well have been used to control river traffic as well, through a system of tolls or customs dues charged for passing through and for use of lifting bridges, as it was in the medieval period (see Chapter 10.1). Goods crossing the river could be subjected to controls at will in the same way.

The importance of the river crossing would have been a major factor in promoting the immediate restoration of the town following the Boudican revolt of AD 60–1, and we know from the excavation of Regis House that new harbour facilities were built next to the bridge as early as AD 63–4. If the bridge itself had been damaged, it would almost certainly have been restored as a priority. Certainly, extensive burning found at the London Bridge Station excavation in Southwark demonstrates that the Iceni had crossed the river, and it would have been militarily advisable to make the bridge unusable to prevent Roman reinforcements being brought up from Kent.

The Regis House evidence, as mentioned earlier, suggests that the army were engaged in reconstruction in the immediate aftermath of the revolt, and it seems likely that any bridgework would have been carried out by the same unit that was indicated by the timber stamp on one of the quay-front beams: a Thracian *cohors* or *ala*. By this time, the civilian procurator (Julius Alpinus Classicianus) is thought to have been based in London (Merrifield 1983, 57–60), and it is likely that his office was responsible for commissioning the restoration work. The military governor was doubtless occupied with pacification in East Anglia and rebuilding work in Colchester and elsewhere.

Thereafter, it is less clear where the responsibility for building and maintenance work lay. The suggested rebuilding of the bridge in the late 1st or early 2nd century may still have been a responsibility of the procurator's office, or of the civic authority whose presence was for the first time confirmed by the construction of a forum basilica in the AD 70s (Marsden 1987). A complicating factor was the arrival *c* AD 70 of the provincial governor, marked by the construction of a range of buildings identified as the Governor's Palace. Arguably, the bridge was of great importance to civil, military and local government at a period marked by enormous expenditure in London and huge expansion of the occupied area as trade burgeoned. Other projects being undertaken at this time included the construction of a timber amphitheatre in the north-western corner of the town, a large baths complex on the waterfront at Huggin Hill and the extension of the harbour facility. The development of Southwark in the late 1st century also reflects considerable public expenditure on the waterfront and the internal infrastructure needed both to protect and extend the occupied area.

It is not impossible that the bridge was maintained jointly, with funds obtained partly from the imperial purse, and partly raised locally from the great concentration of capital which must have accumulated in the town. A third source may have been tolls or customs dues, which for the administrators of the medieval and later bridges proved sufficient not just for maintenance but for rebuilding. A study of the effort and expense required in maintaining the medieval bridge gives some idea of the scale of the work which was required on such a structure (see Chapter 10).

4.14 The end of the bridge

Because of its signal importance in the communications network, it can be argued that the bridge would have survived the apparent demise of many other public works in London in the late 3rd and early 4th centuries. Its timber decking would not, however, have allowed it to remain usable for long periods once maintenance ceased. The demolition of the north range of the basilica *c* AD 300 implies that it was no longer required as the centre of regional government, and it is uncertain how London was governed after this date. A monumental building of 4th-century date, found on Tower Hill, might be a *horreum* (a store for commodities like grain, collected as taxation) or perhaps a cathedral church (Sankey 1998, 81). In the *Notitia Dignitatum*, a list of officals produced in c AD 400–25, the diocese of Britain included the 'director' (*praepositus*) of a treasury in London (Ireland 1996, 138). This reference confirms that some sort of regional government was still functioning in the London region during the late 4th century. A regional treasury could have provided the funds to maintain the bridge. The bridge repair obligations described in the Theodosian code (discussed above, 14.7) are another possible mechanism by which the bridge could have been maintained during the late 4th century. The fact that bridge building is included in the code implies that lack of funds for bridge maintenance was a common problem. The survival of the bridge into the sub-Roman period was, therefore, dependent on the nature of late Roman government in Britain, a subject which clearly lies outside the scope of this study.

The fact that nothing remained of the bridge in the 19th century beyond a mass of tiles and some iron-shod piles strongly suggests that it had already been systematically dismantled at a period when records were either not kept or have not survived. Assuming the bridge became redundant or too expensive to maintain in the 4th century, it is very likely that its clearance was effected by the Romans themselves. This was possibly undertaken to maintain a clear and safe channel for river traffic, although the underwater elements would presumably have had to be removed by grappling. There are parallels for the wholesale demolition of public works including the forum and Huggin Hill baths, both major undertakings. The salvaged building materials could have been reused elsewhere, even at such a late period.

If, on the other hand, the bridge survived in use into the sub-Roman period, the masonry or brick piers may still have been at least partially visible by the time the first Saxo-Norman bridge was constructed, although not, apparently, in a form which allowed even partial reuse as has been argued for the bridges at Rochester and Newcastle, and as was clearly the case for many bridges on the Continent, including the example at Trier. The builders of the late Saxon bridge may even have taken their slightly different line to the east specifically to avoid the ruined piers, since it would be impossible to drive piles through solid timbers and brickwork. A more likely period for the removal of the Roman bridge was during the construction of the late 12th-century Colechurch structure. The area of the medieval starlings covered almost all of the projected line of its predecessor, and it would have been necessary to remove any obstructions to allow construction to begin. The Colechurch bridge took 33 years to build (see Chapter 8), and this slow progress may in part have been due to the need to remove Roman timbers and masonry as well as the remains of earlier medieval bridges. Since the medieval bridge was almost certainly constructed an arch at a time, with each taking an average of 21 months, there would have been plenty of opportunity to break up earlier structures using grapnels and lifting tackle based on the new bridge deck itself, and perhaps dredging the remaining infill. This operation can be compared with clearance following a collapse of two arches of the medieval bridge in 1437, and presumably also in 1281–2, when five collapsed (see Chapter 11.1).

The systematic removal of the Roman bridge, whether in the Roman period itself or subsequently, has ensured that the only fragments surviving to the present day are likely to be those which have been enclosed within the reclaimed area. These mostly consist of the remains of the northern abutment and some of the northernmost piers beneath Fish Street Hill and Lower Thames Street. Perhaps a future opportunity may arise to examine these and to answer some of the many questions that remain about Roman London bridge.

5

The late Saxon bridgehead

Bruce *Watson*

5.1 The military significance of London bridge and the Southwark bridgehead

It is uncertain when the London bridge was first re-established in the Saxo-Norman period (AD 900–1100), and it seems unlikely that there was much need for a bridge until the late 9th or early 10th century when the walled Roman city was reoccupied and refortified. Up to that time the main area of mid-Saxon settlement from *c* AD 600 was Lundenwic, located some 3km upstream in the Strand area (Cowie and Whytehead 1989; Malcolm and Bowsher in prep). Here at Lundenwic, however, the Thames was shallow enough to be forded and no bridge would have been necessary (Margary 1955, 47). At some point during the late 9th century, Lundenwic was apparently abandoned due its vulnerability to attack by seaborne Viking raiders sailing up the Thames.

According to the *Anglo-Saxon Chronicle* the Scandinavian or Viking raids which drove the inhabitants of Lundenwic into the walled Roman city began in 842 and culminated in an over-winter occupation in 872 (*Anglo-Saxon Chron*, 64, 72) followed, apparently, by permanent settlement. The extent of King Alfred's 'restoration' of London after his reoccupation in 886 is uncertain, but was probably confined to the Queenhithe area and, perhaps, the bridgehead (see Glossary) (Dyson 1990, 106–7). It does, however, provide a likely *terminus post quem* for the replacement of the bridge which, it has recently been suggested, may previously have been rebuilt during the period 842–71 as a joint defensive project by the rulers of Mercia and Wessex in response to the Viking raids (Carlin 1996, 11). Had a bridge been built for defensive purposes in the mid-9th century, however, some evidence of activity around the bridgehead – such as pit digging or revetment building – should have been detected in numerous archaeological excavations in the area (discussed later).

It is possible that all or some of the masonry piers of the Roman bridge were still standing in the 10th century (see Chapter 4.14). If this was the case, the Saxons could have simply repaired the piers and then erected a new timber superstructure on top of them, as was apparently the case at Rochester Bridge (Brooks 1994, 21). However, as the Roman and medieval bridges are on slightly different alignments (Fig 15), and in different positions, this suggests that the Saxons were actually avoiding the remains of the Roman bridge. One possibility is that, due to erosion of the Southwark foreshore during the Saxon period (see Chapter 6.1), the Roman bridge abutment and the southernmost piers had either collapsed or had been washed away, so the remaining superstructure was unusable. The earliest archaeological evidence for the Saxo-Norman bridge consists of two *ex situ* timbers dating to *c* 987–1032 (see Chapter 7). The position and alignment of the bridge of which these two timbers formed a part, is not known and, therefore, it is conceivable that these timbers formed part of a bridge which incorporated the remaining piers of the Roman bridge.

One factor especially favouring the rebuilding of the bridge at London at this period was the attention given to river defences throughout northern Europe generally, in which bridges formed an integral part as a means of preventing Viking penetration upstream. A number of *burhs* (see Glossary) or fortified towns and forts of this date were sited so as to command important routeways. Runcorn, for example, was sited to intercept raiders arriving via the Mersey, but at many of these sites there was a closer and more specific association between fortifications and bridges, the building and repair of both having been among the fundamental obligations of land tenure since the 8th century (see Chapter 7.1). Brooks (1971, 84) has postulated that the growth of the common burdens during the 8th and 9th centuries in England was part of the growth in the power and sophistication of royal government. At the same time kings began to grant immunities from secular burdens to the church and other favoured individuals, but theoretically no one was exempt from three core burdens or obligations – building bridges and fortresses, and military service.

The *Anglo-Saxon Chronicle* states that in 912 Edward the Elder built a fort on the north bank of the River Lea at Hertford and in the following year built a second fort on the opposite bank. In 918 Edward built a second fort at Bedford, but on the opposite bank of the Great Ouse to the existing one. In 918 he constructed two 'strongholds' on opposite banks of the Great Ouse at Buckingham. In 924 he constructed a bridge across the River Trent at Nottingham, linking together a fort on either side, as part of his defences against the Vikings (*Anglo-Saxon Chron*, 97–104).

The building of forts to control access to navigable rivers was also undertaken by the Franks. In fact it is possible that this concept was copied by the Franks from the English. Among other defence works of the kind, Charles the Bald built two forts on opposite banks of the River Seine at Pont de l'Arche in 862, linked by a bridge (Hassall and Hill 1970, 192–4; Boyer 1976, 21). The importance of a well-defended bridge and associated landward fortifications was demonstrated by the siege of Paris which lasted from November 885 to February 886. A large Viking army was able to sail up the Seine to Paris by passing the fortified, but undefended, bridge at Pitres. But at Paris the Vikings' route further upstream was blocked by two strongly defended bridges, which they failed to capture despite a siege and numerous assaults. The Vikings finally stormed one bridge after its middle section was carried away by floods, but failed to capture the second bridge, and had to haul their ships overland to bypass it before they were able to continue upstream (Boyer 1976, 22–3). The Vikings themselves were aware of the military importance of bridges as a means of communication.

Such cases provide potential parallels for London, which controlled access to and across the Thames valley, as well as the road network which converged on it, and was reoccupied by Alfred for strategic as well as commercial reasons. But the question of the bridging of the Thames raises in turn the issue of the status of Southwark at this period. Where a bridge existed Southwark would have become a fortified bridgehead and a vital part of London's defences, whereas in the absence of one it could have been little more than a cul-de-sac surrounded by creeks and marsh. It has been suggested that the inclusion of Southwark in the Burghal Hidage, traditionally dated to *c* 915, indicates the re-establishment of the bridge (Biddle and Hudson 1973, 23), but there is no other evidence for the existence of either the bridge or Southwark until the early 11th century. The inclusion of Southwark in the Burghal Hidage of *c* 915 is curious as it is listed as *Suthringa geweorc* or the '[defence] work of the men of Surrey', which is distinct from the simple and universal form of 'South-work' used subsequently. The exact significance of the wording is unknown, but it raises the possibility that the Hidage text may refer to a planned or designated fortress rather than a completed one (Dyson 1990, 110 fn 57).

According to the *Anglo-Saxon Chronicle*, during September 994, in a renewed series of raids and incursions, London was attacked by a fleet of Vikings who burnt down the settlement (*Anglo-Saxon Chron*, 126–7) and harried the surrounding area. This was part of a concerted campaign in which, as the *Anglo-Saxon Chronicle* of the late 10th and early 11th centuries makes plain, the Vikings made extensive use of their ships in navigable rivers such as the Exe, Humber, Thames, Severn and Yare (*Anglo-Saxon Chron*, 126, 131, 133–5), much as they had a century earlier. It is quite possible that it was this attack on London in 994 that first prompted the rebuilding of London bridge and the fortification of Southwark. For, by the time of the next series of seaborne attacks in 1009 on London, the situation was very different according to the *Anglo-Saxon Chronicle*. In the Peterborough (E) *Chronicle* manuscript it states that 'they [the raiders] often attacked London town, but praise be to God that it still stands sound and they always fared badly there' (*Anglo-Saxon Chron*, 139). By 1014, Southwark was a strongly defended southern bridgehead. This earliest reference to the existence of Southwark's defences occurs in the great saga of St Olaf (written down *c* 1220) and in stanza 6 of the *Vikingarvisur* (compiled 1014–15). The saga describes *Suðvirki* as 'a great trading place', defended by 'large ditches' with an internal road and a rampart constructed of 'wood, stone and turf' and defended by 'a great army' (see Chapter 14.12 for detailed discussion and references). According to the great saga of St Olaf, fortifications or 'castles', as well as parapets which reached higher than a man's waist, were built on the bridge. Snorri quotes Sigvatr's stanza as evidence of Olaf's deeds in London, but his knowledge of 11th-century Southwark was presumably nil, so he may have applied a certain poetic licence to the subject, drawing on his knowledge of 13th-century European urban defences.

The only archaeological evidence of the defences of Saxo-Norman Southwark found to date was excavated in 1979 at Hibernia Wharf and consisted of a large ditch, some 4.0m wide, aligned south-east to north-west and apparently of early 11th-century date (Fig 27). The position of the ditch, some 100m west of the Saxo-Norman bridge abutment, suggests

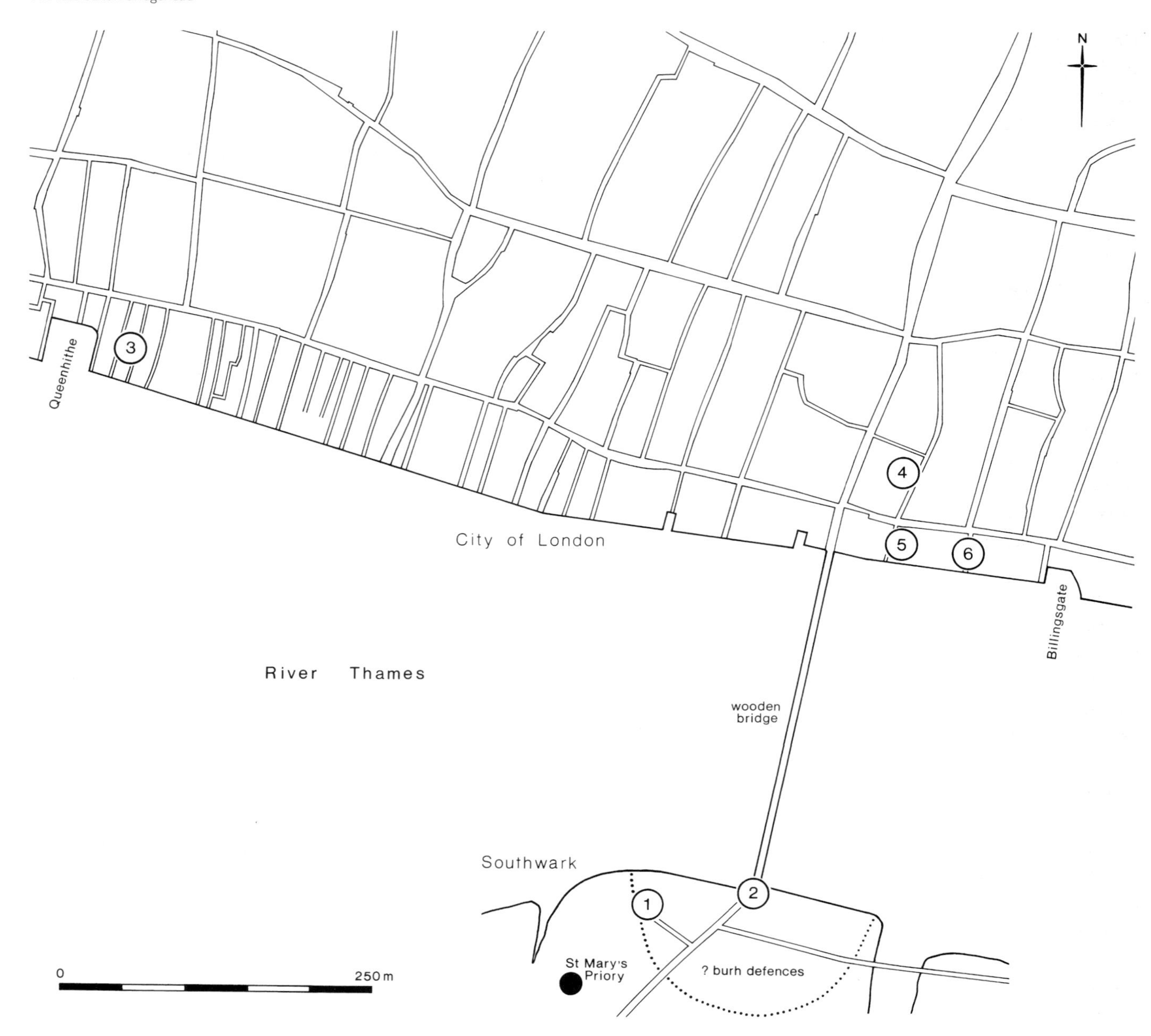

Fig 27 Plan of the London and Southwark bridgehead, in c 1000, showing the position of the bridge, all streets, harbours at Queenhithe and Billingsgate, the location of the Southwark burh defences and their projected alignment, and St Mary's priory
Key to excavations discussed in text: 1 – Hibernia Wharf; 2 – Fennings and Toppings Wharves; 3 – Bull Wharf and Vintry; 4 – Pudding Lane; 5 – New Fresh Wharf; 6 – Billingsgate

that the defended area of the bridgehead could have been fairly small. During dredging around the site of the bridge in 1824–41 there came to light an iron axe and spearhead of 9th- to 11th-century date, while a second spearhead of 9th-century date is probably of Scandinavian origin.[14] A number of iron battle axes, spearheads and a grappling iron were found on the foreshore near the northern end of the medieval bridge during the 19th century and have been dated to c 1000 (Wheeler 1927, 18). Wheeler suggested that these weapons may have been lost during this and other attacks on the bridge during the 10th and 11th centuries. Wilson (1976, 255, 401) pointed out that the decoration of the spearheads and at least one of the axes is stylistically Scandinavian, and believed that a number of the axes were battle axes, although some are now regarded as carpenter's tools that may have been lost during bridge building or repair work.

In May 1016 the Southwark bridgehead was again defended, this time against the forces of King Cnut, who apparently bypassed it by digging a new channel for their ships through the marshes and creeks of low-lying southern Southwark, avoiding the defended bridgehead (*Anglo-Saxon Chron*, 148–9). Clearly Cnut's forces did not dare attack the actual bridge and force their way upstream. The course of Cnut's channel is unknown, but several routes have been suggested (Nunn 1983, 198). Hauling ships overland short distances to avoid obstacles was a fairly common procedure during this period, and in the ninth chapter of *De administrando imperio* written c 950, Viking traders are described as routinely circumventing the rapids of the River Dnieper in the Ukraine by dragging their ships overland (Jannson 1992, 75), and, as has been seen, Viking raiders resorted to a very similar expedient at Paris in 886 (Boyer 1976, 22–3).

5.2 The reoccupation of the London bridgehead

While it is possible that between the early 5th and the late 9th century there was sporadic human activity within the area of the London bridgehead, perhaps connected with a ferry across the Thames, there is very little archaeological evidence to support this thesis. One sherd of 5th- or 6th-century pottery was found in a residual context at New Fresh Wharf, and nearby at the Billingsgate Roman bathhouse a 5th-century Saxon brooch and sherds of 5th-century Eastern Mediterranean wine amphorae (*c* 400–500) have been found (Vince 1990, 7,11).

The archaeological evidence for late 9th-century occupation within the city consists mostly of coin finds (Vince 1991a, 420–4), although it has been shown that soon after the reoccupation of 886 part of the tidal foreshore was used as a landing place for the first time since the building of the riverside wall in the late 3rd century (Brigham 1990b, 139–40). Documentary evidence shows that by 889 a port and a market were functioning at 'Æthelred's Hithe', later known as Queenhithe; the early name records the Mercian ealdorman's involvement in the development of the port in a way paralleled by his role at Worcester, another town under his control (Dyson 1978 and 1990). Recent archaeological work at Bull Wharf has located barge beds dated to *c* 890 for beaching ships at Queenhithe, together with coins of Alfred (Ayre and Wroe-Brown 1996, 20) (Fig 27). Bread Street and probably Bow Lane/Garlickhithe were laid out at the same time to link the port area with Cheapside and the interior of the town, providing a nucleus of the 'grid' pattern familiar at other Alfredian towns which, after the mid-10th century, was extended to much of the rest of the southern part of London (Dyson 1990).

By the beginning of the 11th century London was becoming a significant port, visited by merchants from Flanders, Normandy and Scandinavia. The trade of Saxo-Norman London owed much to geographical factors (see Chapter 4.1) and, although not the only port in south-east England, London was the lowest point where the tidal Thames could be bridged, and where suitable space was available nearby for an estuarine port and a settlement to develop. London was, thus, a prime example of those numerous English towns which developed at river crossings. In this sense, Queenhithe, well above the bridge, was a special case. It appears to have served upstream river traffic from the Midlands (where Æthelred's sphere of activity lay) and, apart from the landing of fish there by the 13th century, there is no evidence that it was used by traffic from the Thames estuary, North Sea or Continent. By the reign of Edward the Confessor (1042–66), however, a port was established for the use of German merchants at Dowgate, between Queenhithe and London bridge, and this raises the question whether the bridge now incorporated a drawbridge to enable large seagoing vessels to pass beneath (see Chapter 7). Dowgate, however, was a special preserve of the emperor's men; and by the turn of the 10th and 11th centuries the generality of overseas shipping that was not intended, or did not wish or need, to pass the bridge was catered for on the waterfront between the bridge and Billingsgate which was embanked at this date to provide a suitable beaching place (Steedman et al 1992, 133–4).

Billingsgate is first documented as a port *c* 1000 (Robertson 1925, IV Æthelred II), and access to this new port just downstream of the bridge from the centre of the city was provided by a network of north–south streets which also appear to have developed during the later 10th and 11th centuries (Steedman et al 1992, 123–8) (Fig 27). These included Pudding and St Botolph Lanes (Steedman et al 1992, 75–6), and the grid probably extended as far north as Fenchurch Street, which appears to have originated as a route along the southern frontage of the Roman forum with which it is aligned at this point. An essential element of this north–south street grid would have been Fish Street Hill on the line of the original approach road to the Roman bridge, and it has been suggested that Fish Street Hill was again in use by the early 11th century (Horsman et al 1988, 114). The northern continuation of the bridge approach road is Gracechurch Street, which was laid out across the remains of the monumental Roman forum and basilica complex; this must have happened by the end of the 11th century, for the church of St Peter Cornhill, first recorded in 1140, stood at the intersection of Gracechurch Street and Cornhill (Marsden 1987, 67, 70). Gracechurch Street in turn led northwards to Bishopsgate on the far side of the walled city from the bridge, thus constituting part of a major trunk route between the Channel and eastern and northern England by way of the bridge. The earliest post-Roman occupation found to date along Bishopsgate consists of a single sunken-floored building, probably of early 11th-century date (Milne et al 1984, 399). To the east of Gracechurch Street, excavations at Leadenhall Court confirm that this area was not reoccupied until *c* 950–1000 (Milne 1992a, 37). This evidence is certainly consistent with the likelihood that Gracechurch Street and Bishopsgate Street were laid out at the turn of the 10th and 11th centuries to provide good communications with the port established at the northern bridgehead (Fig 27) as part of a large-scale development which may well have included the restoration of the bridge itself.

The medieval approach road to the southern bridgehead, Borough High Street, was on a very similar alignment to that of the Roman road (Hinton and Graham 1988) (see Chapter 4.4), and the essential survival of both the northern and southern approach roads to London bridge suggests that the roads remained in use throughout the early and middle Saxon periods (c 400–850) perhaps encouraged, in the absence of a bridge at this period, by the existence of some sort of river ferry. It was the continuing use of these Roman approach roads which ensured that the Saxo-Norman bridge was constructed close to the site of its Roman predecessor. It was claimed by Honeybourne (1969, 33–4) that the Saxo-Norman timber bridge was sited downstream of the stone bridge and that Pudding Lane served as the northern approach road to it (Fig 27), but this theory leant heavily on the incorrect location of Drinkwater's Wharf. Subsequent documentary research and excavation showed that Honeybourne's claim cannot be substantiated (Dyson 1975): excavations in 1975 at the point in question revealed an area of piles dated to c 950–1020; these were too flimsy to have formed the foundations for a bridge abutment and were interpreted as a jetty extending out into deep water (Steedman et al 1992, 23–8, 102).

Judging from the negative evidence of three excavations to the east of Fish Street Hill, and a fourth excavation at Regis House on the west of the street, it is probable that the northern bridgehead area was unoccupied between the 5th and the late 9th or early 10th centuries; environmental evidence from Pudding Lane indicates that the area was waste ground colonised by elder and nettles during this period (Milne 1988, 17). Almost all the dating evidence for the reoccupation of the area is ceramic and imprecise, the earliest features consisting of a number of wells and rubbish pits or cesspits dating from c 850–1000. A number of rubbish pits and wells (later used as rubbish pits) at Regis House date from c 900–1050 (Brigham in prep).

This broad date range allows for some reoccupation by the early 10th century, which fits better with the dating of the earliest revetments found around the northern bridgehead (Steedman et al 1992, 23–9, 48–52) and the earliest timber suitable for tree-ring dating from one of the bridgehead buildings has a felling date of after 912 (Vince and Jenner 1991, 24). At Botolph Lane a series of rectangular wooden-walled and sunken-floored buildings were constructed during the 10th and 11th centuries; these buildings were laid out in rows and sited at right angles to, or parallel with, the adjoining streets (Milne 1988, 12–21).

Nearby at New Fresh Wharf there was no activity on the foreshore until the late 10th or early 11th century, when the jetty and the rubble bank revetment were constructed (Steedman et al 1992, 23–9), in the top of which was found a fragment of a 10th-century clinker-built boat (Marsden 1994, 141–51). At St Botolph's Wharf, closer to Billingsgate, there was a period of silt accumulation from the late 3rd century, when the last Roman waterfront was built, until the port was reoccupied in the 11th century (Fig 27), when planks and other timbers were laid out in c 1039–40 as consolidation before rubble and clay were dumped to form an embankment along both sides of the small inlet (Steedman et al 1992, 48–52). Soon after, the bank was enlarged by dumping more rubble and the addition of timbers and clay. Later the bank was protected by a series of pile, plank and baseplate/stave revetments, the latest revetments dating to c 1150–90 (Steedman et al 1992, 55).

5.3 The reoccupation of the Southwark bridgehead

The earliest artefactual evidence of Saxon activity in the Southwark bridgehead consists of two coins. The first find was a coin of the Byzantine Emperor Justinian I (AD 527–65, alleged date of issue 537), found at King's Head Yard (adjoining Borough High Street) in 1881 (RCHM 1928, 149), and the second one was a halfpenny of Alfred (London monogram, AD 886), found within a Saxo-Norman ditch at St Thomas Street (Stott 1991, 309). However, the earliest archaeological evidence of Saxon occupation at the Southwark bridgehead is represented by an unlined rubbish pit at Topping's Wharf which contained two late Saxon shelly ware cooking pots dating from c 900–1050 (T7.1) (Fig 28). Other Saxo-Norman structural features (1050–1150) around the southern bridgehead at Fennings Wharf consisted of a series of 14 vertical earth-fast posts probably representing the wall lines of a small rectangular building some 3.6m long and 2.7m wide (F17.2, Building 9), its walls were replaced several times. There was also a short length of upstanding cob wall forming part of another building (F17.4), and two unlined and truncated pits of Saxo-Norman date (F17.1, 3) (Fig 28). One of these pits was oval-shaped and had been used for the disposal of organic rubbish, plant remains including charred cereal grains, bread wheat, rye oats and weed seeds, interpreted as waste from crop processing (see Chapter 14.9), and bones of cattle, sheep and chicken (see Chapter 14.8). The other pit (F17.1), which predated the construction of the small rectangular building, was square in section and was used for the disposal of cess and organic rubbish. It contained bones of sheep/goat, cattle and chicken (see Chapter 14.8), as well as mineralised fruit pips of apple and pear. Rushes and sedges had been put into it, perhaps to seal its smellier contents (see Chapter 14.9).

Excavations at Toppings Wharf produced some residual late Saxon or Saxo-Norman pottery from the soil dumps (T8.8) laid down to reclaim the eroded foreshore in the early 14th century (Sheldon 1974, 30) (see Chapter 14.5). The few surviving structural features and the very low number of pits compared with findings around the northern end of the bridgehead (Vince 1991a, 425–6) suggest a considerably lower density of settlement around the southern bridgehead during the Saxo-Norman period.

The Domesday Survey of 1086 shows that Southwark was an unmanorialised settlement without a direct lord of its own, which may perhaps have resulted from the settlement's informal development with minimal outside control or intervention. The settlement at this date was probably confined to the high ground around the bridgehead, and the existence of a port can be inferred from reference in Domesday Book (1086) to tolls collected *in strande* (on the shore) and *in uico aquae* (on the water street) (Carlin 1996, 15–16). Domesday also recorded Edward the Confessor's right to the local tolls in Southwark, which he shared with Earl Godwin, the king taking two-thirds and the earl one-third (Carlin 1996, 106–7).

5.4 The documentary evidence for Saxo-Norman London bridge

What was once taken to be the first reference to London bridge in a charter of 963–84 has been shown to allude to a bridge in Northamptonshire, probably over the River Nene, on the road to London (Hill 1976). That leaves as the earliest reliable mention of the bridge an allusion in a composite manuscript which includes Æthelred II's (979–1013; 1014–16) fourth law code and is usually dated *c* 1000 (Robertson 1925),[15] although it may in part date from the last years of Cnut's reign (1016–35) (Loyn 1962, 93). It also contains some decrees concerning coinage of *c* 995 and some local, London regulations of uncertain date. In the context of a review of London's gates (which also includes the earliest reference to Billingsgate) occurs the statement that 'a merchant who came up to the bridge with a boat containing fish paid one half-penny as toll and for a large ship one penny' (ibid, 93–4).

According to the Olaf sagas, the bridge was successfully attacked in 1014 by King Æthelred's Viking allies (see Chapter 14.12). When considering the sagas as a source of historical evidence it should be borne in mind that they were part of an oral Scandinavian tradition, which was not compiled into written documents until the early 13th century. The *Anglo-Saxon Chronicle* states that in 1016 the Southwark bridgehead and the London bridge was defended against the forces of King Cnut, who apparently bypassed it by digging a new channel along which they hauled their ships westwards (*Anglo-Saxon Chron*, 148–9). According to Osbern's late 11th-century account of the transfer of the relics of St Ælfheah from St Paul's Cathedral to Christ Church Canterbury in 1023, some of Cnut's housecarls occupied the bridge while the relics were in transit in case of public hostility (Lawson 1993, 181). The *Anglo-Saxon Chronicle* records that on 14 September 1052 Earl Godwin's forces sailed unopposed through the bridge, having first waited for the high tide, and keeping close to the south bank (*Anglo-Saxon Chron*, 180–1); the wording of the text implies that this was a difficult manoeuvre, as was sailing under the stone bridge during the 17th and 18th centuries (see Chapter 13.1). In this instance the citizens of London obviously chose to assist the outlawed earl to return and then allowed him to land his forces in Southwark (McLynn 1998, 82).[16]

Unless there was a drawbridge section within the bridge which was opened to allow the ships through (which seems highly improbable), the ships' masts must have been lowered to allow them to pass beneath (see Chapter 7). In 1066 when William, Duke of Normandy, and his victorious army arrived in Southwark after the Battle of Hastings, they found the southern bridgehead defended against them and, having attacked and failed to capture it, burnt down the rest of Southwark as a reprisal (Carlin 1996, 15). William's army returned to London in December 1066, apparently entered it from the west and captured it after a battle within the walls (Mills 1996, 60–1).

Fitz Stephen's 'Description' of London written as the preface to a proposed life of St Thomas Becket some time between 1170 and 1183 makes no mention of London bridge (Butler 1934, 61). This is a curious omission considering that he must have been writing during the early stages of the construction of the stone bridge, and might suggest that fitz Stephen's account was compiled during the early 1170s, soon after Thomas's death and before the construction of the bridge began *c* 1176 (see Chapter 8). It is worth noting that St Thomas was born in London and that the chapel on the bridge was dedicated to him (see Chapter 9.4). However, fitz Stephen's text does state that the walled city of London was entered by 'seven double gates' (Butler 1934, 49): one of these must have been the postern near the Tower, in existence before 1190 (Carlin and Belcher 1989, 83), or part of the bridge.

5.5 The archaeological evidence for the first Saxo-Norman London bridge

The earliest archaeological evidence for the existence of the Saxo-Norman bridge consists of two *ex situ* timbers (F19.4) (Fig 28). The first timber was a large beam found lying within a series of foreshore fluvial deposits. It was made from a squared oak log and was probably a baseplate: 5.30m long, 450mm wide and 420mm thick, its southern end appears to have been squared, while the northern end contained an edge-halved scarf joint within its upper face, and a shallow empty socket (see Glossary). The second timber was a rectangular oak beam, made from a quartered oak log with a possible laft-joint at one end, and found reused as part of the foundations for a 12th-century bridge caisson. Both timbers are from the same tree and were felled *c* 987–1032 (see Chapter 14.2). They are interpreted as part of a very late 10th- or early 11th-century southern abutment (see Glossary) of a timber bridge destroyed during the early 11th century by floods or tidal scouring, a few *ex situ* bridge timbers being buried by the following build-up of fluvial deposits on the tidal foreshore.

The precise date when the timbers used in the first phase of the Saxo-Norman bridge were felled is unknown: it may have been as late as the 1020s in which case it is possible that there had been a still earlier timber bridge of which no trace was found.

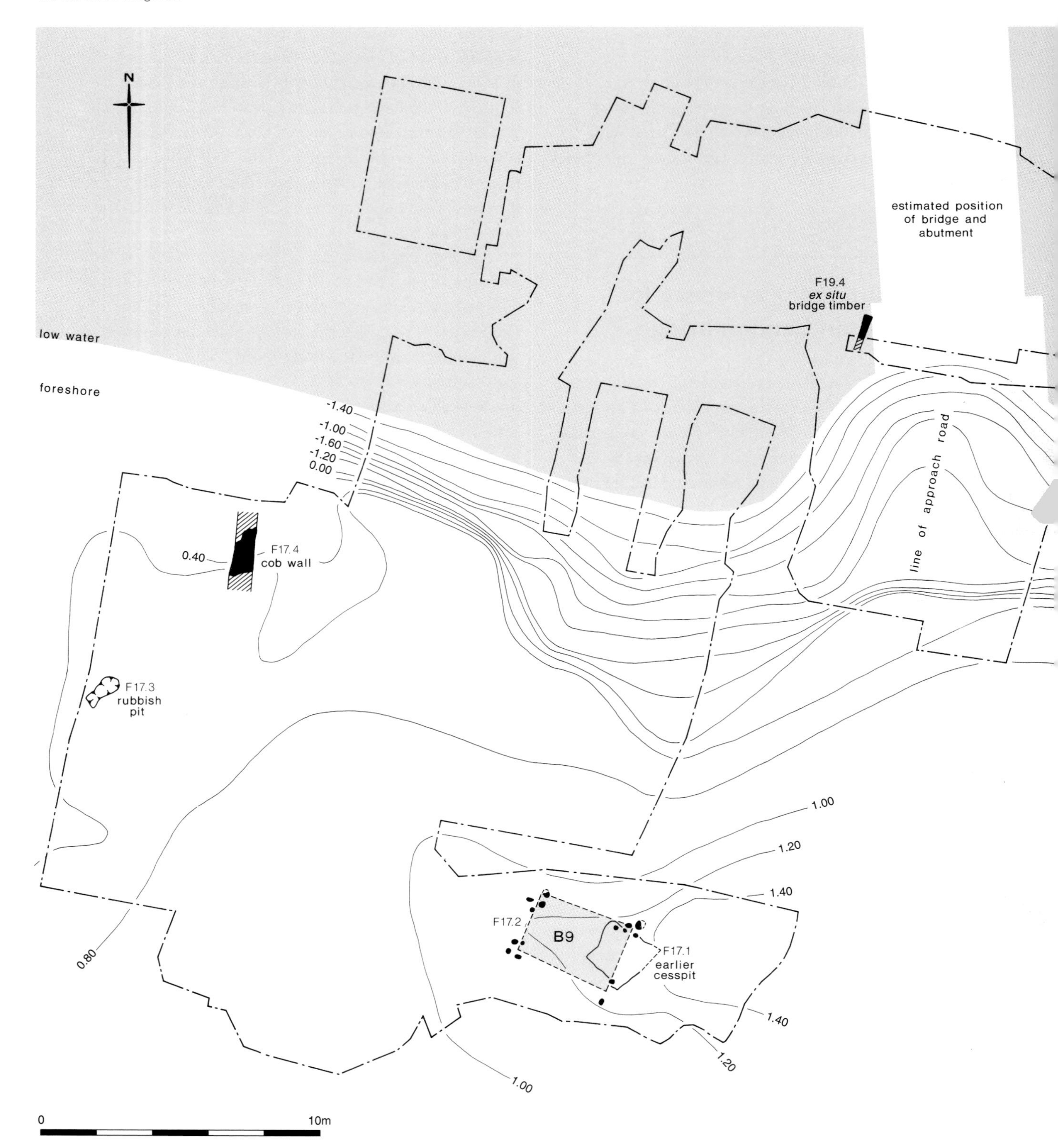

Fig 28 Plan of the Saxo-Norman features and the earliest Saxon bridge structure (F19.4), plus contours of natural geology

The only evidence for the existence of a river ferry which could have been used before the bridge was built, or while it was being rebuilt, is a piece of folklore recorded by Stow c 1600 that before a bridge was first built the adjacent priory of St Mary's in Southwark (known from the 14th century as St Mary Overy) was founded as a house of sisters from the profits of a river ferry, and that the nunnery was later converted into a college for priests and the ferry replaced with 'a bridge of timber' (Stow 1603, ii, 56). As bridge building was considered to be pious charitable work there may well be some truth in this piece of folklore: the late 10th- or early 11th-century jetty found at New Fresh Wharf next to the City bridgehead would have made an ideal northern terminal for a ferry (see Chapter 7). There are other examples of monastic houses having links with local ferries. For instance, a charter of 1130–3 issued by Henry I in favour of the monks of

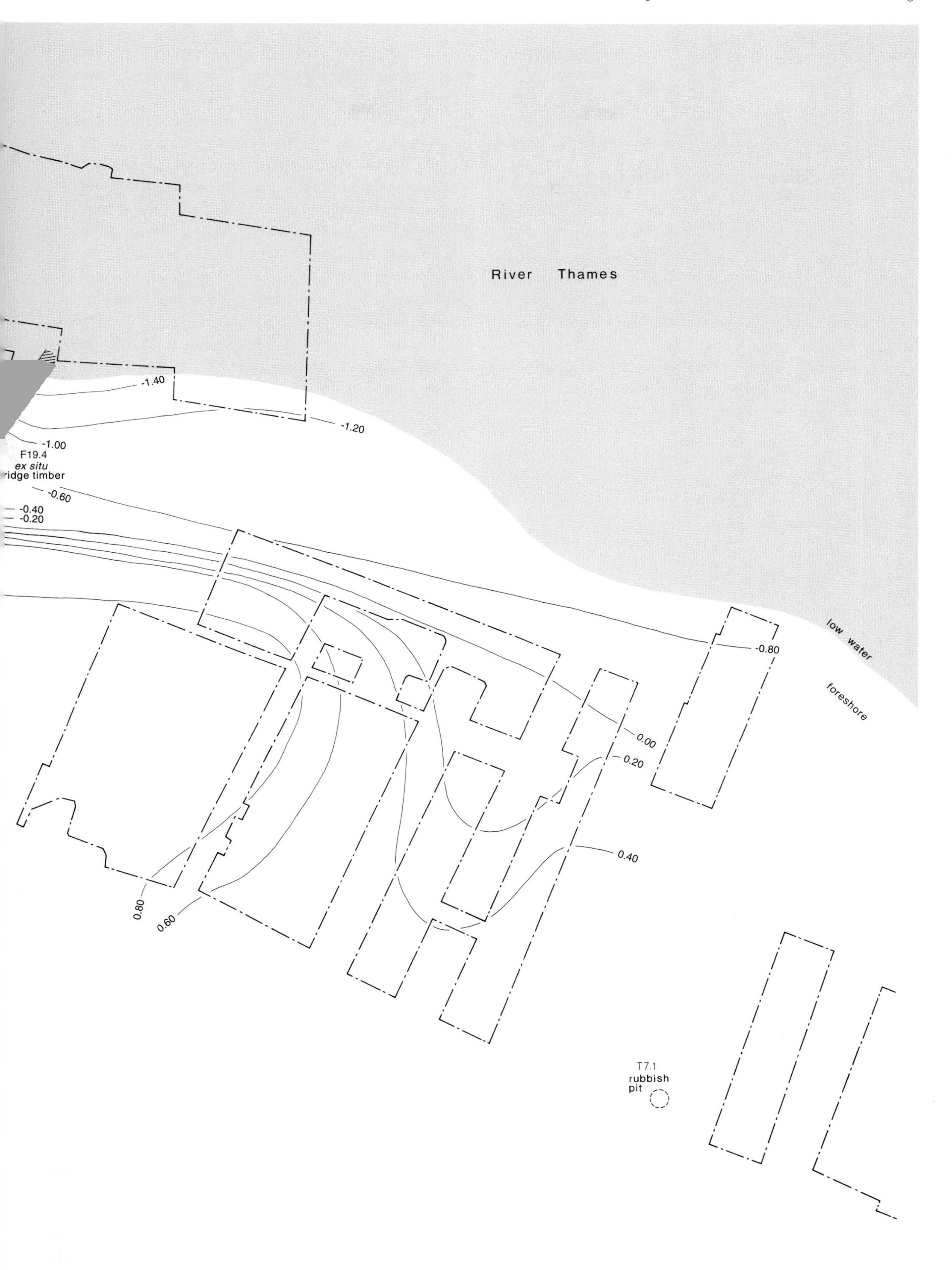
River Thames
-1.40
-1.20
-1.00
F19.4
ex situ
ridge timber
-0.60
-0.40
-0.20
low water
foreshore
-0.80
0.00
0.20
0.40
0.80
0.60
T7.1
rubbish
pit

Rochester stated that their rights included a share of the ferry dues whenever the bridge over the Medway was broken down (Brooks 1994, 35).

5.6 Late Saxon bridge building

Study of authentic 8th-century Anglo-Saxon charters reveals that, of the 91 examples, only one for Crediton in Devon dating from 739 mentions a bridge (Sawyer 1968, S255), while 50 other charters mention fords (Cooper 1998, 20–1). Of the 331 authentic 10th-century charters 20% mention bridges and 57% fords (Cooper 1998, 21). These figures show that few English rivers were spanned by bridges during the 8th century, but that by the 10th century numerous bridges had been constructed (Cooper 1998, 30). Interestingly the charters of the 8th and 9th centuries refer to the building of bridges, while charters of 10th-century date begin to refer to the repair rather than construction of bridges (Harrison 1996, 234).

In France during the 10th–11th centuries there was an identical increase in bridge building too (Boyer 1976, 26–27). What was the impetus for this sudden wave of bridge building in England and France? Cooper (1998, 34–42) listed a number possible reasons – increasing problems with flooding during the 11th century as recorded in the *Anglo-Saxon Chronicle* (see Chapter 6.1); more land clearance and drainage increasing run-off; by the late Saxon period ox-drawn wains were replacing pack animals as a means of freight transport; and also as part of the defences against the Viking raiders (see above, 5.1). This last reason appears to have been the prime mover in the construction of many bridges, including London bridge during the 10th century.

6

The erosion of the Saxo-Norman and medieval foreshore

Bruce Watson

6.1 The erosion of the Saxo-Norman Southwark bridgehead foreshore

At Fennings Wharf, on the upstream side of the bridgehead, the earliest post-Roman foreshore feature was riverine erosion and scouring, represented archaeologically by a steeply sloping truncation horizon (F11.1[17]) cut into the flood plain gravels, its top at -1.42m OD and observed at least as low as -1.72m OD (Fig 29). Above this truncation horizon was a build-up of silt-stained sand and gravel (F11.2). Much of this deposit was later removed by a second period of erosion (F11.3; F13.1) (Fig 29) represented by an irregular, concave truncation horizon, aligned roughly south-east to north-west, its top at 0.51m OD and observed at least as low as -1.42m OD. The later of these truncation horizons is dated to the early/mid-11th century on stratigraphic evidence, but the earlier one is undated. The two truncation horizons could represent either events merely a few years apart within a single phase of general foreshore erosion, or two different phases of erosion decades apart. Contemporary evidence of marine transgression within south-east England is provided by the Essex coast, where the formation of shell banks prior to the deposition of marine sediments has been radiocarbon dated to cal AD 880–1220[18] (Greensmith and Tucker 1969; Wilkinson and Murphy 1986). These sediments were almost certainly deposited during the early stages of a period of transgression, and seem to have coincided with the erosion of the seaward wall of the nearby Saxon shore fort at Bradwell-on-Sea.

Contemporary record of severe flooding survives from this period. The *Anglo-Saxon Chronicle* states that on 28 September 1014 'the great sea-flood came widely throughout this country and ran further inland than it ever did before, and drowned many settlements and countless number of human beings' (*Anglo-Saxon Chron*, 145). The Thames is not specifically mentioned, although its valley must have been one of the areas affected. Catastrophic events such as this could have caused serious erosion along the Thames foreshore. At the end of the same century, in 1097, the *Chronicle* recorded that floods carried away nearly all of London bridge and that workmen engaged in building the Tower of London and the great hall at Westminster were sent to repair it (*Anglo-Saxon Chron*, 234).

There is no doubt that during the later 11th century there was considerable erosion of the foreshore on the upstream side of the southern bridgehead, and that this had possibly been recurring intermittently for some time. It is probable that during this period of erosion any surviving traces of late Roman revetments and bridge were washed away. The Saxo-Norman foreshore deposits found around the southern bridgehead contained a number of residual Roman finds including glassware, brick and tile, early and late Roman pottery (AD 270–400), as well as a copper alloy coin of the house of Constantine (330–75) (F12.13–16) (see Chapter 14.6 and Chapter 3). From 1081 onwards, however, systematic attempts (discussed later) were made for the first

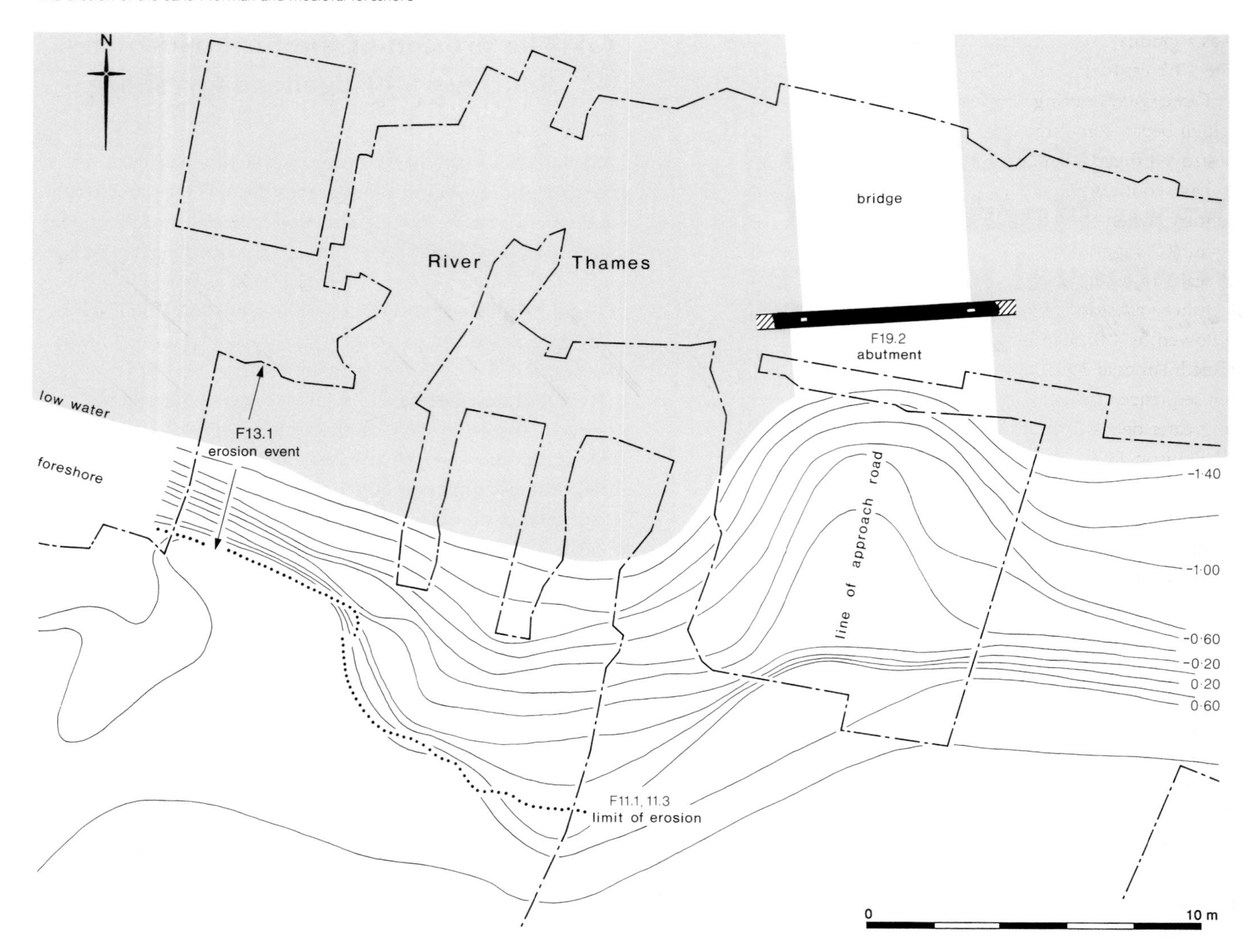

Fig 29 Plan of the Southwark bridgehead foreshore, showing position of the bridge abutments and caissons, the various areas of erosion and contours of natural geology

time to prevent further erosion.

On the downstream side of the southern bridgehead at Toppings Wharf the earliest post-Roman foreshore stratification consisted of a series of estuarine deposits (T6.1): coarse- and medium-grained silt-stained sands and gravels separated by thin bands of laminated clays and silts, their top at -0.87m OD and observed at least as low as -2.10m OD (Fig 29). These deposits indicate a cyclical build-up of sands and clays in succession and repeated three times; a sequence indicating a frequently changing, depositional environment, varying from swiftly flowing to still or standing water. The uppermost sand deposits contained an *ex situ* oak stake with a pointed end and dated to 1122, presumably washed out of some upstream revetment. This build-up was interrupted by a further period of river erosion (T6.2), its top at -0.87m OD and observed at least as low as -1.58m OD. The variable estuarine conditions during this period are emphasised by another entry in the *Anglo-Saxon Chronicle* for 10 October 1114. 'There was so great an ebb-tide everywhere as no-one remembered before, and such that men travelled across the Thames to the east of the bridge by riding and walking' (*Anglo-Saxon Chron*, 244).

6.2 The erosion of the city foreshore

Around the northern bridgehead during the Saxon period there is less evidence of riverine erosion than at the southern end of the bridge. Instead there seems to have been a limited build-up of foreshore sands and silts between the 4th and 10th centuries, probably interrupted by unrecorded phases of truncation caused by riverine erosion (Steedman et al 1992, 22, 49, 98) (Fig 30). The survival of large portions of the 3rd-century Roman quay and riverwall, both upstream and downstream of the northern bridgehead, proves that erosion of this stretch of foreshore during the Saxon period must have been much less severe than that at the southern bridgehead. Perhaps the massive quay structures and riverwall prevented or restricted erosion of the northern foreshore; this could have deflected the downstream current towards the Southwark foreshore, itself apparently not protected in this way (see Chapter 3.3).

To the east of the northern bridgehead at New Fresh Wharf and Billingsgate between the 3rd and 11th centuries there was a slow accumulation of foreshore sediments (Steedman et al 1992, 22, 49) (Fig 30). Further east at Custom House there was a gradual erosion of the foreshore from the 4th until the late 13th century (Tatton-Brown 1974, 128). Evaluation work at Three Quays during 1995–6 revealed a period of erosion which began sometime after the 3rd century and probably destroyed this stretch of the Roman riverwall. This appears to have continued, perhaps intermittently, until the late 12th century, when the accumulation of fluvial deposits began again (Grainger 1996, 25, 37). To the west of the bridgehead at Seal House and Swan Lane, there was an undated period of foreshore erosion between the 3rd and 12th centuries, followed by a limited build-up of silts, sand and gravel (Steedman et al 1992, 78–80, 89). Further west at Bull Wharf the sequence of fluvial deposits has been studied and analysed in greater detail than on any other London foreshore site (Wilkinson 1997 and in prep). This work has shown that, between the construction of the Roman quay in AD 198 and the Alfredian barge beds of AD 890, some 1.5m depth of laminated fine-grained fluvial deposits (silts and clays) accumulated at Bull Wharf. Only after the construction of the first major revetment at Bull Wharf in AD 964 did the depositional environment change dramatically. Laminated organic sediments accumulated on the northern part of the site and were in time sealed by fluvial gravels that accumulated further south on the tidal foreshore (Wilkinson 1997, 32–3, 123–4).

The documentary evidence suggests that long stretches of the late Roman riverside wall were still standing along the city foreshore by the end of the 9th century, but that by 1066 it no longer survived as a continuous and effective defensive feature (Dyson 1980, 9) (Fig 30). Excavation of part of the riverwall at Castle Baynard Street in 1975 revealed that it had collapsed some time after the 8th century (Hill et al 1980, 21). Fitz Stephen in the late 12th century attributed the collapse of the Roman riverside wall to river erosion (Butler 1934, 49).

It is probable that throughout the Saxon period the Thames foreshore on both banks around the bridgehead underwent cyclical phases of erosion and accumulation. The phases of erosion might have been caused by severe winter floods for which there is limited documentary evidence. One factor which would have encouraged the build-up of foreshore deposits between the phases of erosion was the commencement during the late Roman period of a marine transgression (see Glossary) which is still in progress today (Devoy 1980) (see Chapter 3.4, 3.5). The results of this continuing phase of transgression have been most marked in the low-lying areas of the Thames basin, such as the Bermondsey district of Southwark where, east and south of the high ground on which the southern bridgehead was sited, fluvial clays accumulated in some areas until the post-medieval period, when these areas were systematically embanked and drained. It is probable that these clays represent either areas of brackish standing water or areas which were flooded daily by tidal water (Drummond-Murray et al 1994, 252; Thomas and Rackham 1996, 246). The records of Southwark manor make it clear that the flooding of pasture and meadow land was a serious, if intermittent, problem. For instance, pastures were flooded in 1252–3; from 1296 to1301 a pasture called the *Pondfold* was flooded annually and in 1308–9 it was flooded again; more floods of manorial land were recorded in 1325–6 and 1381–2 (Carlin 1996, 38).

6.3 The reclamation and revetments of the Southwark foreshore upstream of the bridgehead during the 11th and 12th centuries

The earliest surviving attempt to stabilise the upstream bridgehead foreshore cannot be identified with certainty as there appear to have been several broadly contemporary piecemeal efforts throughout the late 11th century, following a phase of erosion. One of the earliest revetments (F12.1) consisted of a layer of grey fluvial clay, its top at -0.28m OD, containing a number of horizontal small oak logs or large branches aligned east–west. These logs appear to have been laid down as part of an attempt at consolidation, the clay accumulating around them. Within the upper part of this deposit were a number of logs, anchored by driven stakes, which appear to have been sealed by dumped or waterlaid grey clay. These timbers are interpreted as part of an 11th-century revetment, the rest of which may have been removed by erosion. Nearby another early attempt at foreshore consolidation began with the construction of an east–west revetment (see Glossary), which probably consisted of a line of driven stakes retaining a number of horizontally laid planks, several of them reused (F12.3). This 11th-century revetment collapsed forward, presumably as a result of being undercut by erosion (Fig 30).

The collapsed revetment was buried by a fluvial silt/peat (F12.4), itself overlaid by the next revetment: a mass of horizontally laid, small, split oak trunks and large branches (F12.5) within a mixed grey or blue-grey clay and silt matrix, which probably represented material dug up on the foreshore at low tide and then dumped amongst the timbers. Its top lay at +0.25m to -0.73m OD (Fig 31; Fig 34), and its timber was felled during 1081–2 (see Chapter 14.2). In one place it was sealed by a series of river deposits, starting with a grey clay/silt, overlaid by a further sequence of silts (its top at -0.18m OD) (F12.9). The exact nature of the next revetment is uncertain. It probably consisted of an east–west aligned series of driven piles, retaining horizontal planking, much of it reused (F12.10) (Fig 32). The blue-grey silty clay found amongst the timbers probably represented dumped material derived from the foreshore and intended to form a bank. The multitude of collapsed timbers, many reused, suggests that this revetment was repaired or patched several times, before collapsing forward and scattering, presumably owing to erosion and undercutting (Fig 33). This and its probable repairs date to *c* 1081–94. The collapsed revetment was sealed

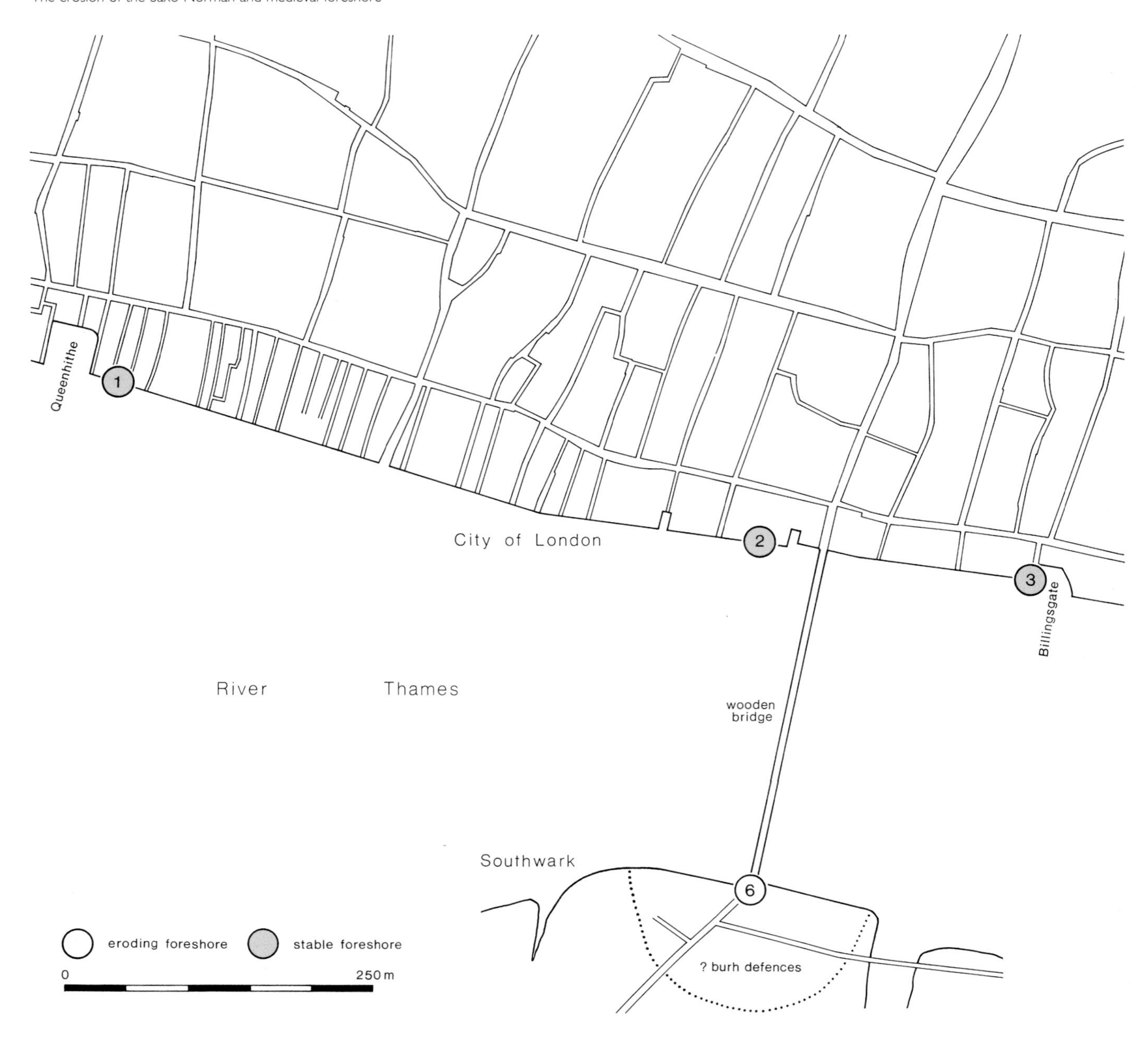

by a build-up of fluvial sediments (F12.11). A sample of this material contained a high frequency and species diversity of plant remains. Wetland species included rushes, sedges, hairy buttercup, tripartite bur-marigold and gipsy-wort. Disturbed ground and arable species included goosefoot, *Polygonum* spp, docks and cornflower, corn marigold and thistles being particularly well represented. Grassland species included self-heal. Some of these plants may have simply washed downstream and been deposited here, while some of the wetland and grassland plants may be residues from the collection of hay or fodder (see Chapter 14.9).

The foreshore was soon reclaimed by the construction of yet another revetment. This consisted of a cruciform arrangement of two overlapping timbers interpreted as the basal portion of a revetment tieback or frontbrace, the rest of the revetment having been either washed away or removed for reuse (F12.12) (Fig 32). To the north of the timbers was a bank of horizontally laid brushwood, branches and small trunks with an average length of about 0.90m and a diameter of about 100mm, its top at 0.20m OD. The timbers were set in a matrix of blue-grey clay, probably dug from the adjoining foreshore and dumped amongst the timbers. All the timbers were oak, except one which was willow or poplar; several date to 1095. Later this bank was repaired or reinforced by the addition of more brushwood, including twigs, branches and several reused timbers (F12.13), their top at 0.36m OD, and at the same time a single stake (F12.14) was driven into the bank. All these revetments date from *c* 1081–1100 and appear to have been part of a continuing effort to stabilise the upstream area of the bridgehead foreshore. During their construction, or immediately after, further attempts were made at foreshore reclamation to the north, presumably to protect the bridge approach further east from persisting erosion.

To the north of the clay and log bank (F12.5, 10) was a sequence of late 11th-century activity (Fig 34) beginning

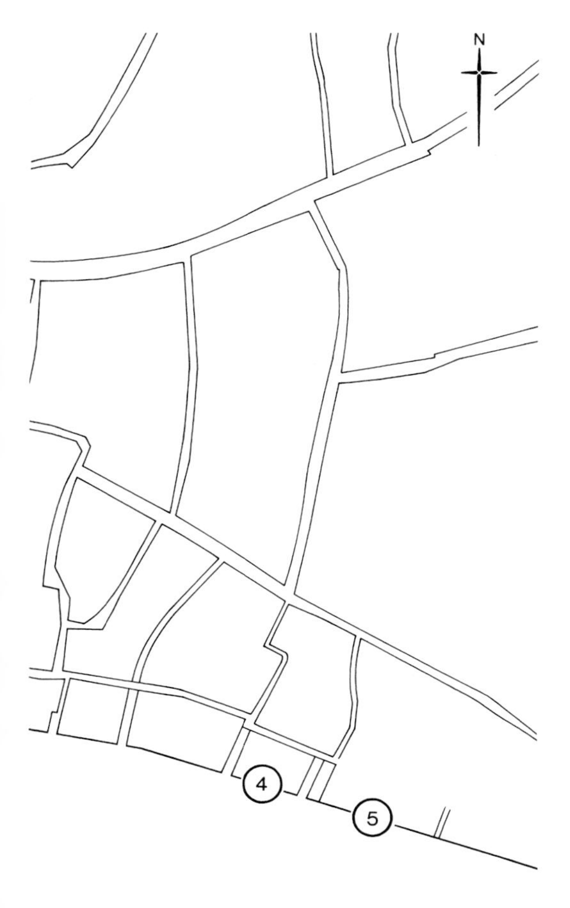

Fig 30 *Plan of the Saxo-Norman Southwark and London bridgehead in c 1000, showing the various locations where the foreshore was eroding and where it was relatively stable*
Key: 1 – Bull Wharf/Vintry; 2 – Seal House; 3 – Billingsgate; 4 – Custom House; 5 – Three Quays; 6 – Fennings and Toppings Wharves

Fig 31 *Clay and timber revetment (F12.5) looking south (2m scale) (MoL)*

with the accumulation of a number of undated *ex situ* timbers or drift wood lying on the foreshore (F14.1: Fig 32); included were two logs and two *ex situ* stakes, possibly deposited on the eroding foreshore by the ebb tide. The next distinguishable revetment was represented by a mass of *ex situ* timbers, comprising the collapsed remains of a line of stakes of uncertain alignment and retaining a number of horizontal planks (F14.2) as if to contain further clay and timber dumped behind it. It dates to c 1077–1100. On top of it the next revetment was built, consisting of a cruciform tieback (F14.3), its top at -0.99m OD. Both surviving elements were made out of reused timbers (Fig 32), the rest of the structure having been apparently washed away or removed for reuse.

The Saxo-Norman pottery from the early revetments and foreshore deposits (F12) is dominated by locally produced handmade cooking pots, identical to the pottery found in the two rubbish pits near the bridgehead. Many of the vessels found in two of the revetments (F12.5 and F12.12) were represented by more than one sherd, suggesting that the material was discarded in bulk by people living nearby, and perhaps partly explaining the scarcity of rubbish pits around the Southwark bridgehead (see Chapter 14.5). The plant remains from the foreshore are indicative of the dumping of waste material from byres, hay or animal fodder, and the processing of cereal crops (see Chapter 14.9). The greatest concentration of animal bone fragments was from F12.2 and F12.5, cattle and sheep/goat being dominant, and other species including chicken, goose, mallard, cod and whiting (see Chapter 14.8). One find of special interest is a goshawk, a high-status find as such birds of prey were used in falconry (Grant 1988, 180). Comparable finds among animal bones from the 12th- or early 13th-century revetments and foreshore deposits (F14.15 and 16: Fig 34) included fallow deer, presumably killed in hunting, and a crane, possibly caught by falconry (see Chapter 14.8).

The next revetment appears to have incorporated another tieback, anchored by two driven stakes (F14.4) (Fig 34). It is dated to 1097 and may represent part of the bridge repairs that followed the flood damage documented in that year (see Chapter 7). A second revetment was constructed to the west (Fig 34), consisting of a substantial north–south baseplate (F14.5) more than 3.60m long, its top at -0.88m OD. In the upper face of the baseplate was a continuous groove to hold the staves (see Glossary), which were made out of lengths of reused clinker-built boat planking (Marsden 1994, 154–5). It is notable that tree-ring dating suggests a *terminus post quem* of 1085 for the boat planks (Tyers 1994, 207), while the baseplate has a felling date range of *c* 1090–1135: this suggests that the boat may only have been in use for ten or 20 years before it was broken up and reused. To the west of the baseplate were driven three clusters of stakes (F14.6–8), one of them dated to 1165 and possibly intended to anchor frontbraces. The stakes were sealed by a build-up of fluvial silty peat containing pottery dating to 1180–1230 (F14.9).

Slightly further east there was another localised phase of foreshore erosion during the early 12th century (F14.10). This was followed during the early or mid-12th century by the accumulation of layers of fluvial sandy silt and silty clay,

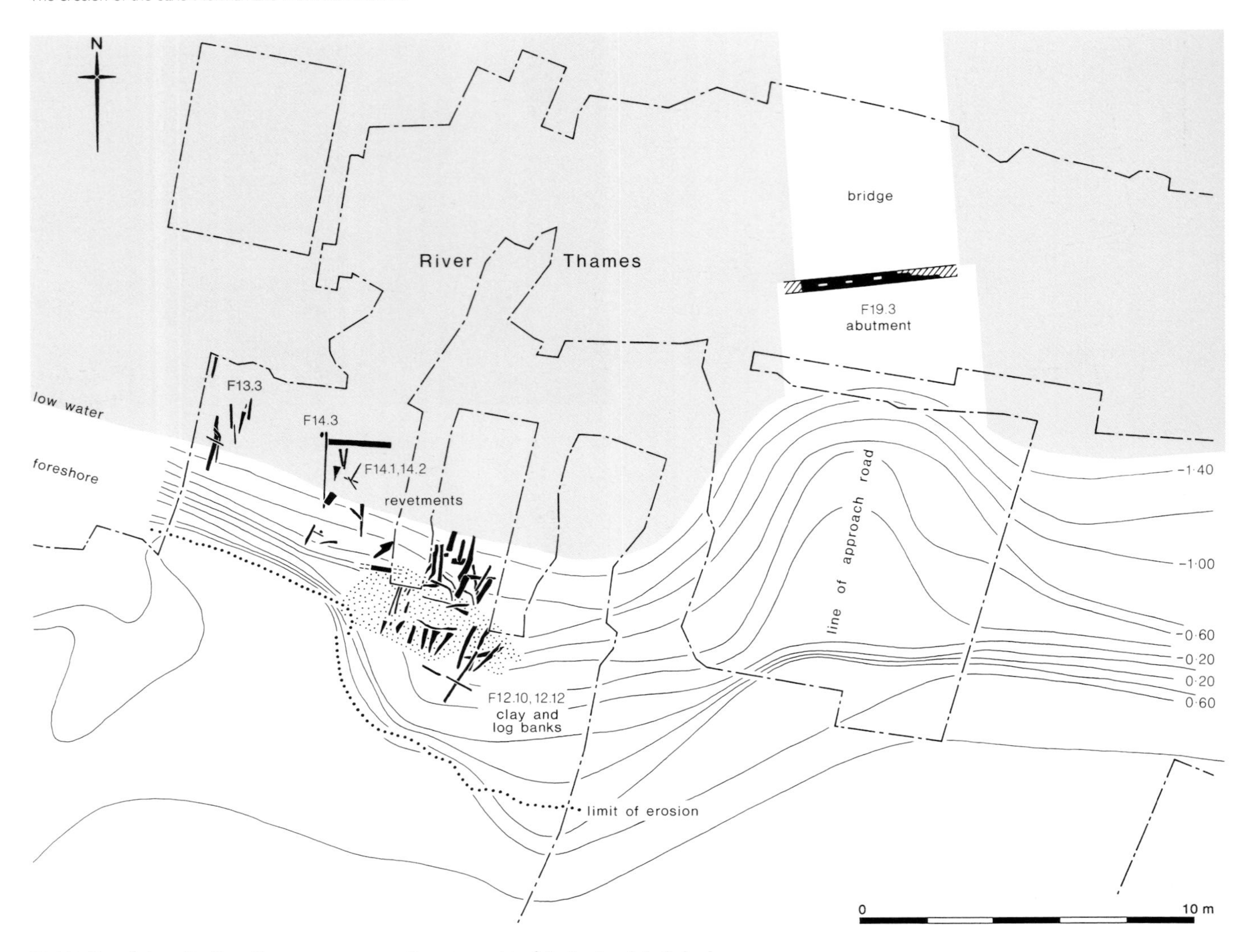

Fig 32 Plan of the earlier Saxo-Norman revetments on the upstream side of the Southwark bridgehead

Fig 33 *View of line of collapsed piles (F12.10), looking north, showing the sharpened tips and providing the clearest evidence of the destruction of the revetment: after the sand and gravel these stakes were driven into was washed away, all the piles simply collapsed forward and were subsequently buried by fluvial sediments (F12.11); the hollow in the foreground marks the southern limit of an earlier phase of foreshore erosion (F11.3) (2m scales) (MoL)*

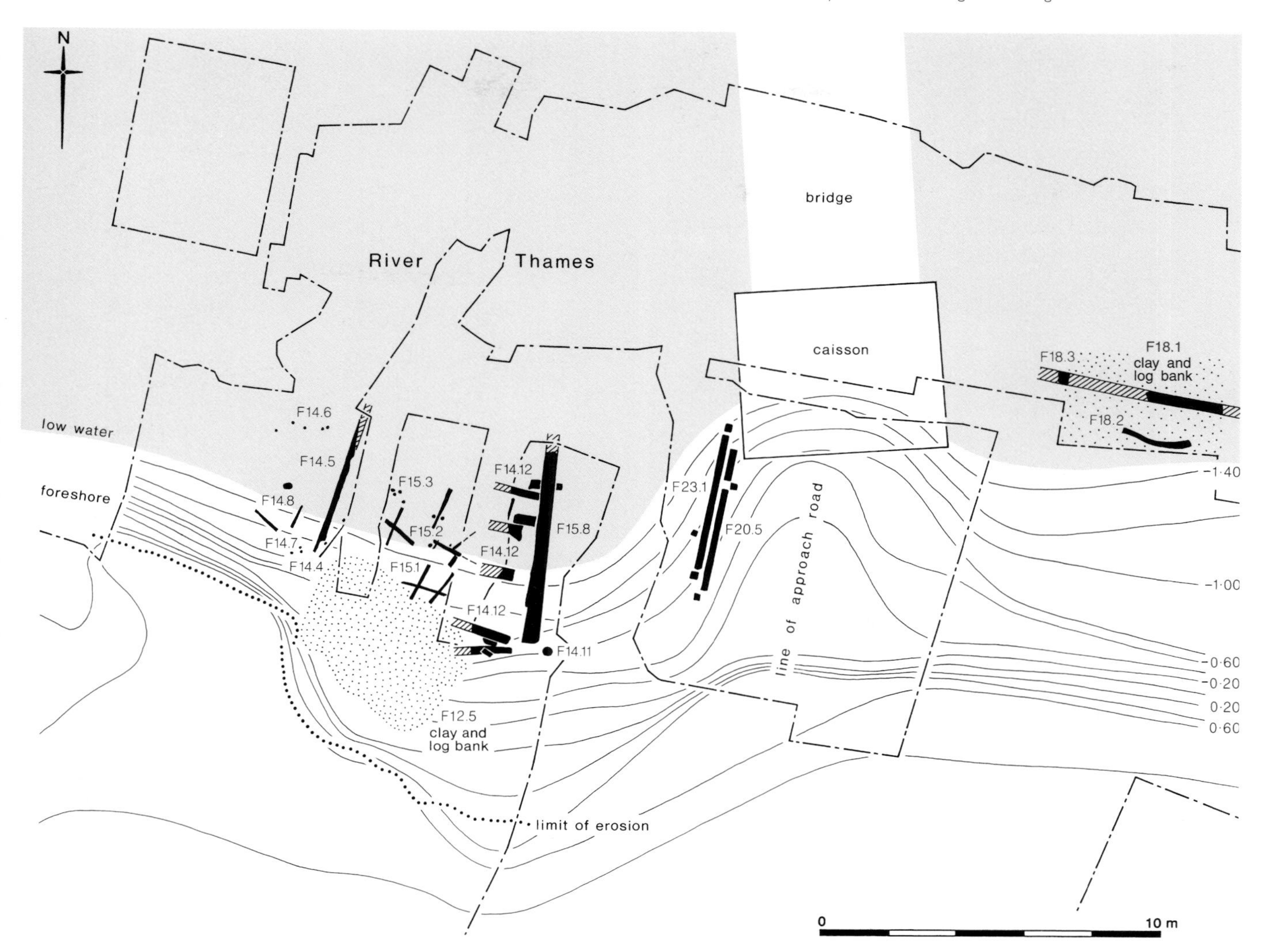

Fig 34 Plan of the later Saxo-Norman revetments on the upstream side of the bridgehead

its top at -0.43m OD, containing a collapsed wooden cask (F14.11: Fig 34). On top of the fluvial sediments laid a series of east–west oriented logs and reused timbers, one of which dated to 1133 (F14.12: Fig 34). It is possible that these timbers were intended to consolidate the foreshore surface, enabling vessels to beach and helping to protect the existing clay and log bank (F12.5, 10) from further erosion. To the east was laid a substantial baseplate, its top at -0.58m OD, with a continuous groove in its upper face to retain staves (F14.13: Fig 32). It was aligned north–south and anchored by driven stakes, all of which were oak except one, which was fruit wood (*prunus*).[19] After the staves in this revetment had been either removed or washed away the area was sealed by an accumulation of fluvial sand and silts (F14.14). These were in turn sealed by the construction of one phase of the upstream bridgehead revetment, dating to 1133 (F15.8: Fig 34).

The sequence of activity to the north and west of the clay and log bank (F12.5, 10) began with the construction of an H-shaped timber structure, interpreted as a tieback or frontbrace of a revetment (F15.1: Fig 34). Later, and slightly further north, two phases of tiebacks or frontbraces (see Glossary) were constructed (F15.2, 3), the earlier phase dating to 1118 and the later to 1137 (Fig 35). The remainder of these revetments were either washed away or more likely taken away for reuse, and the area was covered by a build-up of fluvial clay/silt (F15.4). The next phase of revetment was represented by a single tieback (F15.5), and was sealed by a series of fluvial deposits (F15.6) containing pottery dating to 1080–1150.

Further west of the clay and log bank (F12.5, 10) the sequence of Saxo-Norman revetments began only in the mid-12th century, presumably because everything earlier had eroded away during the late 11th century. After a period of sustained foreshore erosion (F13.1) (Fig 29), a revetment consisting of a north–south line of driven oak stakes was constructed: one of the stakes had been felled in 1148 (F13.2), and the revetment perhaps included a tieback and a fragment of 11th-century boat wale (Marsden 1994, 158–9). It was buried by fluvial deposits and replaced by an east–west line of driven stakes of oak, alder and *prunus* (F13.3), undercut by erosion. Next a series of driven stakes were used to anchor a crosspiece and backbrace for a further east–west revetment (F13.3), most of which was either washed away or

Fig 35 *View of the Saxo-Norman foreshore looking north, showing a mass of* ex situ *piles (F15.3), partly covering an earlier tieback (F15.1) (1m scale) (MoL)*

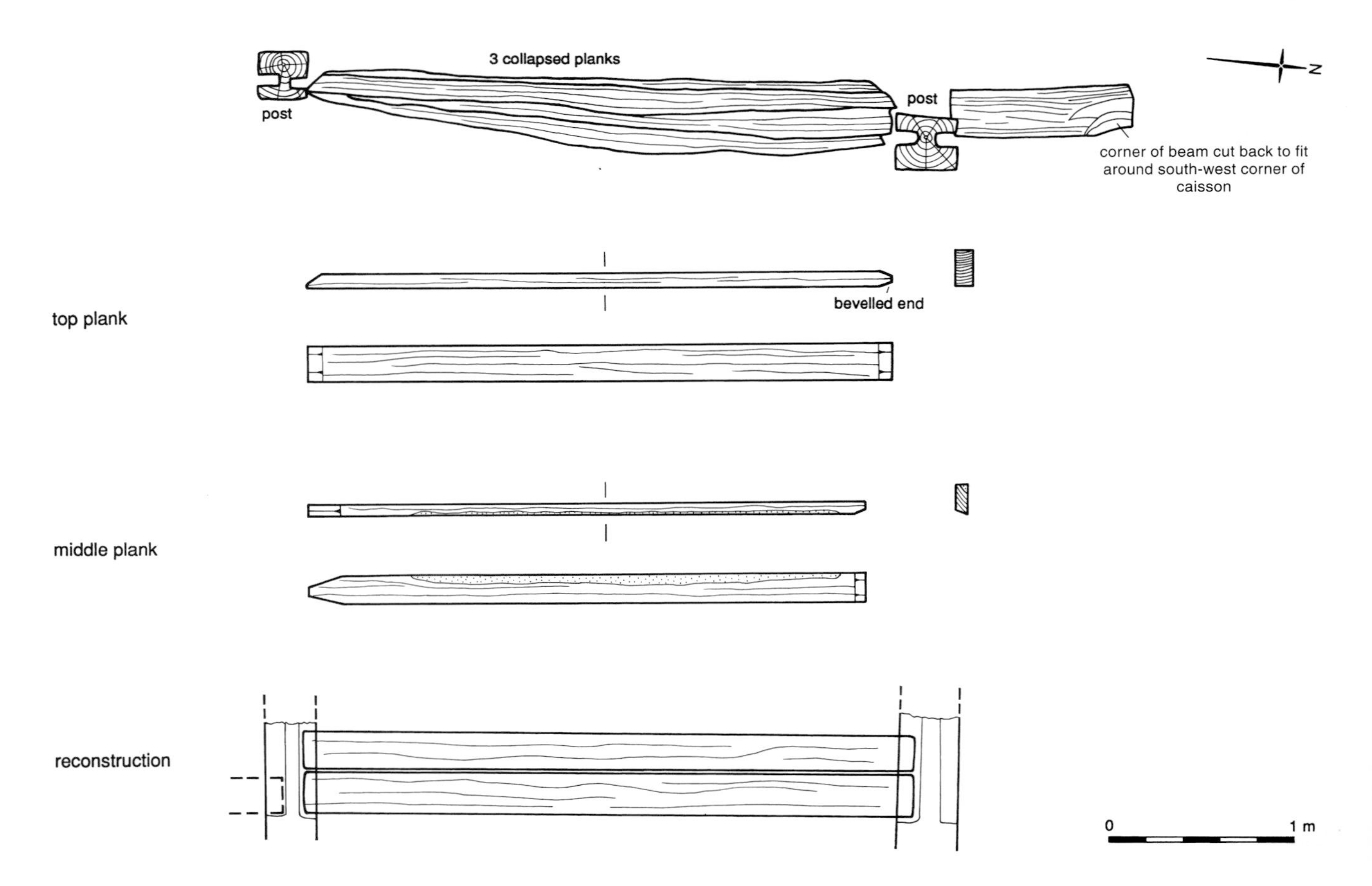

Fig 36 *Plan, elevation and reconstruction of the 12th-century bulwark revetment (F20.5)*

dismantled for reuse (Fig 32). The remnants of the revetment were then sealed by a further build-up of fluvial deposits containing pottery dating to 1180–1230 (F13.4). Two samples from the fluvial deposits included a high frequency and diversity of plant remains. The presence of charred cereal grains suggests waste from crop 'cleaning' or processing, while that of grassland plants and of numerous buttercups (indicative of hay) suggests the dumping of fodder or stable waste (see Chapter 14.9).

Finds from the Saxo-Norman revetments and foreshore deposits included fragments of imported basalt rotary quernstones, a hone stone, a bone knife handle and two bone skates (see Chapter 14.6). Fitz Stephen in his description of 12th-century London mentioned that skating was a popular pastime (Stow 1603, i, 93), and the skates are also a reminder that the Thames froze over on many occasions during the medieval period (Currie 1996, 1–3).

On the upstream side of the bridge approach three phases of substantial revetments were built during the early 12th century. All these revetments were aligned north–south, at right angles to the flow of the ebb tide; they successfully halted the erosion and scouring of the upstream area of the bridgehead and even allowed some of the tidal foreshore to be reclaimed. The first phase of substantial revetment is dated to 1128 (F20.5) and consisted of two vertical, squared oak posts with vertical grooves cut into opposite sides, set in post pits 3.05m apart (Fig 34; Fig 36). Between the posts had been slotted a series of edge-laid, horizontal, radially-cleft oak planks (see Chapter 14.3), a carpentry technique known as 'bulwark' construction (see Glossary). To the east or rear of the planking, soil or rubble would have been dumped to consolidate the reclaimed area. The planking at the southern end of the revetment was later removed for reuse. The reconstruction of the northern portion of this revetment is uncertain, but it was not of bulwark type construction as there was no vertical groove in the north side of the adjoining post for retaining the horizontal planking (Fig 36). A very similar type of revetment, found at the Thames Exchange site, Vintry, on the north bank of the Thames, was dated to *c* 1183–4 (Milne and Milne 1992, 60). It differed slightly from the Fennings Wharf example as it had intermediate front stakes to retain the planking and both front and rear diagonal braces. Possibly these had been removed at Fennings Wharf for reuse. The technique of building with grooved or rebated posts is also demonstrated by the discovery of a number of such posts, reused in an early 13th-century revetment at Billingsgate (Brigham 1992b, 95–7). The same building technique was used in house construction at Husterknupp, Germany, during the 10th century (Chapelot and Fossier 1980, 263).

The second substantial revetment phase dates to 1133 (F15.8). It consisted of a massive, squared oak log baseplate aligned north–south, over 5.8m long with a square cross section and set on two crossbearers (Fig 34; Fig 38). The upper face of the baseplate sloped from south to north and contained four rectangular sockets and a continuous rectangular groove, the top surface at -0.14m OD (Fig 37). Any joints in the southern end of the timber may have been destroyed by decay and machine damage. Evidently the sockets were intended to hold verticals between which staves were slotted into a continuous groove, although the entire superstructure had been removed (Fig 37; Fig 38). The tapering profile of this timber suggests that the wider southern end was probably the butt end of the tree. This revetment was built some 4m to the west or upstream of its predecessor, showing a certain confidence in its attempt to reclaim a large area of foreshore. The confidence was perhaps misplaced as within ten or 20 years (*c* 1136–55) a third revetment was built in almost the same place as the first one.

The third substantial phase of revetment (F23.1) dates to

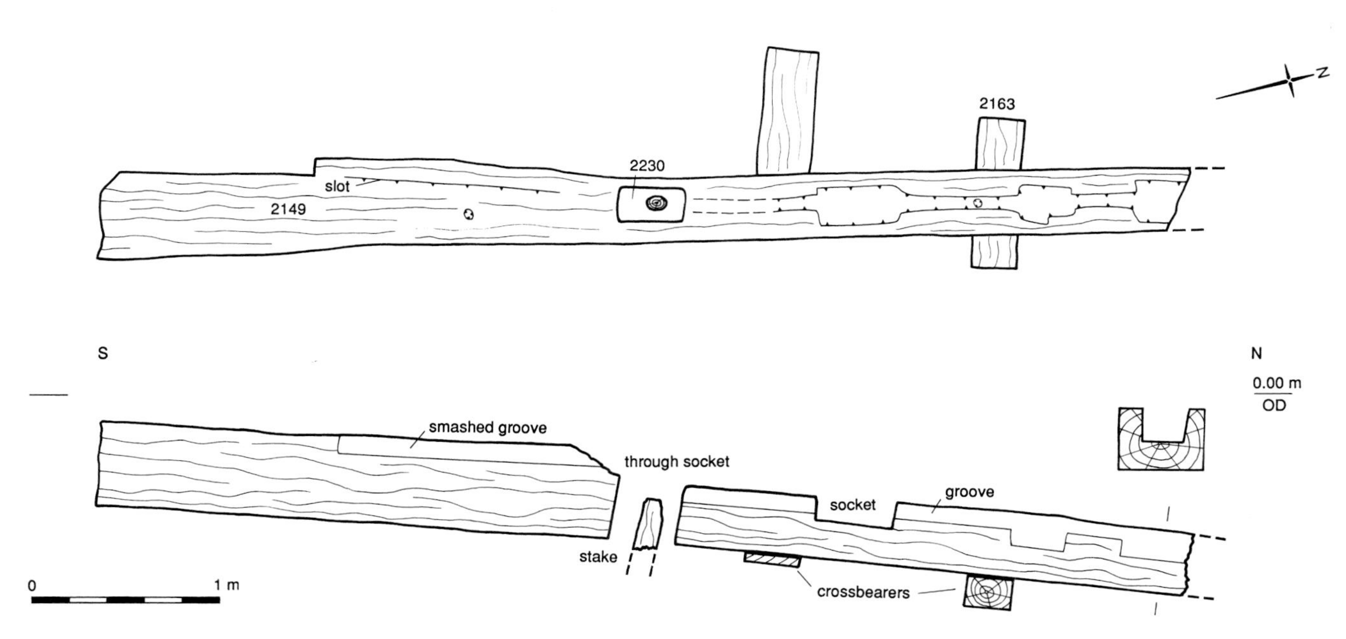

Fig 37 Plan, elevation and cross section of the massive 12th-century revetment baseplate (F15.8)

Fig 38 *View of the 12th-century baseplate (F15.8), looking north; the clay and chalk rubble seen in section are part of the infill of the Colechurch abutment (F26.6) (2m scales) (MoL)*

c 1136–55, making it contemporary with the first phase of 12th-century caisson and probably it was still in use when the second 12th-century caisson was built (Chapter 7). The revetment baseplate was 4.32m long with evidence of earlier use (Fig 34; Fig 39), probably as the top plate of a house, and now turned upside down, so that the relict lap joints would originally have been for diagonal braces. The diagonal slot adjoining the continuous groove in the timber would have served for sliding in the staves between the base- and top plates. Identical gaps can be seen in the 11th- to 12th-century baseplates of the vertical plank wall (*stabbua*) building at Büderich, Germany (Chapelot and Fossier 1980, 266), and the same diagonal slot was found in a reused house baseplate of 11th- or early 12th-century date found in one of the 12th-century revetments at Billingsgate (Brigham 1992b, 91). Other reused house baseplates from Billingsgate had the same pattern of short lengths of a widened, continuous groove in their upper faces, doubtless intended to hold larger than average staves (Brigham 1992b, 91). Within the upper face of the Fennings baseplate was a continuous groove which still contained eight *in situ* staves, some of them secured by wedges (see Glossary) (Fig 39). A small diamond-shaped hole (width 80mm, height 50mm) had been cut into one stave, suggestive of a tiny peephole window – which would indicate that the staves may have been fashioned from reused house planks. At each end of the baseplate was a vertical, square post, set in a post pit. The northern post had a single vertical groove in its southern face for retaining staves, but the southern one had grooves in both the north and south faces, presumably to retain the edge-laid, horizontal planking that would have formed the southern end of the revetment (see Chapter 14.3). This revetment was sealed by the preparatory work for the construction of the Colechurch bridge (see Chapter 8).

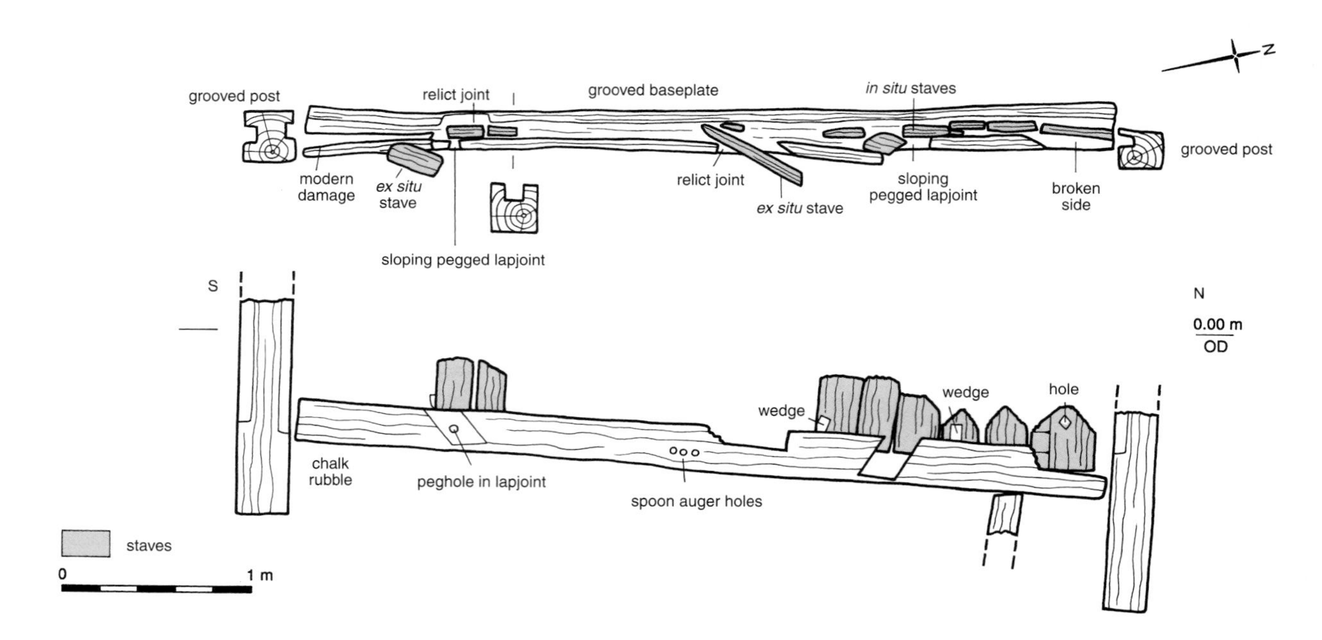

Fig 39 *Plan and elevation of 12th-century baseplate revetment (F23.1), showing the in situ staves and associated posts*

6.4 The reclamation and revetments of the downstream Southwark bridgehead foreshore during the 11th and 12th centuries

Less is known about the sequence of revetments in this area than on the upstream side, as shortages of both time and staff precluded a full investigation. The earliest revetment was an undated clay and log bank aligned east–west (F18.1), its top at -0.95m OD and the timbers within the bank consisting of driven stakes and larger horizontal logs. The next phase of activity consisted of a single log or large branch from which a number of side branches had been cut: it had been dumped on the top of the bank (F18.2: Fig 34). The next phase of revetment consisted of two lengths of scarf-jointed baseplate made from quartered oak logs aligned east–west (F18.3), its top at -1.00m OD. The baseplate was probably anchored by a series of tusk-tenons (see Glossary). In the upper face of the baseplate was a continuous vertical groove, presumably for retaining staves. The later phases of downstream revetment, which are contemporary with the Colechurch bridge, are described in Chapter 8.

It is clear that the bulk of the Saxo-Norman bridge abutments, caissons and landward approach ramp acted as a breakwater and protected the downstream foreshore from the worst effects of riverine and tidal scour. Within this relatively sheltered area of the foreshore, reclamation was obviously successful as the revetments here extended much further north into the river than those found on the upstream side of the 12th-century caissons (Fig 34).

6.5 The Saxo-Norman Fennings Wharf revetments: a discussion

It is clear that during the 11th and 12th centuries riverine erosion on the upstream side of the Southwark bridgehead was a very serious problem, and that to try and prevent it a series of revetments were constructed, many of them destroyed or damaged by flooding.

The most durable type of revetment was the clay and log bank, constructed parallel with the line of the foreshore and containing a number of timbers, either driven stakes or a mass of horizontally laid logs or branches. Very similar clay, rubble and timber banks of Saxo-Norman date have been found in a number of places along the north bank of the Thames within the medieval walled City (Steedman et al 1992, 23–36, 50, 80–1). A number of other revetments were founded on baseplates, normally simply laid on the foreshore but sometimes set on crossbearers or driven stakes. These baseplates usually had a continuous shallow groove in their upper faces, but some also contained sockets to retain upright timbers or staves. In only two instances (F14.5 and F23.1) were *in situ* staves still slotted into the upper face of the baseplates. In one instance (F15.7), the staves from a baseplate revetment were found *ex situ*. These Saxo-Norman baseplate revetments are also known from sites in the City on the opposite side of the river (Milne and Milne 1992, 26–8, 37), where many of them possessed front- or rear braces: any such braces in the Fennings Wharf revetments would appear to have been removed for reuse.

The most difficult to interpret of the Saxo-Norman Fennings Wharf revetments are those incorporating various T–shaped or cruciform arrangements of horizontal timbers. The nature of their superstructure is uncertain but on the evidence of contemporary revetments around the northern bridgehead they probably consisted either of baseplate structures, or of lines of driven stakes incorporating several tiers of horizontal edge-laid planks (Milne and Milne 1992); both types could have required front- or rear braces. Also unclear is the original top height of the various Fennings Wharf revetments, as a result of their incomplete state or modern truncation, but as it is estimated that the highest astronomical tides during this period would have reached no higher than *c* 1.7–1.8m OD (Steedman et al 1992, 119–20; see Chapter 3 above), the average tops were probably *c* 2.0m OD.

The timbers of the Saxo-Norman revetments found at Fennings Wharf contained features which illustrate a number of contemporary woodworking techniques, including the use of baseplates with stavework superstructures and bulwark construction (see Chapter 14.3). Especially notable is an example (F23.1) of the reuse of a house top or baseplate in a waterfront revetment, a practice that was also in evidence at Billingsgate (Brigham 1992b). A number of clinker-built boat planks were reused as staves (Marsden 1994, 154–5), and many other reused timbers were identified by the presence of relict slots and sockets (Fig 39), and it is probable that most of them were salvaged from the earlier revetments that were being replaced.

6.6 The influence of reclamation on the hydrology of the River Thames

Reclamation of both sides of the Thames foreshore along with the presence of the timber bridge must have influenced the hydrology of the River Thames during the 11th and 12th centuries. The stretch of the river from Westminster to Rotherhithe appears as one very large, underdeveloped meander, the development of which has probably been restricted by the area of natural high ground on which the City of London is sited, in contrast to the classic meanders that can be seen further upstream at Chiswick and downstream at the Isle of Dogs.

Both erosion and accumulation are involved in the

formation of meanders. An obstruction swings the current of the river towards one bank, whence it is deflected back towards the opposite bank. Next, the movement of the current causes erosion of the bank on the outside of any slight bend, where turbulence is greatest, and accumulation of deposits on the inside of the bends, where turbulence is least (Judson and Kauffman 1990, 292). If the Thames in the bridgehead area is regarded as part of a large underdeveloped meander, the expected pattern would be for erosion along the northern foreshore and accumulation along the southern foreshore. Instead, extensive revetment and quay construction along the north bank during the 11th and 12th centuries appears to have deflected the current southwards, causing serious erosion of the southern foreshore around the bridgehead, and the erosion was to worsen in the 13th century (see Chapter 12.2).

7

The Saxo-Norman timber bridge

Bruce Watson,
with a contribution by Damian Goodburn

7.1 The documentary evidence

The documentary and archaeological evidence discussed in Chapter 5.4 confirms that there was a bridge over the Thames by the very late 10th or early 11th century (*c* 990–1020). How long this bridge lasted is uncertain. The timber bridge would have demanded almost constant repair and replacement, although there is little direct documentary evidence for such work. In 1097 the *Anglo-Saxon Chronicle* recorded that floods carried away nearly all the bridge and that labour engaged in constructing the Tower of London and Westminster Hall was sent to repair it (*Anglo-Saxon Chron*, 234). Interestingly this reference states that 'many shires, whose work pertained to London were badly afflicted' (*Anglo-Saxon Chron*, 234), which could mean that the obligation to maintain Saxon London's defences and bridge was the responsibility of many counties. The Exchequer accounts in the Pipe Roll for 1130–1 note a payment of £25 for the construction of 'two arches of London Bridge' (Hunter 1833, 144). According to the *Bermondsey annals* a fire in 1135 or 1136 swept London and destroyed the 'wooden bridge' (Luard 1866, iii, 435). The 16th-century London historian Stow stated that in 1163 the timber bridge was rebuilt under the direction of Peter of Colechurch (Home 1931, 19; Stow 1603, i, 22).

There is ample evidence for the use of outside labour for repairs to London bridge during the 11th and 12th centuries: all counties in which the canons of St Pauls held land were liable for labour service to London's castle, bridge and palace (Gibbs 1939, nos 8, 13, 16, 27, 51, 52). The estates of Battle Abbey, Sussex, were excused by Henry I in 1100–23 from certain dues and customs, including work on London bridge and Pevensey Castle (Johnson and Cronne 1956, 1060).

The obligation to perform bridgework occurs in many Anglo-Saxon charters from the 8th century onwards. The first mention of bridgework in an authentic charter is in a grant of privileges to the church by King Æthelbald of Mercia at the Synod of Gumley in AD 749, which stated that the church had the right to hold lands free of royal exactions except for 'the buildings of bridges and the necessary defence of fortresses against the enemies' (Cooper 1998, 45, 276–7 (S92)). The clauses about bridgework generally follow the usual formula of exemption clauses – 'free from all secular services except building bridges, forts and army (*fyrd*) service', these three duties were standard obligations from which very few landholders were exempted (Stenton 1947, 286–7). The later Anglo-Saxon law codes of V and VI Æthelred II (979–1013; 1014–16) and II Cnut (1016–35) all included bridge building as one of a number of required public duties (Cooper 1998, 100).

Bridge building duties are also mentioned in many later documents. For instance, the Cheshire Domesday Survey of 1086 noted that, for the repair of Chester city wall and bridge, the sheriff used to call up one man from each hide in the county (Sawyer 1987, 343). An early 12th-century document known as the 'Rochester bridgework list' records the estates (including those of the king and archbishop) whose holders were responsible for the upkeep of the nine bridge piers at Rochester, and specifies the quantities of beams and planks that each had to provide (Brooks 1994, 16–17). Judging by the Rochester list, and the mention of bridge repairs in the Exchequer accounts for 1130–1, it is probable that the king was only one of a number of contributors to repair work at London bridge during the Saxo-Norman period. It is clear from three sets of evidence – monastic charters, the coronation charter and legal compilations – that the original Saxon concept of bridge building as a communal and unavoidable public duty had changed by the late 12th century to a 'plethora of personal and institutional obligations, exempted through favour, demanded through malice' (Cooper 1998, 125).

Records of the repair of timber bridges in Denmark during the 16th century and later show that major repairs were often necessary only ten to 15 years after building, and that after about 20–40 years a bridge was often in need of complete replacement owing to the rotting of the piles (Jørgensen 1993). The first Saxo-Norman London bridge appears to have been swept away by floods during the early 11th century, perhaps less than 30 years after it was built (see Chapter 5.4).

While the bridge was 'broken' or collapsed, a ferry service was possibly in operation. Just downstream of the bridge on the northern foreshore during *c* 950–1020 a large jetty was constructed at *Rederesgate*, now known as New Fresh Wharf

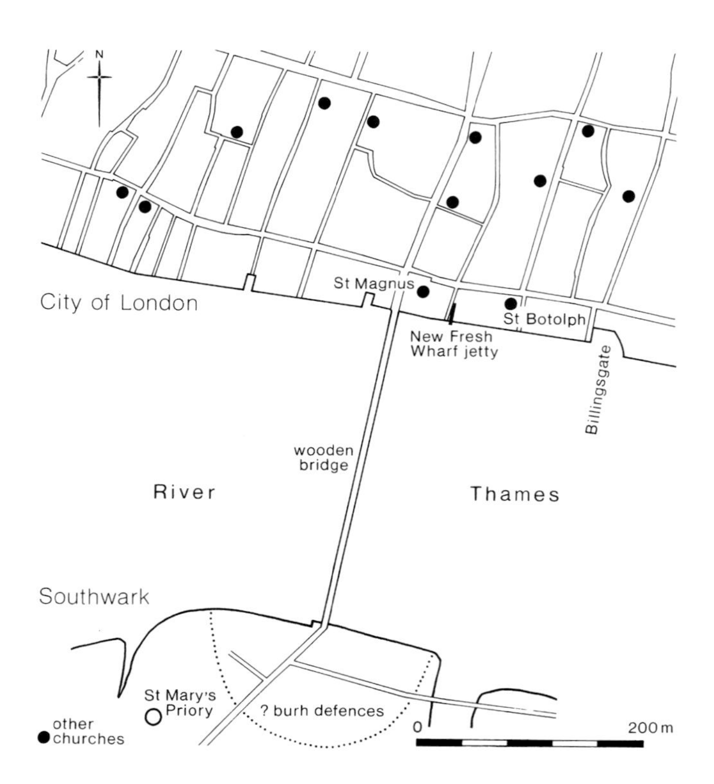

Fig 40 Map of the London and Southwark bridgehead c 1000 showing the extent of foreshore, roads, location of parish churches, position of the bridge and the deep-water jetty at New Fresh Wharf

(Steedman et al 1992, 23–8, 102).[20] This jetty consisted of a rectangular area of piles, arranged in lines to a width of *c* 7.5m and extending out into deep water (Fig 40) perhaps beyond the southern limit of excavation. An apparently unique feature of the Saxo-Norman harbour during this period, when boats normally beached on low-angled embankments or revetments of clay and timber and unloaded during low tide (Marsden 1994, 173), this jetty could have been used for loading and unloading boats, regardless of the state of the tide, and have functioned as a ferry terminal for small boats as a mode of crossing the river when the bridge was 'broken' (see Chapter 5).

7.2 Rebuilding Saxo-Norman London bridge

The second attempt at bridging the Thames at Fennings Wharf began with the consolidation of the foreshore around the bridgehead with dumps of sand and gravel, clay and chalk rubble, their surface lying at -1.12m OD (F19.1). This material contained a length of a clinker-built boat keel with planking made from timber felled after 986 (Marsden 1994, 156–8). The keel fragment was probably either dumped here as part of the consolidation, or was reused as part of a revetment around the dumped material; its original context could not be determined. Further dumping consisted of the deposition of sand/gravel, clay and chalk rubble (F19.2), and the deposits contained three parallel beams aligned east–west, two of which featured relict joints showing that they were reused. Moreover, two of the timbers were from the same tree, which was felled after 925 (see Chapter 14.2). Possibly these timbers served as foundations for the next phase of bridge building.

On the consolidated foreshore a massive, horizontal oak baseplate (see Glossary) was laid, aligned east–west. It was felled after 1056 (F19.2) (Fig 41). The slope of the baseplate from west to east shows that it was not found completely *in situ*, but had been moved slightly by scouring and undercutting, the highest point of the top recorded at -2.00m and the lowest at -1.65m OD. Cut into each end of the upper face of the baseplate were two rectangular sockets (Fig 42). The outer pair, which cut completely through the timber, were 5.58m apart, while the inner pair, 4.68m apart, did not penetrate the full depth of the timber (see Chapter 14.3).

The baseplate is interpreted as part of the northern, riverward side, of a landward bridge abutment. The outer pair of sockets could have held driven stakes to anchor the corners of the structure to the foreshore, while the inner pair could have retained vertical timbers at the corners of the abutment. The groove between the inner pair of sockets presumably held staves (see Glossary), but was full of fluvial sand when found. The width of the roadway can be estimated from the spacing of the inner pair of sockets at *c* 4.6m or 15ft, and it

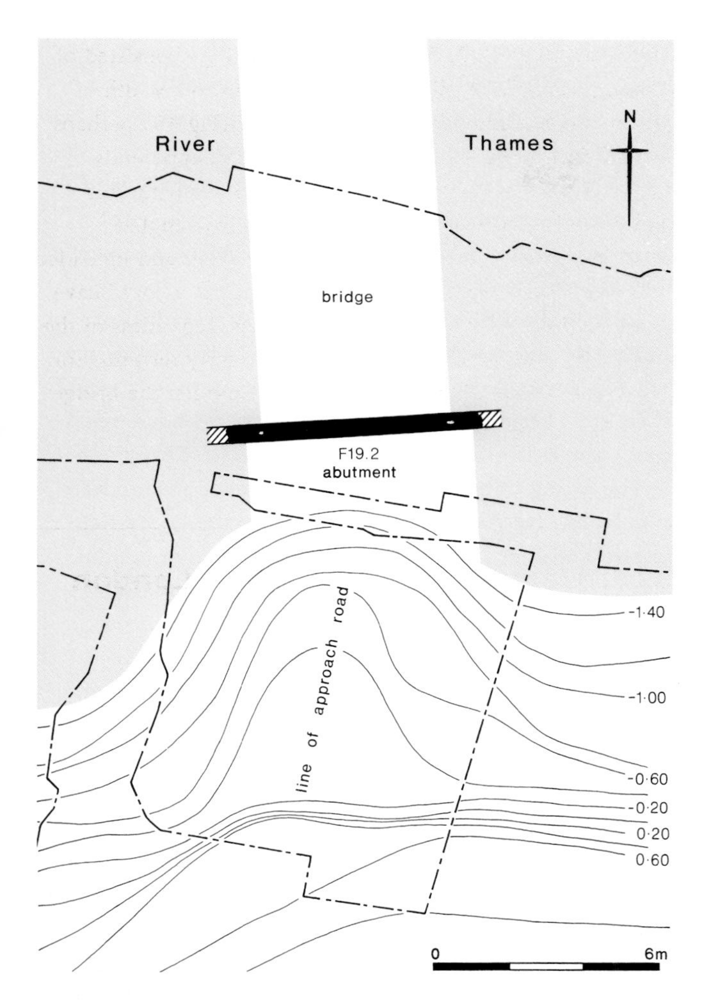

Fig 41 *Plan of the in situ late 11th-century bridge abutment baseplate (F19.2)*

is interesting that the Olaf sagas described the 11th-century London bridge as 'so broad that two wagons could pass each other upon it' (Laing 1964, 123). At the Norman castle of Hen Domen at Montgomery in Wales the spacing of the two sockets in the soleplate (see Glossary) of the late 11th-century motte bridge suggests that the roadway was only 3.5m wide (Barker and Higham 1982, 56–9). At Goltho motte and bailey castle in Lincolnshire, the third phase of the moat bridge dating to *c* 1100–50 was about 3.6m wide (Beresford 1987, 91). The spacing of the sockets in the soleplate of the late 11th-century bridge trestle at Hemington, Leicestershire, indicates that it supported a roadway some 2.7m wide (Cooper and Ripper in prep).

The appearance of the rest of the abutment is uncertain, but probably consisted of a three-sided rectangular timber structure, several metres high and probably infilled with chalk rubble and clay. It is likely to have been open on the southern or landward side, where there would have been a substantial gravel or rubble ramp to facilitate easy vehicular access on to the bridge roadway. The upstream side of the abutment would have been protected from erosion by ebb tides and floodwater by means of a series of frequently rebuilt and repaired revetments dating from *c* 1080–1100 (see Chapter 6.3), and the revetting of this southern abutment was possibly included in the documented repair work on the bridge in 1097 (*Anglo-Saxon Chron*, 234).

A second horizontal baseplate, over 4.25m long and of late 11th- or early 12th-century date on stratigraphic grounds, appears to represent the sole remains of the next phase of work at the southern end of the bridge (F19.3) (Fig 43). Cut through the plate were a series of irregularly spaced rectangular sockets, one of which contained roundwood stakes for anchoring it. The rest of the structure appears to have been washed away or salvaged for reuse, and its fragmentary nature leaves it unclear whether it was part of an abutment or a caisson. It may have rested laftwork-style (see Glossary) on top of the baseplate, since there were no sockets to retain vertical timbers, or grooves to hold staves. If this late 11th- or early 12th-century abutment or caisson was indeed a laftwork structure, it would represent a significant change in design, and would probably have resembled the square (*c* 5.3m) Romano-British rubble-filled laftwork abutment at Aldwincle, Northamptonshire (see Chapter 4.8) (Jackson and Ambrose 1976, 43; Sunter 1976, 50).

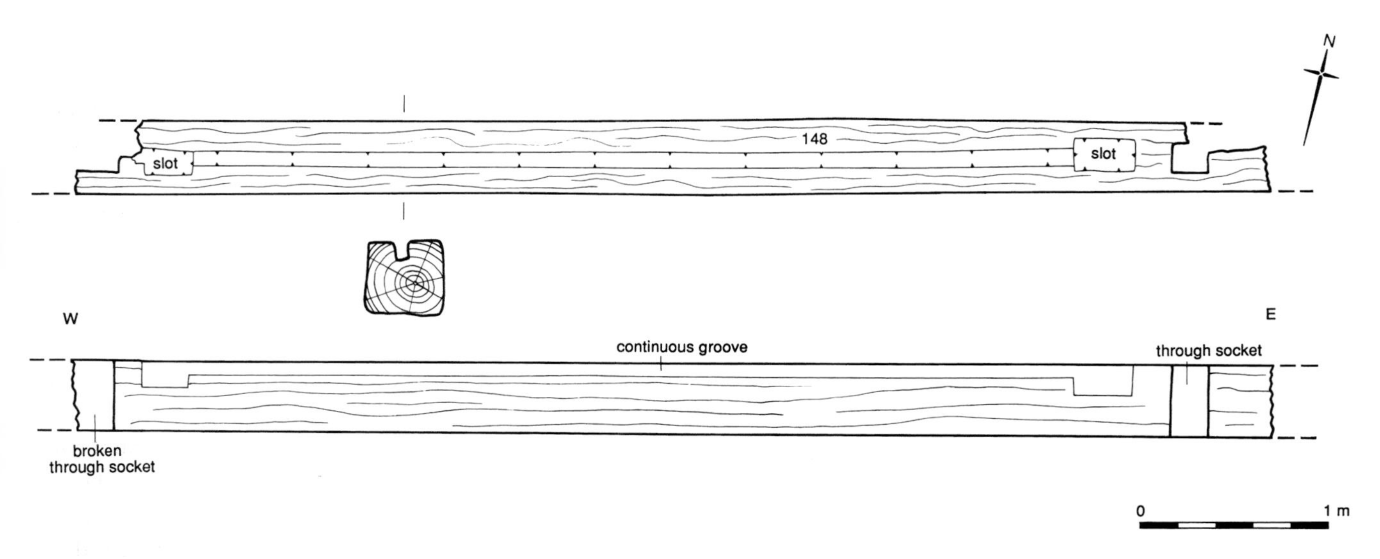

Fig 42 *Plan, elevation and cross section of abutment baseplate (F19.2)*

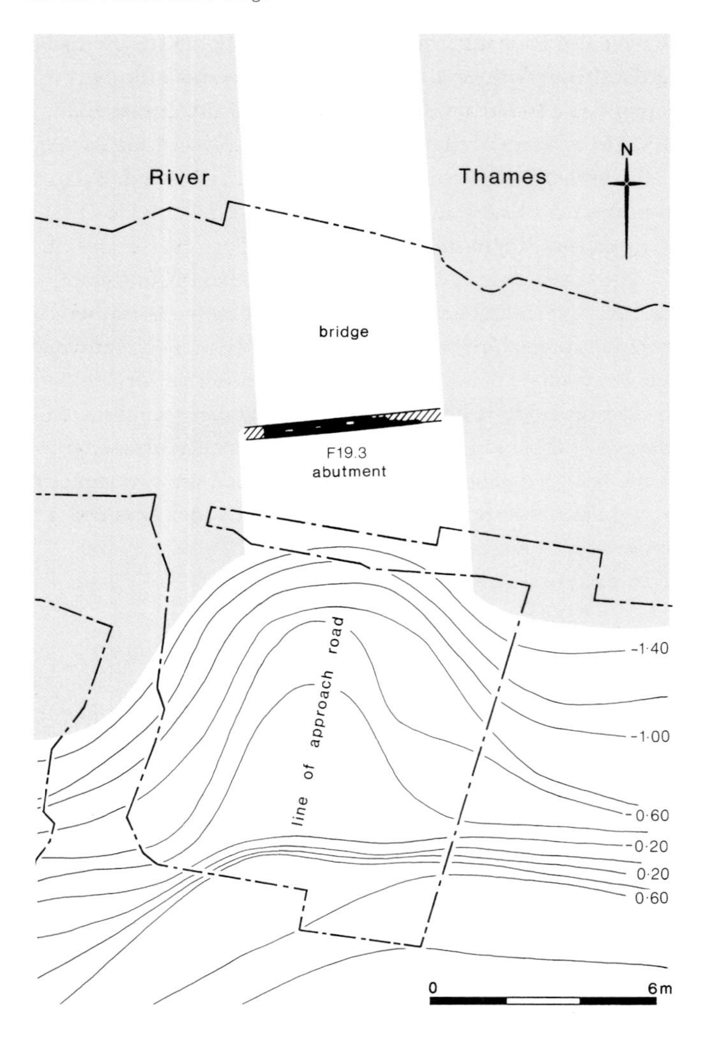

Fig 43 Plan of the late 11th- or early 12th-century bridge abutment baseplate (F19.3)

7.3 The 12th-century caissons

Bruce Watson with Damian Goodburn

For the next phase of landward bridgework a caisson (see Glossary) was constructed during *c* 1158–70. Work started with the digging of a shallow linear trench, aligned east–west and cut into natural gravel (F21.1). This is interpreted as a construction trench for the southern side of the caisson, which was apparently dug into the sloping foreshore. To its north the sloping foreshore was consolidated and levelled up by the dumping of chalk rubble in a grey silt/clay matrix (F21.2), its recorded top at -1.40m OD. Pottery from these deposits dates to 1150–1200. The northern side of the caisson consisted of two horizontal baseplates, aligned east–west, their tops at -1.51m OD (F21.3: Fig 44). These timbers were not fully exposed before removal by machine and their plan and carpentry is uncertain. Cut into their upper face were a number of rectangular through-sockets, one of which contained a vertical stake for anchorage. In the upper face of one of them was a shallow rectangular socket intended to hold a vertical timber. The rest of the caisson was represented by two *ex situ* timbers (F21.4) of different design. The western of these was a rectangular baseplate with a continuous central groove in its upper face for holding staves (Fig 44). In the upper face of the eastern timber was a large, empty, rectangular slot, while the southern side of the upper face featured two continuous stepped grooves of unknown function. Floods probably destroyed the caisson. The nature of its fill is unknown.

Later in the 12th century a second caisson (F22), of similar size and design, was built on top of the earlier one (Fig 45). Its south side (F22.2) contained a timber probably felled in 1145 and the west side two timbers both derived from a single tree felled during 1141–78 (see Chapter 14.2).

The dating of this structure is problematic as it appears from its relatively early tree-ring dates to contain a lot of material (with no signs of reuse) from the earlier caisson, which could suggest that both caissons were of identical size and design. However, judging by the dating of the earlier caisson, the later one would seem to have been constructed during *c* 1160–78, and would have been part of the timber bridge that was rebuilt or repaired in 1163 by Peter de Colechurch (Stow 1603, i, 22). Before building started the site was consolidated and raised by the dumping of ragstone and clay on which were laid a number of planks and beams. Tree-ring dating shows that all these timbers are of late 10th- or early 11th-century date, and they are, therefore, interpreted as residual or reused material, presumably derived from previous phases of bridge or revetment building (F22.1). One beam derived from the same tree as one in the earliest timber bridge (see Chapter 5.5). The caisson was constructed on the consolidated surface. Three sides of it were found (F22.2–.4), and externally the structure measured 5.1m north–south by 6.0m (estimated length) east–west, making it the most complete of the caissons or abutments found. Its northern, southern and western sides were founded on a series of horizontal baseplates anchored by stakes set within through-sockets, their top at -0.67m OD (Fig 46). At the corners of the caisson the baseplates were laft-jointed and pegged together (Fig 47). The superstructure was represented by traces of laftwork beams *c* 200mm thick, standing up to two high (to -0.40m OD). The caisson was infilled with clay and chalk rubble.

Thanks to the relative completeness of this caisson, its carpentry is well understood. It was constructed from a number of oak logs up to 6m long and *c* 600mm in diameter (Fig 48). The baseplates were fashioned from half logs that had been split by clefting with wedges and then hewn into shape. The general absence of working debris on site suggests that all the preliminary work such as clefting and squaring the logs was done off-site. The laft joints and sockets were probably cut with narrow-bladed axes rather than saws and chisels, and after cutting out the joints – perhaps on high ground nearby – the large, heavy baseplates would have been moved into position at low tide. It is probable that they were slid into position on wooden skids, final adjustments being made with levers. On the baseplates were laid a series of rectangular beams made from quarter logs (see Chapter 14.3). All the beams were laft-jointed together at the corners and neatly squared to provide a tight fit. From what is known of tidal river levels during the 12th century it can be estimated that this caisson would have stood at least 3m high above its baseplates. At any lesser height the structure of the roadway that it supported could easily have been washed away by floods (Fig 49).

The superstructure of the two 12th-century London bridge caissons was probably similar to that of the early 13th-century four-sided laftwork caissons (see Glossary) found at Bergen harbour, Norway, although the latter were founded on a solid timber platform, not on baseplates. The Bergen caissons, which were all infilled with earth, sand and stones, are interpreted as quays which had been floated into position and then sunk to provide deep-water moorings (Myrvoll 1991, 152–4). Very similar laftwork caissons (described as 'bulwarks') of 14th-century date found at Oslo harbour in 1992 are interpreted as wharves or deep-water moorings (Krogstad and Schia 1993, 110). In shallow rivers laftwork caissons could be used as piers to support the central portion of the bridge roadway, as at Brogepollen near Vestvågöy in Norway, where the remains of the 11th- or early 12th-century bridge consisted of a line of three fragmentary, stone-filled, square laftwork caissons which lacked cutwaters (Jørgensen 1997b, 54). Very similar laftwork caisson or piers, with cutwaters, are found in the wooden bridges of northern Russia (Opolovnikov and Opolovnikova 1989, 78–9). The rubble-filled, diamond-shaped, timber piers of the late 11th-century bridge at Hemington were constructed of edge-laid planks, which were lap jointed together (Cooper et al 1994, 316).

Fig 44 Plan of the first 12th-century bridge caisson (F21.4) and its contemporary upstream revetments

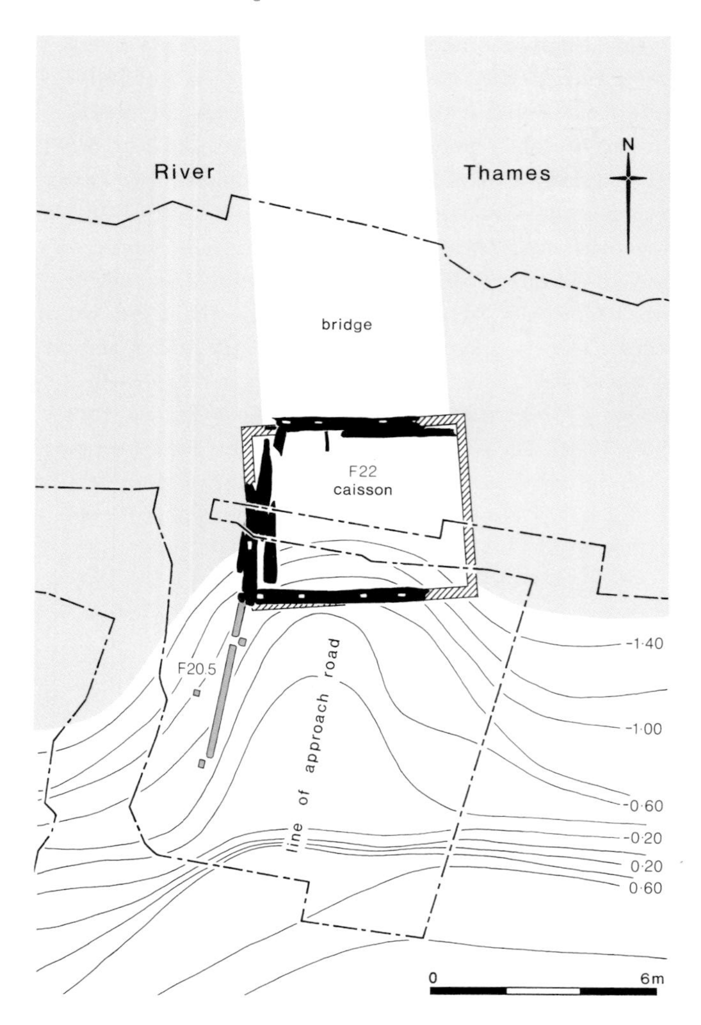

Fig 45 Plan of the second 12th-century bridge caisson (F22) and its associated upstream revetment

Fig 46 View of the north side and part of the west side of the second 12th-century bridge caisson (F22.3) in one of the riverside bays, looking south; under the north side of the caisson fragments of the earlier, first, caisson (F21.4) are visible; in section at the rear of the bay the chalk rubble infill of the caisson can be seen (F22.6) (2m scales) (MoL)

7.4 Reconstructing the timber bridge

The appearance of the remainder of the Saxo-Norman timber bridge is unknown, for none of it was recognised during the demolition of the bridge in 1831–4, and no traces were found in 1984. Investigation of an area up to some 5.0m due north of the later of the 12th-century caissons revealed no evidence of any associated bridge superstructure, although there was a spacing of 3.65m between the front of the abutment and the first timber piles at the Romano-British timber bridge abutment at Aldwincle, Northamptonshire (Sunter 1976, 50) (Fig 23). The possibility that fragments of the pre-Colechurch bridge went unrecognised during the demolition of the stone bridge in 1831–4 is worth considering, as, when one of the piers of the medieval stone bridge over the Tyne was demolished in 1871, three phases of construction were identified, the earliest of which was all thought to be of Romano-British date (Bruce 1885, 6) (see Chapter 4.8).

The only description of the timber bridge occurs in the Olaf sagas, first written down during the early 13th century, which describe how in 1014 Olaf Haraldsson and the Vikings

Fig 47 View of the south-west corner of the second 12th-century bridge caisson (F22), looking north, showing the arrangement of laft-jointed baseplates; the end of the revetment baseplate in the foreground has been cut back (F23.1) to allow the insertion of the caisson timbers (0.5m scale) (MoL)

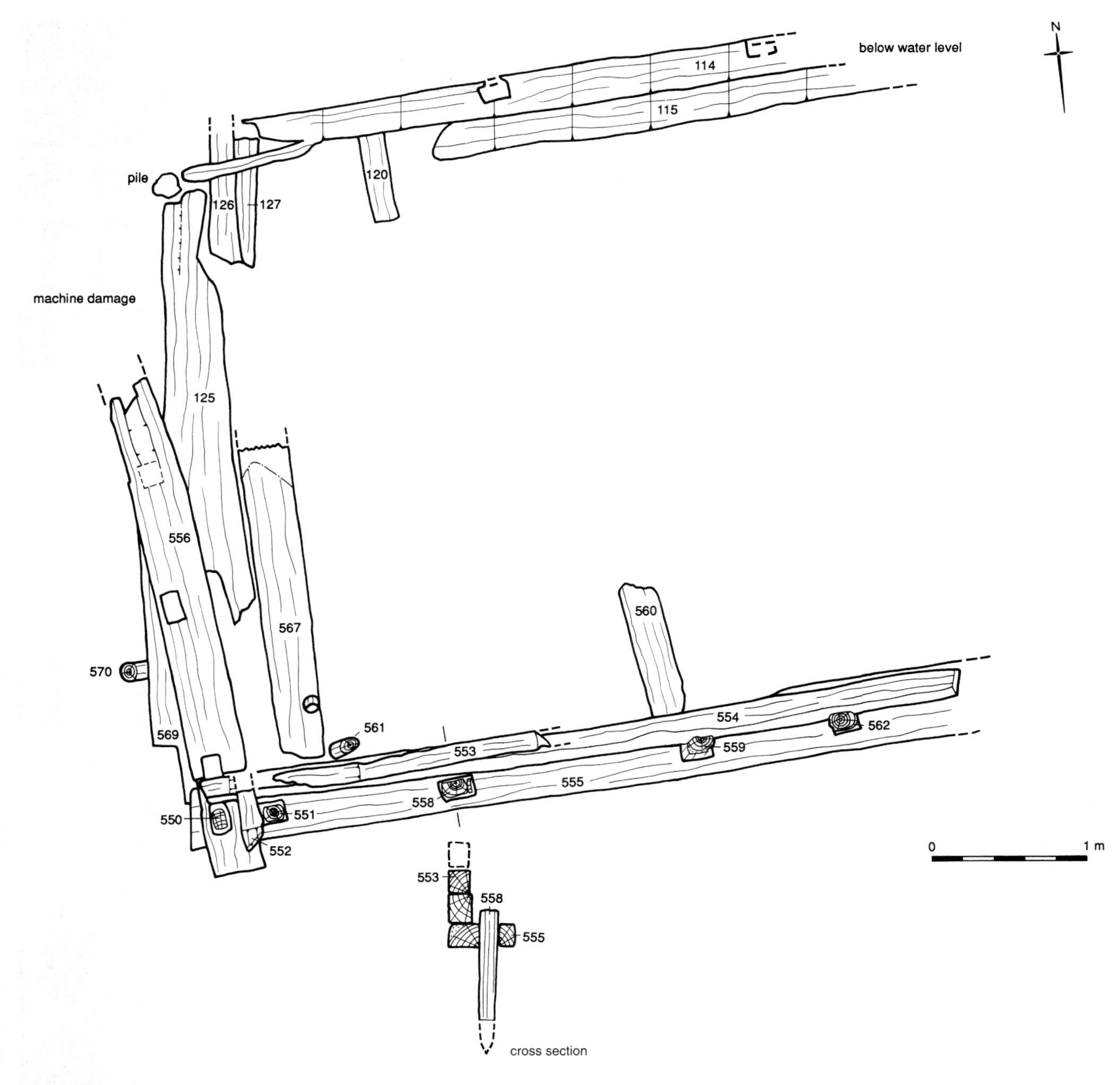

Fig 48 Plan of the second 12th-century caisson (F22)

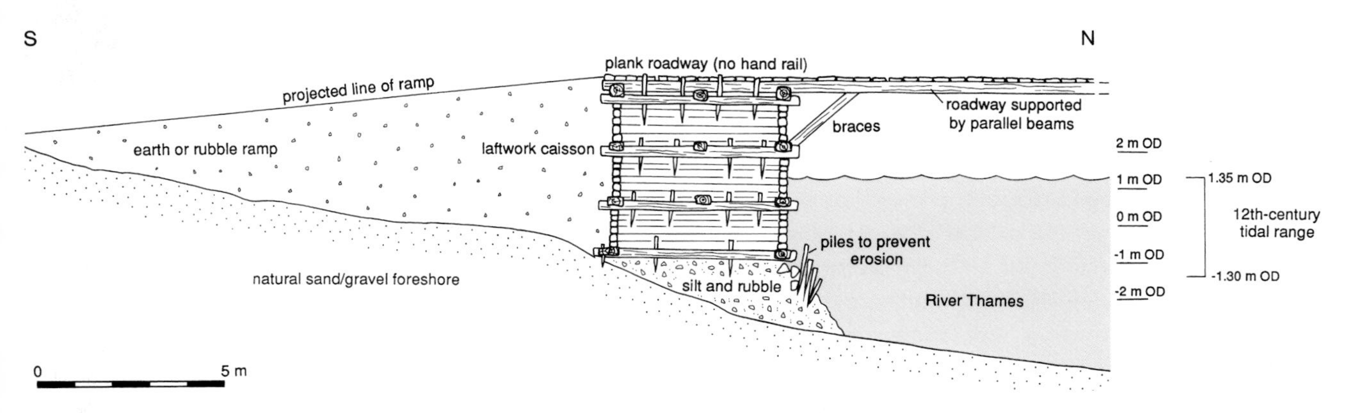

Fig 49 Reconstructed elevation of caisson F22; the height of the caisson, the design of the roadway and the slope of the ramp are all conjectural

attacked the Anglo-Danish forces defending London bridge (see Chapter 14.12 for a detailed discussion and references). The Vikings, protecting their ships from attack from above with timber and wattle screens, rowed upstream to the bridge and fixed ropes to its superstructure. Then they rowed downstream, allegedly demolishing much of the bridge and compelling the defending forces to surrender (Laing 1964, 123–4). A new translation of part of the description of this event in the *Heimskringla* by Jan Ragnar Hagland reads: 'and under the bridge were staves [stafir, ie vertical timber poles or pillars] and those [they] stood down on the shallow [parts] in the river'. The wording of this passage in the legendary saga of St Olaf reads: 'And the bridges [plural] were [constructed] thus that they stood out in the River Thames, and [there] were poles [stolpar] below [undir] down in the river which held up the bridge [singular]'. These passages appear to describe a trestle, not a caisson, bridge. See Chapter 14.12 for details.

Although these sagas were biographies of King Olaf of Norway (reigned 1016–29, died 1030) and intended to glorify his deeds, so that the success of these ploys and the extent of the damage inflicted on the bridge were doubtless exaggerated, the details of the bridge structure deserve serious consideration. It is perfectly possible that at the edges of the river, where the water was shallow, the roadway of the mid-12th-century Saxo-Norman bridge was adequately supported by piers formed of laftwork caissons, but that this technique was not practical in the deeper water at the centre of the river. It is possible that if the later phase of the Roman bridge possessed masonry piers (see Chapter 4.11), and some of them still remained above water level, they could have been reused during the Saxo-Norman period, but this seems unlikely in this instance as the Roman and medieval bridges were apparently on slightly different alignments (see Chapter 5 and Fig 15). It has been suggested that during the late Saxon period the bridge at Rochester was founded on Roman piers (Brooks 1994, 21).

However, a trestle-built Saxo-Norman London bridge is a further possibility with ample parallels in both the Roman and medieval periods. The Rhine bridge built by Caesar's army in 55 BC is believed to have been of pile and trestle construction (O' Connor 1993, 140–1). It is clear from the pictorial evidence provided by the columns of Trajan (Fig 22) and Marcus Aurelius that the Roman army was capable of building large timber bridges of many different designs, but the most common one was apparently of pile, trestle and girder construction (O'Connor 1993, 139).

An important source of evidence for the design of medieval English trestle-built bridges is provided by the study of castle moat bridges (Rigold 1975, 56–9). Most of these bridges were simple trestle and soleplate structures (see Glossary). Fragments of a medieval timber bridge of soleplate and trestle construction were found in the River Cashen in Co Kerry, Ireland, during dredging in 1953–7 (O'Kelly 1961, 137–45). The soleplates were spaced about 2.7m apart, and each contained seven through-sockets: three for short stakes to anchor the structure to the riverbed and four to accommodate upright timbers. It is estimated that the roadway was *c* 2.0m wide, with handrails at each side (ibid, 140). Currently the Cashen River bridge is undated but, from the design of its soleplates, it is considered to be of later medieval date (Aidan O'Sullivan and Charles Mount, pers comm 2000). In 1989 a fragment of a collapsed 11th-century bridge was discovered spanning the River Fleet in London. It appeared to have consisted of a number of piles supporting a series of horizontal, parallel beams, lying edge to edge and forming a roadway *c* 2.5m wide (Fig 50) (Goodburn 1993).

The late 11th-century timber bridge found in 1993 across a former course of the River Trent at Hemington in Leicestershire consisted of two small diamond-shaped and rubble-filled wooden pier bases (see Glossary). Associated with these piers was a collapsed bridge trestle, dated by tree-ring dating to 1089–1124 (Cooper and Ripper in prep). Over time these piers were displaced by floodwater and were reinforced or superseded during the early 12th century by a double row of vertical piles (Cooper and Ripper 1994, 152–7; Cooper et al 1994). During the early 13th century another bridge, consisting of a series of irregularly spaced, paired piles was constructed at Hemington, and this would have been very similar to the timber bridge over the River Aveyron (France) recorded in 1120 and described as consisting of 'planks on piles' (Boyer 1976, 83). A recent underwater survey of the River Shannon, Ireland, has located an early 9th-century bridge structure at Clonmacnoise, which consisted of two lines of vertical piles 4–5m apart, to secure the soleplates (A O'Sullivan, pers comm 1997). The spacing between the pairs of posts varied from *c* 5–7m (Boland et al 1996, 15–20).

An exceptionally long medieval timber causeway bridge was constructed during AD 979 or 980 at Ravning Enge over the River Vejle (Denmark) and its floodplain. This bridge, 760m long, consisted of groups of four squared piles arranged in lines, each of the outer piles being braced by an additional timber. The spacing between the individual lines

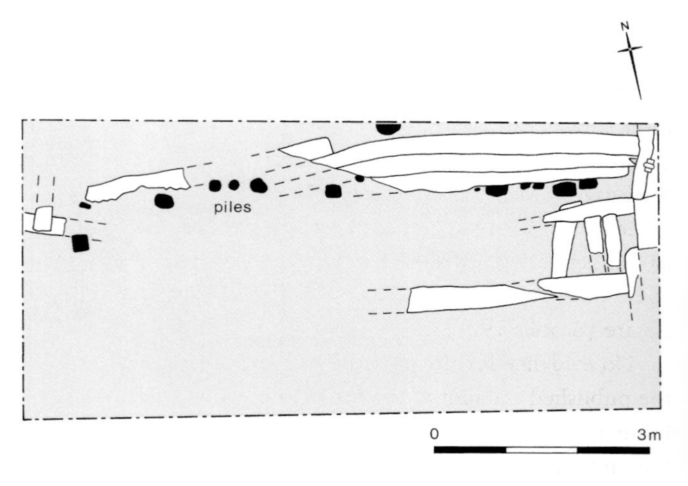

Fig 50 Plan of Fleet Valley bridge (McCann 1993, fig 19)

of piles was 2.4m, and it is estimated that the bridge roadway was *c* 5m wide (Crumlin-Pedersen et al 1992, 48). Although only fragments of the superstructure survived, it consisted of crossbeams set on the lines of piles and on which planking was laid to form the roadway (Jørgensen 1997a). During the 7th century a wooden causeway bridge some 750m long was constructed across Teterow Lake (Mecklenburg-Vorpommern, Germany) (Brooks 1995, 20), its decking supported by pairs of regularly spaced vertical piles, each supported by a brace. The bridge was rebuilt during the 9th century.

The decking or roadway of Saxo-Norman London bridge was probably composed of small, squared or split logs, laid edge to edge on a series of stout horizontal beams, aligned at right angles to the decking logs and supported by the piers or trestles. In northern Russia the roadway of many contemporary timber bridges consists of a single layer of short logs laid edge to edge and supported by tiers of three or more log beams (Opolovnikov and Opolovnikova 1989, 78). Radially cleft planking (logs split into one eighth of the full trunk) are only likely to have been used for a bridge roadway if the supporting beams were very closely spaced, as was the case at the 11th-century River Fleet bridge (Fig 50). The spacing of the piers or trestles can be estimated from engineering principles at a maximum of 9m to 12m; if the trestles were placed further apart the roadway would have been dangerously flexible (Brooks 1994, 24). However, the rigidity of the roadway would have been greatly improved by the use of diagonal braces or struts between the trestles and beams supporting the road planking.

A significant parallel can be seen in the remains of a bridge of uncertain date found at Wallasey Pool (Birkenhead on Merseyside) in *c* 1850: a little-known discovery of great importance as the most complete ancient bridge known in England (Fig 24). Probably of Romano-British date (Watkin 1886, 81; Dymond 1961, 155; see Chapter 4.8), it consisted of two collapsed piers some 9.0m apart, above which were four lines of horizontal, parallel beams which would have supported the timber roadway, traces of which also survived (Fig 24). The beams were squared oak logs 460mm wide, 230mm thick and 10.0m long, all arranged in tiers of three. Interestingly, the total width of the river from bank to bank was only 22.3m although the beams measured 30.5m, allowing for a substantial overlap on each bank (Massie 1849–55, 55–6). Moreover, the tiers of beams supporting the roadway of the Birkenhead bridge are much more substantial than in the postulated reconstruction of the roadway beams of the late Anglo-Saxon bridge at Rochester, where it has been suggested that a hypothetical interval of up to 9m between stone piers for a wooden roadway *c* 6m wide could have been spanned by three oak beams each 300mm square (Brooks 1994, 25).

No evidence for the presence of handrails is included in the published account of the Birkenhead bridge, but the two joints (a through-socket and a diagonal halving), shown on the only detailed drawing of a beam fragment, could have been intended to support a handrail (Massie 1849–55, 76). A timber handrail can be seen in the representation of the Apollodorus bridge on Trajan's column (Fig 20). It is possible that the Saxo-Norman London bridge had no handrail, a feature that appears to have been absent from some medieval timber bridges (Boyer 1976, 3), and in a few instances absent from stone bridges such as the 12th-century Pont-Saint-Bénézet (Avignon, France) (Boyer 1976, 125). This deficiency could certainly explain the number of people and horses that fell off Rochester Bridge during the 12th and 13th centuries (Brooks 1994, 35–6). There are two medieval stone bridges in Surrey – at Eashing and Tilford East – which have no stone-built parapets, and they were presumably both built without parapet walls (Jervoise 1930, figs 8 and 10). The wooden handrails on both these Surrey bridges look like post-medieval additions or replacements. A contract dated 1485–6 to build a 12-arch timber bridge at Newark (Nottinghamshire) stated that it should have handrails (Salzman 1952, 546).

It is clear from the documentary evidence that in the early 11th century ships were sailing under or through London bridge (see Chapter 5.4) and, although the Saxo-Norman harbour at Queenhithe upstream of the bridge seems to have catered only for up-river traffic, the establishment of Dowgate and Vintry by the mid-11th century for the use of German and French shipping confirms that it did not restrict river traffic. This would imply that ships either lowered their masts to pass under the bridge or that there was a drawbridge section within it, which could also have formed part of the City's defences. It is also probable that the Romano-British bridge possessed a drawbridge section to allow ships to pass upstream (see Chapter 4.12). The existence of a drawbridge could explain why in Æthelred's fourth law code (see Chapter 5.4) tolls were being collected at the bridge by *c* 1000.

The exact length of the timber London bridge is uncertain, but as the southern masonry abutment was built over the remains of the earlier timber caissons it is probable that the timber bridge was of a similar length to its stone successor, which measured 276.09m long from abutment to abutment; a good discussion of this topic is provided by Home (1931, 340–1). Precise information on the length of other medieval timber bridges is limited, but the longest known example is the 10th-century Ravning Enge causeway bridge at 760m (Jørgensen 1997a). The River Cashen bridge was at least 184m long (O'Kelly 1961, 145), while the Clonmacnoise bridge was some 160m long (Boland et al 1996, 23; A O'Sullivan, pers comm 1997). From the width of the relic channel of the River Trent it can be estimated that 11th- and 13th-century timber bridges at Hemington were both over 70m long (Cooper and Ripper 1994, 155). From documentary evidence it is calculated that the roadway of the 12th-century timber bridge at Rochester was 133.3m long (Brooks 1994, 21), and a timber bridge across the Rhine (at Mainz, Germany), which burnt down during the reign of Charlemagne (768–814) (Boyer 1976, 3), would have been *c* 450m long judging from the width of the modern river. This brief survey of medieval bridge dimensions implies that Saxo-Norman London bridge was one of the longest timber bridges in western Europe.

The timber bridge spanning the River Thames was replaced in stone during *c* 1176–1209 (see Chapter 8), only one of a number of instances of the process in England and Wales for which there is archaeological evidence. At Hemington a masonry bridge replaced an earlier timber one *c* 1235–40 (Cooper and Ripper 1994, 160–1; Cooper et al 1994, 319). In 1988 fragments of a soleplate and trestle bridge of late 12th-century date were found incorporated into the late 13th- or early 14th-century Monnow Bridge, Monmouth (Rowlands 1994, 78–81). During 1949–55 two large timbers were found in the riverbed under the bridge at Rievaulx, Yorkshire. One timber was 4.5m long and 0.5m wide with six sockets in its upper face to hold stakes or piles (Weatherill 1963, 75–8). These two timbers appear to have been soleplates for a trestle bridge, presumably predating the medieval stone bridge.

It is possible that the 14th-century stone bridge at Waltham Abbey, Essex, was built on the remains of an earlier timber bridge (Huggins 1970, 133), but the evidence is uncertain.[21]

8

The construction of the Colechurch bridge

Bruce Watson

8.1 The decision to build in stone

Construction of the stone bridge was begun in *c* 1176, according to the annals of Merton, Southwark and Waverley (Brett 1992, 302). The work was directed by Peter, chaplain (*capellanus sacerdos*) of St Mary Colechurch, the parish church where, according to Stow, St Thomas Becket was baptised (Stow 1603, i, 23, 264; Home 1931, 21–4). Peter appears to have lived in a house in Southwark, to the east of St Olave's church, while the bridge was under construction (see Chapter 10.4). He died in 1205, four years before the bridge was finished and, as he was buried in the chapel on the bridge, it is possible that this portion of the bridge was already complete by that date (Brett 1992, 305; Home 1931, 23–4). The discovery in 1730 of a stone bearing the date '1192' in the cellar of one of the houses above the northern arches of the bridge would certainly suggest that this portion of the structure was completed during Peter's lifetime (Brit J 1730).

Little is known about how the construction of the stone bridge was funded. Bridge building and repair were clearly an expensive undertaking and a duty that could be burdensome to small communities. One of the clauses of Magna Carta (1215) expressly stipulated that 'no town or person shall be forced to build bridges over rivers except those with an ancient obligation to do so' (HMSO 1965). However, it is possible that this clause was only intended to avoid the imposition of new obligations, as it addresses a very specific problem, and not the feudal duty of bridge building (see Chapter 7.1). The failure to repair bridges was the subject of many medieval commissions and inquisitions. The commissioners invariably asked who was responsible for the repair of the bridge in question (Harrison 1996, 240), which shows that the concept of personal or collective liability was deeply ingrained, but difficult to enforce.

The most obvious way for Peter of Colechurch to have raised the money to build the stone bridge would have been by seeking gifts from the king, the bishops and the Archbishop of Canterbury, the aristocracy and the richer London merchants. Stow claims that funds were donated by the king (presumably Henry II, 1154–89), a visiting papal legate and Richard, Archbishop of Canterbury (Stow 1603, i, 23). Another early benefactor of the bridge was London's first mayor, Henry Fitz Ailwyn (mayor *c* 1189–1212) (Welch 1894, 263). It is probable that Peter of Colechurch formed the first of a series of confraternities or guilds to raise money for the purpose as a pious charitable work, comparable with oblations to parish churches and religious houses, and holding out similar prospects of spiritual reward. The religious aspect of the works was emphasised by the presence of a chapel on the bridge (see Chapter 9.4), to which donations were made throughout its existence. The Bridge House accounts for 1461–2, for example, record the receipt of a sum of 26s 2d as 'the offering of the faithful to Christ in the chapel of St Thomas the martyr on the bridge' (Harding and Wright 1995, 123) and, when in 1383 money was required to rebuild Rochester Bridge, Archbishop Courtenay granted indulgences in return for contributions to bridge funds (Britnell 1994, 85). The existence of the charitable guilds is documented. For instance, in 1179–80, while London Bridge was under construction, five bridge guilds owed the king fines totalling 42 marks (£28) (Brooke 1975, 44). The size of this debt is best illustrated by comparing it with the 1212 London list of maximum daily wages in the building trade. This source states that carpenters, masons and tilers were paid 3d plus food, or 4d without (Salzman 1952, 68).

Two bridges, linked by a causeway, were constructed over the River Lea at Stratford-by-Bow (Essex) during the early 12th century with funds donated by Maud, queen of Henry I (1100–35). She bestowed certain lands on Barking Abbey to maintain the bridges, a responsibility that was later transferred to Stratford Abbey and was the subject of a dispute between the two houses, which was resolved in 1315 (Fowler 1907, 116). In France from the middle of the 12th century onwards there were instances of bridges being built not by single rich individuals, but by groups of citizens organised into pious fraternities or guilds. For instance, St Bénézet, the builder of the bridge at Avignon, by 1181 had founded a fraternity, *fratres donati*, both to raise and to manage funds for bridge construction (Boyer 1976, 39).

After the bridge was completed in 1209, by which time Peter of Colechurch had died, there would have been a clear need for an organisation to maintain both the bridge and its endowments. It seems to be implied in a document of 1265 that prior to that date the bridge revenues were being administered by the 'brethren and chaplains ministering in the chapel of St Thomas' on the bridge, for in that year the following grant was made: 'Commitment for five years to the master and brethren of the hospital of St Catherine by the Tower of London of the wardenship of London Bridge, so that they apply the rents, tenements and other things belonging thereto within and without the city to the repair of the bridge; with mandate to the brethren and chaplains ministering in the chapel of St Thomas upon the said bridge and other inhabitants on the bridge and all others to be intendant to them' (Cal Pat R, 1258–66, 507). It appears that by 1284 the control of London Bridge passed from the chaplains to the bridge wardens (see Chapter 11.1). The most likely date for this handover is 1282–3, when after some years of royal control and neglect, which only ended when part of the bridge collapsed, the citizens of London regained control of their own bridge (see Chapter 11.1).

As Peter of Colechurch had previously repaired the timber bridge in 1163 (see Chapter 7.3), he would have been keenly aware of the practical difficulties involved in constructing a new bridge. There are several possible reasons why he might

Fig 51 View of the south end of the Elvet Bridge, Durham, showing the former bridge chapel of St Andrew; only the oval arch is original, the two Gothic style arches (spanning the River Wear) are part of a later rebuild (B Watson)

have decided to build in stone. The timber bridge may finally have been considered beyond repair, and his guilds must have now raised or secured enough money to build a stone bridge. This would have been the most important and difficult consideration of all. Gautier, an 18th-century French authority on bridges, stated that a stone bridge was worth constructing even if it cost ten times as much as a wooden one, because the difference in maintenance costs would justify the initial outlay (Boyer 1976, 145).

London Bridge was certainly not the earliest stone bridge erected in England in the 12th century. The Durham bridges over the River Wear appear to be the earliest documented examples. The 'Old' or Framwellgate Bridge, believed to be stone-built, was erected in *c* 1120 by Bishop Flambard (1099–1128). The 'New' or Elvet Bridge was built by Bishop le Puiset (1153–93) (Roberts 1994, 62–3) (Fig 51), and these two examples may at least partly be explained by the exceptional wealth and prestige of the Prince Bishops of Durham. Queen Maud constructed two bridges, both apparently stone-built and linked by a causeway over both branches of the River Lea, at Stratford-by-Bow (Essex) during 1100–35 (Fowler 1907, 116) (see above). In north-west Europe during the period from *c* 1050 to 1350 numerous stone bridges were constructed, and it is worth noting that, after churches, bridges are the most common form of surviving medieval monument in north-west Europe (Brooks 1995, 12). A total of 119 bridges are recorded as being built in France during the 12th century (Boyer 1976, 162). London bridge was rebuilt in stone during a period when London was increasing in population, wealth and civic independence. The construction of the bridge can, therefore, be seen as a monument to the prosperity and growing self-confidence of the citizens of 12th-century London, whose charitable donations funded it.

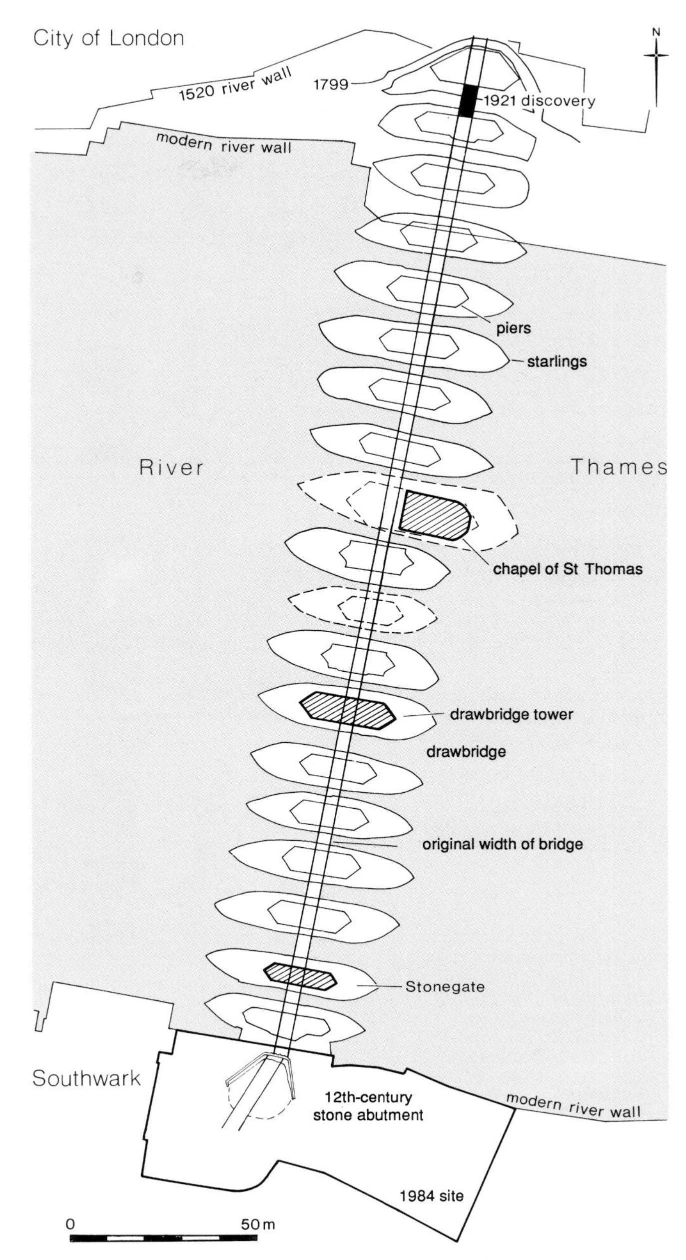

Fig 52 Plan of the Colechurch bridge, showing the extent of the Southwark abutment and various features of the bridge

8.2 Building the stone bridge

The stone bridge was 905ft 10in (276.09m) long, and was supported by 19 piers set on starlings (see Glossary).[22] Between the piers were 19 stone arches and one other spanned by a drawbridge (Fig 52; Fig 53). The original width of the roadway is unclear but according to Stow, writing *c* 1600, it was 30ft (9.14m), effectively reduced to *c* 12ft (3.66m) by the presence of buildings on both sides of the bridge (Stow 1603, i, 26). In 1633 the roadway at the northern end of the bridge was said to be 15ft (or 4.57m) wide (Chandler 1952, 19), and the original width of the bridge masonry was said to be 7.31m (Dodd 1820, 8). During the demolition of the bridge superstructure in 1831–2, it was noted that the cellared buildings at the northern end of the bridge flanked a roadway which was 20ft (6.09m) wide, while at the southern end the roadway was only 12ft (3.66m) to 14ft (4.26m) wide (Knight et al 1832, 203). It is probable that the cellars found on the northern end of the bridge were part of the post-Great Fire of 1666 rebuilding of the buildings here (see Chapter 9.1 for the buildings and Chapter 13.2 for road widenings).

Once the scale of the bridge is appreciated it can be seen to have amounted to a major construction project comparable with the building of a large castle or cathedral in terms of costs, manpower and materials required. As 17 English cathedrals were either built or rebuilt during the 12th century there was clearly a great deal of regional expertise and skill available for large-scale construction projects. In France, at about the same time that construction was starting on London Bridge, work was also beginning on a massive bridge: the 22-arch Pont-Saint-Bénézet, spanning the Rhône at Avignon (Boyer 1976, 38).[23] During the late 12th century there were a number of innovations in English carpentry, which

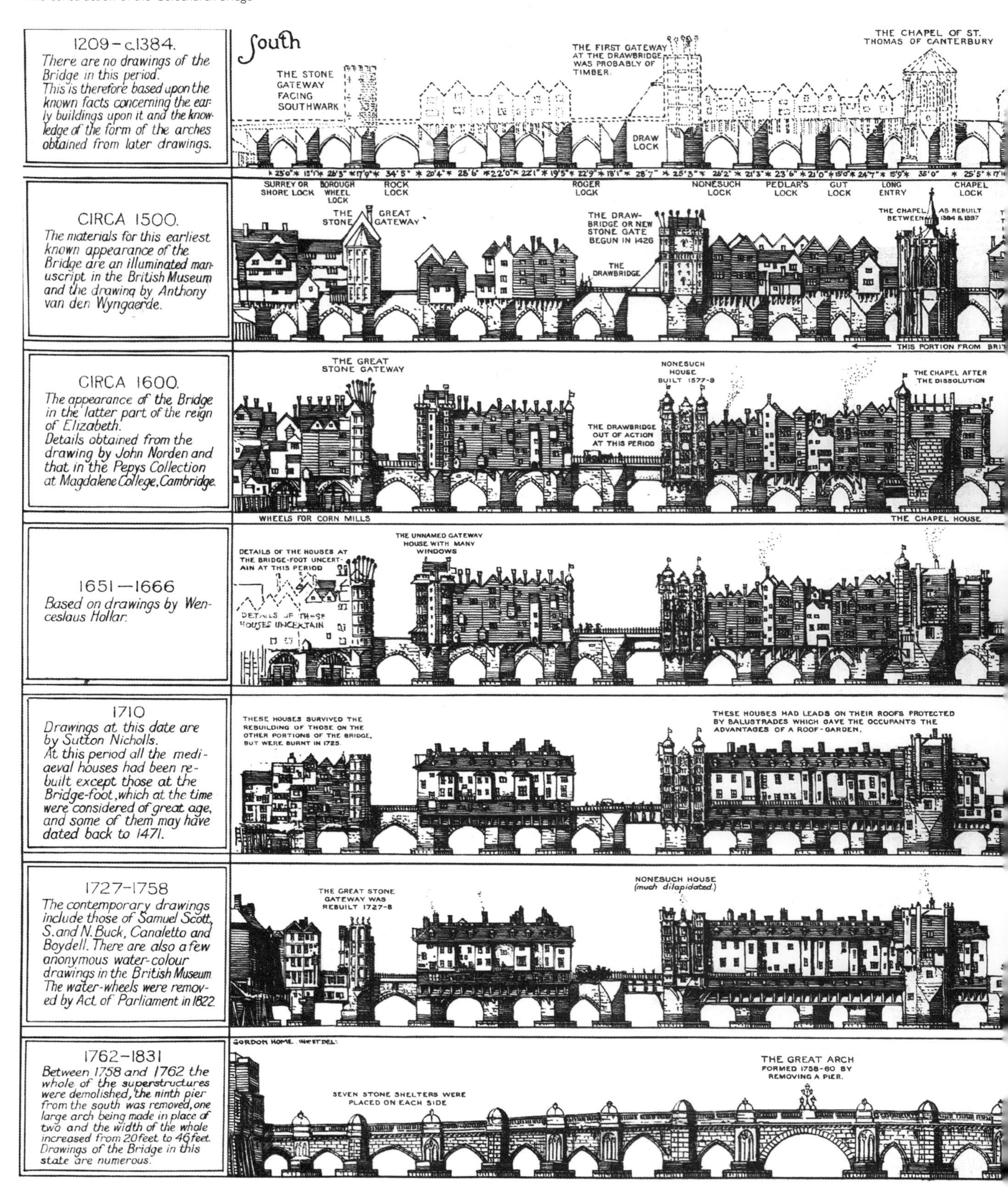

Fig 53 Home's 'Seven phases in the evolution of old London Bridge, 1209–1831' (reproduced from Home 1931) (MoLPL)

could well have emerged in response to the new technical requirements of such major projects (see Chapter 14.3).

The construction of the stone bridge took about 33 years, a slow process which meant that on average one pier would have been built about every 21 months. Construction during the winter months was almost certainly hindered by high river levels, which would have halted work on starlings and foundations. Research into French medieval bridges shows

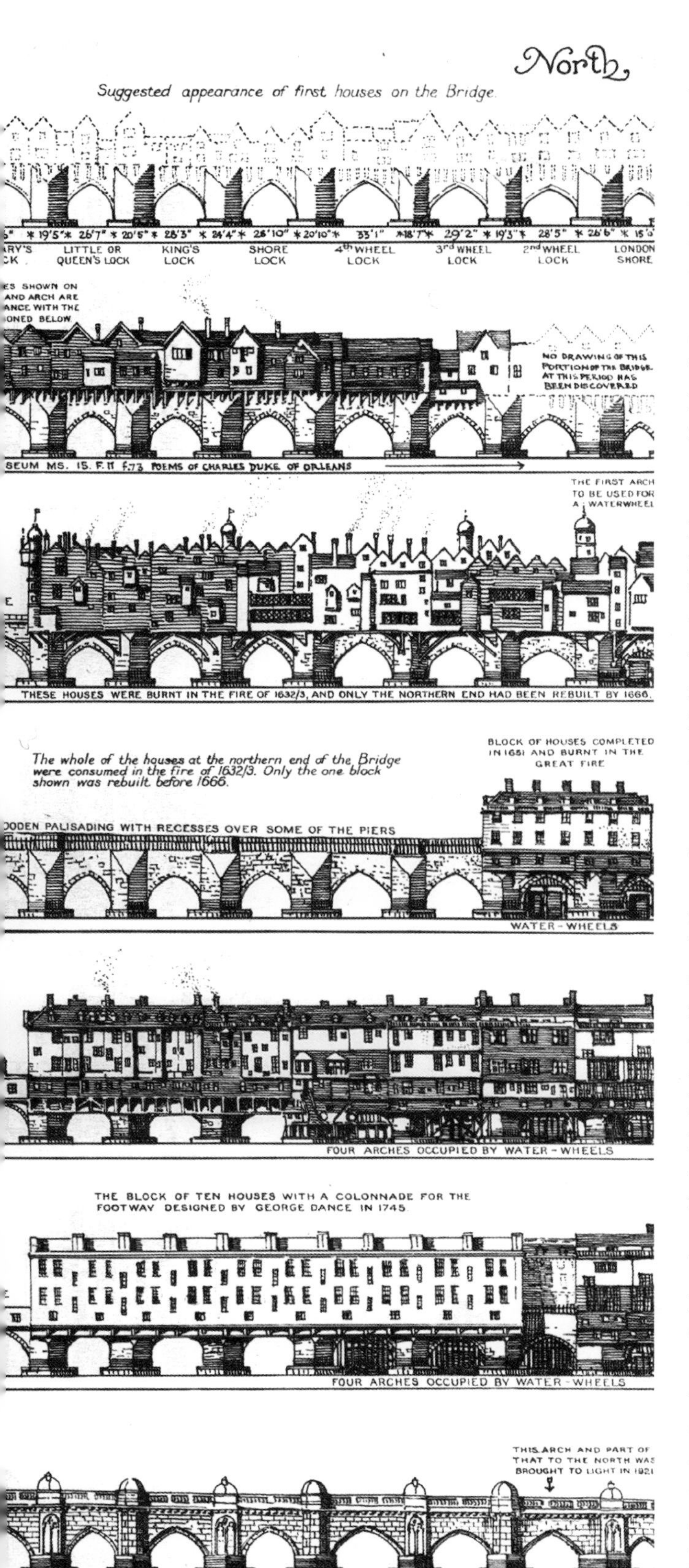

that construction and repair work were largely restricted to the period June to October (Boyer 1976, 145), and the weekly accounts for London Bridge in 1381–2 show that piling work on the starlings was restricted to a period from mid-April to December, most of the activity taking place between May and July (Harding and Wright 1995, 11–15, 21–9).

How were the starlings and pier foundations constructed in standing water? It can be estimated that there would have been about 1.0m depth of water at low tide at the edges of the River Thames during the 12th century, and probably as much as 4.0 or 5.0m in the central portion of the channel (see Chapter 3.5). One suggestion is that the starlings were built as cofferdams (see Glossary), to create a dry working area in which to build each pier (Hopkins 1970, 38), but this seems most unlikely. The technique of building cofferdams was known to the Romans and the methodology was described by the 1st-century BC architect and engineer Marcus Vitruvius Pollio (Vitruvius, 162–4), and put to use in constructing the stone piers of the 2nd-century bridge at Trier (see Chapter 4.8). There is, however, no evidence that it was used in 12th-century England. It has been claimed that the 14th-century bridge at Waltham Abbey (Essex) was constructed using a cofferdam, but the evidence is not conclusive and it is possible that the timbers in question are part of an earlier bridge structure instead (Huggins 1970, 129, 133).[24] One of the earliest post-Roman documented instances of cofferdam usage is at the Albi bridge (France) in 1408 (Boyer 1966, 30–1). The 1420–1 contract to build a three-arch stone bridge at Catterick (Yorkshire) states that the abutments and piers were to be constructed using timber *branderethes* or cofferdams (Salzman 1952, 497–8). It is perhaps more likely that the pier foundations of London Bridge were built during low water by setting up a piling rig or engine within a boat to drive in a number of short piles and produce an enclosure into which rubble could be dumped to raise the interior level above low water level (Knight 1831, 119). On this platform the pier could then be founded and sillbeams laid (Fig 54) before a larger piling rig could be set on the pier to drive in the longer piles of the surrounding starling. At the bridge over the Loire at Orleans during the 14th century a piling rig was set up regularly within a *chaland* or barge to maintain the starlings (Boyer 1976, 152–3). The 18th-century French authority on bridge construction, Gautier, stated that in deep water bridge piers could be founded on an enclosure of piles infilled with dumped material (Ruddock 1979, 201). Hutton (1772, 103–4) described the sequence of bridge pier construction without cofferdams. Work started with the creation of an enclosure of piles, lined with planking into which rubble was dumped to infill the space, then the pier constructed. He stated that 'this method was formerly much used, most of the large old bridges in England being erected that way, such as London Bridge, Newcastle Bridge, Rochester Bridge etc'.

The stone piers of the new bridge were founded on platforms of rubble, reinforced with short piles, on which sillbeams (see Glossary) were laid. The only practical way of constructing the piers in deep water would have been to have first constructed a ring of short piles using a pile driver mounted on a platform between two barges at low tide (Fig 54, a). This piling technique is illustrated in a 16th-century manuscript, depicting Caesar bridging the Rhine (BL, Harley

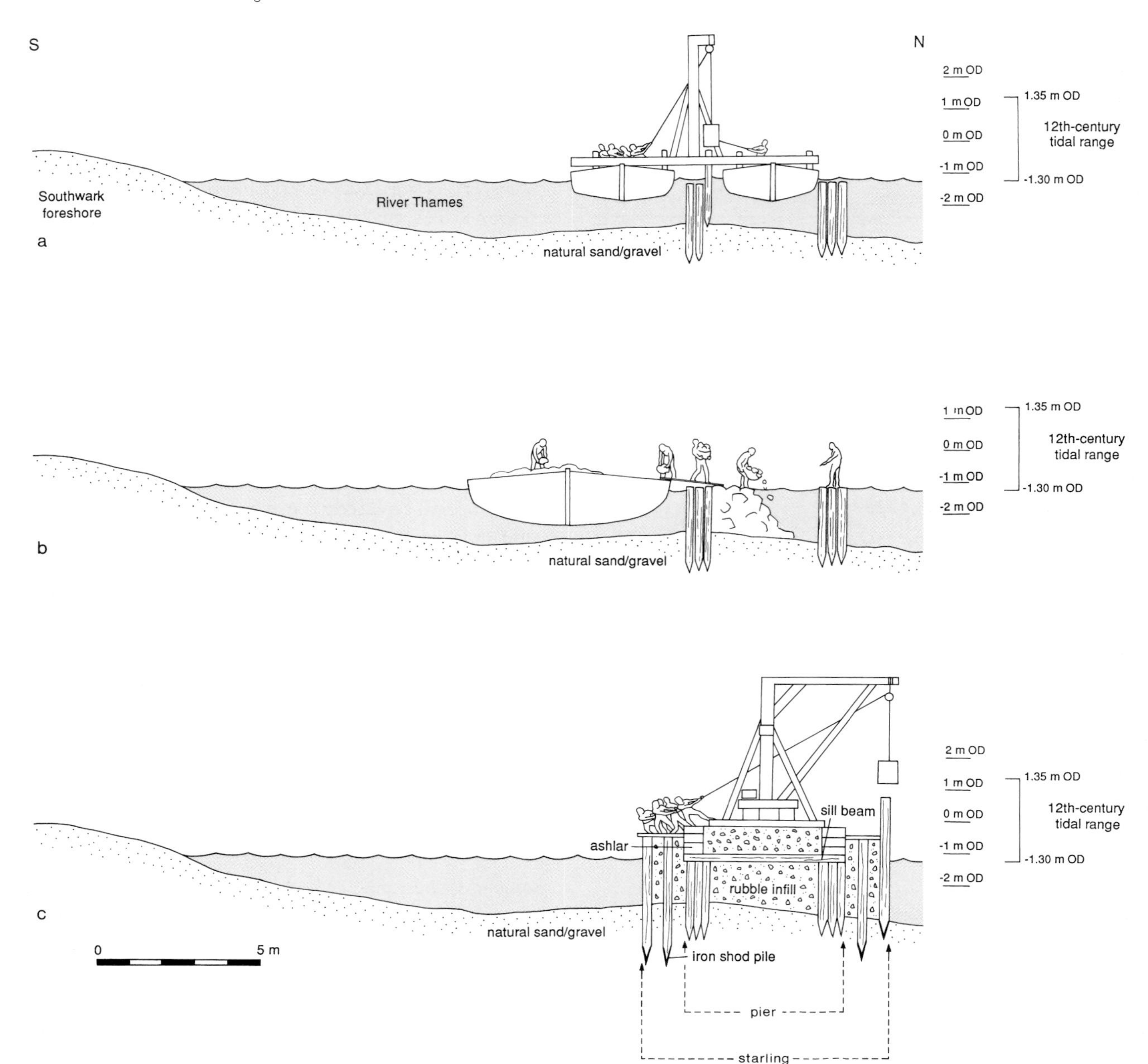

Fig 54 Cross sections showing the various stages of pier construction: a – ring of short piles and water level; b – ring infilled with rubble; c – after the addition of the sillbeams and standing masonry the starling was constructed around the pier; notice how its deep piles protected the shallow pier foundations

MS 6205, fo 21). Still at low tide, the ring of piles would be infilled with rubble (Fig 54, b), and on this the foundations of the pier could be constructed. Then, on the partly constructed pier, a large pile driver could have been set up to drive the many long shod piles which made up the starling (Fig 54, c). Once the starling was constructed the lower portion of the pier could then have been safely completed.

Ashlar facing walls were constructed next and then the interiors of the piers were infilled with rubble. Next the centring for the adjoining arches could be constructed on the starlings. On these centrings the voussoirs of the arches were laid. The piers were equipped with cutwaters on both the upstream and downstream sides to help them withstand both floods and incoming tides. The foundations of each pier were protected by boat-shaped starlings (Fig 52), the edges of which consisted of several lines of long, closely spaced elm piles, infilled with chalk rubble and braced across the top by a series of horizontal timbers. It is probable that many starling piles were iron-shod or -tipped, as three examples of iron-shod piles date from the preparatory work on the Colechurch bridge (see Chapter 14.6). The starlings were completely covered by the high tides, so that repair work on them was restricted to low tide (Harding and Wright 1995, xxii).

It has been claimed that a number of major 12th-century French bridges 'had piers founded on starlings rather than piles' (Boyer 1976, 84). This method of construction consisted, it is said, of driving short piles into the riverbed to a depth of perhaps half their length, so as to form an enclosure. Into this enclosure rubble was dumped to a height above water level, then sillbeams were laid on the rubble, and finally the pier was constructed (ibid). It appears that the author is confusing bridge piers founded on piles and piers protected by starlings.

The function of all the starlings was to protect the relatively shallow pier foundations from scouring and erosion, although if the pier foundations had been as deeply founded as the piers of the Ponte Vecchio, built across the River Arno at Florence (Italy) in 1345 (Hopkins 1970, 40–1), no starlings would have been required. The 17-arch stone bridge at Exeter (constructed c 1196–1214) had no starlings, and its deep-water piers were instead founded on a network of short oak piles, unshod and hand-driven, and layers of dumped rubble enclosed within lines of stakes; a process repeated several times to increase both the area enclosed and its height. The Exeter bridge pier foundations were protected from erosion, scouring and ice flows by lines of stakes, interlaced with brushwood and wattles. The enclosed space was then infilled with gravel or brushwood. This technique may have been appropriate at Exeter where the River Exe was shallow enough to have been crossed by ford in the medieval period. Nevertheless, these pier 'defeynes' needed constant repair according to the Exeter bridge warden's accounts (Brown 1991, 6, 22). The late 14th-century bridge at Rochester was founded on masonry piers, set on a mass of closely packed iron-shod elm piles, which were surrounded by starlings, edged with further piles and infilled with rubble (Britnell 1994, 49).

One factor which certainly influenced foundation design was the depth of water within which the pier foundations were constructed. For instance, excavation of four masonry piers of the late 12th-century bridge spanning the River Thames at Kingston-upon-Thames (some 35km upstream from London Bridge) showed that the first pier, built just above normal river level, was founded merely on sillbeams, while the piers built in the body of the river were founded on rings of short piles within which unmortared flint foundations were constructed (Potter 1991, 140–1). None of these piers was accompanied by starlings. Later, at least one pier was reinforced by the driving of a number of piles to protect its foundations from the effects of erosion and scouring. The foundations of the Exeter bridge piers also varied with the depth of water, in shallow water consisting of rubble dumps on which masonry (without sillbeams) was constructed, while in deeper water unshod oak piles and rubble, enclosed by lines of stakes, were used to raise the ground level to the required height for building on (discussed above) (Brown 1991, 5–6). The foundations of the early 13th-century stone bridge at Hemington also appear to have varied according to water depth. One of the deep-water stone piers excavated in 1993 was founded on a hexagonal enclosure of 3m-long oak piles, infilled with rubble: there were no starlings (Cooper et al 1994, 316–17). Dispensing with starlings did not prove particularly successful at Exeter, where parts of the bridge seem to have collapsed in 1286, 1351 and c 1384. In 1386–8 the collapsed sections of the bridge were replaced in timber (Brown 1991, 2).

Peter of Colechurch evidently decided to build the stone bridge directly on top of the old timber one, judging from the evidence of the superimposed southern abutments. There are several probable reasons for this decision. Firstly, the building of the new bridge in any position significantly different from the previous one would have involved the major task of realigning the approach roads at each of the bridgeheads. Secondly, it must almost certainly have been considered necessary to keep the timber bridge open to traffic as much as possible during the construction of its successor, and that would also have left more scope for the piecemeal rebuilding that the seasonal aspect of much of the early work would have demanded. There is also a third possibility to consider: just as there are a number of documented examples of medieval stone bridges being supplemented with timberwork as temporary repairs (Boyer 1976, 83; Britnell 1994, 72), so Peter could have done the reverse and replaced the timber bridge on a piecemeal basis with masonry. In fact the bulk of bridge piers, compared with the relatively short spans of the arches, suggests the medieval bridge was probably built one arch at time. Constructing the bridge this way would have meant that the latest arch to be erected would not be balanced on both sides by the opposing thrust from the next span until the bridge was completed, so massive piers would have been required to support the arches during construction.

8.3 The design of the stone bridge

Dance's survey of London Bridge in 1799 (after the removal of the central pier in 1757–62) revealed that the spans of the 17 remaining arches varied in width from 4.57m to 10.49m, the mean being 8.10m (Dance 1799).[25] The 18 remaining piers varied in width from 4.60m to 10.67m, the largest occupied by the chapel (see Chapter 9.4). The mean width of the other, more typical, piers was 6.45m. A striking feature of the various arches and piers would, therefore, have been their general lack of symmetry, which was doubtless increased over the years by piecemeal repair and replacement (Fig 53). It is probable that the Gothic style arches recorded by Norden in c 1600 (Fig 67) all date from the mid-15th-century rebuilding of the bridge that took place after the 1437 collapse of two arches (see Chapter 11.1). However, it is possible that the semi-circular arches shown on Norden's engraving are in the same style as the original ones. Knight observed that 'although this bridge had so unsightly and irregular an appearance it is nevertheless fair to conclude that the architect originally intended to make his work symmetrical', and he noted that 'under the arches of the old bridge several phases of construction or repair could be identified' (Knight et al 1832, 205–6). Many surviving medieval bridges feature arches of varying spans. For instance, the 11th- or 12th-century bridge at Albi in France has eight arches, the spans of which vary from 4.1 to 16.0m (Boyer 1976, 26). It has been suggested that Peter of Colechurch's original intention may have been to construct arches with spans of c 8.5m and piers

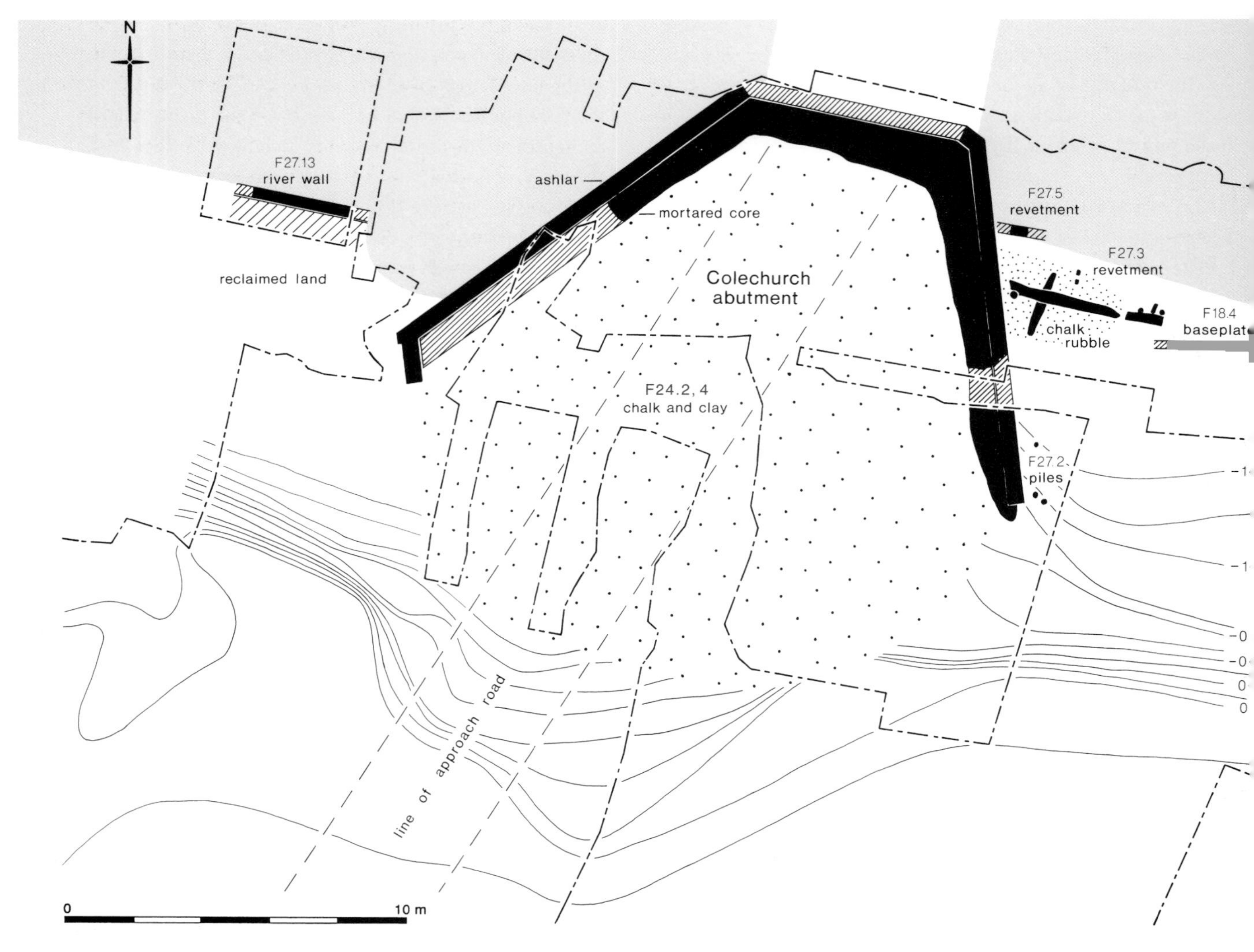

Fig 55 Plan of Colechurch bridge abutment, with the contemporary downstream revetments (F27) and the contemporary upstream riverwall (F27.13)

c 4.2m wide (Hopkins 1970, 38).

Hawksmoor (1736, 38) was of the opinion that the arches were too narrow and the piers too wide, with the result that the bridge restricted the flow of the river (see Chapter 13.3). He suggested that the large central pier (the site of the chapel) (Fig 52) was intended as a buttress to ensure that the adjoining structure would withstand sudden and drastic changes in lateral pressure that might have been caused by any collapse of arches (ibid).

8.4 The construction of the southern abutment

Peter Colechurch's first action at the southern abutment was to consolidate and raise the ground level to the south of the pre-existing 12th-century timber bridge (see Chapter 7). A shallow hollow within the natural sand/gravel was infilled with a mass of flint cobbles set within a matrix of sand/clay (F24.1). This levelling was followed by two separate but probably contemporary dumpings of chalk rubble set within a grey clay matrix, its top at 1.06m OD (F24.2, 4) (Fig 55; Fig 56). Associated pottery dates to 1150–1200 (see Chapter 14.5). These deposits also sealed the southern side of the 12th-century caisson and its associated upstream revetments, confirming that by this time these structures had been dismantled down to foundation level. The eastern side of the chalk rubble dumps was retained by a short length of revetment consisting of horizontal planking retained by stakes (F24.3).

Soon after the deposition of the chalk rubble dumps there was an undated period of erosion which removed some of the rubble and created a hollow (F25.1). The primary fill of this hollow was a build-up of fluvial silt/peat containing driftwood (F25.2). Further accumulation of deposits included a layer of slumped chalk rubble and river silts, as well as the dumping of chalk rubble and sand/gravel intended to raise the ground level (F25.3), its top at 0.66m OD. Pottery from these deposits dates to 1170–c 1200 (see Chapter 14.1 and 14.5), but on stratigraphic evidence this activity dates after c 1160 and before 1187.

The construction of the abutment started with the driving

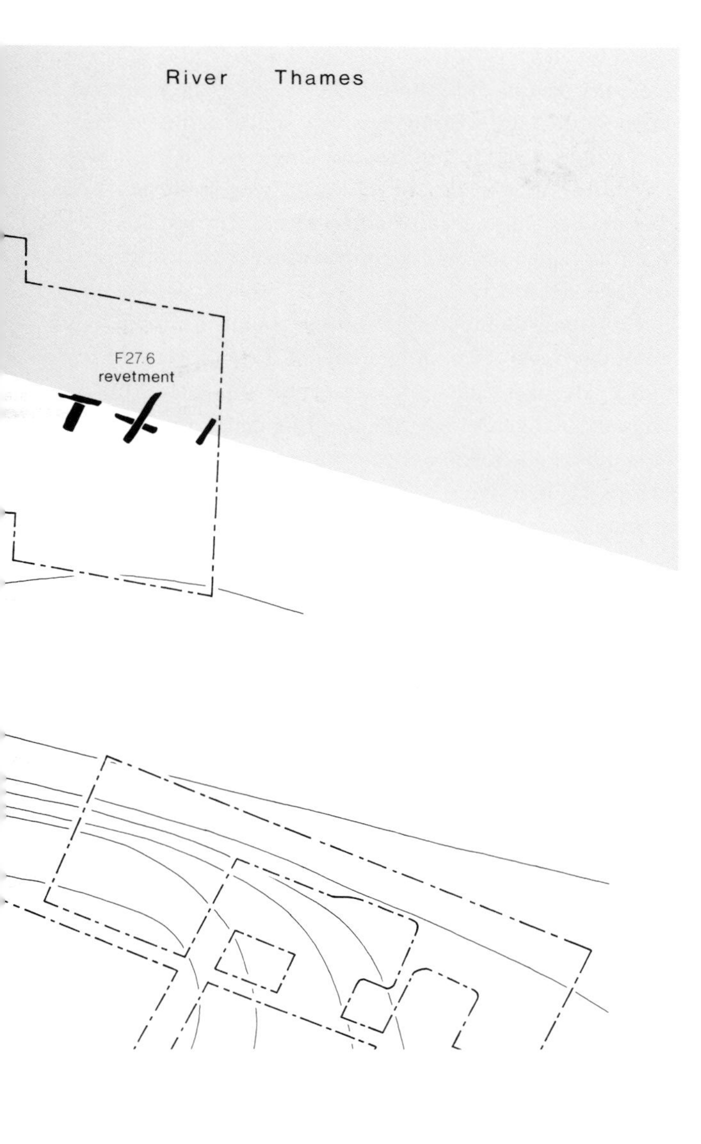

in of a number of piles, at least two of which were iron-shod (F25.4) (see Chapter 14.6). This activity may have been accompanied by further dumping of clay and clay rubble to make a compact level surface on which a single course of oak sillbeams was laid, except along the west side of the abutment where there was a double course of oak sillbeams. Along the east side of the abutment the sillbeam was founded on a number of short beech piles (F26.1) (see Chapter 14.2). The western side of the abutment, which would have received the full force of the current, incorporated two courses of sillbeams, the top of the lower at -1.22m OD and of the upper at -0.80m OD, separated by a single course of ashlar masonry (F26.4). Within the rubble under the western sillbeam was a London Ware dripping dish, packed full of concreted shells and food remains. This type of vessel was used as a receptacle to collect the juices from spit-roasted meat (see Chapter 14.5). The sillbeams of the north side of the abutment, their top at -1.35m OD, were founded on chalk, rubble and clay, and there was no trace of piles (F26.2). All the sillbeams were made from halved oak logs. The beams were smoothly dressed and rectangular in cross section, with a width of 580–610mm and a thickness of 180–230mm. The only complete sillbeam recovered was 6.35m long (see Chapter 14.2 and Table 1). All the beams were through-splayed, scarf-jointed and pegged together. A number of the timbers used as either sillbeams or piles were felled during 1187 or 1188, confirming that construction of the abutment took place during the summer of 1189 or perhaps 1190, when the lower water levels would have made access easier (see Chapter 14.1 and 14.2). This variation in date range could be due to the stockpiling of materials for a year or two, a practice which tree-ring analysis has detected on other major medieval construction projects (English Heritage 1998, 12). The relatively small size of the beech timbers used for the piling argues against the long-term stockpiling of this type of material (see Chapter 14.2).

On the sillbeams were constructed the ashlar facing walls of the abutment (F26.5) (Fig 55; Fig 57). The ashlar consisted of dressed rectangular blocks made from a variety of materials. The basal course of the east and west walls consisted of a white quartzite sandstone, probably from the North Downs or the Chilterns, while the standing masonry consisted of a mixture including Kentish ragstone, Purbeck marble, Purbeck stone and Wealden 'marble' (see Chapter 14.11). The blocks varied in length from 380 to 700mm, width from 300 to 440mm and height from 160 to 210mm (Fig 58), and all were fixed together by U–shaped iron cramps set in holes in the upper faces of the blocks and fixed by pouring molten lead into the holes (see Chapter 14.6). The various courses of blocks were all offset and the frequent use of short blocks created a 'header and stretcher' type pattern (Fig 57). All the blocks were bonded by the same pale brown, sandy lime mortar as was the solid mass of uncoursed ragstone wall core about 1.0m wide, some of which adhered to the rear of the ashlar. Pitch had been mixed with mortar used on some of the ashlar facing (see Chapter 10.4). The full height of this ashlar was concealed by modern damage, but surviving masonry stood up to 2.99m OD. At an uncertain height above the sillbeams there was a chamfered offset in the external face. The resulting structure was an irregular trapezoid, open on its southern or landward

Fig 56 View of the truncated chalk rubble infill of the Colechurch abutment (F24.1), area 4, looking north; in the foreground is natural gravel (1m scales) (MoL)

Fig 57 *View of the facing blocks of the west side of the Colechurch abutment seen in elevation, looking north-east in bay 16; the chalk rubble to the left is the infill of the 15th-century enlargement of the abutment (0.2m and 0.5m scales) (MoL)*

Fig 58 *View of the north-west facing of the Colechurch bridge abutment, seen from the corner of a riverside bay (no. 16), looking north-west; the chalk rubble dump on the far side of the ashlar facing is part of the infill of the enlarged 15th-century abutment; this illustrates the piecemeal nature of the investigation of the bays during ground reduction (2m scales) (MoL)*

side where there would have been an access ramp (Fig 55; Fig 94). The abutment measured 19m long from east to west and 11m long from north to south, and would originally have stood some 4.5m high above its sillbeams.

The finished masonry abutment was filled with a series of dumps of chalk rubble all set in a brown clay matrix (F24.2, .4: Fig 55; Fig 94). These dumps extended much further south on the landward side of the abutment than did the masonry of the abutment, and were intended to infill an existing hollow as part of the construction of the access ramp up to the roadway. No trace of any access ramp was found further south during the recording of the abutment, probably because of truncation caused by the construction of the basement of the 19th-century warehouse and its foundations. The design of the southern abutment of London Bridge was very similar to that of the two abutments of the 14th-century single-arch stone bridge at Waltham Abbey. The two abutments were founded on elm piles and oak sillbeams 40–50mm thick, 0.2m and 0.4m wide, on top of which the ragstone superstructure was constructed (Huggins 1970, 133–5).

How the building of the southern abutment related to the rest of the construction programme of the 12th-century bridge is uncertain, as it is the only element for which tree-ring dates are available. Interestingly, the tree-ring dating confirms that the southern abutment was constructed during 1189 or perhaps 1190, roughly halfway through the *c* 33-year building programme. This means that by this date a lot of the superstructure must already have been built, but presumably not any of the masonry arches adjoining the southern abutment or there would have been problems with their lateral thrust. Therefore, it seems likely that the first part of the stone bridge to be constructed was the northern portion of the bridge and that its large central pier (discussed earlier) was being used as a buttress to withstand the lateral pressure from the northern masonry arches. Perhaps the erection of the southern portion of the bridge began with the abutment, and the remaining arches were then constructed one at a time from south to north (Fig 52).

8.5 Downstream revetments and foreshore contemporary with the Colechurch bridge

A number of developments were broadly contemporary with the construction of the Colechurch abutment. The newly built abutment stood further into the river than the wooden caisson had done, and it acted as a breakwater, with the result that soon after the construction of the bridge there was further reclamation on the eastern or downstream side of the abutment. The earliest downstream revetment which is broadly contemporary with the Colechurch bridge consisted of a single, squared oak log baseplate, aligned east–west with its top at -1.20m OD (F18.4), and made from timber felled *c* 1167–1212 (Fig 55). Cut into the upper face of this baseplate were four closely spaced mortices (Fig 59), a type of joint used in London carpentry only from *c* 1180 onwards (see Chapter 14.3).

Immediately after the construction of the bridge there was a build-up of fluvial silts and peats around the eastern side of the abutment (F18.5 and F27.1). These silts were sealed by dumps of chalk rubble set in a sandy silt matrix. Probably contemporary with these rubble dumps was the dumping of more chalk rubble in a grey sand/clay matrix further north on the foreshore, its top at -1.35m OD (F27.3) (Fig 55). On

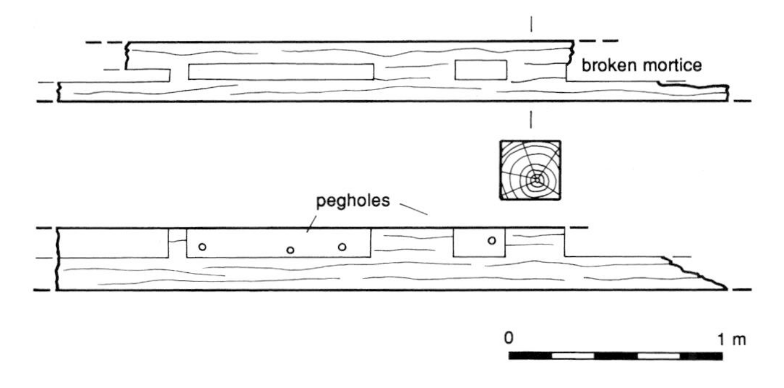

Fig 59 *Plan, elevation and cross section of the late 12th- or early 13th-century revetment baseplate (F18.4)*

this newly consolidated foreshore an east–west revetment was constructed containing timber felled in 1197 (F27.3) and confirming that this revetment was built some nine years after the construction of the Colechurch bridge abutment. The revetment consisted of two baseplates set on crossbearers and retained by a number of short stakes, their top at -1.03m OD (Fig 55). Cut into the upper face of the eastern baseplate were three empty mortices (Fig 60), the westernmost of them partly removed by diagonally cutting back that end of the baseplate so it butted right up against the east side of the abutment. As the beech baseplate in this revetment was made from timber felled in 1197, it is unlikely to be have been reused (see Chapter 14.2). Thus, the cutting back of the western end of this baseplate, removing part of a joint, appears to represent a crude adaptation of the prepared timber, which had presumably turned out too long for the available space. This revetment is of particular interest as an early example of the use of the mortice and tenon joint (see Chapter 14.3).

The reconstruction of the superstructure of this waterfront is uncertain. It probably consisted of a series of vertical timbers set into the upper face of the baseplate and retaining a series of horizontal planks laid on edge, similar to the frontbraced post and plank revetment dating to *c* 1220 found at Billingsgate or the early 13th-century revetment from Seal House (Milne and Milne 1992, 32, 40), except that no evidence of frontbraces was found.

This revetment was probably soon dismantled as it was sealed by a dump of coarse sand mixed with clay and ragstone rubble (F27.4) which was itself sealed by a fluvial silt containing a number of stakes and plank fragments, possibly the *ex situ* remains of stake and plank revetments destroyed by floods. Several metres to the south of the former revetment, there was a further build-up of fluvial silts/clays during the 13th century, into which were driven a number of oak piles (F27.2) sealed in turn by a further build-up of fluvial deposits. Pottery from these later deposits dated to 1230–70.

The next revetment to be built consisted of a baseplate aligned east–west (top -0.68m OD) (F27.5) and one isolated stake (Fig 55). The baseplate lay 1.2m north of the former baseplate (F27.3) and had originally butted up against the eastern side of the abutment. Cut into the upper face of the baseplate was a rectangular socket, which suggests that its

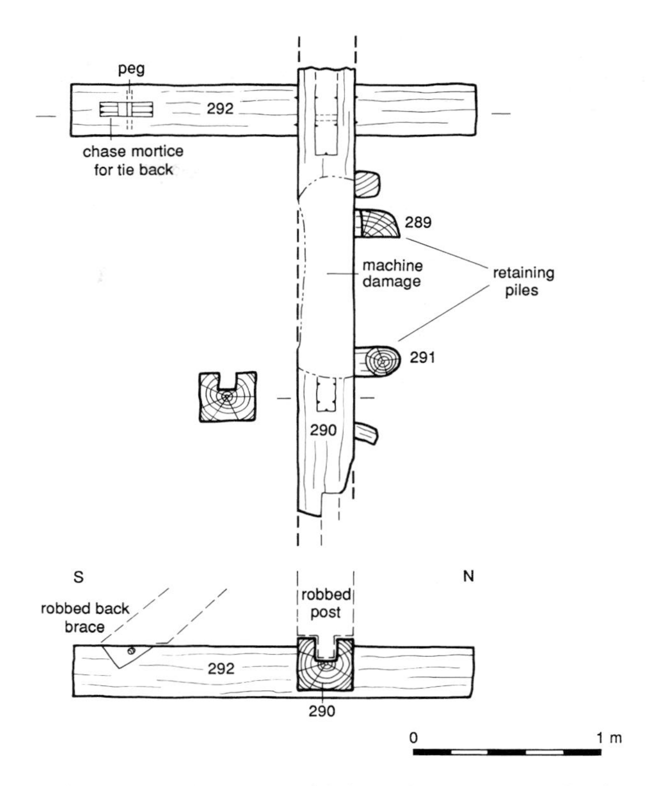

Fig 60 *Plan, elevation and cross section of the late 12th-century revetment baseplate and its crossbearers (F27.3)*

superstructure was probably very similar to that of the former revetment. On stratigraphic grounds this revetment was of late 13th- or 14th-century date.

The easternmost downstream revetment, its top at -0.65m OD, is undated, but its alignment suggests that it was of either 13th- or 14th-century date (F27.6: Fig 55). It consisted of a series of four cruciform-shaped timber structures, all half-lapped together. The individual structures were regularly spaced at 1.4m apart. The timbers probably represent the basal remains of a series of frontbraces or tiebacks. The complexity of the joints in some of them suggests that they were reused. The revetment was sealed by a build-up of fluvial sands and silts of uncertain date (F27.7).

8.6 Upstream riverwall and foreshore contemporary with the Colechurch bridge

The 12th-century upstream foreshore consisted of reworked natural sand/gravel overlaid by fluvial silts (F27.8). Into these deposits were driven a succession of piles, interpreted as four phases of activity. The first phase of piling (F27.9) dated to 1190 and served as foundations for a masonry riverwall. The riverwall was faced with ragstone or limestone ashlar blocks, not cramped together (Fig 57). The ashlar facing was fused with wallcore, composed of uncoursed ragstone rubble and occasional fragments of chalk held together by light brown sandy mortar (F27.13). In front of this riverwall, and intended to stop it being undercut by the scouring of the ebb tides, were driven three phases of oak elm and beech piles, one dating to 1274 (F27.9–.12). Pottery from the foreshore in front of the riverwall dated to 1270–1500.

8.7 Buildings of the southern bridgehead broadly contemporary with the Colechurch bridge

No traces of any buildings likely to be contemporary with the erection of the stone bridge were found along the projected line of the bridge approach road. This is presumably because any such buildings were either uncellared or not deeply cellared, so that their remains did not survive truncation of the archaeological deposits caused by the construction of a 19th-century warehouse basement.

It was during the 12th century, at about the time the bridge was rebuilt in stone, that cellared buildings with stone foundations became common in both Southwark and London. They are often represented archaeologically by fragments of trench built, unmortared chalk rubble foundations (Schofield et al 1990, 49–52, 83, 119–21), and at the northern end of the bridgehead during the late 12th and early 13th centuries a series of rectangular, uncellared, masonry buildings were constructed at New Fresh Wharf and Billingsgate (Steedman et al 1992, 109–11). The church of St Magnus the Martyr, adjoining the northern end of the bridge, was in existence by 1128–33 (Carlin and Belcher 1989, 88).

At the south end of the bridge one nearby building which would have been in existence by the time of Peter of Colechurch's operations was the church of St Olave, close to the east side of the Bridge Foot (Fig 61). Before becoming King of Norway (c 1015–30) St Olaf was closely linked with London, assisting King Æthelred to regain his throne after the death of Swein Forkbeard in 1014 (see Chapter 5.4). 'St Olave' later became a popular English saint with three church dedications within medieval London, and ranking 49th in a national list of popularity ratings (Oxley 1978, 117). It is, therefore, interesting that the nearest parish church to the southern bridgehead, which was in existence by 1096, should be dedicated to him. By the late 11th or early 12th centuries it had been granted to the Priors of Lewes, who built a house nearby in the late 12th century (Carlin 1983, 365) (see Chapter 12). The burial ground lay to the north of the church, adjoining the river, and was partly lost to erosion during the early 14th century (see Chapter 11).

The medieval church of St Olave's is clearly shown on Wyngaerde's panorama of c 1544, featuring a western tower topped by a steeple (Colvin and Foister 1996, 31). The medieval church was rebuilt in 1740 and demolished in 1926

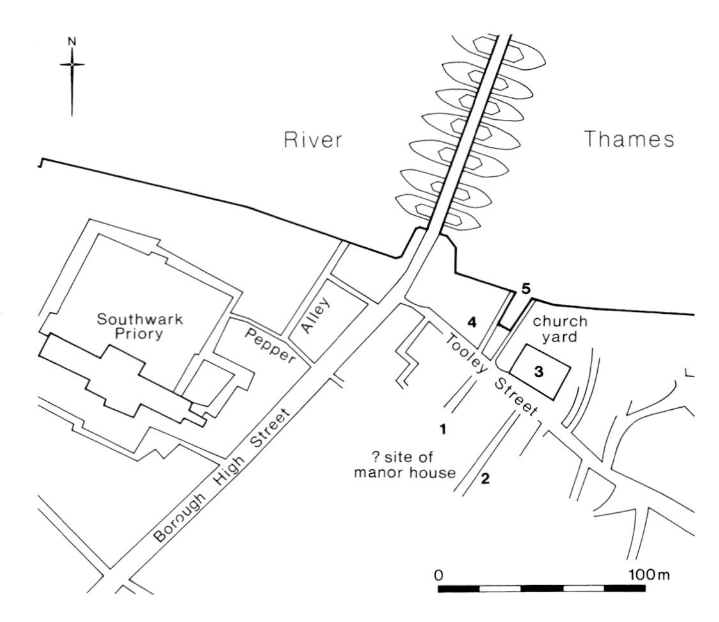

Fig 61 Map of the Southwark bridgehead in the late 12th century showing the southern end of the Colechurch bridge
Key: 1 – stone-built hall found c 1830; 2 – stone-built undercroft found 1832: this undercroft and the hall (no. 1) may be part of the same manorial complex; 3 – St Olave's church, the burial ground was on the north side of the church; 4 – a cellared building (T7.3) on Toppings Wharf; 5 – a hypothetical inlet or natural creek on the site of St Olave's watergate

when St Olaf House was built on the site. A fragment of the south wall of the church was located in Tooley Street in 1972–3 (Graham 1988, 48), and during redevelopment on the site in 1985 the foundations of the 18th-century hexagonal tower were located (Thompson et al 1998, 217).

Only one excavated building at the southern bridgehead was broadly contemporary with the construction of the Colechurch bridge. This was situated east of the Bridge Foot on St Olave's Wharf and was constructed during the late 12th or very early 13th century (T7.3) (Fig 61) (see Chapter 14.1). The rectangular building was represented by three substantial wall foundations, the northern wall having been eroded during the 13th century (see Chapter 12). The surviving portion of the structure had an internal area c 5.0m by 5.0m, containing clay floors, the top of which was at 1.10m OD. The three walls were c 1.0m wide and constructed of mortared, coursed, chalk rubble blocks, founded on alternate layers of rammed gravel and chalk rubble (Sheldon 1974, 24). They are interpreted as the foundations of a large cellared building fronting on to Tooley Street. The same type of unmortared chalk foundation was used nearby on the 12th-century phases of the hall at Winchester Palace (Seeley in prep).

Broadly contemporary with the building found on St Olave's Wharf are fragments of two substantial residences found nearby during the early 19th century on the south side of Tooley Street (Fig 61). In c 1830 redevelopment on the opposite side of Tooley Street from St Olave's revealed a stone-built hall with a vaulted undercroft below. The style of the masonry, particularly the use of scalloped capitals, confirms a mid- to late 12th-century date for the construction of the hall

(Gage 1831, pls 20–4). It has been suggested that the hall was part of the manor house of the Earls of Warenne and Surrey, which by the late 14th century was occupied by the bailiff of Guildable Manor in north Southwark (Corner 1860, 39). Nearby, a second vaulted undercroft or cellar of similar date was found in 1832 (Gwilt 1834), and interpreted as part of the house of the Priors of Lewes, Sussex, who are known to have had a residence here from the late 12th century onwards (Gage 1831, 299–300). Recently Carlin has reviewed these interpretations and concluded that 'Corner's identification of this complex as the original Southwark residence of the Warennes is possible, but all that can definitely be said of this property is that at one time it formed part of the Southwark estate of the Warennes' (Carlin 1983, 270).

9

The buildings and spaces on the medieval bridge and their use

Bruce Watson with Tony Dyson,
and contributions by Vanessa Harding,
Peter Jackson and Richard Lloyd

9.1 The buildings on the bridge

Today bridges are seen as little more than an easy route across an obstacle of one sort or another, but in medieval towns they were much more than that. They often served a number of other functions, such as forming part of the civic defences, providing sites for housing, shops or chapels, and making obvious locations for watermills. Like its medieval counterparts at Bristol (Home 1931, 255), Exeter,[26] Newcastle (Bruce 1885, 11) and the Ouse bridge at York,[27] for example, London Bridge was lined with buildings (Home 1931, 256), as indeed many major urban English and French bridges still were as late as the 18th century. In part the phenomenon arose from the feeling expressed by the Italian architect and town planner, Leone Battista Alberti, in 1485, that 'the bridge is undoubtedly a main element of the street ... and, in the very heart of the city, ought to be accessible to everybody' (Alberti, 76). Bridges were seen as natural places for shops and businesses, no doubt on account of the volume of passing trade which was funnelled across them; and to people crossing over it – as was pointed out by several commentators at various periods – a bridge lined with buildings several storeys high would have looked much like any other thoroughfare (Fig 62). During the 12th and 13th century at Provins (northern France), the bridge over

Fig 62 Reconstruction showing the houses and congestion on medieval London Bridge, by Peter Jackson

Fig 63 The Pulteney Bridge, Bath, the last bridge lined with buildings to be constructed in England (B Watson)

the River Durteint, like the rest of the approach road to the walled town, was lined with stalls selling 'bread and fish' (McCormack 1999, 109). Thirty-five medieval French bridges had houses built on them, and by 1141, when Louis VII decreed that money changers should work only on the Grand Pont in Paris, teachers and scholars already occupied houses on the Petit Pont (Boyer 1976, 76–7). There are several further recorded instances of French bridges featuring houses in the late 12th century: the bridge at Angers was lined with them before 1175, and before 1202 they were erected on the bridge at La Rochelle to raise money for bridge maintenance (Boyer 1976, 75). It is more than likely that such examples provided models and precedents for house building on London Bridge.

The concept of the habitable bridge went out of fashion during the 18th and 19th centuries, when such structures were either replaced or their buildings demolished, often in order to widen the roadway. The only surviving example of a habitable medieval bridge today in England is the High Bridge at Lincoln. The western side of the High Bridge is lined with timber-framed shops and until 1762 the eastern side of the bridge was occupied by the chapel of St Thomas the Martyr (Pevsner and Harris 1989, 523).[28] The Pulteney Bridge at Bath, constructed in 1769–74 and rebuilt in 1804, is probably the last habitable bridge to have been erected in England (Beard 1978, 48) (Fig 63). Today there are only a few such surviving medieval bridges in northern Europe, including the Ponte Vecchio in Florence, the Ponte di Rialto in Venice and the Krämerbrücke at Erfurt in Germany (Eaton 1996), although many more still retain chapels. Very recently the notion of the urban bridge as a place for living, shopping, entertainment or as a public open space has been rediscovered by architects, and it is now proposed to build such an amenity (but without housing) over the Thames not far from the site of medieval London Bridge (Murray 1996).

The buildings which were to be such a striking feature of medieval London Bridge were first mentioned in a letter of King John in 1201, referring to the imminent completion of the bridge and pointing out that the rents and profits of 'several houses' which would or could be erected on it would contribute towards its maintenance (Home 1931, 42). There is, however, no direct evidence that such houses had been built until 1221, when a grant to London Bridge of land in the parish of St Olave, Southwark, mentions 'the houses on the bridge' as a landmark (Welch 1894, 78), although the fact that the great Southwark fire of July 1212 or 1213 'destroyed' the southern portion of the newly completed bridge, including the chapel (Home 1931, 44; Riley 1863, 3), strongly implies the presence of a superstructure of buildings that could sustain a fire that far across it (Fig 53). In 1281 a royal writ concerning the ruinous state of the bridge mentions 'almost innumerable people dwelling thereon' (Cal Pat R 1272–81, 422), while a rental survey of Bridge House properties in

Fig 64 The earliest known view of medieval London Bridge from an illuminated manuscript of the poems of Charles, Duke of Orleans, dating from c 1480; in the foreground is the Tower of London, where the Duke was imprisoned 1425–40; in the background can be seen the chapel and the central portion of the bridge (BL, Royal MS 16, F.11, fo 73)

c 1358 shows that there were then 62 shops on the east side of the roadway and 69 on the west side (Welch 1894, 258–9). According to Stow, writing in *c* 1600, 20ft (6.09m) of the bridge's overall width of 30ft (9.14m) was occupied by buildings on either side (Stow 1603, i, 26).

The earliest artistic representation of the bridge appears as an illustration to a manuscript of the poems of Charles, Duke of Orleans, which dates from *c* 1480 and shows him confined within the Tower of London (1415–40) with the central portion of the bridge revealed in the background (Fig 53; Fig 64). The portion of the bridge to the north of the chapel was covered with houses, except for the arch immediately adjoining. The extreme northern end of the bridge was obscured in this view by the vaulted arcade of the Custom House buildings (Home 1931, 144). A cross section of the buildings on the Pont Notre Dame in Paris in the 16th century shows that the central roadway was flanked by two lines of four-storey cellared buildings, the ground floors of which were used as shops (Eaton 1996, 59). At the attack on the Drawbridge tower during Fauconberg's rebellion of 1471, 14 (13, according to Stow) houses on the south side of the tower were burnt down (Fig 73), and in 1471–2 the bridge carpenters framed and erected wooden rails (*lez rayles ligneos*) where the 14 tenements destroyed in the rebellion had stood (Stow 1603, i, 25, 60; CoLRO, BHR 3, fo 198v). Mancini, writing about London in 1482–3, described London Bridge as 'a famous bridge built partly of wood and partly of stone. On it there are houses and several gates with portcullises; the dwelling-houses are built above the workshops and belong to diverse sorts of craftsmen' (Armstrong 1936, 125): the wooden portion of the bridge was presumably the drawbridge.

The first complete representation of the bridge is the sketch which Wyngaerde produced as part of his London panorama (*c* 1544) (Colvin and Foister 1996, 28–9). Despite the fairly steep perspective of this northwards view across the river, it shows much detail (Fig 65). The variation and complexity of the rooflines of the many properties on the bridge suggest much piecemeal building, except for the block at the far northern end whose uniform roofline suggests a single phase of construction. Of the many shops and houses on the bridge depicted by Wyngaerde only the backs are generally shown, and it is difficult to tell how many of their fronts, facing inwards towards the carriageway, included jettied facades (Fig 65), although the one shop front which is visible, on the west side of the Stonegate, had no jetty. By this date the houses were arranged in four blocks (Fig 66), which appears to correspond with the Orleans depiction. The area south of the Stonegate was built over, but the roadway over the arch to the north of it was open, after which interval there was a further block of houses – some spanning the roadway at upper storey level – before the drawbridge, where one arch remained unbuilt upon and the drawbridge spanned another. The portion between the Drawbridge tower and the chapel was completely built over, although there was another open arch on the north side of the chapel (Fig 66). A number of properties included small gabled structures, interpreted as privies, overhanging the river (Colvin and Foister 1996, 12). According to Stow (1603, i, 25) 'the common siege' (a public privy on the bridge) fell into the river in 1481, drowning five men. The height of the properties on the bridge was generally three or four storeys and their size varied, suggesting that some may have been amalgamated with others and rebuilt as one.

In the 'Agas' woodcut map of London (*c* 1562–3),[29] only a little later than Wyngaerde's, most of the blocks of buildings on the bridge are shown with their roof ridges aligned parallel with the roadway, although eight buildings clearly spanned the roadway at upper floor level (Fisher 1981). The complex roof-lines of the properties shown in the later Norden view of *c* 1600 reveal that the 'Agas' representation of the same buildings is much simplified, implying it is an artistic representation of the bridge rather than a true to life cartographic image. Norden's view of the bridge shows that some of the buildings on it were three or four storeys high (Fig 67). The five most significant differences between Norden's depiction of the buildings and Wyngaerde's are: the presence of the waterworks tower (see below); the demolition

Fig 65 *View of London Bridge in c 1544, looking north-west, as shown on Wyngaerde's panorama of London (© Ashmolean Museum; reproduced from Colvin and Foister 1996, 28–9)*

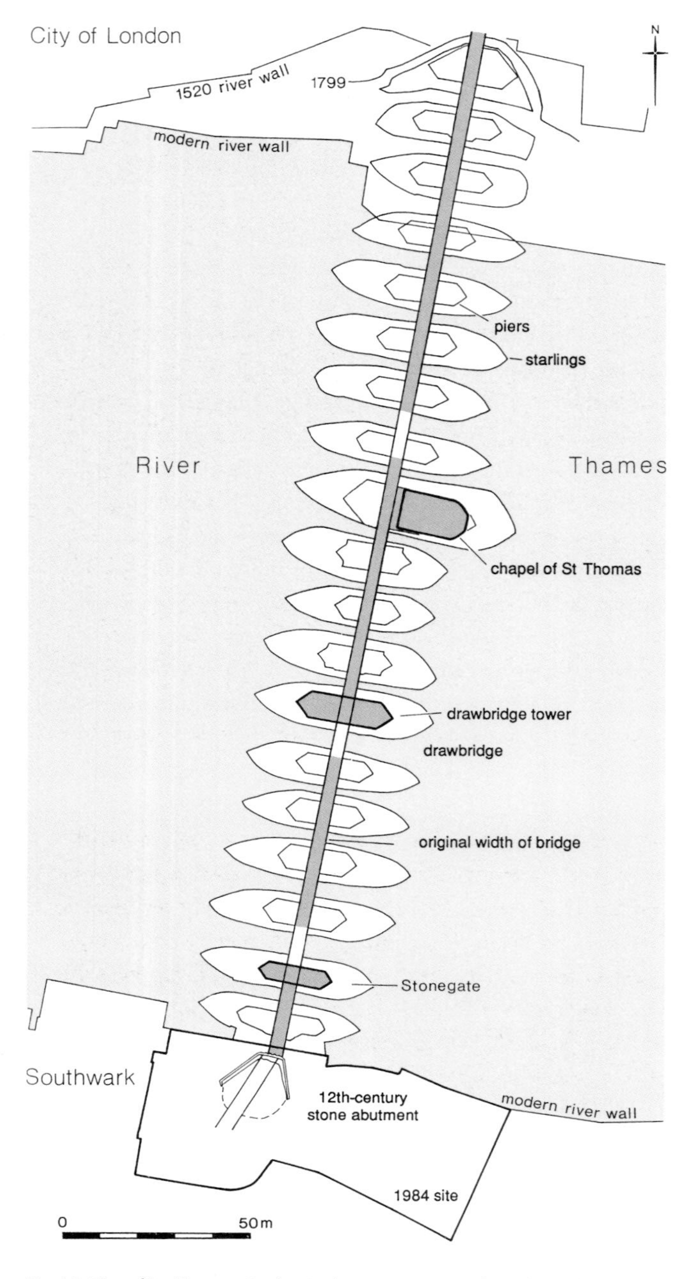

Fig 66 Plan of buildings on London Bridge c 1544; notice how almost the entire bridge was covered with buildings; at this scale it is not possible to show the narrow central roadway, which was flanked on both sides by overhanging buildings

of the upper portion of the chapel and its replacement by a timber-framed building; the replacement of a number of the buildings south of the Stonegate; and the addition to a number of buildings north of the Drawbridge Gate of small, domed towers which would have functioned as observation towers, and of stair turrets giving access to upper storeys (one of these new towers was clearly a rebuild of the former chapel stair turret) (Fig 53). To the north of the chapel another mock gatehouse with domed, twin towers had also been added by c 1600. Some of the buildings south of the Stonegate were possibly damaged or demolished during Wyatt's rebellion in 1554 (see 9.2 below). The replacement block of buildings shown by Norden was four storeys high with twin observation towers at its northern end, giving it the appearance of a gatehouse, and a central upper-storey block apparently spanning the roadway; portions of it overhung the river (Fig 67).

In the text accompanying his engraving, Norden described the bridge as 'adorned with sumptuous buildings ... and beautiful houses on either side, inhabited by wealthy citizens and furnished with all manner of trades ... [the] buildings are artificially contrived, and so firmly combined, as it seemeth more than an ordinary street, for it is one continuall vaute or roofe, except certain voyde places reserved from buildings, for the retire of passengers from the danger of carres, carts and droves of cattell, usually passing that way' (Soane, drawing no. 105, p 187).

Norden's engraving has been the main inspiration for the reconstruction (Fig 68) by Jackson who gives the following account of his sources.

'This reconstruction of London Bridge from the east shows it at the zenith of its greatness in the late 16th and early 17th centuries. We are fortunate in having a splendid engraving of the eastern elevation of the bridge made at this time which, with its meticulous detail, has all the appearance of considerable accuracy (Fig 67). It is this engraving which has formed the basis of my reconstruction. The engraving was produced by John Norden (1548–1625), Surveyor to Henry, Prince of Wales, elder son of James I. Norden engraved the plate towards the end of Elizabeth I's reign (1558–1603), but it was stolen or, as Norden puts it in his dedication, "the plate had bene neare these 20 years imbezeled and detained by a person till of late unknown". The plate was finally printed and published in 1624, and bears the arms of Sir John Gore, then serving as Lord Mayor of London. Another useful source of information was the anonymous drawing on parchment in the Pepysian collection at Magdalene College, Cambridge. Although it is of the same period as the Norden view it shows the western side of the bridge and the corn mills on the southern starlings (reproduced in Home 1931, 136), which do not appear on Norden's view.

Nonesuch House (the replacement for the Drawbridge Gate tower, built 1577–9; see Chapter 9.3) presents a problem since its appearance varies in both views and also conflicts with those of Hollar (Home 1931, 224, 241) and Visscher (Home 1931, 215). The result was a compromise using elements from all sources. Its colouring was based on a description on a manuscript of 1558 which is a record of the payment to "Durram, the paynter, to bye Coulors to paynt the *Vawte* [balcony] at the Maior's palace, in parte of payment of xxx s. to ley the vawte in oyle Colers substancially, the greate posts in jasper Collur [reddish-brown], as *the newe house on London Bridge ys:* all the rayles in stone Coulor, the smale pillors in white leade Coulors, the great pillars in perfect greene Coullor xiij.s. iiij.d." (Thompson 1827, 254).

The appearance of the starlings was not completely conjectural; their erratic shapes are based on George Dance's survey of 1799, which was also used to produce Fig 62. However, it is possible that in 1600 some starlings were of slightly different sizes and shapes to when Dance surveyed the bridge, but one can assume that their shape had not changed significantly, except in the case of the central starling which

Fig 67 *Norden's engraving of London Bridge* c 1600, *looking west; notice the river boats attempting to pass under the bridge and the overturned boat in the foreground* (MoLPL)

was removed during the modifications of 1757–62 (see Chapter 13.2).'

Many of the properties mentioned in the *c* 1358 rental had a *hautepas* – an upper storey external passageway or gallery. Home (1931, 87) believed that some of the *hautepas* may have spanned the roadway to create a structural link at third storey level between buildings on either side of the street. However, Wyngaerde's panorama shows that, apart from the entire block of houses at the far northern end of the bridge, all of which spanned the roadway at upper storey level, few of the other houses did and certainly none of those south of the Stonegate. This would suggest that some of the *hautepas* were either galleries or portions of properties built out over the river beyond the edge of the bridge superstructure; evidence of this practice is clearly shown on Norden's view of the bridge (*c* 1600) (Fig 67). Many of these small structures overhanging the river were probably privies (Fig 69).

Stow described these buildings as 'compact and joined together with vaults and cellars; upon both sides be houses built, so that it seemeth rather a continual street than a bridge' (Fig 62; Fig 66; Fig 69), and stated that some of these shops were held by mercers and haberdashers (Stow 1603, i, 81). It is plain that many of these properties featured narrow cellars built into the superstructure of the bridge on each side of the roadway, either within or between the various piers (Knight et al 1832, 203). When these cellars were encountered during the demolition of the bridge superstructure in 1832, those at the southern end of the bridge were observed to be smaller than those at the northern end, perhaps a consequence of the 17th- and 18th-century rebuilding of the latter properties (see below).

According to the martyrologist John Foxe, a woman was imprisoned in the 'cage' on the bridge in April 1555, and told to 'cool herself there', for refusing to pray at the church of St Magnus the Martyr for the recently deceased Pope Julius III (Madan 1760, 162). This appears to be the only reference to the 'cage' on the bridge. Although 18th-century engravings of the incident imply that the 'cage' was positioned near the Stonegate, and was possibly one of the small, privy-like structures shown by Wyngaerde's panorama on the south-eastern side of the gate (Colvin and Foister 1996, 29) (Fig 65), in reality (assuming that the story can be relied upon), she is more likely to have been simply hauled over the bridge and imprisoned in the Tooley Street cage, shown on the Southwark map of *c* 1542 (Carlin 1996, 39).

A fire which broke out in a house near the church of St Magnus during the night of 11 February 1633 spread the next day to the northern end of the bridge and burnt down 42 houses to the north of the chapel pier, plus nearly 80 houses in the parish of St Magnus the Martyr (Chandler 1952, 19). A single house on the eastern side of the bridge was rebuilt during 1639, but the site of the rest was simply clad with a fence of 'deal' (softwood, probably pine) boards. During 1645–7 a large block of timber-framed, three-storey houses was built at this end of the bridge and in 1648 a further three houses were erected (Chandler 1952, 20–1) (Fig 53). Pepys noted in January 1666 that the wooden fencing around the vacant plots on both sides of the bridge had been blown away by gale-force winds (Latham and Matthews 1972, 22). These new houses and all the other buildings on the northern end of the bridge were destroyed during the Great Fire of London on 2 September 1666 (Bell 1923, 26) (Fig 53; Fig 70). The Great Fire had started earlier that day in a bakehouse nearby in Pudding Lane. More buildings on the bridge would probably have been destroyed, including those above the five or more

Fig 68 Reconstruction of London Bridge in c 1600, looking west, by Peter Jackson

piers between the newly rebuilt houses and the chapel, had all those burnt in 1633 been replaced. The heat of the fire caused some damage to the masonry of the bridge, which had to be strengthened at a cost of £1500 (Home 1931, 243). Thereafter, house plots were leased out on an individual basis for the tenants to rebuild, and by 1677 a block of low, probably single-storey, buildings had been erected on the northern end of the bridge (Fig 53).

In 1685, to improve the traffic flow on the bridge, the street was widened from 12ft (3.65m) to 20ft (6.09m) and all the houses on the bridge, except those adjoining the drawbridge, were pulled down and rebuilt in 'a new and

regular manner' (Seymour 1734, 48) (Fig 71). On 8 September 1725 the houses on the bridge south of the Stonegate were destroyed by a fire which spread north from Tooley Street (ibid, 49). In 1745 George Dance the elder rebuilt a large four-storey block of ten houses (five on each side) on the northern end of the bridge (Fig 53), the ground floor partly occupied by a colonnaded walkway (Home 1931, 320). All the houses on the bridge were demolished during 1757–61, when the bridge roadway was further widened (see Chapter 13.2).

When Zacharias Conrad von Uffenbach visited London in 1710 he walked over part of the bridge without actually

Fig 69 London Bridge, c 1600, view looking west of model from the Living Bridges exhibition (Royal Academy; photograph by A Chopping)

Fig 70 The Great Fire of London September 1666, looking north-west (Dutch School c 1666); notice the damage to the northern end of the bridge and the gap in the houses on the bridge remaining from the 1633 fire (MoLPL)

Fig 71 London Bridge in 1749 looking north-east, showing Nonesuch House and the adjoining blocks of houses; this engraving was published in September 1749 as part of Samuel and Nathaniel Buck's Prospect of London (1st edition) (Hyde 1994, pl 44) (MoLPL)

realising that he was doing so, commenting that 'one does not take it for a bridge because it has on both sides large and handsome houses, the lower storeys of which are all shops' (Quarrell and Mare 1934, 56). It is clear that until the buildings on the bridge were finally demolished in 1757–61 there was a wide of range of shops and trades represented. A list of residents on the bridge in 1734 included a cooper, a dyer, a farrier, two fishmongers, two glovers, a goldsmith, two lorimers (makers of metal fittings for horse harness and trappings), a needlemaker and a painter-stainer (*Daily Post Extraordinary* 1734). From 1749 until 1759 one of the block of ten new houses and shops was occupied by the printmaker William Herbert (Heal 1931, 315), while from 1751 until 1759 the former chapel was occupied by the stationers Wright and Gill. Two former taverns on the bridge – the Bell and the Blue Boar – were occupied by a stationer and a toyman (selling needles, buttons and small metal objects) during the early 1750s (Heal 1931, 321, 323). A somewhat unsympathetic impression of what it might have been like to live on the bridge in the 18th century is provided by Pennant (1813, 447), who claimed that 'nothing but use could preserve the repose of the inmates, who soon grew deaf to the noise of the falling waters, the clamours of watermen, or the frequent shrieks of drowning wretches'.

Medieval finds contemporary with the houses on the bridge

A number of medieval finds contemporary with the houses on the bridge were recovered during dredging in 1824–41 and 1967. They appear to have been either thrown into the river or lost, perhaps by being inadvertently thrown out with rubbish. Objects found include iron axe heads, bronze fish hooks, coins, jettons, a crucifix, three signet rings, lead cloth seals, daggers and knife blades, strap-ends, buckles, forks, garter hooks, horse shoes, iron keys, pins, pottery vessels, and padlocks, spurs and silver spoon bowls (Knight 1834, 600–1; Ward Perkins 1940, 132, 142; Marsden 1971, 12).[30]

On the foreshore near the bridge abutment (G28.1) was found a medieval gold finger ring set with a garnet. The 1967 dredging also produced a medieval finger ring, which had originally contained a gem stone (see Chapter 14.6). Two of the late medieval rings in the British Museum collection were found during the demolition of the bridge in the 1830s. The first ring has an octagonal bezel with a trefoil, on which is engraved *cest mon ure* (it is my destiny). The second ring is an early 16th-century signet ring with the letter 'I' on the bezel. A third ring in the collection was apparently found during dredging around the bridge, and is made of gold and contains a gem engraved with Venus (Dalton 1912, nos 439, 471, 223). One of the many finds from dredging on the site of the bridge is a late medieval stone (Reigate?) corbel; decorated with the head of a man, it is believed to have come from one of the buildings on the bridge, possibly the Stonegate (Fig 72).

Fig 72 A Reigate or Greensand stone corbel decorated with a human head; it was found during dredging on the site of medieval London Bridge and possibly formed part of the medieval Stonegate which collapsed in 1437; length 555mm, height 345mm and width c 310mm (MoL acc no. 36.62; MoLPL)

9.2 The Stonegate

A number of English medieval bridges possessed gatehouses as part of the local town defences, including Old Bedford Bridge which possessed two gatehouses (Simco and McKeague 1997, 33), Bristol (Frome Bridge), Chester, Durham, Newcastle, Norwich (Shepherd 1998, 40–1) and Shrewsbury (Welsh Bridge).[31] The only two bridge gatehouses still to survive in England and Wales are at Warkworth Bridge, Northumberland, and the Monnow Bridge at Monmouth, Gwent (Rowlands 1994, 100).

At London the second pier from the southern end of the bridge accommodated a gatehouse or barbican known as the 'Stonegate tower' or the 'great gate'. Stow referred to it as the 'bridgegate', and suggested that it represented one of the seven principal city gates listed by fitz Stephen in the later 12th century (Stow 1603, i, 42) (Fig 52), although as fitz Stephen does not mention London Bridge the validity of the claim is difficult to assess. It is nevertheless probable that the Stonegate was an original feature of the Colechurch bridge, although its existence is not documented until 1258 (Riley 1863, 42). In the Bridge House accounts for 1381–2 a new 'latten pulley' was purchased for the portcullis there (Harding and Wright 1995, 21). A large portion of the gatehouse collapsed in January 1437 (see Chapter 11.1), some of the collapsed masonry apparently falling as a single mass into the gully between the second and third pier from the southern end. It was never removed, and gave the name 'Rock Lock' to the partly obstructed gully. The Stonegate was rebuilt between 1437 and 1465–6 (CoLRO, BHR 5, fo 107), perhaps as a consequence of damage by fire during the Cade rebellion of 1450 (discussed below), when the houses on the southern end of the bridge were burnt down (Home 1931, 128).

The newly rebuilt Stonegate was burnt down during Fauconberg's abortive attack on the bridge in 1471 (detailed below). Just before the attack, carpenters had hung 'two great leaves upon the gate' which were burnt and were afterwards replaced once more (CoLRO, BHR 5, fos 180v, 184v). The paving of the bridge between the Drawbridge tower and the southern end of the bridge during 1471–2 appears to represent the completion of the rebuilding (CoLRO, BHR 3, fo 199v). In 1472–3 the tower on the west side of the new gatehouse was rented out for the first time. In 1475–6 a 'hanging lock' was purchased for closing the gate (CoLRO, BHR 5, fos 255v, 257).

The Stonegate is shown in some detail in Wyngaerde's view of London Bridge c 1544, when it had a large central gate with the City arms above and flanked by a pair of heraldic beasts, probably lions (Fig 65). On each side of the gate were flanking hexagonal towers, following the exact shape of the bridge pier beneath (Colvin and Foister 1996, 12). The number of buildings adjoining the southern side of the Stonegate by this date suggests that it no longer served a military function.

The Stonegate and the buildings on the southern end of the bridge were destroyed on 8 September 1725 by a fire which had spread northwards from Tooley Street. A survey of the Stonegate made at the time shows it to have consisted of two circular crenellated stone towers with a stair turret, one on each side of the gate. The gatehouse was four storeys high with a pedestrian or postern gate on its east side (Fig 73). In 1728 the damaged gatehouse was rebuilt for the last time, with the gate widened from 11ft (3.35m) to 18ft (5.48m) (Seymour 1734, 49). The rebuilt, crenellated, gatehouse also featured two postern gates for pedestrians and bore the arms of George II (1727–60) (Fig 73; Fig 74). There are a few discrepancies between the two elevations, probably representing differences between the proposed and completed designs. The gatehouse was demolished in 1760 as part of the roadway widening scheme (see Chapter 13.2). The royal arms from the Stonegate were salvaged and set in the facade of the King's Arms public house, established at Newcomen Street, Southwark, in 1787, by which time they had been converted into the arms of George III (Home 1931, 273). When the pub was rebuilt in 1890 the royal arms were retained in the facade of the present building (Fig 75).

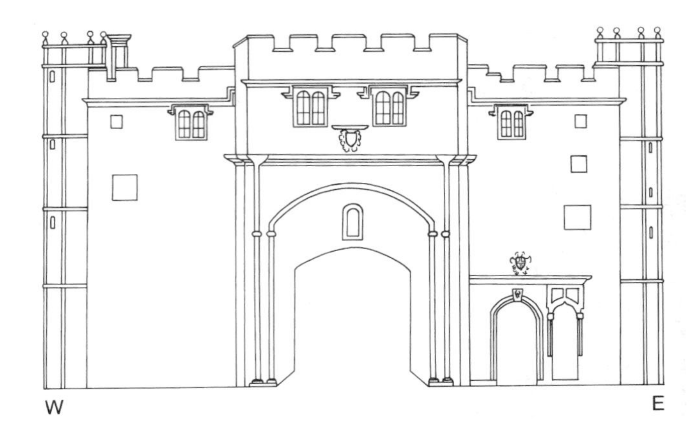

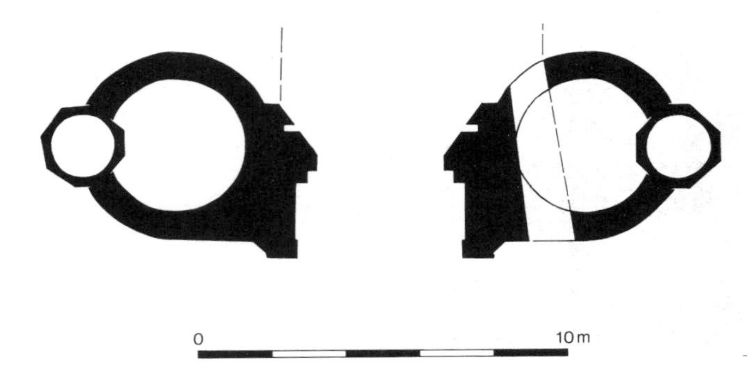

Fig 73 Plan and elevation of pre-1728 Stonegate (CoLRO, CD 203–35)

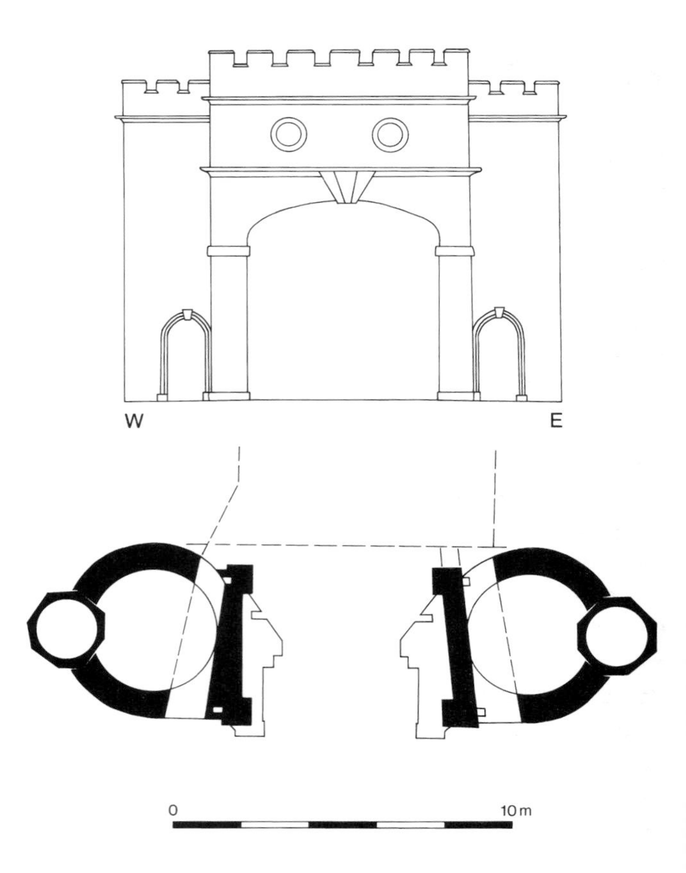

Fig 74 Plan and elevation of post-1730 Stonegate (CoLRO, CD 128Q)

Fig 75 *View of the royal arms of George II (1727–60) from the Stonegate, now reset in the facade of the King's Arms public house, Newcomen St, Southwark; the arms have been altered to those of George III (1760–1820), probably in 1787 (MoL, R Bartkowiak)*

9.3 The Drawbridge Gate

The seventh opening from the southern side was spanned by a drawbridge lowered from a masonry gatehouse which accommodated a portcullis on its southern side (Fig 52; Fig 76). The Drawbridge Gate is believed to have been an original feature of the Colechurch bridge, as it was already in existence by 1258 (Riley 1863, 42). It served as an integral part of the City's defences and allowed large ships to pass upstream. Tolls incurred at the lowering of the drawbridge formed part of the income of the Bridge House until 1475–6, after which the accounts recorded that it could not be raised because the stonework was in need of repair (Harding and Wright 1995, xxi; see Chapter 10.2), a frequent if not continuous state of affairs thereafter.

Between 1405 and the collapse of the Stonegate in 1437 (see above) the Bridge House accounts record frequent building work on the Drawbridge tower, from which it is clear that it was extensively repaired during the early 15th century. Despite all this work, the tower still remained a cause for concern, bridge wardens expressing anxiety in July 1481 about the 'great perils and jeopardies' in which the structure stood, along with the adjoining arches and piers which were all thought to be at risk from the shaking caused by the passage of carts with iron-shod wheels, and the raising and lowering of the drawbridge (CoLRO, JCCC 8, fo 248). In March 1497 a committee of the Court of Aldermen was appointed to view the Drawbridge Gate and consider whether it should remain in its present position or whether the tower should be taken down and rebuilt (CoLRO, REP 1, fo 13). The military importance of the Drawbridge Gate and Stonegate stemmed from the fact they represented London's only civic defences along its entire river frontage. The tidal river, too deep to ford at this point, was about 270m wide during the medieval period, making it a formidable obstacle for any army to cross unless it possessed a lot of ferry boats.

On five recorded occasions the Drawbridge Gate fulfilled a military role. Firstly, on 11 December 1263, Henry III's supporters in the City raised the drawbridge and locked and chained the gates to prevent Simon de Montfort and his baronial army in Southwark crossing the bridge and entering the City. Then Prince Edward's army at Merton and Henry's army at Croydon both converged on Southwark, intending to trap and defeat de Montfort's army (Williams 1963, 223). However, the *populares* of the City seized the bridge and allowed de Montfort's army to cross and evade their pursuers (Home

Fig 76 *View of the Drawbridge Gate in c 1550, looking north; notice the decapitated heads of five traitors displayed on poles (Home 1931, frontisplate)*

1931, 56; Williams 1963, 223). It was probably this action that encouraged Henry, after the defeat and death of de Montfort in 1265 at the Battle of Evesham, to seize the bridge revenues (see Chapter 10). Henry also punished the Londoners for rebelling by confiscating the property of many of de Monfort's supporters and also by arresting numerous people (Williams 1963, 232–3). In October 1265 Thomas Fitz Thomas, the mayor, and 40 leading London citizens travelled to Windsor, under the royal promise of safe conduct, to see the king to try to resolve matters, and were all thrown into prison instead. Mayor Thomas was only released in 1269, when he paid a redemption of £500 (Williams 1963, 232, 241).

On 13 June 1381 Wat Tyler and his Kentish followers, protesting against the poll tax, entered the City from Southwark, and three days of arson, looting, murder and rioting ensued, until Mayor William Walworth mortally wounded Tyler at Smithfield and earned himself a knighthood. It is uncertain why London Bridge was not defended against the rebels. According to the the Anonimale chronicle, Thomas Walsingham's and Froissart's accounts of the revolt, the mayor intended to defend the bridge (Dobson 1970, 156, 188). However, it appears that a combination of threats by Tyler to burn down Southwark and London's other suburbs, if he was not allowed to cross, and mass support for the rebels amongst the common people of London were enough to make the City capitulate and lower the drawbridge and open the Stonegate and admit the rebels (Dobson 1970, 168). Other accounts allege that the surrender of the bridge involved treachery. The jurors' returns to the inquisitions held before the Sheriffs of London in the autumn of 1381, stated that two aldermen – John Horn and Walter Sybyle – conspired with the rebels to discourage resistance within the City and then assisted them to gain access to the City by leaving the Stonegate and a number of other City gates open and undefended (Dobson 1970, 212–18).

On a third occasion the Drawbridge Gate failed to keep insurgents out of the City. On 3 July 1450 Jack Cade's Kentish forces obtained the keys to the doors of the Stonegate and seized the drawbridge before it could be raised, so enabling them to enter the City from Southwark unopposed (Home 1931, 125–6). Jack or John Cade was 'a soldier of fortune' who exploited the unpopularity of the government, led by the feeble-minded Henry VI, by successfully organising a rebellion. Cade's revolt differed from that of 1381 in so far as it was aimed less at the landowners and more at a number of unpopular and tyrannical crown officials (Jacob 1961, 496–7). On 2 July the City of London Common Council considered the matter of defending London and decided to admit Cade's army; one member of the Council was imprisoned for opposing their decision (Lyle 1950, 12). However, Cade's actions after he entered the City soon ensured that he lost public support. He executed Lord Say, Treasurer, in Cheapside after a mob trial (Harvey 1991, 93), and then executed Lord Say's son-in-law, William Crowner, Sherriff of Kent. Cade had the heads of the two unfortunate men displayed on London Bridge. During the night of 5 July forces composed of the Tower of London garrison and Londoners, led by Lord Scales and Captain Matthew Gough recaptured the bridge (Harvey 1991, 95–6). The rebels, who were encamped near the bridge in Southwark, realised that their success depended on controlling the bridge; they immediately counter-attacked and succeeded in recapturing all or most of it, only to be driven back again. Gough's forces appear to have been unable to raise the drawbridge due to damage caused by Cade's men. After heavy fighting a truce was arranged the following day, both sides agreeing not to try to cross the bridge. On 7 July the rebels were then offered a general pardon, so returned home a few days later (Harvey 1991, 97), while Cade was pursued and captured. After execution, the heads of Cade and 23 of his followers were displayed on the Drawbridge tower (see Chapter 9.6). All the houses on the southern end of the bridge were burnt down on this occasion, but, because of a gap in the Bridge House records between 1445 and 1460, the repair of the Drawbridge tower and houses after the rebellion is undocumented.

In May 1471 the Drawbridge Gate was defended by forces under the command of Ralph Joslyn, a former mayor (1464) and sheriff (1458), against Thomas Fauconberg and his rebel army.[32] The drawbridge was closed and three holes cut in it for 'sending out gun shot' (CoLRO, BHR 5, fo 182v). The defenders' weapons included longbows and several cannon. While the rebels attacked the bridge, some of their force crossed the river by boat and attacked Aldgate. Neither attack succeeded, and Joslyn's forces then lowered the drawbridge and counter-attacked, routing the rebels (Home 1931, 135–6).

The situation was potentially very serious as Fauconberg intended to release Henry VI from his imprisonment in the Tower of London and depose Edward IV, who had only regained his throne earlier that year after being victorious at the Battles of Barnet and Tewkesbury (Seward 1997, 60). After the rebellion Fauconberg was offered a free pardon on 10 June 1471 and he joined Richard Duke of Gloucester's army and marched north to help pacify the border region. However, Richard had him beheaded at Middleham and sent his head south to be displayed on London Bridge (Seward 1997, 61).

The last time the Drawbridge tower was instrumental in the City's defence was during Sir Thomas Wyatt's rebellion. Wyatt's rebellion like so many European conflicts in the 16th century was sparked by denominational intolerance. Queen Mary (1553–8), a devout Catholic, was about to marry Philip of Spain, a match that was not popular with her predominantly Protestant subjects. 'The situation was serious and the City might have fallen to the rebels had not Mary, with her accustomed courage, ridden to the Guildhall, where she appealed to the citizens to remain faithful to her' (Lockyer 1964, 125). As Wyatt's forces advanced through Southwark on 3 February 1554, it proved impossible to raise the drawbridge, which was broken down into the Thames (Holinshed *Chronicles*, 1097). According to one account, the lifting ropes had been cut by the rebels, while according to another the tower may have been gutted by fire – although there may have been confusion here with the Stonegate tower (Home 1931, 126–30). At any rate, the rebels occupied Southwark, dug trenches at the Bridge Foot and set two pieces of ordnance against the gate (Holinshed *Chronicles*, 1097; Stow 1603, i, 25–6). Wyatt appears to have been unsure what to do next, so he simply waited in Southwark; presumably he was hoping for support from London, which did not materialise (Fletcher and MacCulloch 1997, 86). After three days of waiting Wyatt, having resolved not to attack the bridge, decided to march his forces westwards in an attempt to cross the Thames at Kingston instead. To try and prevent Wyatt's force crossing Kingston Bridge, a 30ft length of the bridge was broken down.[33] However, Wyatt's forces were able to repair the damage and crossed over, and then marched east towards London. En route at St James Fields Wyatt's forces were attacked by cavalry, but other government troops put up a craven performance and the rebels marched onwards, to reach Ludgate; the City gate here was closed and defended by the London militia. Wyatt, knowing he was now defeated, surrendered (Fletcher and MacCulloch 1997, 86–7). As much of the resistance to Wyatt appears to have been rather half-hearted and dilatory, it suggests that he had considerable public support within London, but not enough to win the City over to his cause.

Of the five occasions that Londoners were called upon to defend their stone bridge, only once, in 1263, were they opposing Crown forces. It is clear that in 1381 and 1450 there was either a degree of sympathy with the rebels, or perhaps a lack of resolve shown by members of the citizen's militia, not to defend the bridge on the first occasion and on the second to allow its capture without a fight. It is probable that treachery played some part in the failure to defend the bridge against the rebels in 1381 according to some contemporary accounts (Dobson 1970, 156, 212–18). In 1450 the common council decided not to defend the City against the rebels, an act which showed civic support for the rebellion (Lyle 1950, 11–12). However, a few days later public opinion had clearly changed as members of the London militia fought alongside royalist troops to recapture London Bridge. In complete contrast during 1471 and 1554 the City of London showed a disciplined resolve to defend its bridge and on both occasions this refusal to support the rebels was the pivotal act which ensured the failure of these ventures. London, as the capital city, was seen by all rebels as the key to England. However, if rebel forces were advancing on the capital from the south-east, then gaining the bridge was the key to London. William Duke of Normandy realised this in the autumn of 1066 and, having failed to capture the wooden bridge, chose to march on London from the north instead, outflanking the bridge he could not capture (see Chapter 5.4).

On Wyngaerde's panorama (*c* 1544), dating from a few years earlier than Wyatt's insurrection, the Drawbridge Gate is shown as a massive masonry structure several storeys high, the elaborate central archway flanked by polygonal turrets (Colvin and Foister 1996, 12) (Fig 65; Fig 73). On top were displayed the decapitated heads of traitors, a practice first recorded at the expense of Sir William Wallace in 1305 (Home 1931, 78). However, after the replacement of the Drawbridge tower with Nonesuch House in 1557, there was no space to display the heads of traitors. Instead they were displayed on top of the adjoining Stonegate. The last person to receive this punishment was Thomas Venner (Home 1931, 231), a leader of the Fifth Monarchists, who in January 1661 threatened to overthrow the government and was executed later in the year.

The demolition of the Drawbridge Gate is recorded as beginning in April 1557 and finishing before August (Welch 1894, 67). No doubt it was the sheer dilapidation of the tower as much as its military value that prompted its final rebuilding between August 1577 and September 1579 (Stow 1603, i, 60). The new tower was known as Nonesuch House, and featured a stone-built ground floor. The three upper storeys and four domed corner towers were all timber-framed and clearly not intended as a fortification (Fig 53; Fig 69). Above the southern arch of the gate were the royal arms of Queen Elizabeth; two sundials were added during the mayoralty of Sir Patience Ward (1680–1) bearing the motto 'Time and tide stay for none' (Seymour 1734, 48). In 1685 the height of the gateway was raised as a road improvement (Seymour 1734, 48). Nonesuch House was demolished in 1757 (Home 1931, 186). The fixed drawbridge was rebuilt in 1672 and 1722 (Seymour 1734, 49), and replaced by a stone arch during the 1757–62 modifications (see Chapter 13.2).

9.4 The chapel of St Thomas the Martyr

On the large central pier of London Bridge stood the chapel of St Thomas the Martyr (Fig 52; Fig 53). Thomas Becket, a Londoner, was Archbishop of Canterbury from 1162 until his murder in 1170, which was carried out at the prompting of Henry II, whom he had previously served as Chancellor. He was immediately hailed as a martyr and canonised in 1173, and his shrine at Canterbury Cathedral rapidly became a popular place of pilgrimage. The popularity of St Thomas is demonstrated by the number of English churches or chapels which were once dedicated to him: he is credited with 80 examples, making him the 22nd most popular saint (Bond 1914, 17). An inventory of articles in the bridge chapel in 1350 mentions an 'image of St Thomas of Canterbury before the altar' (Cal L Bk F, 263). There was also an image of St Thomas 'over the bridge' in the form of a wooden signboard, repainted in 1420–1 (Harding and Wright 1995, 107).

During the 12th century, when the concept of building bridges as a work of piety became popular in England and France, chapels such as St Thomas's at London were evidently regarded as an appropriate means of ensuring divine protection. A number of surviving English medieval bridges still preserve them: the bridges at Bradford-on-Avon (Wiltshire), Cromford (Derbyshire) (Pevsner 1978, 159), Derby (St Mary's at Bridge Gate), St Ives (Cambridgeshire) consecrated in 1426 (Savill and Craven 1999), Rotherham, and Wakefield licensed in 1356 (Taylor 1998) all still possess chapels. The Elvet Bridge at Durham formerly had two chapels, one at each end, but only one survives (Fig 51). Attached to the new (1383–91) bridge at Rochester was a 'bridge chapel' finished in 1392 (Britnell 1994, 49). Exeter Bridge possessed St Edmund's church and a chantry chapel; the salary of the chapel priest was funded by the Exeter Bridge Trust (Brown 1991, 2; Juddery 1991, 23). Fragments of St Edmund's church survive and now form part of the Exe Bridge Park. Bridge chapels were not exclusive to major bridges; the only bridge chapels within the medieval diocese of Worcester were both situated on relatively small bridges at Bewdley and Droitwich (Houghton 1919). The 1485–6 contract for building a timber bridge at Newark (Nottinghamshire) stated that there should be a cross set up within the centre of it (Salzman 1952, 456).

The presence of chapels on medieval bridges would have served as a reminder to travellers that bridge building was pious charitable work, but, for the theologically sophisticated, bridges were symbolic as well as practical structures and emblems of Christian life. For instance, the Pope and his bishops each had the title of *pontifex* (bridge-builder) (Duffy 1992, 367). Bridges were used as images in allegorical literature as a link between heaven and earth. The 15th-century English translation of the work of St Catherine of Siena (c1347–80) known as *The orchard of Syon* devoted five chapters to an elaborate exploration of the notion of Christ as a bridge (Hodgson and Liegey 1966).

Dredging of the riverbed and work on the foreshore in the vicinity of London Bridge has produced numerous lead and pewter pilgrim badges and ampullae of St Thomas and other saints, mainly of 14th- or 15th-century date (Ward Perkins 1940, 256–258). It has been suggested that the chapel of St Thomas on London Bridge could have been the 'starting point of innumerable pilgrimages; and of the many pilgrim signs recovered from the Thames nearby, it is notable how many are those of St Thomas' (Ward Perkins 1940, 257). These badges were small talismans or souvenirs sold at saints' shrines for pilgrims to wear as visible tokens of their devotion (Alexander and Binski 1987, 218–25). 'Their popularity rested on the knowledge that they had been pressed either by the owner or by a shrine keeper against the saintly relics, shrine or image, thereby, it was believed, absorbing some of the shrine's virtue and transforming the badge or ampulla from mere souvenir into a touch relic' (Spencer 1998, 16). The ampullae or phials were filled with water collected from St Thomas's well at the shrine and held in great esteem for its miraculous healing and protective properties. A lead ampulla of probable 13th-century date, found at Toppings Wharf (T8) (Fig 77), depicts the Saint's murder (Spencer 1974). Another find from dredging near the site of the chapel is a lead seal or bulla of Pope Urban VI (1378–89) (Knight 1834, 600).

The highly popular shrine at Canterbury was immortalised

Fig 77 The lead ampulla from the Toppings Wharf erosion deposits (T8); it depicts Becket's murder by two knights clad in mail suits, wearing conical helmets and armed with swords; four knights were implicated in Becket's murder, but normally only one or two are shown on ampullae (Spencer 1974, 114)

in Geoffrey Chaucer's *The Canterbury tales*, the earliest version of which was written in 1387–92 (ibid, 17). The Prologue tells of a group of pilgrims who met at the Tabard Inn, Southwark, and travelled together to St Thomas's shrine. There is no mention of the chapel of St Thomas on the bridge, although there is good reason for supposing that it served as the finishing, as well as starting, place of many pilgrimages to Canterbury. This is suggested by local finds of pilgrim badges, described by Langland in *Piers the ploughman* (1377–9) where, in Book Five, Piers meets a pilgrim who had pinned around his hat many badges and dozens of phials of holy oil as souvenirs of his travels (ibid, 115). The vast numbers of such badges and ampullae recovered from the Thames around London Bridge strongly suggest that these were not casual losses, but were thrown into the river on a regular basis by returning pilgrims; an interpretation reinforced by the discovery of pilgrim badges at many major European river crossings (Spencer 1978, 250). These discoveries suggest that the badges may have been discarded as a thanksgiving for a safe homecoming. It is even possible that an extra badge was purchased at the shrine in order to invoke the saint's protection on the return journey, and then discarded on arrival home as a token of gratitude (Merrifield 1987, 109).

St Thomas's chapel appears to have been built as an original feature of the bridge and Peter of Colechurch, the builder of the stone bridge, was buried there (Brett 1992, 305) (see Chapter 8). In 1265, before Henry III seized the bridge revenues, they were apparently being administered by 'the brethren and chaplains ministering in the chapel of St. Thomas upon the said bridge' (Cal Pat R, 1258–66, 507). This reference implies the existence of a fraternity or guild of citizens to administer the bridge by the mid-13th century. The chapel was maintained and run by the bridge wardens, who paid the chaplains and clerks, and other expenses. The number of chaplains varied widely: in 1381–2 there were at

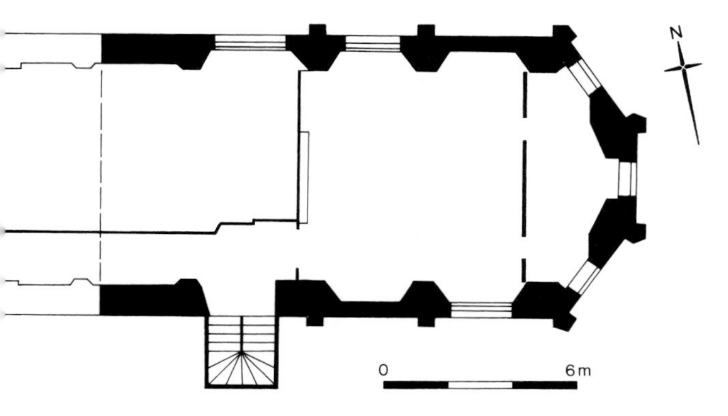

Fig 78 Plan of the bridge chapel undercroft (CoLRO, CD 203.54)

different times three, four, and five chaplains and one clerk; in 1460–1 four chaplains and three clerks; in 1501–2 two chaplains and four clerks; and in 1537–8 three chaplains and six clerks (Harding and Wright 1995, xvi). The chapel services appear to have comprised a daily mass and the seasonal observances of the church, together with placebo, dirige and a requiem mass for the benefactors of the bridge four times a year, in addition to the celebration of the feast of St Thomas the Martyr (29 December) and other saints (ibid, xvi). From the Bridge House rentals a lot is known about the provision of music in the bridge chapel (see below, 9.5).

During 1384–96 the chapel was rebuilt in the perpendicular style (Home 1931, 99–100). One of the leading English exponents of this new architectural style was Henry Yevele (c 1320–1400), a London mason and architect who was elected a bridge warden in 1365 and held the post until 1399 (Harvey 1944, 28). Although it is unlikely that Yevele was closely involved in the rebuilding of the chapel, as from c 1383–91 he was directing the construction of the new bridge at Rochester (Britnell 1994, 44), it is more than likely that he influenced its design. The degree to which Yevele might have influenced the chapel design would have depended partly on the amount of time he spent at Rochester during 1383–96, as well as his other commitments including work on Canterbury Cathedral and Canterbury city walls. As the bridge building at Rochester was presumably seasonal Yevele could have easily spent the winter months working in London or elsewhere. The new chapel had a five-sided, apsidal east end built above an undercroft or lower chapel and a small stair turret in the south-east corner (Colvin and Foister 1996, 12) (Fig 78; Fig 79). It

Fig 79 The bridge chapel undercroft revealed during demolition in 1832, view looking north by E W Cooke; notice the roadway above the remains of the undercroft dating from the 1757–62 modification of the bridge (MoLPL)

appears probable that the chapel had always functioned as a 'double chapel' with altars both on the ground floor and within the undercroft. The accounts for 1397 mention the purchase of white and stained glass for the chapel windows (Welch 1894, 72).

In 1539–40 the dedication of the chapel was changed to St Thomas the Apostle in deference to Henry VIII's condemnation of Becket as a traitor in 1538 (Home 1931, 167). At the same time a Southwark painter was hired to 'deface and mend divers pictures of Thomas Becket in our Lady Chapel' (Welch 1894, 77). In 1543 a representation of the martyrdom of St Thomas Becket was changed into an image of Our Lady (Welch 1894, 77). But the bridge wardens' efforts on behalf of the idolatrous fixtures in the chapel were unavailing. The date of the chapel's closure is uncertain. According to Welch (1894, 77) it was decided in 1548–9 that the chapel should be closed and converted into a house. On 16 May 1549 arrangements were made for the 'divising of the new frame or alteration of the chapel to be made and transposed there where the said late chapel was' (CoLRO, REP 12/1, fo 86), and in 1550 the chapel organ was sold (Welch 1894, 74) (Fig 53). But as Hugh Bowell, chapel clerk since 1543, was employed continuously by the Bridge House wardens until 1551–2 (see section 9.5 below), it appears that either the 1549 alteration was not immediately carried out or that at least part of the chapel (perhaps the undercroft or lower chapel) still remained open. Home suggested that this closure and conversion may not have taken place until 1553, when the building was leased to William Bridger, a grocer (Home 1931, 168).

The remains of the chapel superstructure were demolished in 1757–62. During the demolition of the chapel pier in 1832 (see Chapter 13), the undercroft of the chapel with its groined vault was rediscovered (Fig 79). The chapel walls were built of 'firestone' (a type of Reigate stone from the Godstone area of Surrey) with Caen stone mouldings. An adult burial of 'middle stature', found in a stone-lined tomb under the floor, may well have been the remains of Peter of Colechurch (Knight 1832, 98). There is a close parallel here with the remains of St Bénézet, who is credited with starting the construction of the stone bridge over the Rhône at Avignon in *c* 1177 or 1178 and who, when the bridge was completed a few years after his death in 1184, was reburied in the chapel on the bridge there (Farmer 1992, 48).

The fate of the bones from the London chapel is uncertain: according to one source they were thrown into the river by the workmen during the demolition of the medieval bridge in 1832, although according to another the lower jaw and three other bones were sold as the remains as Peter of Colechurch in 1833, as part of the effects of the late William Knight, a keen antiquary and Clerk of the Works during the bridge rebuilding. It was claimed at the time that the bones were not from Colechurch's tomb but had been found in Southwark.[34] A small wooden casket now on display in the Museum of London contains five fragments of bone, reputed to be Peter's remains. Their examination in 1997 revealed that only one was human (a fragment of an adult humerus), the others being from a goose and a cow (see Chapter 14.10). It seems probable that the claim that these bones were actually those of Peter of Colechurch was fraudulent. As the relics of saints or famous people are saleable, there is a long tradition of fraudulent claims (Bentley 1985, 89–100, 170–5).

It is probable that two medieval wooden statues recovered from the Thames during dredging on the site of the medieval bridge in 1824–41 were originally part of the chapel fittings (Fig 80; Fig 81), presumably thrown into the river during the mid-16th century and thus representing rare survivals of pre-Reformation English religious art (Smith 1854, 111). The statue of God the Father, however, may have been concealed within one of the starlings during the Reformation rather than thrown into the river (see Chapter 14.7). The concealment or disposal of the two statues perhaps took place during

Fig 80 Wooden statue of a monk, found during dredging on the site of the medieval bridge during the early 19th century; height 480mm (British Museum)

Fig 81 *Wooden statue of God the Father, found during dredging on the site of the medieval bridge during the early 19th century; height 590mm (Sudeley Castle, photograph by K Upton)*

September 1547, when the royal visitation of the City of London churches was accompanied by assiduous image and stained glass window breaking (Duffy 1992, 453).

The first statue, dating from c 1480–1550, was a standing male figure (Fig 80), the dress and tonsure indicating that he was a Benedictine monk, possibly St Benedict himself (see Chapter 14.7). In the breast was a small orifice intended to hold a relic or the Easter Day sacrament. In the second statue God the Father is personified as a crowned and seated king or emperor (Fig 81). Both hands were missing in 1847 and have since been replaced, with the addition of an orb in the right hand, although the figure probably originally held either an orb and sceptre or a representation of Christ on the cross surmounted by a dove symbolising the Holy Spirit. This statue dates from c 1450–1500 (see Chapter 14.7). Since the mid-19th century it has been in the possession of the Dent-Brocklehurst family at Sudeley Castle, Gloucestershire, and can be seen there today by visitors to the castle. Both wooden statues from the Thames would originally have been painted, although neither retains any trace of paint or gilding today.

It is clear from the Bridge House accounts that the chapel was well furnished with statues. In 1420–1 six 'images' in the chapel are recorded as having been cleaned and painted (Harding and Wright 1995, 102). A statue of the Virgin in the chapel was repainted in 1426. The location of these statues is not stated, but in 1531 the crucifix, along with figures of Mary and John and two angels on the rood loft, are all recorded as gilded (Welch 1894, 72, 76). In 1489 the accounts mention the construction of 'two stone walls with two images in tabernacles [probably canopied niches] thereupon standing in the void room on the north side of the said chapel' (Welch 1894, 72).

9.5 Music at the chapel of St Thomas, London Bridge, in the later Middle Ages

Richard Lloyd

One of the most significant influences on the rise of music in the late medieval church was purgatory (the place in which the souls of those who die in a state of grace are believed to undergo a limited period of suffering to expiate their venial sins before they go to heaven). It was purgatory that inspired many benefactions to parish churches, in the belief that certain provisions could be undertaken to reduce the amount of time spent there (Duffy 1992, 338–76). The very rich endowed chantries with full-time chaplains to reduce their time in purgatory, while the less wealthy endowed obits or annual remembrance services for the same reason. Often these remembrance services were sung by a chaplain and many London parishes had a musical chantry priest, and such singing priests were employed too at the chapel on London Bridge. From an early date the bridge chapel attracted such endowments. John Hatfield (died 1363) left a tenement in Gracechurch Street for the maintenance of a chantry in the church of St Benet Gracechurch Street. In the event of the parish church not maintaining this chantry, it was to pass to the bridge chapel (Sharpe 1890, 79, 81). Hatfield's obit first appears in the Bridge House rentals of 1392–3, when the chaplain was paid 2s for 'celebrating for the soul of John Hatfeld' (CoLRO, CH 91, 125). It is perhaps significant that the chapel commissioned two antiphonals (books of liturgical material such as psalms) in 1396–7, which were enlarged the following year. It is likely that these books contained the antiphons customarily performed as part of the *Officium defunctorum* (the Office of the dead), and the Requiem and the Mass of the dead. Also in 1396–7 the chapel purchased a missal (prayer book); whether or not it was notated or used at funeral obsequies is not certain. However, it does appear that the transfer of Hatfield's chantry from St Benedict's to the bridge chapel may have prompted the chapel to purchase two antiphonals and a missal, and to start the practice of singing mortuary services (Lloyd 1995).

The first volume of the Bridge House rentals (1404–21) contains little detail of the musical activities of the chapel, other than noting its total annual expenditure. It is not until the start of the third volume in 1460 that a more detailed picture of the employment of the singing clerks, in addition to the chantry priests, begins to emerge. In that year two clerks, John Beller and William Holford, were both paid for 52 weeks' service and a third clerk, William Clotton, for 31 weeks. The following year six clerks were retained by the chapel, but not all were fully employed. It seems that only two clerks were in residence at any one time. During 1461 three clerks were employed for the whole year and one for four weeks. This pattern of employment continued until 1470–1. The staff turnover was high: during the course of twelve years from Michaelmas 1459 to 1471, some 21 clerks were employed by the chapel (Lloyd 1995, appendix 1). The number of clerks retained by the chapel remained at four until 1508–9, when it was increased to five. The number of clerks remained at this level until 1538–9, when only John Feres was retained as clerk. In 1543–4 Feres was replaced by Hugh Busewell, who remained sole clerk until 1551–2. The chapel was closed in *c* 1553 (CoLRO, BHR rolls 12, 14 and 15, vols 3, 4, 6 and 7).

The Bridge House rentals list the chapel's purchase of plainsong and, later, polyphony. The chapel's first book of polyphonic masses and anthems was purchased in 1473–4 and at least eight more such books were purchased over the course of the next 40 years.

Despite the number of endowments being made to, or claimed by, the bridge chapel, it was not until 1478–9 that priests and clerks were recorded as 'syngyng placebo diriges and masses of Requiem for ale the benefactors of the Bridge iiij tymes this yere'. In fact it is difficult to discern from the Bridge House rentals the number of chantry priests involved in the music of the chapel. It must be understood that the bridge chapel was an extra-parochial institution serving those who crossed the bridge. It was not a parish church and those who lived on the bridge were served by the church of St Magnus the Martyr, situated at its north end. At this church a number of chantries were founded by bridge residents (Kitching 1980, 15–16).

Many of the chantries operated by the bridge chapel had only been claimed when the first choice of the donor, normally the parish church, defaulted on the terms of the endowment. The usual stipulation in such endowments was for musical chantry priests, and this would have certainly affected the size of the singing body in the chapel. However, examination of the chantry mechanism of the bridge chapel demonstrates a further distinction which may have had significant implications for the development of music in the chapel. Chantry endowments were almost always made to the donor's own parish church. From the 15th century onwards bequests only reached the bridge chapel when the parishes defaulted. Therefore, it appears that the development of music at the bridge chapel was critically dependent upon the default of endowments in first-choice parishes.

9.6 Pageantry on London Bridge

Vanessa Harding

As a link with the outside world, the bridge had a major role to play in ceremonies that presented the interaction of the City and the monarchy, but it was little used in civic ceremonies such as the midsummer watch or the Lord Mayor's riding, since these focused on the internal spaces of the City. Its liminality also made it the appropriate location for the display of executed traitors' heads, a public statement of the state's victory over treason, and of the City's loyalty.

English kings, like many other European princes, used the urban landscape as a backdrop for pageants and processions of national importance. Most medieval and early modern monarchs made at least one ceremonial entry into the City of London; one of the complaints against the Stuarts was that they neglected these rituals. There were obviously several coronation entries, but royal weddings (Katherine of France, Margaret of Anjou and Katherine of Aragon) and visits of foreign princes (the Emperor Sigismund, the Emperor Charles V, the Queen Mother Marie de Medici) also provided splendid occasions (Anglo 1962, 1963 and 1969; Bergeron 1971; Kipling 1982; Smuts 1989). The entry into the City via the bridge was also used for more overt political purposes: the victorious return of Henry V after Agincourt was marked by a procession through London that both elevated the monarchy and complimented the City, whose supply of money and materials had been crucial in the Agincourt campaign of 1415, while London's quarrel with Richard II in 1392 was in some sense healed by the staging of a sumptuous 'reconciliation' pageant that figured Richard II as Christ and London as the new Jerusalem (Taylor and Roskell 1975, 101–9; Kipling 1998).

We owe much of our information on royal entries to the London chroniclers. From the later 14th century, it is clear that royal entries to the City were normally following a fixed route, that included entry to the City over the bridge, and a traverse of the main streets north and then westwards via Gracechurch Street, Cornhill and Cheapside, to St Paul's Cathedral. On the processional route, there were a number of stages or locations at which a pageant or *tableau vivant* would be presented: these included the conduits and crosses in Cheapside, but also the Drawbridge Gate and Stonegate or barbican on the bridge (Chapter 9.2 and 9.3). Occasionally there was an entry or procession starting from the Tower of London, but for the most part it is clear that the crossing of the bridge was an essential part of the proceedings. The structure of the bridge supplied three appropriate spaces: before the Southwark Bridge Foot, where the wide street of the market narrowed into the bridge roadway; at the Stonegate itself, where the tall gatehouse towers offered vantage points for figures to stand and a large frontage for displaying decorations; and at the Drawbridge Gate, another tower and open space. The bridge itself was evidently thronged with people during these events: nine people, including the prior of Tiptree, were crushed to death on the bridge during the procession for the marriage of Richard II and Isabel of France in 1396 (Thomas and Thornley 1983, 47).

The processions as a whole were carefully scripted, with each pageant or tableau conveying its own message, and culminated in a major pageant at the west end of Cheapside by St Paul's. The bridge pageants, therefore, could set the tone for each event; in some cases they gave the entrant a sketch of what was to come. Thus in 1432 the young King Henry VI was greeted on the bridge by three empresses, alias Nature, Fortune and Grace, who gave him the gifts of grace and kingship, a recurrent theme in all of the entrance pageants. A second theme of the entry, the union of the crowns of England and France in Henry's person, was also initiated on the bridge with the arms of England and France displayed together (Thomas and Thornley 1983, 156–70). As the first of the pageants seen, the ones on the bridge may have been especially impressive; on the other hand, there was a need to keep the procession going, and Katherine of Aragon was apparently hurried on from the bridge before everything had been said, so that she could get on to the next one at Gracechurch Street in time (Anglo 1963).

Some of the bridge pageants seem to have had a more martial character than those at some other locations. There are references to giants there from at least 1413; there were figures of lions and giants for Henry V in 1415 and again for his entry with Katherine of France in 1421; there was a giant armed as a champion against the king's enemies for Henry VI in 1432; even the wedding entry of Katherine of Aragon in 1501 was greeted by Hercules and Samson (Anglo 1963; Harding and Wright 1995, 82–3; Kipling 1998; Taylor and Roskell 1975, 103–5: Thomas and Thornley 1983, 156–70). The entry of Henry V after Agincourt was understandably very military in character, with a figure of St George on the bridge, and much play was made of his arms, which were very similar to those of the City (Taylor and Roskell 1975, 105–9).

The symbolism of the open gates and welcoming crowds was always important, but it must have been especially pointed when the gates had been recently closed against rebels or invaders. Equally striking was the use of the Drawbridge Gate to display traitors' heads and quarters; Henry V on his return from Agincourt must have passed under the heads of the Oldcastle rebels and the Cambridge conspirators. After the collapse of Cade's rebellion (Chapter 9.3), contemporaries must have appreciated the grisly irony of displaying the executed leader's head on the bridge that his rebel forces had captured. Twenty-three other rebels were similar treated, and given a fairly generous interpretation of the statute of treason – invoked for offences such as coining – there can have been few times when the bridge was not so adorned.

9.7 The corn mills and waterworks at London Bridge

The water flow between the starlings on a number of medieval bridges was utilised by undershot waterwheels as a source of energy: a corn mill on the Grand Pont in Paris is documented as early as 1070 (Boyer 1976, 75). In 1497 a Flemish hydraulic pump in brass was purchased by the Bridge House and installed on the bridge (Welch 1894, 86), presumably to extract drinking water for supplying people living on or near the bridge. It was only in 1580, however, that the first waterwheel on London Bridge was built at the northern end, on the western side of the starlings, by the Dutchman or German, Peter Morris. The intention was to power a pump for extracting water from the river, channelling it into a network of lead pipes and conveying it to the Leadenhall area. The scheme proved so successful that in 1582 a second arch was leased for the installation of another waterwheel (Home 1931, 193–4; Stow 1603, i, 188). The tower which housed the waterworks mechanism is shown on Norden's view of the bridge *c* 1600 (Fig 53; Fig 67). Both waterwheels were destroyed by fire in 1633 and 1666 (Hulme and Jenkins 1895, 261–3). The destruction of the waterworks during the first day of the Great Fire in 1666 was a major misfortune, as this pumping engine was one of the main sources of water in the City, and was the major source of water in the Pudding Lane area where the fire started (Porter 1996, 37). By 1745 'London Bridge Waterworks' operated five waterwheels powering 16 'engines' (pumps for raising water) set in the channels between the first, second and fourth arches from the northern end of the bridge (Fig 53). These wheels were powered continuously by both flood and ebb tides (Dearne 1745, 4–5) and were protected by wooden barriers from damage by driftwood or boats. Naturally, the erection of waterwheels would have increased the partial dam effect the sheer bulk of the bridge had already created (see Chapter 13.3).

With the construction of a new and wide central arch in 1757–62 the head of water on the upstream side of the bridge was reduced, diminishing the efficiency of the wheels. To remedy this John Smeaton suggested in 1763 that rubble

Fig 82 The corn mills on the starlings at the south-west corner of London Bridge, c 1600; view of the Living Bridges exhibition model, looking south (Royal Academy; photograph by A Chopping)

should be dumped in the channel under the central arch to act as a weir and improve the head of water (Smeaton 1763, 461–542). In 1767–8, however, Smeaton was required to design and install a new wheel and pump for the fifth arch from the northern end of the bridge, leased by the waterworks company in 1767, to compensate for the loss of power (Moher 1987, 62). In 1786–96 the waterwheels and pumps of the waterworks were rebuilt under the direction of John Foulds, millwright and engineer, large quantities of oak, elm, hornbeam and other water-resistant woods being purchased for the new wheels. In 1786 a steam engine was installed to power an atmospheric pump that provided stand-by power while the wheels were rebuilt (ibid, 65).

During the late 16th century waterwheels were also fitted under the two southernmost arches of the bridge, again on the upstream side (Home 1931, 198) (Fig 82). These initially powered corn mills but by the 18th century were pumping water to supply Southwark. In 1767 the waterworks company was granted the right to site a wheel in the 'Borough Wheel Lock', the channel under the second arch from the southern end of the bridge (Anon 1767). By 1789 the 'Borough Waterworks' had a steam engine as a stand-by power supply. In 1792 the wheel was rebuilt by John Foulds (Moher 1987, 65). All the wheels were removed from the bridge in 1822 when the two waterworks were demolished.

9.8 London Bridge and the Thames fisheries

Until the early 19th century the Thames teemed with fish, including salmon, and fishing was one of the activities carried out on the medieval bridge. In 1381–2 the bridge wardens received a rent of 20s from the rent of the 'fishery under London Bridge', presumably an allusion to fish trapped or netted within the channels between the starlings. In 1461–2 a fine of 12d was exacted for having a 'fishery beside the staddles [starlings] of the bridge contrary to the order made' (Harding and Wright 1995, 1, 122). The sheer bulk of the piers and starlings of the medieval bridge acted as a partial dam (see Chapter 13.3), restricting the tidal head and increasing the salinity further downstream. This would have had a considerable impact on the distribution of fish, and those, like barbel, that could not tolerate salt water were not found below the bridge. Upstream, freshwater species such as barbel, roach and dace were found in plenty (Wheeler 1979, 67). During the early 19th century a toxic combination of raw sewage and industrial pollution poisoned the Thames estuary, Thames salmon becoming extinct in 1833 and the entire tidal river being described by 1957 as 'virtually lifeless' (Wheeler 1979, 89). Since then due to the proper treatment of sewage, and the effects of the Environment Agency and its predecessors, the water quality has greatly improved and the Thames is now home to 115 species of fish – including salmon, which returned to the river in 1974 (Environment Agency 1998). In July 2000 nine sea lampreys were found in the Thames for the first time within living memory. All the lampreys had died after spawning (Nuttal 2000), so hopefully a breeding population of this popular medieval delicacy has now been re-established in the Thames.

9.9 Conclusion: the importance of London Bridge

Vanessa Harding

London Bridge was, therefore, one of the most important structures and spaces in medieval and early modern London, rich in historical events, contemporary activities and symbolism. It was surely the biggest and most successful civic enterprise in the Middle Ages, a striking architectural and technical achievement. It was home to a thriving residential and working population, and a feature in the daily lives of many more whose business brought them across it; but it was also the scene of momentous historical events. Gaining the bridge crossing was vital to the success of protesting crowds and rebel armies, and it was an integral part of rituals of acknowledgement and welcome; it was also, evidently, a significant moment for returning pilgrims. Part of its importance lay in the fact that it was both connection and barrier. However, unlike the bridges of Paris, let alone that of Venice, London Bridge was not an arterial link between equal parts of the same urban entity. Much the greater part of London's population, commercial activity and significance was concentrated in the City north of the river, and Southwark on the south bank was only a minor and satellite settlement. In 1377 the City's recorded population, at 27,000, greatly exceeded that of Southwark at c 1000 (Benton 1978, 104). In the mid-16th century, late medieval growth in the suburb had been more rapid than in the City, but Southwark's population was still only about 8000, while that of the City north of the river was over 50,000. By 1630, Southwark's population may have increased to c 20,000, but that of the City north of the river was over 120,000, while the spreading eastern and western suburbs may have contained nearly as many again (Harding 1990, 112, 115–18). Only in the 18th century, and still more in the 19th, did south London's population really expand, and this was of course tied into the ending of London Bridge's monopoly of the river crossing, with the building of several more bridges and new road and rail systems.

The City and Southwark were unequal in other ways. The bridge marked a transition: to cross it was to cross a number of important boundaries. Southwark was a separate jurisdiction from the City of London, in a different county and diocese, once a borough in its own right, and subsequently a congeries of special jurisdictions and manorial rights. It was subject to some control from the City from 1327, and in theory full control was obtained in 1550, but the two were never integrated (Johnson 1969). The relationship was unequal but symbiotic: Southwark's economy was dependent on and complementary to, that of the City, in that it housed activities and individuals not welcome in the latter. Southwark accommodated noisome trades and foreign craftsmen, and became known for its brothels and theatres. It also offered space for institutions found in other London suburbs such as hospitals, prisons and the London homes of nobles and ecclesiastics. It had very little of the mercantile, professional and high-class retail activities that characterised the City (Carlin 1996).

The bridge was obviously fundamental to this symbiosis, insofar as Southwark residents needed to sell their wares in the City's shops and markets, and City dwellers needed to cross to Southwark to partake of facilities offered there. It was also London's gateway to the southern road network, including the roads to the channel ports. Stane Street and Watling Street converged on Southwark and the Bridge Foot, bringing with them traffic from radiating roads through Kent and Surrey (Fig 16), and incidentally encouraging the development of another of Southwark's characteristic industries, innkeeping and victualling. However, the bridge was only one, and not the most important, of the gates through which the City's supplies flowed. Recent work on the grain supply of medieval London suggests that lines of communication other than London Bridge were of greater significance: most of London's food probably came either overland to the northern gates of the City, or up and down the Thames to the riverside markets of Queenhithe and Billingsgate (Campbell et al 1993).

As well as its importance as a connection, the bridge also had a real impact as a barrier, both when the crossing was denied, and in other ways. It held back the natural flow of the river and altered the tidal flow. This may well have helped the use of the river above the bridge, making it easier and safer for small boats; a less vigorous tidal regime probably made it easier to build out the upstream waterfront as well, a key feature of the medieval City's development. On the other hand, the dam effect, and perhaps the low salinity of the river above the bridge, must have contributed to the freezing of the river. The bridge was a major impediment to east–west communications by river, because of the fall in levels and the force of the current. It limited the access of ocean-going shipping in the lower Thames to downstream of the bridge and this restriction, therefore, really determined the eastwards shift of maritime and port-related activities from the 14th century onwards.

The bridge was an important icon of the City, and of civic pride and enterprise, and is one of the most obvious and recognisable visual images of London between the 13th and the 17th centuries – far more so than St Paul's or the

Guildhall. The London chronicles demonstrate that news of the bridge was seen as news worth recording, and praise of London Bridge was a common theme in early modern descriptions of London and of England. The two main points of the numerous encomia, repeated to the point of plagiarism, are that it was lined with fine houses, making a street, and that it was all thought to be very beautiful (Manley 1986, 30–2; Groos 1981, 175). The importance of the structures on the bridge has already been examined, and something of their architectural character and appearance indicated. However, given the vicissitudes of the bridge's structural history, and the best known 'fact' or folklore about London Bridge, there is some irony in the comment of a 17th-century writer that it was 'the admirablest monument and firmest erected structure of that kind in the universe' (Manley 1986, 45).

10

The maintenance and repair of London Bridge

Tony Dyson, Vanessa Harding and Bruce Watson, with a contribution by Laura Wright

10.1 The Bridge House archive

Vanessa Harding

The Bridge House records are the muniments of title and administrative documents of London Bridge as a major civic enterprise. From an early date the bridge was endowed with lands and rents in and near London as a work of charity, to provide an income from which the structure could be maintained and repaired, and the chapel's activities supported. Muniments of title date from the 12th century, rentals and detailed accounts for the management of the Bridge House estate from the mid-14th century, and very full week-by-week accounts of building works from the late 14th century. From the 16th century we have further records about materials and supplies, and the series continues into the 19th century (Deadman and Scudder 1994, 22–3; Harding and Wright 1995; Keene and Harding 1985, 9–11).

The particular historical value of the Bridge House accounts lies in their combination of scale, detail and almost unbroken continuity from the later Middle Ages. They compare favourably with those of such enterprises as Rochester Bridge, where accounts were kept in roll form and the series is incomplete (Becker 1930), or the almost complete series of 345 account rolls for Exeter Bridge (Juddery 1991).

The London enterprise was substantial, with an annual turnover of income and expenditure rising from some £600 in the late 14th century to over £1500 in the mid-16th century, and employing a significant workforce for most of the year. Although the format and, to some extent the quality, of the records varied in the later Middle Ages, for many years there are detailed week-by-week accounts of expenditure, and their continuity over a long period allows important changes in the focus and scale of the bridge's activity to be charted. The records illustrate the changing property and labour markets of later medieval and early modern London and yield an immense amount of data on building techniques, materials, wages and other work practices (Harding 1995; Harding and Wright 1995). No other aspect of London's civic administration is so well documented for this period, so the records also illuminate the activity of local government, and the diligence and reliability of the municipality and its officers.

The documents are very well preserved, kept among the muniments of the City of London, and appear to have suffered no significant loss or depredation (unlike the City's Freedom Registers or the accounts of the Chamber). It is clear that the series of London Bridgemasters' account rolls for the late 14th century started some years before the first surviving roll for 1381–2, and it is possible that the earlier rolls, if they were stored at the Bridge House yard, were destroyed in the Peasants' Revolt of 1381 when much of Southwark was burnt down. However, there is no independent evidence to support this suggestion (Harding and Wright 1995, xxv). Once the accounts were kept in book form from the early 15th century (in the series known as 'Rentals') the sequence is almost complete, apart from a period of poor administration in the middle of the 15th century (Harding and Wright 1995, xii, xxvi). The visual, artistic and palaeographical quality of the main record series has also been noted (Christianson 1987).

The language of the medieval Bridge House estate archive

Laura Wright

The Bridge House estate archive is linguistically very valuable for two reasons. Firstly, its entries are minutely detailed and, secondly, because the archive runs uninterruptedly for many centuries, it enables us to see how the language changed and developed over time. The minute detail gives us a knowledge of the names of all sorts of artefacts, ranging from the machines that were used for pile driving to the tinfoil that was used to make the angels' wings of the figures used in the ceremonial processions on the bridge (Harding and Wright 1995; Wright 1996).

Before the accounting year Michaelmas (29 September) 1479 to Michaelmas 1480, the Bridge House estate accounts were written in medieval Latin. This form of Latin was not a fixed or uniform language, in the way classical Latin tends to be taught, but one that varied depending on where it was written and for what purpose. The Bridge House accounts (the rentals and the weekly payments books) were written in a medieval Latin that incorporated a large number of Anglo-Norman and Middle English words. This was not due to the Bridge House clerks' ignorance of Latin, but the usual English way of keeping accounts at this period. The English and French elements were confined to certain parts of the language (nouns, verb stems and adjectives). However, there is one exception to this rule: the entry just before 1479–80, when other parts of the language (articles and prepositions) also appeared in English. After 1479–80 the accounts were produced entirely in English.

The vocabulary of the Bridge House estate archive contains information on a whole range of topics, including the maintenance of the bridge, construction and maintenance of property on the estate. For instance, the various machines or piling engines used to repair the starlings and piers were called 'the five man beetle', the 'gibbet ram', the 'hand ram', the 'running ram' and the 'wilkin'. The generic term for these was not a machine but a 'gin'.

These piling engines 'consisted of an apparatus in the nature of sheer-legs, carrying a pulley by means of which a heavy block of wood or iron, the ram proper, could be hoisted and let fall on the head of the pile' (Salzman 1952, 86). The five man beetle was used for driving short piles and was worked by five men. The gibbet ram was used to drive long piles around the edge of the starlings, deep into the riverbed. According to the accounts of 1480–1, 22 labourers worked 'the gibbet ram of brass'. There was a 'greater' and 'lesser gibbet ram' and in 1475–6 two labourers were paid 4d a tide (a penny extra) to act as 'hookholders' to direct the ram. The name gibbet is probably due to the fact it had a 'sword' (blade or ram), which was greased with animal fat or tallow. It took ten labourers the time of one tide to dismantle it and convey it to the nearby Bridge House yard. In 1437–8 a smith was paid for supplying 'pileshoes, cramps, hooks, bolts' and, again, 'cramps' for the running ram. Horse hides were purchased to make harnesses for the gins, and wooden poles were purchased to act as sheer legs. This level of detail allows us to comprehend the complexities and the cost of maintaining medieval London Bridge.

10.2 Maintaining London Bridge – gifts, tolls and rents

Vanessa Harding and Bruce Watson

London Bridge was maintained by funds from a number of sources including the rents from the houses on the bridge, and gifts and bequests from pious Londoners. These bequests included money and numerous properties, amounting to a sizeable landed estate which helped provide the regular income needed to finance the endless cycle of bridge maintenance. The most substantial source of income for the repairs and maintenance of the bridge was the landed estate, which it had acquired in the main by the mid-14th century. The estate included various tenements in the City of London, including an important concentration in Old Change and Paternoster Row; quit-rents (annual rents charged on property owned and occupied by others), similarly acquired, were also widespread. The actual houses on the bridge itself made an important contribution to the rental income. Further afield there was property in Southwark as well as water mills in Lewisham and Stratford. At the latter site there were two mills, one for grinding corn and the other for fulling. In 1381–2 the fulling mill at Stratford was rebuilt at a cost of £50 (Harding and Wright 1995, 29).

Over the course of the later Middle Ages the value of the estate increased significantly, although only a few new properties were acquired, some of them by default or reversion (Chapter 9.5). The rent actually received at the end of the 14th century was probably around £400, while that in 1537–8 was over £800. Rents in the City of London as a whole may have been declining (Keene and Harding 1985, 20), so this improvement was partly due to good fortune and the advantageous location of important elements of the estate. The houses on the bridge and in the periphery of London accounted for much of the increase in value before the mid-16th century, although other parts of the City estate were in trouble. It was also due to good management, however, including the Bridge House's investment and policy of keeping properties in repair and lettable when the market was sluggish. The quit-rents, about which the bridge wardens could do little, declined by about one third in value between the end of the 14th century and the mid-16th century (Harding and Wright 1995, xviii).

In the second half of the 16th century, the property market in the City of London recovered, and in the 17th century, at least until the Civil War and the Great Fire, property values rose rapidly (Keene and Harding 1985, 20). Most of the Bridge House estate, like that of most institutional landowners, was let leasehold, with the principal return to the landlord coming in the entry fine at the start of the lease, rather than in a variable rent. The landlord's profit was further increased by passing on the cost and responsibility for repairs to the leaseholder. For instance, a large house in Cheapside, rebuilt by the Bridge House in 1536–8, was let at £13 6s 8d rent from 1538. In 1544 the tenant bought a new 21-year lease at the old rent for a fine of £20, but in 1595 the fine paid for a new lease of the same term and rent was £200 – a tenfold increase. In 1633 a new lease of *c* 31 years was sold for £500 (Keene and Harding 1987, property no. 11/1). This property, like much of the Bridge House estate, was destroyed in the Great Fire of 1666. The Bridge House, like many institutional landlords, preferred to extend the length of leases and reduce or remit the rent for a period, in return for the tenant's bearing the cost of the rebuilding.

The landed estate made up a large proportion of the Bridge House's total income in the later Middle Ages, rising from about 75% in 1381–2 to 87% in 1537–8 (Harding and Wright 1995, xvii), and still more thereafter. However, this increase was partly due to actual decline in other sources of income, including the rents from the Stocks market, the tolls from carts passing over the bridge and dues from ships passing under the bridge. The Stocks market was built during the late 13th century on the initiative of mayor Henry le Waleys, in the centre of the City of London (Stow 1603, i, 225). It was principally a meat and fish market, and the rental from stalls and selling places there was granted to the Bridge House to help maintain London Bridge. At the end of the 14th century these rents came to over £100 annually. However, by the early 15th century the rents had declined sharply despite the rebuildings of the market during 1406–11. By the end of the 15th century the market rents were farmed or leased for a total of £56 19s 10d, at which they remained until the early 16th century, although they may have improved by the 1540s (Archer et al 1988, 33, 88–9; Harding and Wright 1995, xx). The market place was enlarged in 1668 after the Great Fire and in 1737 it was demolished to make way for Mansion House.

The Bridge House estate also received a variable amount of income each year in the form of tolls paid by some carts passing over the bridge. Carts owned by citizens of London were exempt from tolls and members of other communities claimed exemption too. When, in 1280–1, it was suddenly realised that owing to years of neglect the bridge was in a ruinous condition (see Chapter 11.1), tolls for travel from Southwark to London or vice versa were charged at the rate of a farthing for every pedestrian, a penny for every horseman and one halfpenny for every pack carried on a horse (Home 1931, 64–5); all wheeled vehicles were presumably banned from the bridge until it was repaired. The customs of the Bridge House, compiled in *c* 1266, list a range of tolls in money and in kind, mostly on the bringers of fish for sale but also on packhorses and horse-drawn carts (Riley 1861, 205–7), but the Bridge House accounts from the 14th century list only money payments, probably at the rates of 2d per cart and 1d per ship passing under the bridge. The receipt from cart tolls declined from £24 8s 5d in 1381–2 to £7 6s in 1420–1, which must reflect either a decrease in traffic or an increase in vehicles claiming exemptions as they were owned by Londoners. An ordinance of the 1420s banned iron-shod carts from using the bridge on account of the damage they were causing to the fabric, and they were similarly prohibited in 1481. By the end of the 15th century the cart toll was farmed out for £23 a year (Harding and Wright 1995, xxi). At a rate of 2d per cart the figure of £23 per annum would have been raised by collecting tolls from less than ten carts per day, a very low figure (Harrison 1996, 280). In contrast, at Rochester Bridge during the 15th century, tolls raised annually a figure varying between 1s and £6, compared with an average income of *c* £107 from rents (Yates and Gibson 1994, 314).

Other sources of income included tolls from ships passing through the drawbridge. The toll on passing ships rose from a probable 1d in the late 13th century, to 2d by 1460 and to 6d by 1463. However, in 1481 the bridge wardens petitioned that the drawbridge should only be raised for the defence of the City and the receipts from ship tolls were last recorded in the 1475–6 accounts (Harding and Wright 1995, xx–xxi). In conclusion it appears that tolls only contributed a small proportion of the income of the Bridge House estate.

10.3 The Bridge House workforce

Vanessa Harding

The Bridge House estate employed a large workforce, partly for the essential repair of the bridge structure, partly to repair and maintain the estate on whose rents the enterprise depended. As the labour market changed in the later Middle Ages, employment and work practices evolved, but there was still a need for a core team of masons, carpenters, 'tidemen' (who worked on the bridge starlings), and general labourers. There was also a need for a number of building craftsmen such as sawyers, daubers, tilers, bricklayers, plumbers, paviours and glaziers. In order to bring in and remove materials from their Southwark yard (see 10.5 below), the bridge wardens for a time kept a team of carthorses, and later they hired carters. They also had at least two boats and at times may have kept up to seven (Spencer 1996a), for which they employed a boatman continuously and shipwrights occasionally (Harding and Wright 1995, xiv–xv).

The masons and carpenters were the most regularly employed, each group headed by a chief or master mason or carpenter, who was paid a higher rate or given extra bonuses. These individuals may also have supervised the supply or procurement of materials for the enterprise, another opportunity for profit. There were normally four to six other masons, working almost entirely on the stonework of the bridge , and a larger number of carpenters, often eight or ten, working both on the bridge and its houses, and on the estate, depending on the work in hand. In the later 14th century masons and carpenters were paid by the week and employed throughout the year, suggesting an effort to retain and reward skilled workmen; later in the 15th century, more and more of them were employed by the day for part of the year, although several clearly worked almost all the possible working days in the year for the Bridge House. Others, however, must have needed to take employment elsewhere (Harding and Wright 1995, xv–xvi).

The other building craftsmen (listed earlier) were similarly employed at first by the week for the whole year, but later by the day and only as required. Over the same period, however, wages for both craftsmen and labourers rose, although there was always a range of wage rates at any one time. Whereas the masons seem to have worked as individuals in coordinated teams, other craftsmen were employed with their own servants or apprentices, sometimes by the task rather than by the day. Usually only one master of any of these auxiliary crafts was employed at any one time, but often the same man was employed year after year.

The labourers for the Bridge House, paid by the day from the 15th century, had a varied range of tasks, mostly relating to the handling and transport of building materials. They unloaded timber and other materials arriving by boat at the Bridge House yard on the Southwark waterfront (see below, 10.5) (Fig 114). Materials such as stone and waterproof cement from the yard were transhipped to the nearby bridge (Harding 1995, 115).

For exceptional necessities, such as the work of preparing the pageants for the entry of Henry V and Katherine of France on 14 February 1421 (see Chapter 9.6), many extra workmen were drafted in, sometimes at higher wages and clearly working day and night to get the task completed. For several weeks before that event the bridge wardens hired a carver, a plasterer, at least five joiners, up to 13 other carpenters, and nine, rising to 16 stainers (painting wood and other absorbent materials) and four (cloth) painters. They also paid an unspecified number of 'virgins' to appear in costume and take part on the day (Harding and Wright 1995, 77–86). Specialist work was also needed on the Bridge House's properties outside the City of London, which included two mills at Stratford and fields in Southwark, Peckham and Lewisham, so the bridge wardens occasionally employed millwrights, hedgers, ditchers and haymakers.

A unique element in the labour force of the Bridge House estate was the army of tidemen, working in gangs to insert elm piles in the starlings and infill the starlings with chalk rubble (see Chapter 8.2). This work was as continuous as tides and weather would allow (see below, 10.4). The impression is of a coherent group or gang of men working together regularly. Unlike other craftsmen, individual tidemen are not normally named in the accounts, and were paid as a group. There were rarely more than five workable tides in one week, so it seems likely that the tidemen also had other employment. At 3d to 4d a tide, for working a shift of perhaps four to five hours, it was fairly well paid, but was also irregular, and must have been wet, uncomfortable and often dangerous (Harding and Wright 1995, xvi).

In the later 15th and early 16th centuries, there seems to have been a progressive 'casualisation' of the labour force, in the move towards more day labour and the reduction in the fixed workforce. In the 16th and 17th centuries the size of the workforce directly employed by the Bridge House at any one time was significantly reduced as a result of the growing practice of leasing property, with the obligation to repair it passing to the lessee. Some of the carpenters, and almost all the miscellaneous house-building craftsmen, largely disappear from the accounts from the mid-16th century. However, a core team of masons, carpenters and labourers was still employed, as the bridge wardens continued to repair and maintain the bridge itself and the numerous buildings upon it (CoLRO, BHR 3, *c* 1550–1650). This development does have an impact on the value of the Bridge House records as sources for wages and work practices, as well as rents and the property market. It appears that the bridge wardens kept a small but regularly employed workforce, paid year-round at something less than the market rate, possibly by way of a retainer rather than in relation to the amount of work actually done, although the workmen argued otherwise in 1689. It does seem to be the case that, when other workmen were employed on a casual or occasional basis by the day, they were paid higher daily wages (CoLRO, BHR 3; Boulton 1996, 274).

10.4 Maintaining the starlings

The Bridge House accounts show that maintaining the bridge starlings and purchasing the materials for the task, chiefly elm logs and iron pile shoes, was a perennial commitment. The piles were driven in by several engines described as 'gins' or 'rams' (Harding and Wright 1995, xxii). These medieval engines would have consisted of a wooden tripod frame from whose apex a heavy weight – the 'ram' – was suspended by a rope, and then allowed to fall on to the top end of the pile, driving it into the ground (see above, 10.1).

The accounts from the week ending 19 April 1381 to 4 January 1382 show an almost continuous programme of up to ten hours' work per week (presumably several shifts or tides on the 'rams' (Harding and Wright 1995, 11–15, 21–9) (Fig 83). During 1420–1 up to five or sometimes six shifts or 'tides' (of probably four to five hours at low tide) were being worked per week by the piling crews between the end of February and the end of November, excepting only July (Fig 84) (Harding and Wright 1995, xxii, 65–113).

An inventory of the stores at Bridge House in 1350

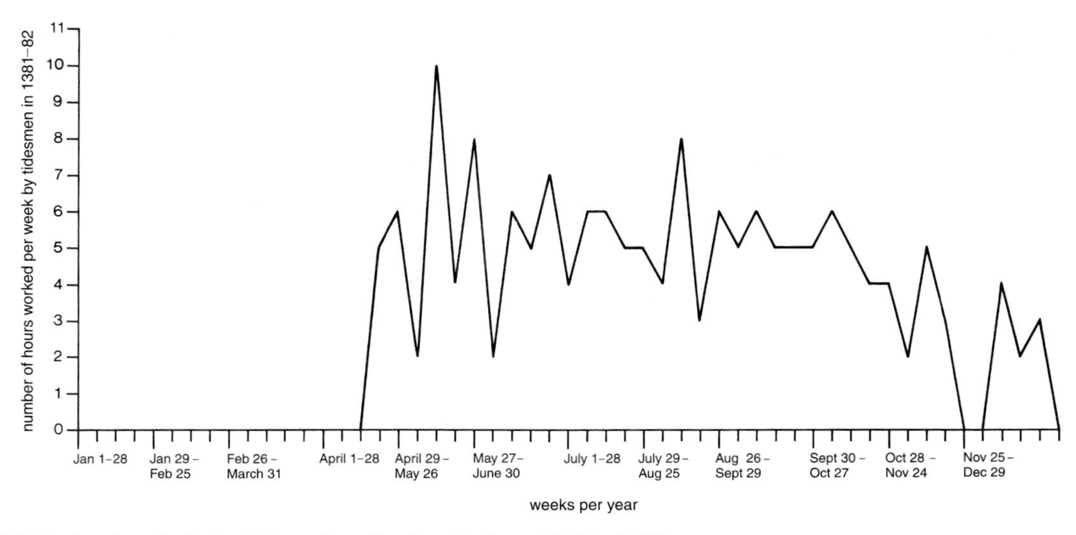

Fig 83 *Graph of 1381–2 work on the London Bridge starlings (data from Harding and Wright 1995)*

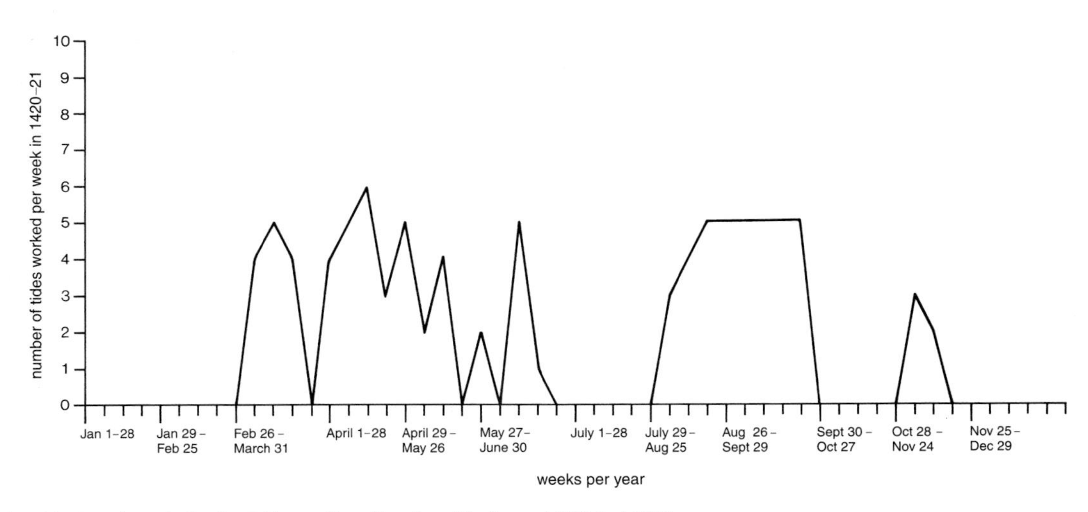

Fig 84 *Graph of 1420–1 work on the London Bridge starlings (data from Harding and Wright 1995)*

included '400 great pieces of oak timber .... Also timber for 14 shops, fully wrought and framed for immediate building ... 120 pieces of elm for piles ... 110 irons for pile tips value 4d per iron ... two engines with three *rammes*, for ramming piles of the said bridge' ... 'one great boat and one small boat and one *shoute* [barge]' (Welch 1894, 260). The boats were principally used for fetching stone from north Kent and assisting with the piling operations (Spencer 1996a, 209).

10.5 Maintaining the bridge: purchase of materials

Tony Dyson and Bruce Watson

The Bridge House accounts contain a great deal of information about the range of building materials purchased for the maintenance and repair of the bridge and the other Bridge House properties. The day-to-day running of the bridge and the purchase of building materials was the responsibility of two bridge wardens. As detailed accounts of income and expenditure only date from 1381, it is not certain how things were organised before this date (Harding 1995, 111–12). However, it is documented that in 1282–3 Henry le Waleys started to reorganise the bridge funds by constructing the Stocks market hall and building houses near St Paul's Cathedral; the latter scheme ended unsuccessfully in an Exchequer suit in 1293. The documentation for these ventures mentions the posts of bridge wardens (Williams 1963, 86–7). Williams (1963, 87) states that after 1311 the bridge wardens (two in number) were to be elected from among the citizens of the City of London. A charter of the reign of Edward II (1307–27) includes the statement: 'Also, that the keeping of the Bridge shall be entrusted unto two reputable men of the city aforesaid, other than the Alderman thereof' (Riley 1861, 128).

The wardens had sufficient funds to buy materials in bulk and space to store them in their yard in Southwark (see below), downstream from the bridge (Fig 114). In the years for which the weekly accounts survive it can be seen that the wardens bought very little during the winter months, when transport would have been difficult and, from March each year, they started to accumulate materials for use during the summer months (Harding 1995, 115).

It should be stressed that, while the accounts often state that particular materials and labour were exclusively earmarked for the bridge or some other named premises, there were also many occasions when the destination of materials and work was not specified. As a result it is not possible to determine the total expenditure on any individual undertaking. The accounts show that the craftsmen employed by Bridge House worked on a wide variety of jobs. During 1461–2, for instance, John Chambre worked on the bridge and helped construct one timber-framed building in Southwark and another in Wood Street in the City (Harding and Wright 1995, 142).

The timber supply

The two main types of timber mentioned in the accounts are elm and oak. During the 15th century elm was obtained from various places in Buckinghamshire, Berkshire, Essex, Kent, Middlesex and Surrey (Fig 85), while oak was obtained from Berkshire, Essex, Hertfordshire, Kent, Middlesex, Surrey and Sussex (Fig 86). The Bridge House records do not always specify the source of timber, so that it is only possible to determine the source of a fraction of the material purchased.

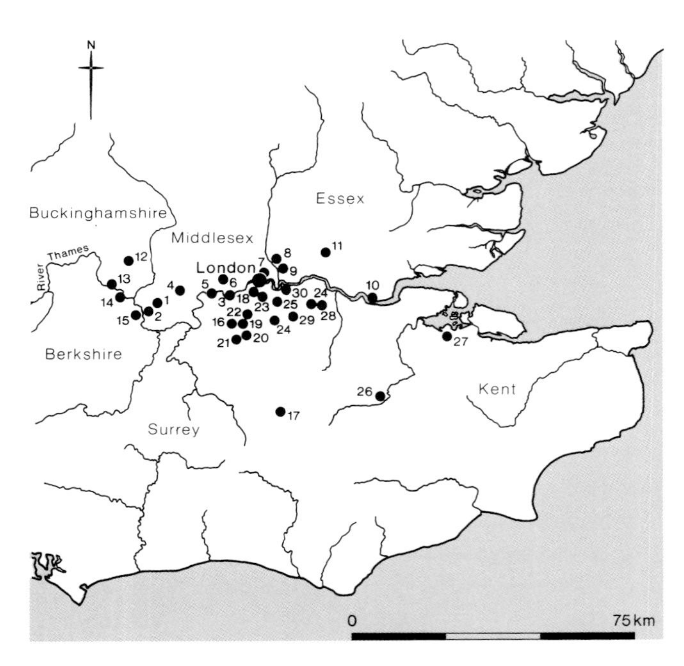

Fig 85 Map of south-east England showing major rivers and all documented sources of elm timber mentioned in the Bridge House archive during the 15th century
Key: Middlesex: 1 – Stanwell; 2 – Yeoveney (Staines); 3 – Fulham; 4 – Heston; 5 – Holgilles/Holyryng Grove and Longford (Chiswick); 6 – Gey near Westminster; Essex: 7 – Mile End; 8 – Leyton; 9 – Stratford; 10 – Tilbury; 11 – Brentwood and Romford; Buckinghamshire: 12 – Ankerwyke and Langley; 13 – Boveney; Berkshire: 14 – Southlegh by Windsor; Surrey: 15 – Egham; 16 – Merton; 17 – Lingfield; 18 – Camberwell; 19 – Mitcham; 20 – Beddington, Wallington and Wodecote; 21 – Carshalton; 22 – Streatham; 23 – Peakham; Kent: 24 – Bexley; 25 – Lewisham; 26 – Peakham; 27 – Newington; 28 – Lesnes and Crayford; 29 – Chiselworth; 30 – Greenwich

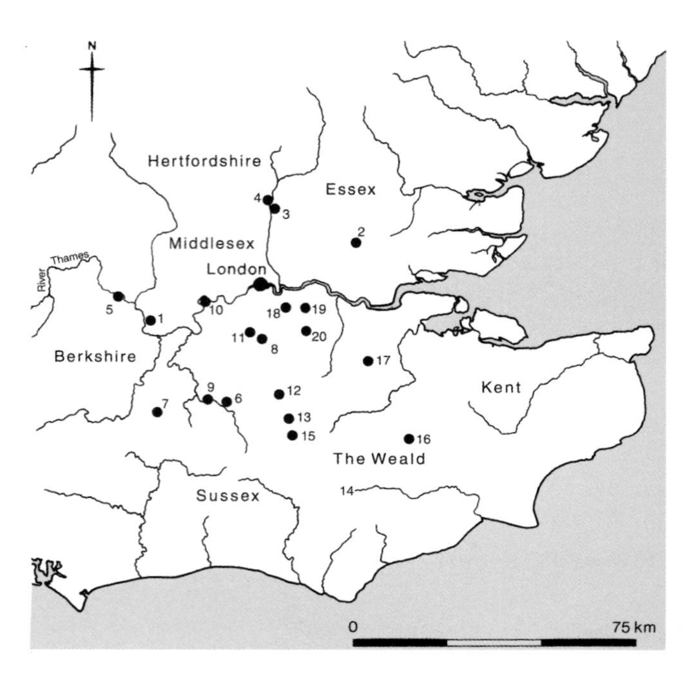

Fig 86 Map of south-east England showing major rivers and all the documented sources of oak and 'timber' mentioned in the Bridge House archive during the 15th century
Key: Middlesex: 1 – Yeoveney (Staines); Essex: 2 – Brentwood; 3 – Waltham; Hertfordshire: 4 – Cheshunt; Berkshire: 5 – Windsor and Shottesbrook; Surrey: 6 – Betchworth and Reigate; 7 – Farleypark; 8 – Croydon; 9 – Brockham by Dorking; 10 – Sheen; 11 – Carshalton and Mitcham; 12 – Tandridge. 13 – Lingfield; Sussex: 14 – the Weald; 15 – East Grinstead; Kent: 14 – the Weald; 16 – Goudhurst; 17 – Wrotham; 18 – Lewisham; 19 – Eltham Park; 20 – Chiselworth

Most of the elm was obtained from places close to the River Thames, suggesting that they were normally moved to the Bridge House yard by barge. In March 1385, 39 elm trees were purchased at Windsor, the felled and lopped trunks being hauled by horses to the river and then shipped downstream by barge (Spencer 1996a, 209). Oak trees were often purchased from the same places as elm, but oak was also obtained from a much wider area (Fig 86). The Bridge House accounts often include the travel expenses of officials who evidently scoured the Home Counties and even further afield in search of timber, implying that it was in relatively short supply. The cost of carrying timber overland or by barge is often mentioned in the accounts. In some cases the oak must have been hauled overland considerable distances. Croydon appears to have served as an important storage depot and collection point for oak and other 'timber' (as distinct from elm) from Surrey and the North Downs. There is no evidence in the Bridge House accounts, or from the growth patterns of any of the 15th-century oak timbers recovered from the enlarged bridge abutment (see Chapter 11.2), that any imported timber such as Baltic oak was used or purchased.

There are occasional references to other species of timber. For instance, ash trees were purchased twice in 1461–2 (Harding and Wright 1995, 128, 140), and in 1422 beech was purchased from Tandridge, Surrey: beech piles were used in the construction of the abutment in 1187–8 (see Chapter 8.4).

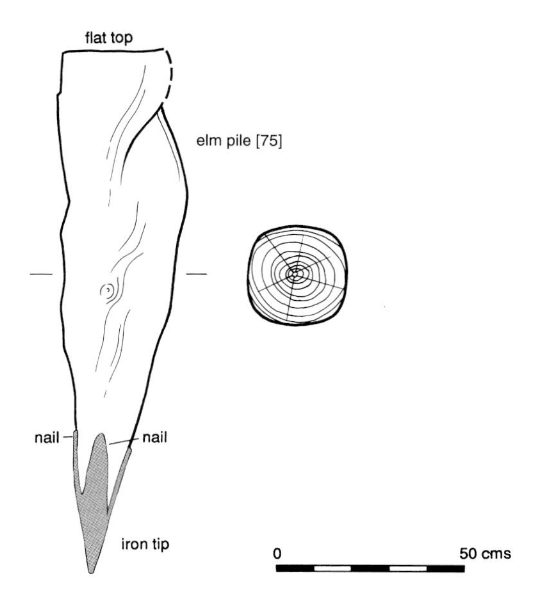

Fig 87 Drawing of one 15th-century iron-shod elm pile [75] <77> (F28.3)

Most of the elm logs were used as piles, although in the 15th century some of the logs may have been sawn up into boards. Numerous iron-shod elm piles were used in the 15th-century rebuilding of the abutment (Fig 87), for which elm as well as oak was used for sillbeams and for the upstream breakwater (see Chapter 11.2); the oak was clearly used as building timber and is sometimes described as 'quarter-cut' (Harding and Wright 1995, 128). Halved oak logs felled during 1187–8 were used as sillbeams in the abutment and iron-shod oak timbers were used as piles (see Chapter 8). A variety of oak and softwood (*estrichesborde*) boards are recorded, as well as a number of types of lath, both 'heartlaths' (laths with heartwood) and 'eaves-laths' (Harding and Wright 1995, 128). Both wooden treenails (*traversnails*) and metal nails are mentioned in the accounts.

The Rochester Bridge accounts (1398–1479) also describe the purchase of timber, mostly elm for use as starling piles. Both large and small trees were purchased, sometimes 260 a year. The trees were stripped of branches where they were felled and even sharpened before being moved to the bridge by timber-tug: dragged by horses or oxen. The main sources of timber were Allington, Aylesford, Botley, Boxley, Halstow, Islingham, Monkdown Wood and Tottington, all in Kent. The principal source of elm piles was apparently Monkdown Wood in the Kentish downs, some 9km south-east of the Rochester Bridge (Becker 1930, 68–70). Interestingly, the buyers of timber for London and Rochester Bridges obtained their material from two different areas. The London buyers purchased from a very wide geographical area (which excluded east Kent), while the Rochester buyers obtained all their timber from central and eastern Kent.

The stone supply

There is no surviving documentary evidence for the construction of London Bridge during c 1176–1209, but examination of geological samples from the basal course of the ashlar facing of the 12th-century abutment (F26.3 and F26.4) revealed that it was built of a quartzite sandstone of uncertain source. While the standing masonry (F26.5) of the 12th-century abutment consisted of Purbeck marble, Kentish ragstone from the Maidstone area, and a quartz-rich sandstone and a shelly limestone of uncertain source (see Chapter 14.11).

Almost all the stone purchased by the Bridge House during the period 1409–1509 came from Kent, although the source of the stone was often unspecified. The three most common stone types were chalk, ragstone and 'Reigate' (Fig 88). Chalk rubble was purchased by the boatload and used to infill the starlings and the gullies created between the starlings by the current, as well as for lime burning. The bridge abutment was also infilled with chalk, and the piers infilled with a mixture of chalk, flints, firestone (part of the Reigate stone beds near Dorking, Surrey) and ragstone. Chalk was purchased at Northfleet in Kent, during 1465–6 and 1470–1. Kentish hassock (a medium-grained sandstone) was also used as rubble infill for the starlings. Kentish ragstone or 'rag' often purchased at Maidstone was a general purpose building and paving stone. The ashlar of the 15th-century additions and rebuilding of the abutment (F28) included Kentish ragstone, Purbeck marble and Purbeck stone, while Kentish ragstone was also used as the rubble infill of the abutment (see Chapter 14.11). Possibly all the Purbeck marble and stone was reused from the earlier bridge abutment, as the purchase of Purbeck ashlar during the 15th century is not documented (see Chapter 11.2).

The ashlar facing of the piers demolished during 1826–7 was described as 'chiefly of Kentish ragstone' (Knight 1831, 118). The geology of a number of fragments of bridge masonry reused as garden walling at Wandsworth Common

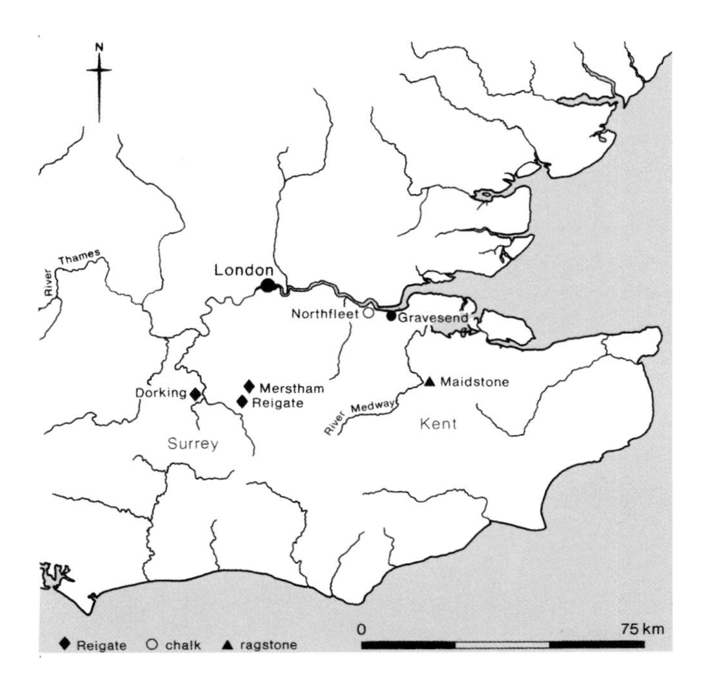

Fig 88 Map of south-east England showing rivers, the location of London and all sources of stone mentioned in the Bridge House archive during the 15th century

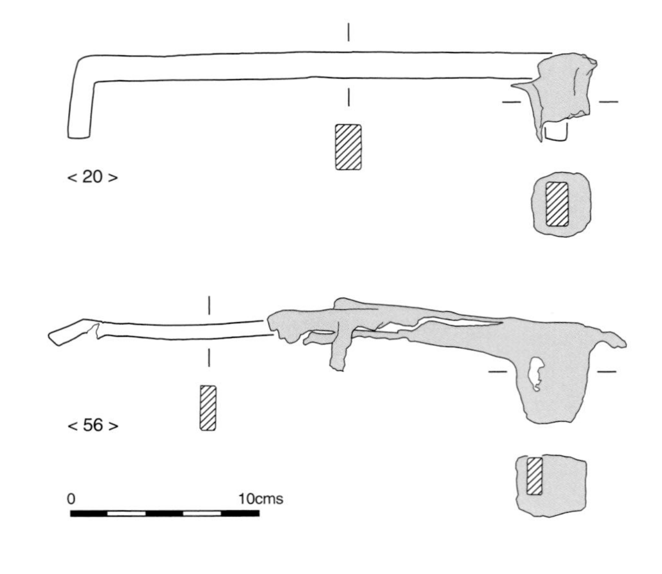

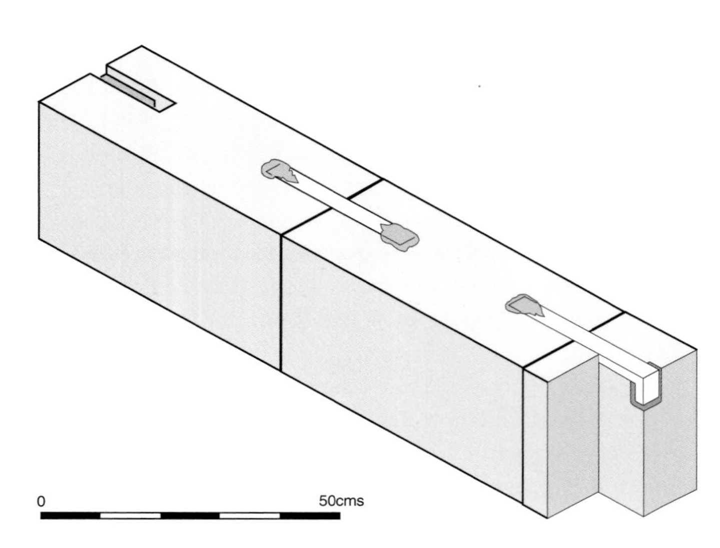

Fig 89 Drawing of two stone cramps from ashlar facing of the bridge abutment; <20> is from the 12th-century bridge abutment (F26.5) and <56> is from the 15th-century enlargement of the abutment (F28.2); the diagram shows how the cramps were used to secure each ashlar block

was mainly Kentish ragstone from the Medway area (Home 1931, 339). 'Stone of Kent called Bridge ashlar' was often purchased at Maidstone, Kent, and carried to London by barge, and occasionally at Merstham, Surrey (Fig 88). Technically it was indeed Kentish ragstone, although it was often referred to in the accounts as 'Reygatestone', no doubt loosely; there is no geological evidence that it actually came from Reigate.

A number of specialised stones such as springers, corbels and voussoirs were purchased already dressed (Fig 101; Fig 103). Occasionally stone was purchased from distant sources, including Quarr, Isle of White (1405–6), 'Shireburn in Yorkshire' (1463, 1467–8: either Sherburn in Elmet, West Riding, or Sherburn near Scarborough, East Riding) and Caen stone from Normandy (1445, 1466–7). The 1350 inventory of Bridge House stores included '690 feet of stone of Portland, handworked and squared', presumably ashlar from Portland, Dorset, and '18 great stones of Bere' (Welch 1894, 259), possibly Bere Regis in the same county. Much of the stone purchased was brought to the yard by barge, details of expenditure on transport being recorded, but freestone and mouldings from Reigate were brought to the bridge by cart (Spencer 1996b). Study of the Bridge House records for the period 1382–98 shows that the Bridge House boats were principally used to ship cargoes of chalk or ragstone (Spencer 1996a, 209).

The Rochester Bridge accounts (1398–1479) include numerous references to the purchase of stone. Large quantities of chalk were needed to infill the starling, much of it purchased locally at Burnham and Cuxton. Chalk was also burnt in a lime-oast or kiln to produce lime for mortar (Becker 1930, 72–3). Kentish ragstone was purchased locally, it included ashlar or 'freestone' blocks, and paving stones or cobbles for the bridge roadway. There are mentions of 'ashlar' (probably ragstone) and voussoir, wedge-cut stones; the former were purchased regularly in large quantities, the latter purchased only when an arch had collapsed. Caen stone was an occasional purchase (Becker 1930, 74–5).

Lead was used for fixing the U-shaped iron cramps which secured the ashlar blocks of the bridge piers and abutments to each other (see Chapter 14.6), and for such plumbing purposes as making pipes, cisterns and gutters (Harding and Wright 1995, 183). In the 1537–8 accounts 'four sows [ingots] of Peak [Derbyshire] lead' are mentioned (Harding and Wright 1995, 184). Ironwork purchased in 1461–2 included cramps (see Glossary), pile shoes, hooks, hinges, nails for pile shoes and a variety of other nails and spikes (Harding and Wright 1995, 130). A number of iron cramps set in hot lead were recovered from the ashlar of the abutment (Fig 89). The Bridge House accounts do not describe the type or number of pile shoes that were purchased in any detail. However, it is clear from the 15th-century examples recovered during the 1984 excavations (Fig 87; Fig 90), that there were two weights of pile shoes, a lighter one weighing 3–3½ lb, and a heavier one weighing about 9lb (see Chapter 14.6).

Pitch was purchased by the barrel and tar by the gallon. In August 1382 nine barrels of pitch were purchased 'for the work of the Bridge masons' (Harding and Wright 1995, 27). The 1350 inventory of Bridge House stores includes 'two cauldrons for melting pitch for cement' (Welch 1894, 260). Pitch was added to the mortar used in the pointing of the bridge abutment (Chapters 7 and 10), presumably in an attempt to make the mixture waterproof. The ashlar of the piers demolished in 1826–7 was described as 'cemented with a composition of pitch and rosin' (Knight 1831, 118).

The Bridge House and yard

To the east of St Olave's church, on the north side of Tooley Street and fronting on to the river, was the original Bridge House and yard (Fig 114). The earliest portion of this property may have been acquired by Peter of Colechurch, the builder of the stone bridge, as there is a grant dating from the mayoralty

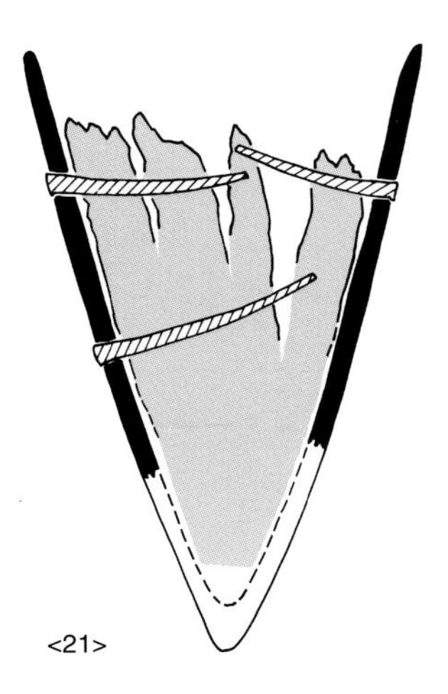

Fig 90 *Drawing of two iron-shod pile tips from the 15th-century work around the bridge abutment; <21> is from the foreshore (F28.1) and <77> is from the upstream starling and breakwater (F28.5) (scale 1:4)*

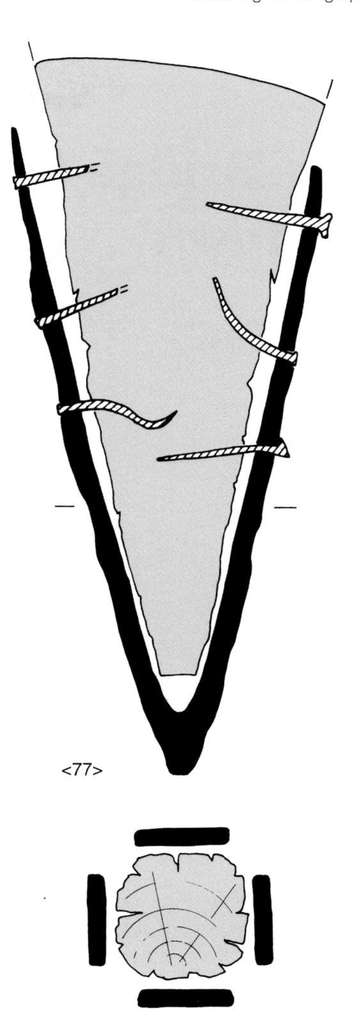

of Henry fitz Ailwyn (*c* 1189–1212) which refers to his house to the east of St Olave's church, and this was perhaps the nucleus of the Bridge House property (Carlin 1983, 354–5); while a deed of *c* 1222–3 refers to the *domum de ponte* or Bridge House (Carlin 1983, 355). The property was entered from Tooley Street via a narrow lane and a single gate. By 1381–2 there were several guard dogs kept in the yard and the 'cook' or janitor was paid to feed them (Harding and Wright 1995, xiv). This yard contained a number of buildings which were used as workshops by carpenters and masons. The chaplains of the bridge chapel also had a house within the yard. Stow, writing *c* 1600, described the Bridge House as 'a storehouse for stone, timber, or whatsoeuer pertaining to the building or repairing of London Bridge' (Stow 1603, ii, 65).

In 1514 an empty house in the Bridge House yard was leased out for nine weeks for the storage of wheat, and within three years a timber granary was being built. In 1519 12 bays for granaries were provided, and in 1522 a new oven was built for the use of the City (Carlin 1983, 358), probably following a bequest of £200 from Sir John Throstone. According to Stow, ten ovens were built, of which six were described as 'very large', and were to be used with grain stored at the yard; the bread was to be distributed to the poor (Stow 1603, ii, 65). By the late 16th century the grain could have been milled in the nearby corn mills beneath the southernmost two arches of the bridge (see Chapter 9.7 and Fig 82).

11

'London Bridge is broken down'

Tony Dyson and Bruce Watson

11.1 London Bridge: the nursery rhyme

London Bridge is broken down,
Broken down, broken down,
London Bridge is broken down,
My fair lady.

Build it up with wood and clay,
Wood and clay, wood and clay,
Build it up with wood and clay,
My fair lady.

Wood and clay will wash away,
Wash away, wash away,
Wood and clay will wash away,
My fair lady.

The above lines constitute the first three verses of a traditional English nursery rhyme (Opie and Opie 1955, 76) of unknown antiquity. Consequently the collapse to which it refers is undated, but if the 'lady' in question was Eleanor of Provence, queen of Henry III, who had custody of the bridge revenues from 1269 until *c* 1281 and may well have contributed to the ruinous state into which the bridge fell by that date, it is possible that the rhyme is alluding to the events of 1281–2.

In 1265 the bridge revenue was apparently being administered by 'the brethren and chaplains ministering in the chapel of St. Thomas upon the said bridge', which was presumably a fraternity or guild of citizens. However, in that same year, perhaps to punish the Londoners for supporting Simon de Montfort (see Chapter 9.3), the king granted the revenue for the next five years to the hospital of St Catherine, near the Tower of London (Cal Pat R, 1258–66, 507). Then, on 10 September 1269, Henry transferred the 'custody of our bridge at London' to Queen Eleanor, who appears to have retained it until *c* 1281 (Home 1931, 53, 60–3). Queen Eleanor and the Londoners plainly did not get on, and the bridge would have held particularly unpleasant memories for her since July 1263 when, attempting to pass it while travelling upstream by barge, she was pelted with stones and garbage until she abandoned the attempt and returned to the Tower (Home 1931, 55; Stow 1603, i, 51).

Whatever the date of the collapse it refers to, the rhyme serves as a reminder that a medieval bridge needed almost constant maintenance to keep it in good repair, without which there was a real danger of collapse. If the starlings of London Bridge were not repaired on a regular basis (see Chapter 10), then there was nothing to protect the shallow pier foundations from the constant scouring effects of the river flow, tides and periodic floods (Fig 54; Fig 95). If the scouring of soft river sediments created a void below a pier, then the superstructure would crack and eventually collapse (discussed later). According to the London Bridge accounts for 1461–2, 14 boatloads of Kentish chalk were purchased for infilling the 'straddles' (piers and starlings) and the 'gullys' (probably scour channels between the starlings or holes within the starlings) (Harding and Wright 1995, 127).

Edward I, in a letter of 8 January 1281 to 'all bailiffs, for the keepers of London Bridge or their messengers', noted that the bridge was in a ruinous condition (Cal Pat R, 1272–81, 422), and during the severe winter of 1281–2 the accumulation of ice around the piers caused five of the bridge's arches to collapse (Stow 1603, i, 24). On 4 February 1282, after this 'sudden ruin of the bridge of London', the king instructed the mayor and the wealthier citizens to contribute £200 or £300 towards the cost of building a temporary wooden bridge, to be erected before Easter, on the understanding that parliament would repay the money and fund the proper repair of the bridge in due course (CoLRO, SBHR, fo 86). Stow states that in 1289 a subsidy was granted for the reparations of London Bridge (Stow 1603, ii, 160), presumably to repay the Londoners who had financed the bridge repairs a few years before.

Between that date and the early 15th century there is no evidence of any comparable damage to the bridge, although (and perhaps in part because) the Bridge House accounts show that there was a continuous programme of maintenance. New piles were frequently added to the starlings to ensure the continued protection of the pier foundations from scouring (Harding and Wright 1995, xxii). This stable state of affairs suddenly and rapidly changed during the week ending 5 January 1425, when an examination of the arches on the west side of the central portion of the bridge (between the chapel and the Stonegate) was reported to have revealed the bridge to be so badly cracked that the river could be seen through the fractures (CoLRO, BHWP 3, fo 123v) (Fig 91). During the following week payments were made for repairs (CoLRO, BHWP 3, fo 124), although the work cannot have amounted to much more than patching on the evidence of the weekly payments records: there the main focus of attention at this time was the rebuilding of the Drawbridge tower, on which there had been frequent work since at least 1405. All the same, an ordinance by the mayor and aldermen in the same month forbade the driving of carts with iron-shod wheels over the bridge (Cal L Bk K, 38). In November and December 1429, a number of aldermen were appointed to undertake a further survey of the fabric of the bridge with the help of masons and carpenters. This survey recognised some new unspecified peril (CoLRO, JCCC 3, fos 67–8), perhaps more financial than structural, for there was much concern at this time about the administration, or maladministration, of the bridge's finances. Such problems would certainly not have facilitated the repair of the bridge at a critical time.

On 27 July 1435 it was noted once more that the bridge had fallen into a ruinous condition (Cal L Bk K, 191), and during the first week of January 1437 damage to the bridge, caused by a build-up of ice around the piers, was recorded. Then, between 5 and 12 January 1437, the Bridge House weekly payments recorded that Stonegate and its two adjoining arches had collapsed (CoLRO, BHWP 4, fo 218) (Fig 91), the *Chronicles of London* for 1436–7 reporting that the 'tower' of London Bridge and two arches had collapsed into the river (Kingsford 1905, 142). A large quantity of masonry fell into the gully between the second and third piers from the southern end of the bridge, later known as 'Rock Lock' (Dance 1799) (Fig 91). A very low tide in 1715 revealed a mass of masonry lying there, held together by mortar and iron cramps and apparently part of a collapsed arch (Seymour 1734, 49).

The collapse set in train a salvage operation to recover masonry and timber (CoLRO, BHWP 4, fos 218–19). Some of the salvaged timber derived from collapsed tenements and the tower, and was recovered from the Thames, often by the boatload, over the next few weeks. All of it was moved to the Bridge House yard, a short way downstream of the bridge on the Southwark side, east of St Olave's church (Fig 105). This

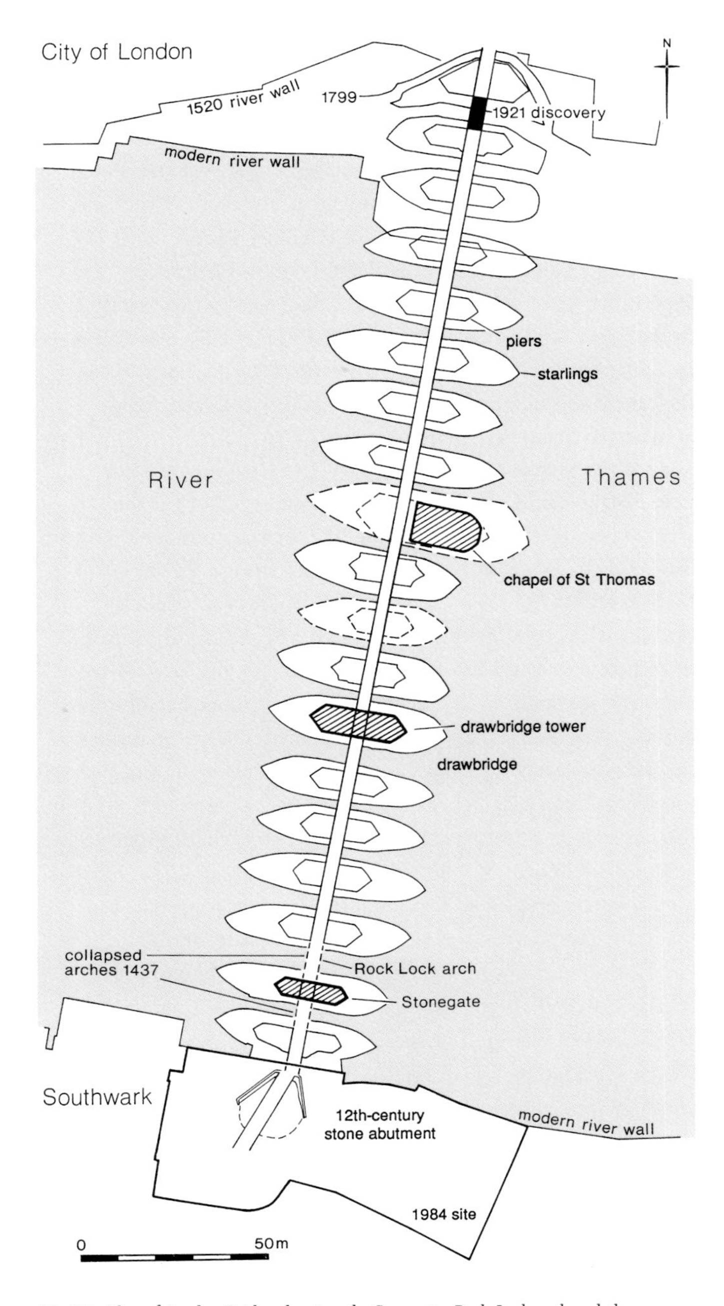

Fig 91 Plan of London Bridge showing the Stonegate, Rock Lock arch and the position of the two arches which collapsed in January 1437

operation appears to have continued until mid-February (CoLRO, BHWP 4, fo 223). Immediately after the collapse, a gang of 17 carpenters were paid to work day and night on the bridge (CoLRO, BHWP 4, fo 218). They were assisted by three other carpenters and one labourer by day. Their task is not stated but it is probable that they shored up and consolidated the remains of the fallen tower, to prevent further collapse. A retrospective payment made in the week ending 4 May for rope, bought for binding elms upon the bridge at the time of the collapse (CoLRO, BHWP 4, fo 232), suggests either an attempt to stabilise the damaged masonry or even the construction of an emergency pontoon. Elm logs were normally used as starling piles and may only have been used for emergency repairs because they were readily available in the Bridge House yard.

The seriousness of the situation was such that a former chief mason of the bridge, Richard Beke, currently clerk of works to the Dean and Chapter of Christ Church Canterbury, was paid 4s in expenses to inspect the bridge during the week after the collapse (CoLRO, BHWP 4, fo 219). Beke was doubtless summoned because of his expertise and knowledge of the bridge, having been employed as a mason by the Bridge House as early as November 1409 (CoLRO, BHWP 1, fo 256), becoming chief mason on the death of John Clifford in September 1417 (CoLRO, BHWP 2, fo 266) and remaining in that post until as recently as April 1435 (CoLRO, BHWP 4, fo 163). Beke would have known more than anybody about the fabric and defects of the bridge, and was to stay on in Southwark to supervise the repairs.

For the week ending 2 February 1437 payments were made for the carriage of 'two great posts' bought for the wooden bridge (*pro ponte lignea*) at 12s per post and for 'two cart loads of braces and rails for the said bridge' (CoLRO, BHWP 4, fo 222). This is the first mention of a wooden bridge, which was evidently intended as a temporary repair to restore communications between the City and Southwark, although fees paid in late March show that work began in mid-January (see below). By the end of February large quantities of timber were arriving from Croydon and Lewisham, and during the week ending 23 February as many as 23 carpenters were on the pay roll and presumably engaged in constructing the timber bridge (CoLRO, BHWP 4, fos 224–5). By 9 March the number of carpenters employed had risen to 30 and a payment was made for the guarding of the gate on the bridge, so that those entering should not hinder the labourers at their work (CoLRO, BHWP 4, fo 226r–v), and further rails, planks and other timbers were purchased. This may have resulted in the wooden bridge being complete enough to enable people to cross on foot: during the last week of March, John Crofton, the chief carpenter in charge of the construction of the new bridge, was paid 60s for 60 days work at 12d a day, with a bonus of 40s from the Mayor and Bridgemasters as a reward for his labour (CoLRO, BHWP 4, fo 228v). At the same time Richard Beke was paid 66s 8d for his good counsel and supervision of the masons and labourers working on the new wooden bridge during the previous ten weeks (CoLRO, BHWP 4, fo 229).

Most of the information about the financing of these repair works is provided by the records of Bridge House expenditure in the 1430s, at that date compiled weekly. From these weekly totals yearly figures were produced, not always accurately, around Michaelmas (29 September) when the new Bridgemasters were elected or re-elected. These annual figures were organised under five headings: wages and expenses, 'tides' (mostly maintenance work on the starlings), quit-rents (rents payable from properties belonging to the bridge), and bridge chapel expenses (the last two categories are not included here) (Fig 92).

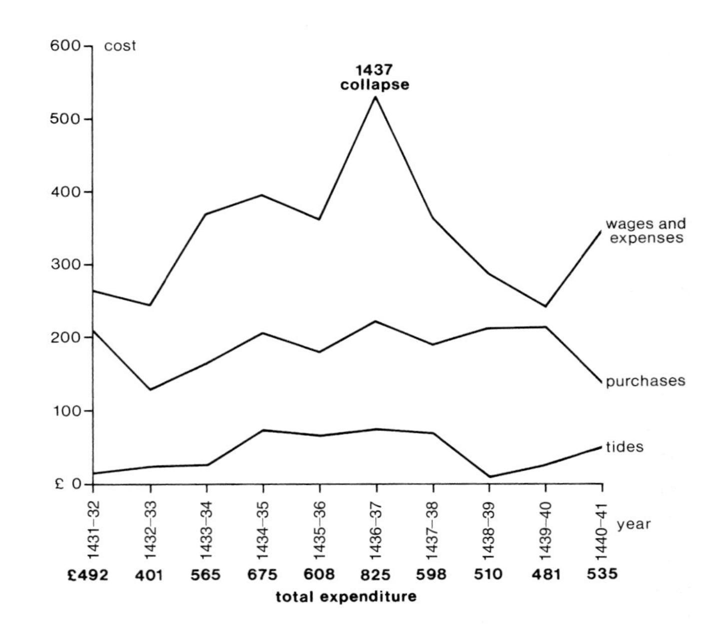

Fig 92 Graph of the annual Bridge House expenditure on maintenance from 1431–2 to 1440–1

Excluding the last two categories, the average annual expenditure for these ten years was £569, compared with an average of £521 for the years 1404 to 1445. Unsurprisingly, the highest recorded expenditure figure for the entire 15th century was for the year 1436–7, the year of the Stonegate collapse. Much of this extra expenditure was due to the costs of salvaging materials and building the wooden bridge. However, the figures suggest a more complex pattern, for although the purchases totals were erratic throughout the entire period, wages and expenses had been steadily rising in two out of the three years preceding the collapse, only to fall progressively during the three subsequent years after the collapse, and rise again in 1440–1 (Fig 92). It is particularly interesting that the figures for the years 1434–5 and 1435–6 show that a serious repair programme was underway before the collapse, even although this clearly included extra work on the starlings, as increased expenditure on 'tides' was maintained until 1437–8. It cannot, therefore, be said that the bridge was completely neglected in the three years immediately before the collapse of the Stonegate. Work had started in earnest, but evidently too late to avert disaster.

Perhaps the scale of the disaster would have been worse without the accelerated repair programme of 1434–6.

How was this extra work on the bridge funded? The normal revenue of the Bridge House estates could not cover extraordinary expenditure on this scale. Instead, the additional costs seem to have been met from levies, loans and gifts (Harding and Wright 1995, xxi). Stow states that a number of wealthy citizens donated money towards the cost of repairing the bridge after the 1437 collapse, including Robert Large, mayor in 1439, who gave 100 marks (£66 13s 4d), Stephen Forster, mayor in 1445, who gave £20 and Sir John Crosby, sheriff in 1470, who gave £100 (Stow 1603, i, 42). The Common Council raised 500 marks (£333 6s 8d) for the repair of the Gatehouse tower in 1439 (Barron 1970, 190).

There is a regrettable gap in the Bridge House records between 1445 and 1460, the only information about bridge repairs during this period being provided by a few references in the journals of the Court of Common Council. These include details of the repair and strengthening of the wooden gate (*portum ligneum*) on the bridge in January 1448 (CoLRO, JCCC 5, fo 28v) and of the supervision of repair work on the bridge funded by a bequest of Cardinal Beaufort, late Bishop of Winchester (d 1447) in February 1453 (CoLRO, JCCC 5, fo 103v). This was also the general period of the enlargement of the southern abutment as revealed in the excavations at Fennings Wharf (discussed later). Although not precisely dated, pottery associated with these works belongs to the years *c* 1450–1500 (see Chapter 14.1 and 14.5), and it is possible that it was in these sparsely documented years, 1445–1460, that the abutment and the adjoining southernmost arch of the bridge were both rebuilt, for, unlike most of the others, this arch is not specifically mentioned in the accounts after they reappear in 1460. They may have been included in the general operations at the southern end of the bridge, which are recorded on and off throughout the 1460s and beyond, but one reason for rebuilding this portion of the bridge first would have been to secure good landward access to Bridge House yard on the Southwark side of the river, from which the building materials would need to be brought for the duration of what was to amount to a rebuilding of the entire bridge (Fig 105).

It is clear that the rebuilding of the Stonegate tower and the two collapsed arches was still in progress by 1460 (CoLRO, BHR 3, fos 12, 13). In 1462–3 two great hooks were procured 'for the gate of the new stonework of the bridge' and a pier on the south side of the drawbridge was being repaired (CoLRO, BHR 3, fo 56). In 1463–4 a total of 163 loads of Reigate stone were purchased at Mersham for use in making a 'pendant' for two arches for the new stonework of the bridge, while 40 tons of 'Kent stones called hassok' was acquired for 'making the walls' of the new stone work as well as filling the 'gulys' between the piers and paving within the City (CoLRO, BHR 3, fo 74). Building work continued on the southern end of the bridge during 1464–5, and included the closing of the arch on the southern side of the Stonegate (CoLRO, BHR 3, fos 92v, 96), and the completion of this arch suggests that the abutment and the southernmost arch of the bridge may already have been rebuilt by this date. In 1465–6 the arch in the new stone tower was complete and the temporary wooden bridge (over the collapsed spans) was dismantled. In addition 71 loads of gravel were purchased to surface the roadway of the southern end of the bridge in 1465–6 (CoLRO, BHR 3, fos 107, 111, 112), which suggests that the rebuilding of the Stonegate tower and its adjoining arches was largely complete before the tower was damaged during the Fauconberg rebellion of 1470 (discussed earlier).

During the demolition of the superstructure around the Stonegate pier in 1832 (Fig 91) it was noted that the masonry here consisted of several phases of construction. It was also recorded that the Stonegate pier 'had evidently been rebuilt', as there were piles beneath the pier – in contrast to a number of other piers which were not founded on closely spaced piles and were considered to be part of the original bridge (Knight et al 1832, 204). Knight's observations, made during the demolition of parts of the bridge in 1825–32, thus imply that the Stonegate pier was wholly rebuilt during the mid-15th century and that this rebuilt pier was founded on a mass of long piles, rather like a starling, instead of the rubble-filled enclosure of short piles on which the foundations of the original piers were seated (see Chapter 9.2). The same technique of founding new masonry on a dense mass of piles was also used during the enlargement of the southern abutment at this time (see Chapter 11.2). The totality of this pier reconstruction could explain why the rebuilding of the Stonegate pier and tower, for which special funds were raised in 1439, was probably not completed until 1465–6. It is assumed that the Stonegate pier was rebuilt using the same techniques as they used to build the original 12th-century one (see Chapter 8). However, as a cofferdam was in use at the Albi bridge in France during 1408 (Boyer 1966, 30–1), it is possible that the same technique was used in London for some of the 15th-century reconstruction of the bridge.

After the completion of the Stonegate and its adjoining arches, building work continued on other portions of the bridge during the next decade, when many other arches were rebuilt. The impression is of a fairly orderly progression from one arch to the next. It is also noticeable that there was a tendency to return to the same work after an interval of up to as much as ten years. Whether this was because parts of it had been left incomplete (superstructural work on the two towers, for example, conceivably being deferred until the basic work on the arches was complete) or because of recurring structural problems or unsatisfactory workmanship (an explanation invited by the recurrent attention given to the Drawbridge tower throughout the 15th century and later) is not known. A prime example of this apparent repetition is the Stonegate tower itself and its adjoining stone arches, which were the main priority in the early 1460s but were still receiving attention in the 1470s and 1480s.

From 1469–79 the main focus of work was the northern end of the bridge (CoLRO, BHR 3, fos 166–71) and, from that year until 1476–7, these arches were restored at the approximate rate of one each year, working from north to south (CoLRO, BHR 3, fos 257v, 259). This work involved the construction of a wooden framework by carpenters to serve as a 'former' for the arch, then the rebuilding of the arches and often the adjoining piers by stone masons. Whether this work represented the partial or complete rebuilding of the arches is not entirely clear from the details of the accounts, but there can be little doubt that the scale of operations which took place in the four decades or more after the collapse of 1437 was so extensive as to amount to a virtual rebuilding of the bridge. Throughout this process, as ever, the vital and endless task of repairing the starlings continued almost independently (discussed earlier).

In 1468–9 Malteswharf, a short distance upstream of the northern approach to the medieval bridge (and on the site of the approach road for the modern bridge), was used as a depot for stone used to repair the bridge (CoLRO, BHR 3, fo 157). This wharf had been bequeathed to the Bridge House by Richard Malt, a Thames Street stockfishmonger in 1456 (Sharpe 1890, ii, 529), and would have made a useful works depot while the rebuilding of the northern end of the bridge was proceeding. In 1468–9 the Bridge House staff also repaired the starlings at Malteswharf (BHR 3, fo 157).

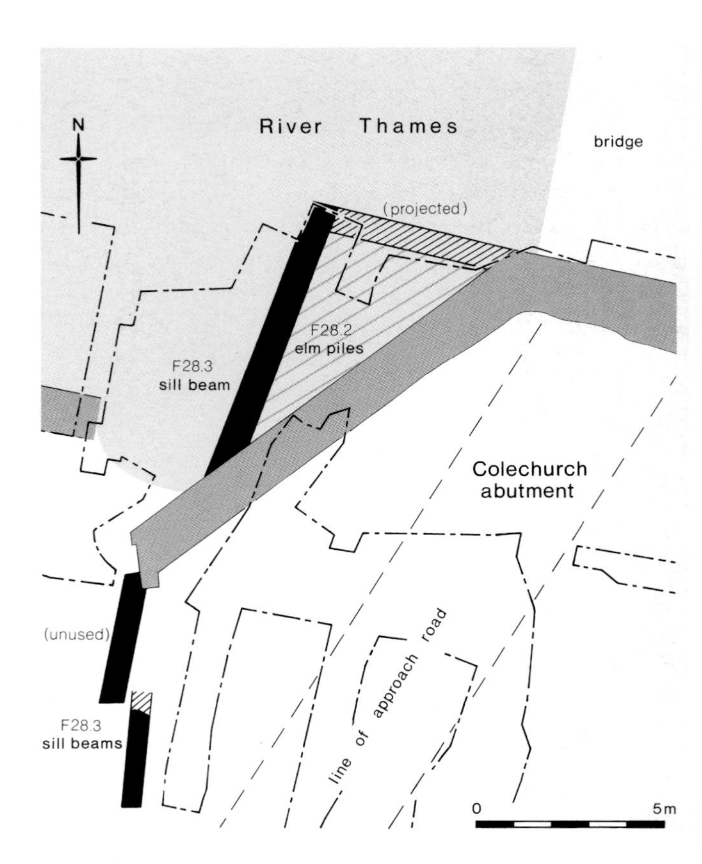

Fig 93 Plan showing the foundations (piles and sillbeam F28.2) for the 15th-century enlargement of the north-west corner of the Colechurch abutment

11.2 The enlargement of the southern abutment

The 1437 collapse of the Stonegate tower and its two adjoining arches clearly precipitated a major rebuilding of London Bridge, beginning with the enlargement and partial rebuilding of the southern abutment, and it is unfortunate that most of this work appears to have taken place between 1445 and 1460, when there is a gap in the Bridge House records.

Work on the abutment started with the driving of a large number of elm piles into the area of foreshore adjoining the north-western and south-eastern sides of the existing 12th-century abutment (Fig 93; Fig 94). The existing foreshore on the upstream side of the abutment consisted of fluvial silts, mixed with dumps of chalk rubble and sand/gravel (F28.1). Finds from these deposits included lumps of ferrous and non-ferrous slag, copper alloy pins, an irregular Claudian coin (AD 45–56), a gold finger ring containing a garnet (see Chapter 14.6 and Chapter 9.1), pottery dating from c 1400–1500 and fragments of medieval roof tile. Into the foreshore were driven a dense mass of elm piles, their top at 0.02m OD (Fig 94). Across the western side of this area of piles was placed a series of rectangular sillbeams made out of halved oak logs (Fig 95). These beams were 710mm wide and 200mm thick, their tops at -0.30m OD (F28.2). In their undersides were a number of peg holes, probably for use in securing the beams to the underlying piles. The method of jointing the beams together is not known. One of them was cut from a tree felled after 1364 (see Chapter 14.2).

On top of the sillbeam was constructed a new ashlar wall in stone types including Kentish ragstone, Purbeck marble and Purbeck stone (a white oolitic limestone almost certainly Portland stone) (F28.2) (Figs 95–8 and see Chapter 14.11). It was constructed of close-jointed rectangular blocks, 430mm–1.1m long, 200–340mm wide and 160–260mm high, mainly laid stretcher-fashion, although alternate courses featured header blocks at irregular intervals, extending back into the wallcore. The facing blocks were bonded together by brown/pink sandy mortar containing traces of black pitch, added as waterproofing (see Chapter 10.4). The blocks were fixed together by iron cramps (see Glossary), set with molten lead (see Chapter 14.6 and Fig 89). The cramps were set in 25mm-square holes, all 60mm deep, and the bars fixed into a carved slot in the upper face of the blocks. Three ashlar courses above the sillbeam ran a 100mm-wide chamfered offset which terminated at the southern end of the sillbeam, where the ashlar ran over the top of the existing masonry (Fig 95; Fig 97). South of this point the ashlar stood up to nine courses high, its top at 2.50m OD. Behind the ashlar facing was a band of mortared rubble wallcore over 1.3m wide and consisting mostly of uncoursed chalk rubble together with fragments of ragstone and flint cobbles, bonded together by a light brown sandy mortar containing frequent chalk flecks.

The southern continuation of this wall lay to the south-west of the abutment. Work here started with the driving of a series of unshod elm piles (F28.3), around which were dumped ragstone rubble blocks, probably intended to consolidate the

Fig 94 The mass of elm piles (F28.2) used to consolidate the foreshore to the north-west corner of the Colechurch bridge abutment (visible on the left), view looking south (MoL)

ground surface. The tops of the piles had been sawn off to create a level surface, those at the southern end amounting to little more than unshod, squared tips (Fig 95). On top of these piles was laid a sillbeam of composite construction, consisting of two rectangular halved oak logs, each 250–260mm wide and 120–130mm thick. These two beams were held together by a series of concealed crosspieces, morticed into the inner face of each half of the sillbeam. The assembly was fixed together by a series of circular vertical pegs 15mm in diameter and extending through the entire width of the beams. None of these oak timbers was datable (see Chapter 14.1). On top of the sillbeam was constructed the southern continuation of the ashlar masonry. The types of stone used in this wall included Kentish ragstone and Purbeck stone (see Chapter 14.11). This

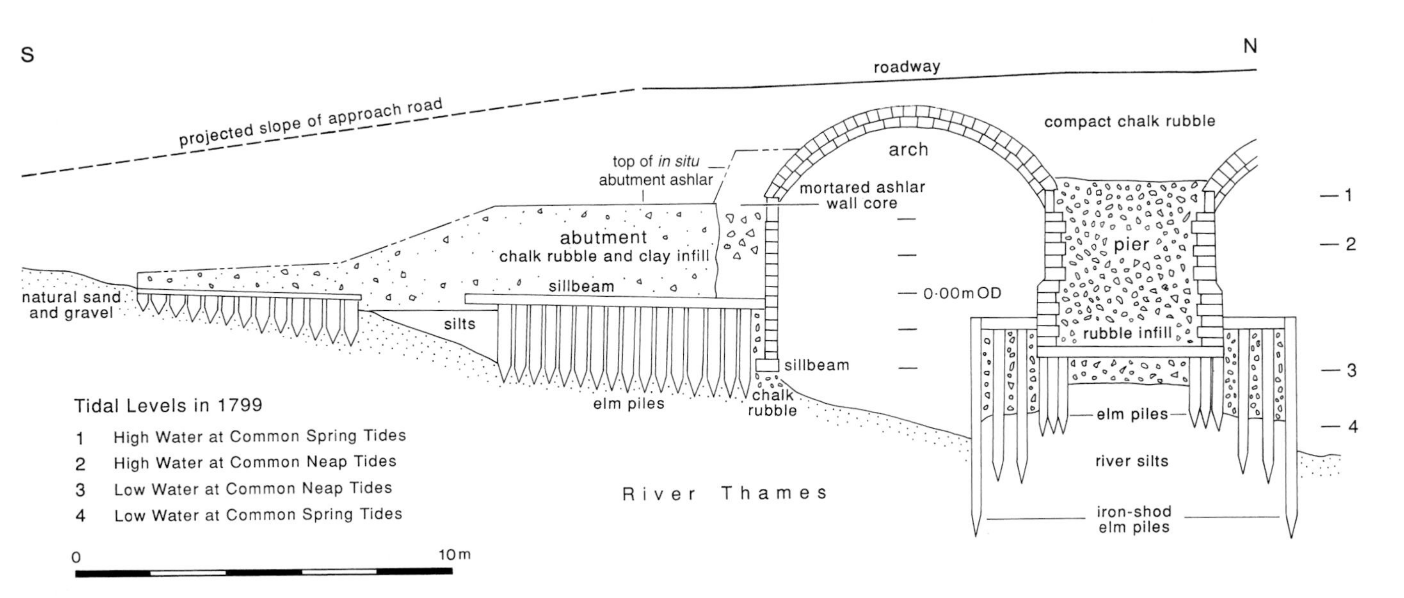

Fig 95 Cross section of the 15th-century bridge abutment and southernmost arch of bridge (for location see Fig 96); composite elevation based on excavated data, the 1799 Dance survey and Knight (1831)

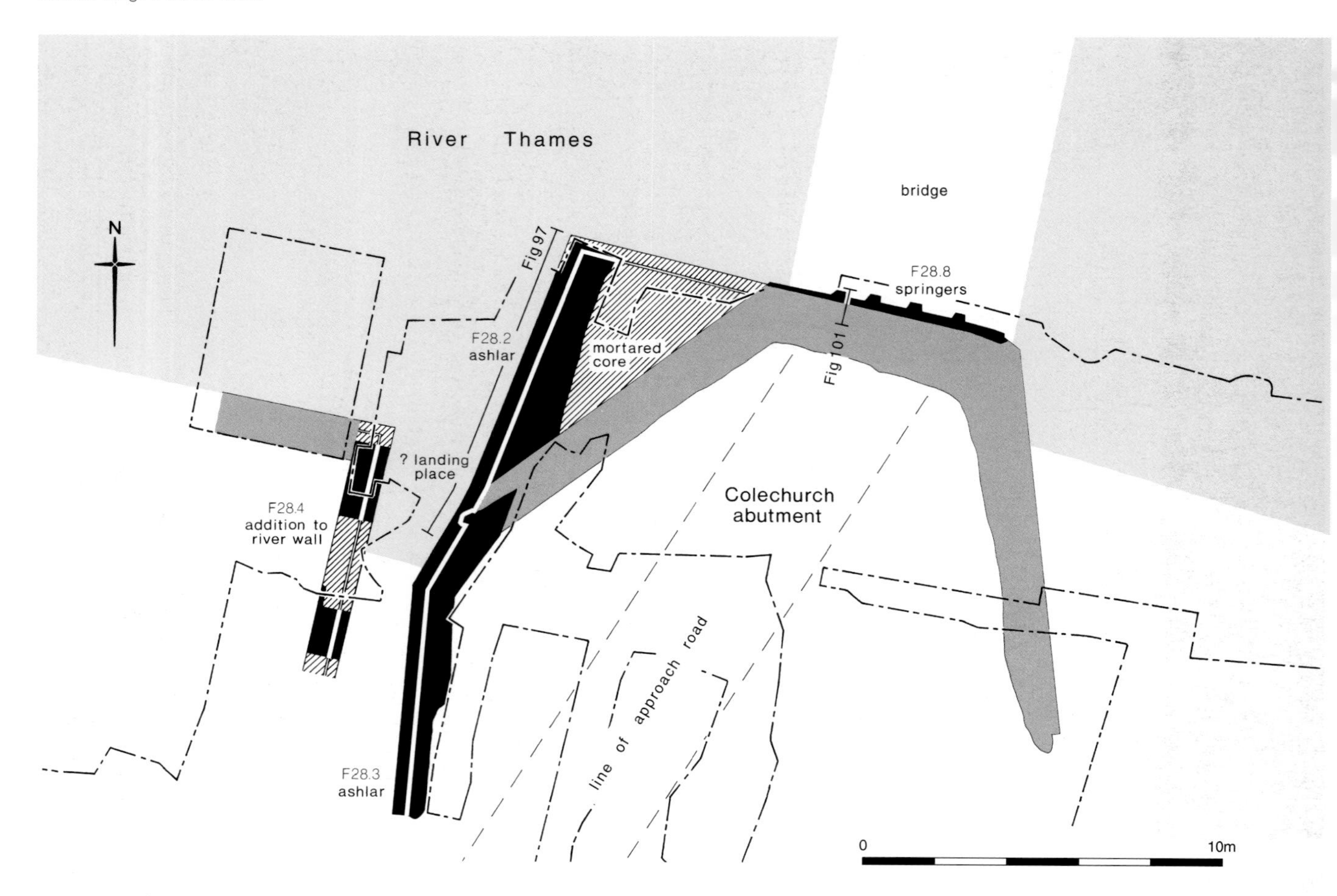

Fig 96 Plan of the enlarged 15th-century bridge abutment showing the new masonry and the associated upstream riverwall

masonry formed the new western face of the abutment and it extended back 6.1m further south than the same side of the original abutment, almost into the natural sand/gravel.

On the east or downstream side of the abutment a series of rubble dumps were deposited on the foreshore, their top at -0.92m OD, to raise and consolidate the ground surface, on which were placed a number of planks (F28.6). Into this raised ground surface were driven a number of unshod elm piles up to 2.0m long with pointed tips, and on them in turn was laid a single course of ashlar blocks on which were laid a series of oak sillbeams (Fig 99). The design of these beams was not recorded in detail, but they appeared to consist of two adjoining rectangular oak logs, each 350–380mm wide and 60–140mm thick. The varying thickness of these beams reflects their apparent function as a levelling course, with its surface at -0.52. OD. The northern or external side of the beams and basal course of ashlar was covered by lead sheeting. The extreme western end of the southern sillbeam had been

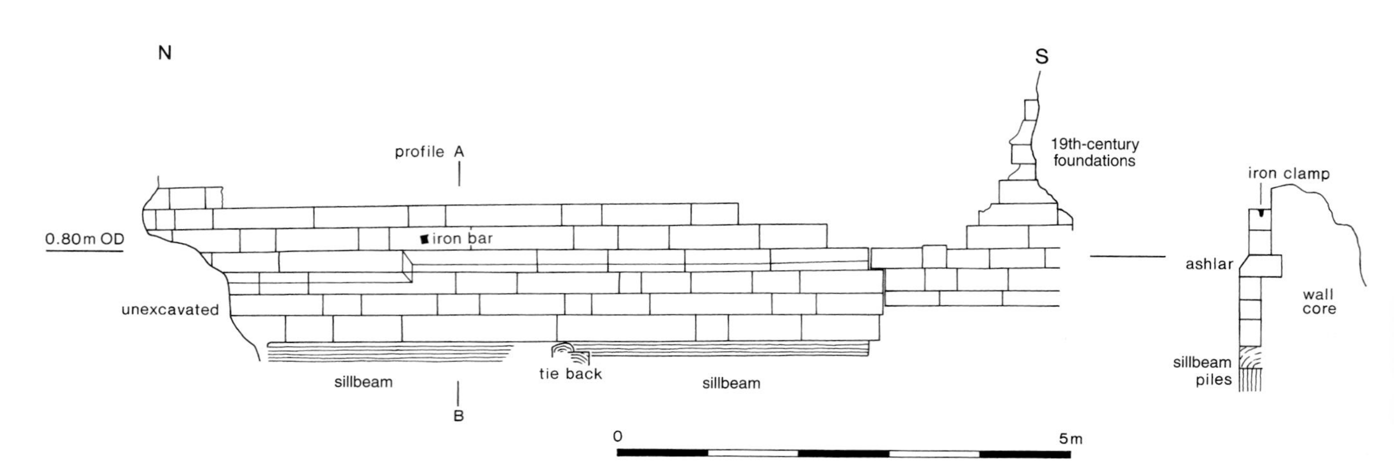

Fig 97 Elevation of the truncated western face of the enlarged 15th-century bridge abutment

Fig 98 *View of the ashlar facing of the west side of the enlarged 15th-century bridge abutment, looking east (2m scale) (MoL)*

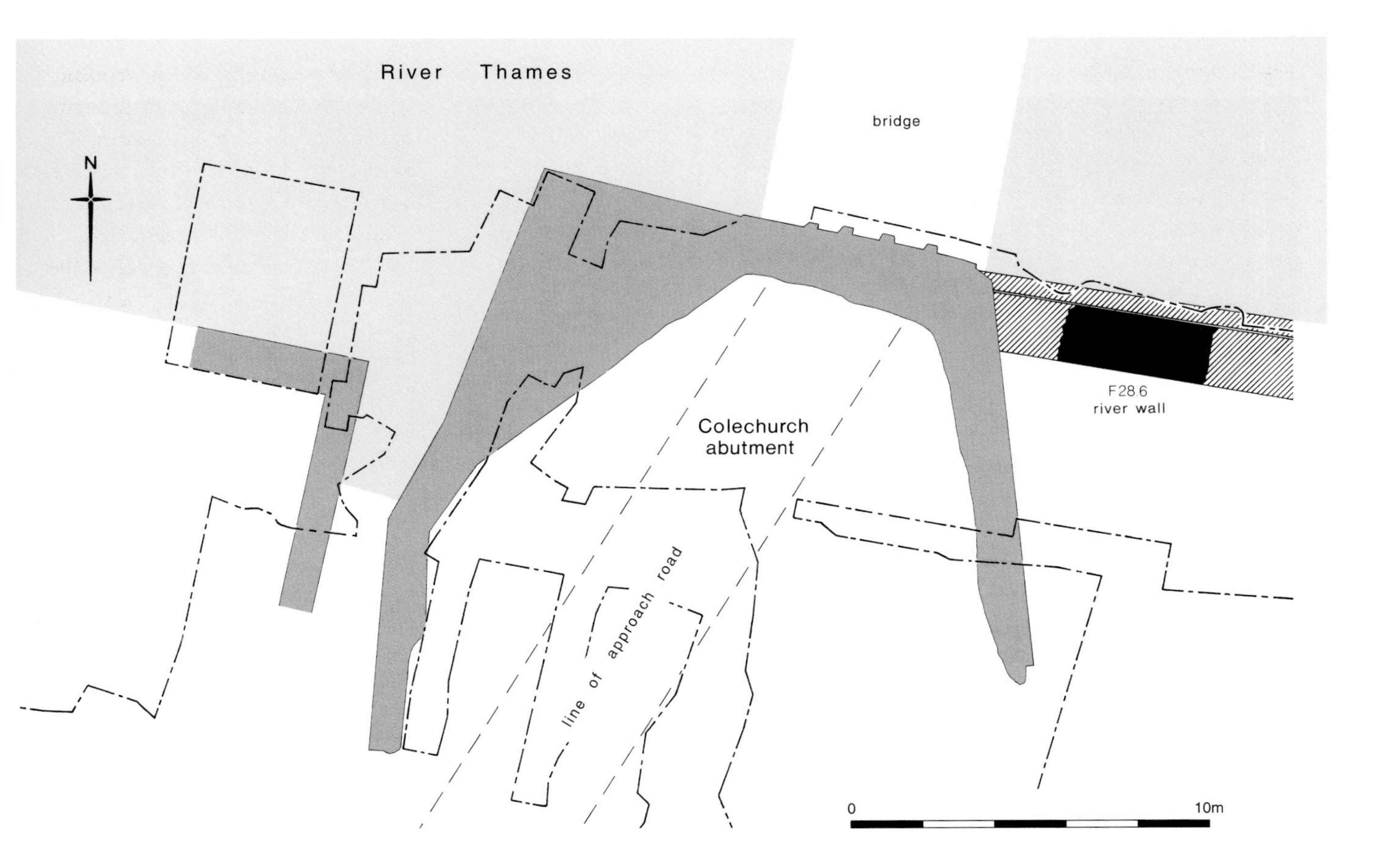

Fig 99 Plan of downstream riverwall (F28.6); this length of wall is probably of 15th- or 16th-century date

carefully cut to fit around the abutment, while at the western end of the northern beam was a gap, infilled with a triangular block of timber.

On top of the sillbeams was built a facing course of close-jointed, rectangular ashlar blocks of Kentish ragstone measuring some 500–700mm long (and 250mm thick, width unknown), standing up to nine courses high (Fig 97). These blocks were not cramped together. To the rear of the ashlar was a band of truncated rubble wallcore, 1.70m wide and reaching 2.33m OD at its highest point. It consisted of a mass of uncoursed chalk rubble and ragstone fragments, bonded together by a light brown sandy mortar. There is no dating evidence for this masonry, but it was part of a substantial downstream riverwall of probable 15th-century date, built as part of the post-1437 bridge repair and rebuilding programme.

While the abutment was being enlarged, its northern side and the southernmost arch of the bridge were also rebuilt (Fig 95). The rebuilding of the north wall involved the placing of another tier of elm sillbeams on top of the original sillbeams (F28.7), the two sets of sillbeams being separated at one end by a thin layer of broken ragstone rubble blocks and by a number of thin wedge-shaped pieces of wood at the other, material which served as either a levelling course or infill. The elm sillbeams were *c* 200mm thick and consisted of a single timber at the west end and of two separate timbers at the east end (Fig 100); the eastern and western sillbeams were not jointed together but simply lay end to end.

On top of the elm sillbeams was built the new northern wall of the abutment (F28.8). It was constructed of rectangular, close-jointed ashlar blocks of Kentish ragstone, secured by iron cramps. Set within the upper portion of the ashlar was a series of four springers for the supporting ribs of the voussoirs (Fig 96; Figs 101–3), set some 800mm apart. There were probably another two extra springers further west which were destroyed during the demolition of the arch in 1831–2. The use of ribs to strengthen stone vaults in buildings, or arches in bridges, was fairly common during the late medieval period. For instance, the arches of the late 13th- or early 14th-century Monnow Bridge at Monmouth are each braced by three sets of wide ashlar ribs (Rowland 1994, 97). The arch of the mid-14th-century bridge at Waltham Abbey, Essex, was also braced by three sets of ribs (Huggins 1970, 137). However, the medieval single arch Stony Bridge at Waltham Abbey originally possessed five sets of ribs (Huggins 1970, 137). It is not known if such ribs were an original feature of the Colechurch bridge, or were added during the 15th-century rebuilding. Mill Lock arch at the northern end of the bridge, discovered in 1921, had three wide ashlar ribs added in 1703 (see Chapter 13.7). Above the westernmost springer were three *in situ* voussoirs, with traces of mortared rubble wall above, their top at 3.52m OD (Fig 101). To the rear of the ashlar blocks was a band of rubble wall core. It consisted of coursed Kentish ragstone rubble, bonded by the same light brown sandy mortar as the core above the arch.

Fig 100 *View of the sillbeams below the north-east corner of the abutment, showing the two phases of foundations, looking north; on the left are two sillbeams – the upper one is elm and of 15th-century date and the lower one is oak of 12th-century date; a large stone block, which served as a quoin stone, is sandwiched between the two phases of beam; the elm piles on the right are the foundations for the downstream riverwall (F28.6) (1m scales) (MoL)*

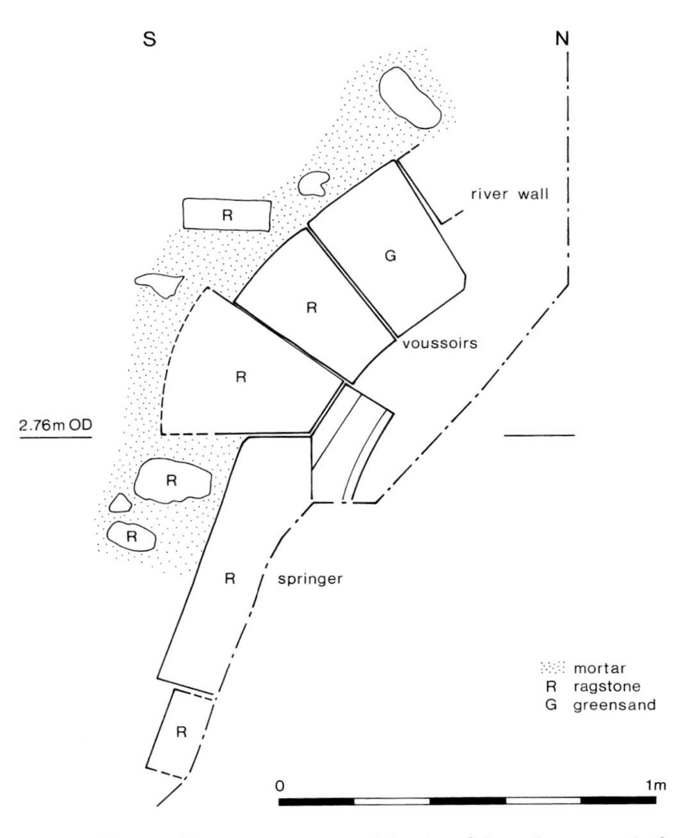

Fig 101 Elevation of the in situ springers and baseplate of the southernmost arch of the 15th-century rebuild of the bridge

On the upstream side of the bridge, during the 15th century or later a series of large and undated oak sillbeams, aligned north–south and with their tops at -0.09m OD, were laid on top of the elm piles. On top of these sillbeams was built a facing course of close-jointed Kentish ragstone ashlar blocks (F28.4) (Fig 96). The blocks were up to 940mm long, 420mm wide and 220–270mm high, none of them cramped together. Behind them was a mortared core of coursed chalk rubble, bonded by light brown sandy mortar. There was no sign of an ashlar facing on the west side of the wall core. This wall is interpreted as the western side of an inlet or landing place on the upstream side of the abutment (Fig 96). To the northern end of the wall was later added a small polygonal block of masonry consisting of close-jointed rectangular facing blocks with a mortared rubble core, its highest point at 2.80m OD. This was probably a breakwater and it was founded on a massive elm sillbeam 3.86m long, 1.19m wide and 140mm thick, its top at -0.04m OD (Fig 102). The width of this timber shows that it had been obtained from a very large elm tree.

On the downstream side of the abutment there was a substantial east–west wall (F28.6, Fig 99) with a mortared rubble core and traces of ashlar facing on its north side; the space to the south of the wall was infilled with a massive dump of chalk rubble and clay.

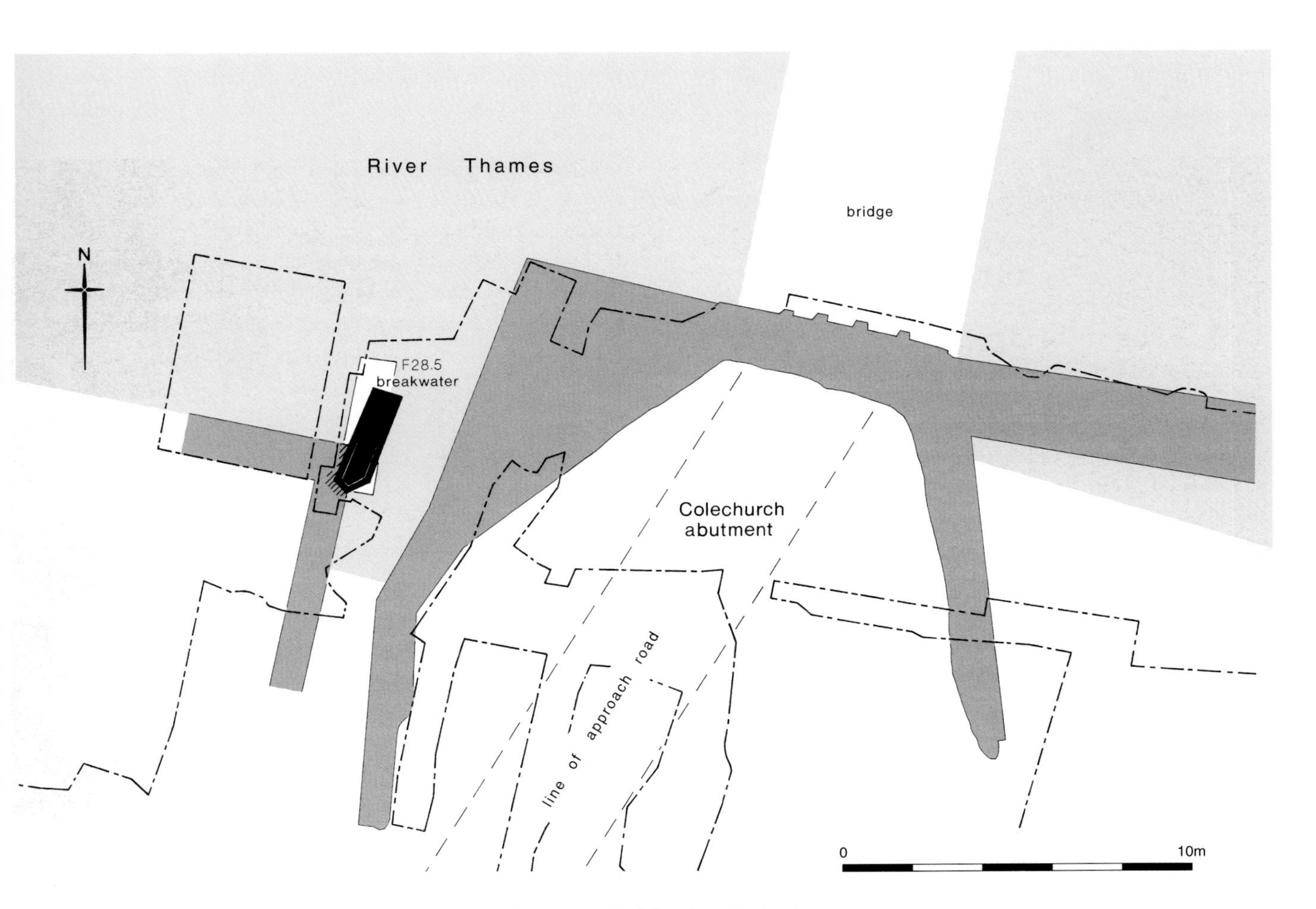

Fig 102 The breakwater (F28.5), which was added to the inlet on the upstream side of the enlarged bridge abutment

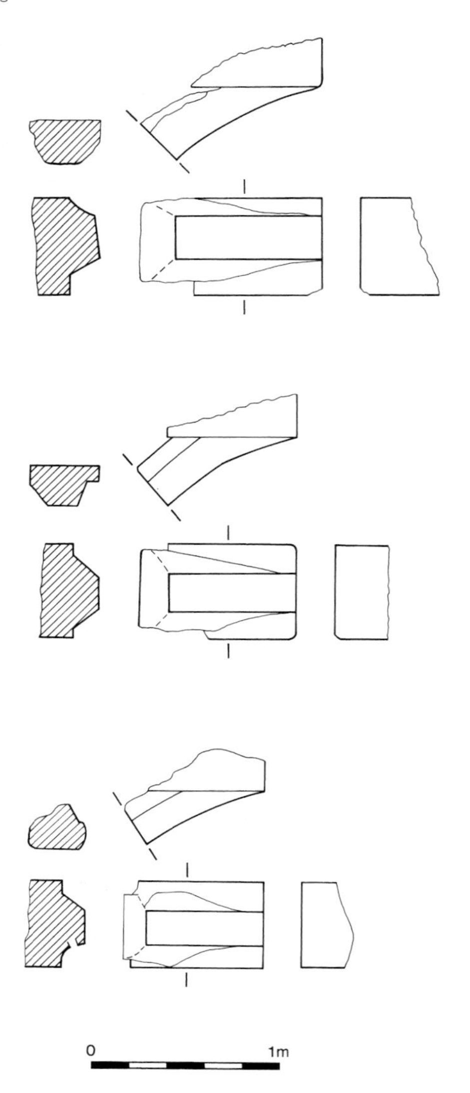

Fig 103 *Drawings of the* ex situ *springers for the ribs below the southernmost arch of the bridge*

11.3 Broken bridges

A perpetual problem of medieval bridge-builders was that their bridges often could not withstand floods and ice floes. The archaeology of the first three phases of Saxo-Norman London bridge shows that all of them were apparently washed away or badly damaged by floods, and in 1097 chronicle sources specifically recorded that floods had carried away nearly the entire bridge (see Chapter 7.1). The late 11th-century timber bridge at Hemington was also apparently destroyed by flood water (Cooper and Ripper in prep), and there is no doubt that floods frequently destroyed some medieval bridges. At Vizille in south-eastern France, for instance, the bridge over the Romanche was swept away in 1336, 1350, 1388, 1399, 1453 and 1457 (Boyer 1976, 6), while Rochester Bridge, over the Medway estuary in Kent, was broken down by floods and ice floes on 19 occasions between 1271 and 1381, when it was finally decided to build a new bridge, the old one being beyond repair (Brooks 1994, 38–40). A new stone bridge was erected at Rochester in 1383–91, but this was not the end of the problem: within 20 years the new bridge had developed serious cracks, two arches collapsing in 1423–4 and the bridge being 'broken' or unusable for more than seven months during 1445 (Britnell 1994, 70–1). The frequency of the structural problems at Rochester Bridge suggests that the technology in use was inadequate. Leone Battista Alberti, an Italian architect, wrote in 1485 that bridges should not be sited on 'elbows' or bends in rivers, as here the banks were more likely to erode and there was a greater risk of logs and debris catching on piles and blocking the arches, in effect damming the river and causing the bridge to collapse under the weight of the pent-up water. He also recommended siting a bridge away from areas of strong current, and constructing substantial abutments 'to bear against the weight of the arches, to keep them from opening in any part' (Alberti, 76–7). It is worth remembering that the same forces have also washed away many bridges during the post-medieval period. For instance, in 1771 part of the medieval bridge over the Tyne at Newcastle was pushed over by the sheer volume of flood water, while at Whitney-on-Wye, Herefordshire, floods washed away three bridges between c 1774 and 1795 (Anon 1994) (Fig 104).

The Colechurch bridge appears to have had fewer collapses than many contemporary medieval bridges, notably Rochester. The builders of medieval London Bridge faced the triple problem of spanning a wide, tidal, estuarine river and contending with the hazards of floods and ice, so that fairly frequent collapses might have been expected (see Chapters 1 and 12). But apart from the two collapses of 1281–2 and 1437, both of them very serious, the bridge seems to have withstood the sort of conditions which destroyed other bridges. The difference may have been due to the costly and almost constant maintenance programme carried out by the Bridge House staff. It is interesting that before both collapses this maintenance appears to have lapsed or been seriously reduced, or been reintroduced too late on the latter occasion, and that both followed periods of especial financial difficulty. But although relative neglect of the bridge's superstructure, and especially of the arches, was a prime cause of the collapses, maintenance of the starlings seems never to have been skimped and to have been carried out constantly. They, rather than the arches, seem to have been regarded as the bridge's weak point, and with good reason. If the starlings were not properly maintained, the tidal scour would have undermined the shallow pier foundations and caused their masonry to crack. The real danger was if these cracks filled with water, which then froze – a process which increases the volume of water by 9.9%; this rate of expansion is enough to cause serious structural damage within a restricted space. An additional hazard was the build-up of ice around the piers and starlings which, by restricting the water flow and trapping any ice floes coming downstream, would initially act as a partial dam when the melt water started to flow. Any substantial damming of the melt water could have resulted in a build-up of water sufficient in volume and weight to push the bridge over.

Fig 104 *View of the Whitney-on-Wye toll bridge, Herefordshire; during 1796–8 the central portion of the bridge was replaced by two oak trestles after the central two stone arches were swept away by floods in February 1795 (B Watson)*

The impression is that Bridge House estate income was sufficient to fund all the regular maintenance work on London Bridge, along with a small building programme. This programme could include the rebuilding or improving of the various properties owned by the Bridge House, such as the six new shops built in the parish of St Dionis Fenchurch in 1381–2 (Harding and Wright 1995, 1). However, Bridge House appears to have possessed no financial reserves, and could not cope with major disasters without the help of outside resources.

12

The medieval Bridge Foot

Tony Dyson and Bruce Watson

12.1 The buildings of the southern Bridge Foot

Excavation of a large portion of the area to the south of the bridge abutment during 1984 revealed the truncated remains of a number of the medieval buildings that flanked the bridge approach road known as the 'Bridge Foot' (Fig 105; Fig 109), a term never applied to the northern, City, end. The Bridge Foot consisted of all the buildings on the bridge south of the Stonegate (see Glossary) and apparently some of those adjoining the bridge (Home 1931, 328). South of Tooley Street, the bridge approach road continued as Borough High Street, referred to as 'High Street' in 16th-century deeds. At each end of the bridge was a pair of posts known as 'the staples' between which a chain could be fixed to close the bridge to traffic (Fig 105). The southern 'staples' are shown on the first map of Southwark produced in *c* 1542 (Carlin 1996, 39) and were mentioned in a legal dispute of 1258, when it was established by inquiry that the legal jurisdiction of the City of London sheriffs extended across the entire bridge as far south as the bridge staples (Riley 1863, 42).

Properties on the west side of the Bridge Foot: archaeological evidence

On the western side of the bridge approach road were the trench-built foundations of a cellared building (F29.1: Fig 106), situated to the south of Pepper Alley (Fig 105). The walls were constructed of uncoursed chalk rubble blocks, with flint cobbles and ragstone blocks bonded by light brown sandy mortar. On stylistic grounds the building was of *c* 13th- to 16th-century date. The cellar, with a brickearth floor at 0.94m OD, seems to have formed the north-eastern portion of a second property to the south, that fronted the west side of the Bridge Foot. The reconstruction of the upper portion of this building is uncertain, but it is probable that it had a stone-built ground storey, with several timber-framed upper storeys. Excavations at New Fresh Wharf, part of the northern bridgehead, revealed a number of stone-built, cellared, quayside buildings of 14th- to 16th-century date, with internal brick-lined cesspits and probably with two jettied upper storeys (Schofield and Dyson 1980, 56). No trace was found of the medieval property to the north of the cellared building (Fig 106). The walls of the two properties were not realigned during the 1761 road widening. Truncation caused by the redevelopment of the site during the 19th century has left no trace of any of the other medieval buildings that occupied the western side of the Bridge Foot, but it has been possible to reconstruct the pattern of the properties here from documentary evidence (Fig 105) (see also below).

Post-medieval occupation of the western side of the Bridge Foot is represented by a number of buildings and features. The earliest feature appears to be a small brick-lined soakaway constructed sometime between the late 15th and early 17th century and not backfilled until the 18th or early 19th century

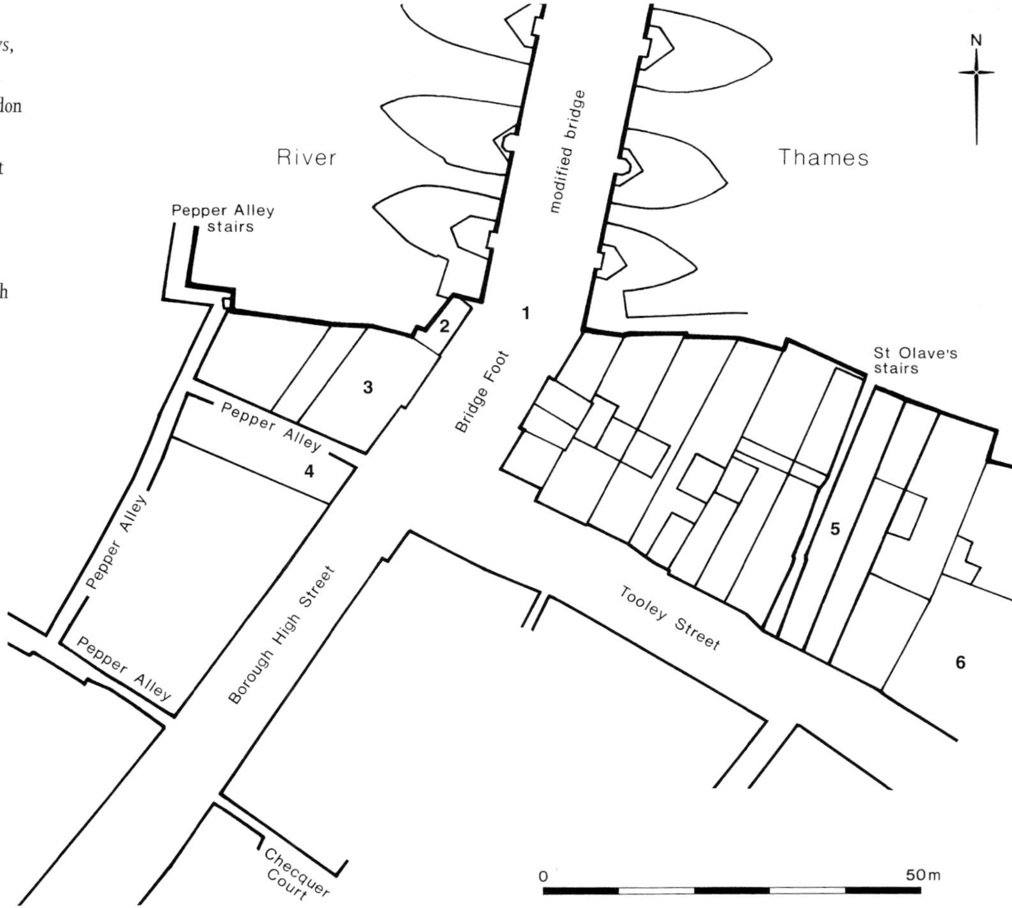

Fig 105 Map of the post-medieval Bridge Foot, showing the position of roads and alleys, and the layout of tenements in August 1821 (CoLRO, CP, drawer 1, roll 3, plan of London Bridge approaches)
Key: 1 – site of bridge staples; 2 – tenement occupied by two shops in the 13th century; 3 – site of the Bear tavern; 4 – site of the Dolphin tavern; 5 – site of St Olave's Dock, later Toppings Wharf; 6 – St Olave's church

(F30.14) (Fig 106). Over this a cellared building was constructed in the 16th or 17th century (F30.15), consisting of three lengths of mortared, rubble-built wall foundation (Fig 106) and incorporating a bricked-up staircase (F30.16) in its north wall. This cellar appears to have been deeper than most of its contemporaries as it extended southwards under Pepper Alley, and it is clear that its east wall lay several metres further east than the post-1761 street frontage. It was probably part of the easternmost of two properties documented in 1603 as fronting the north side of Pepper Alley (see below). Within the cellar was later constructed a square, brick-lined cesspit (F30.17), systematically backfilled with brick rubble during the 19th century, probably in the 1820s (F30.18): the kind of feature that Samuel Pepys described in July 1663 as a 'vault for turds' (Latham and Matthews 1971, 220).

On the south side of Pepper Alley one or more, brick-built, cellared buildings were constructed between the late 15th and early 17th century on the plot between the alley and the existing masonry cellar (F30.11, .13) (Fig 107). On the eastern side of this plot, and presumably within the cellar, a circular, brick-lined well was constructed between the late 15th and early 17th century, and systematically backfilled during the early 18th century (F30.12).

No trace of Pepper Alley itself was found, owing to the truncation of the medieval ground surface caused by the construction of the 19th-century warehouse basement on the site (see Chapter 12.1), but its line can be projected from the 1821 survey of the Bridge Foot (Fig 105). This shows that the alley extended back westwards from the street frontage, providing access to a number of properties: a northern branch, in existence by 1513 (see below) and shown on the Southwark map of *c* 1542 (Carlin 1996, 39), led to a set of river stairs.

On the west side of the enlarged abutment a small building with stone foundations was constructed during the post-medieval period; perhaps it was either part of the house that stood here or a fragment of the Borough waterworks (see Chapter 9.7) (F31.2: Fig 107). An internal brick wall and an area of flagstone floor were later added to the building (F31.3); pottery from the ash and clay dumped over the flagstone floor (F31.4) is dated to 1740–70, which suggests that the building was demolished as part of the 1761 widening of the approach road. It is probable that the inlet or drain to the west side of the abutment (F31.6) was infilled and blocked at the same time, or during the 1830s.

Properties on the west side of the Bridge Foot: documentary history

By the mid-16th century three properties adjoined the southern Bridge Foot, in addition to a house on the abutment itself which belonged to the Bridge House (Fig 105; Fig 107). It is possible that the public toilets or 'jakes' on the Bridge Foot, which were rebuilt in 1542, were sited near this property (Carlin 1996, 237).

Fig 106 Plan of the various medieval and Tudor wall foundations and other structural features on both sides of the Bridge Foot and on Toppings Wharf; the outline of the various properties and the streets is taken from the 1821 survey, after the widening of the approach road and Tooley Street in 1761; the exact widths of the medieval roads are not known and their position is conjectural

River Thames

Toppings Wharf
Fig 113
T9.3
T9.2
cellar
crane base
T10.2
dockside
structure
St Olave's dock
T9.1
cellar
T9.5
cellar
T10.4
Tooley Street
T10.1
dockside
structure
T10.3
drain
T9.4
cellar

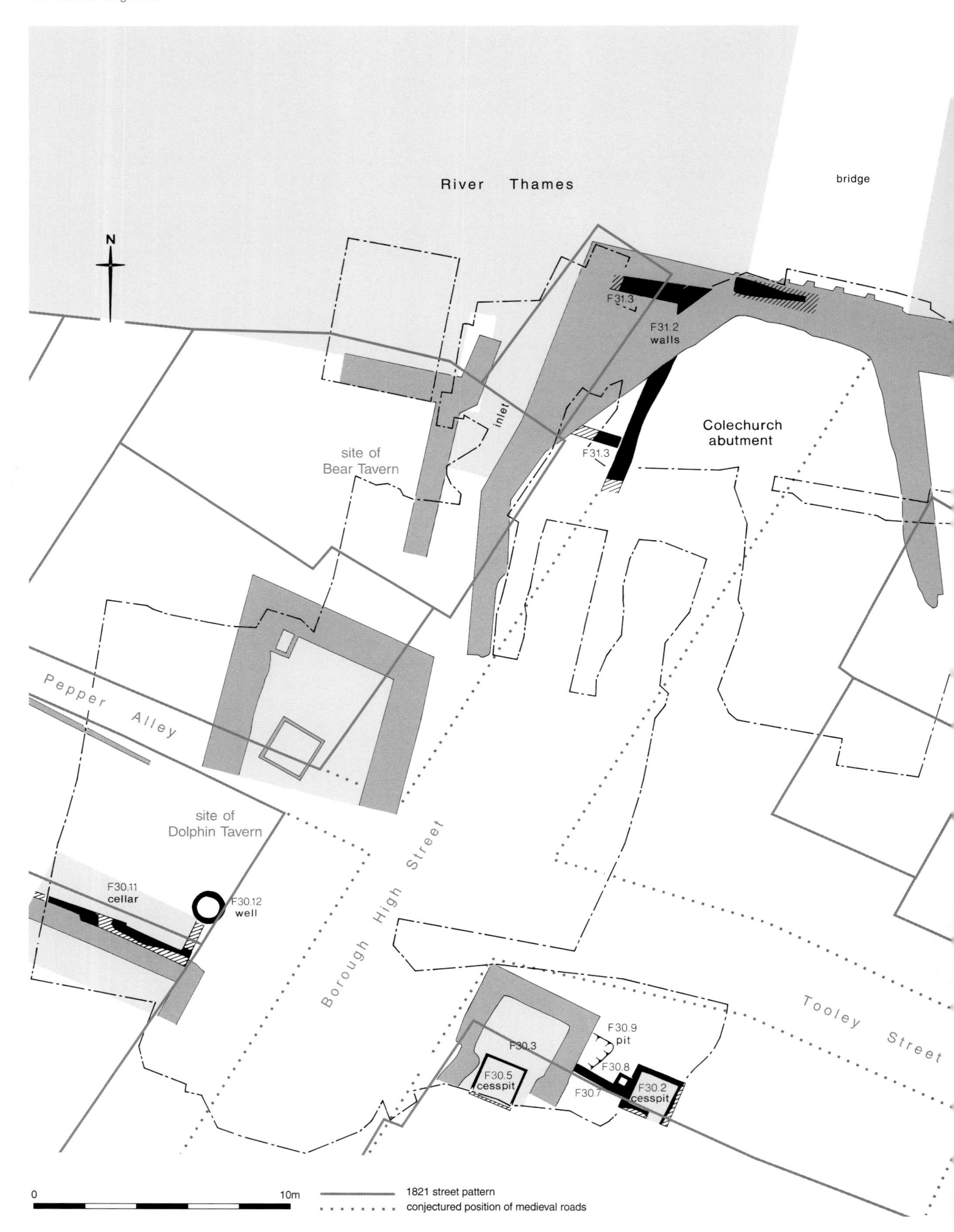

Fig 107 Plan of the various post-medieval buildings and features on both sides of the Bridge Foot and on Topping Wharf; by c 1716 most of St Olave's dock was infilled and built over, the remaining portion of the dock is conjectured from documentary evidence; access to the dock was via an alleyway; the extent of the property boundaries and streets is taken from the 1821 survey, after the widening of the streets in 1761; all the earlier cellars and features which remained in use during this period are shown

River Thames
Toppings Wharf
St Olave's dock
Fig 113
southern portion of dock infilled by c 1716
T11.1 cesspit
T11.2–3 cellar floors
T11.1 cesspit
T11.6 cellar
Tooley Street

The northernmost property, fronting on to the Thames, was the Bear Inn. To the south lay the Dolphin Inn and a third property, documented from the mid-14th century, which belonged to St Thomas's Hospital and fronted south on to 'the king's way of Southwark'. It has been suggested by Home (1931, 76) that the Bear Inn may have been identical with the newly built property described in 1319 as being in the parish of St Olave and at the head of London Bridge (Cal L Bk G, 81). The only one of these properties which is clearly documented during the medieval period was the southernmost, belonging to St Thomas's Hospital. This was recorded in the Bridge House rental of *c* 1358, as a tenement situated on the corner at the staples, between the tenement belonging to the Bridge House on the north and the high street of Southwark to the south (CoLRO, SBHR, 47). In a second, roughly contemporary, version of this rental the same property was described as 'two shops' (BL, Harley MS 6016, fo 152). It is apparently represented in the St Thomas's cartulary by three property grants of early 13th-century date, including a shop at the head of London Bridge. References in the Bridge House weekly payments up to 1445, and in the rentals after 1460, to an annual gratuity paid to the porter (*ianitor*) of St Mary Overy priory, for providing access to the bridge for workmen undertaking repairs, imply that St Mary's owned property here, no doubt that to the west of the Bear tavern, as recorded in 1513. The 1461–2 accounts record repairs on the waterfront either side of the bridge carried out by the Bridge House on behalf of St Mary's (CoLRO, BH accounts 3 (1460–84), fo 29v). These references imply that St Mary's owned waterfront property on both sides of the Bridge Foot, although no other evidence has been found for the downstream lands.

On 12 June 1513, when John Cooke sold the Dolphin tavern to William Seyntpier, the property consisted of a brewhouse and three adjacent tenements built under one frame. To the east was the road to the bridge, to the west Pepper Alley, to the south the tenement of St Thomas's Hospital, occupied by one Anthony, a pouchmaker, and to the north a second tenement of Cooke's, occupied by one Derewik, a shoemaker. This second tenement, not named but plainly the Bear tavern, was also included in the grant to Seyntpier and lay between the Thames and the Dolphin, with a stable and hayhouse adjoining the water and with free access to Bear Alley. To the west of the Bear was a property belonging to St Mary's (BL, Stowe MS, fo 75). Bear Alley appears to have been a passageway within the boundaries of the Bear and leading from the street directly to the tenements north of the Dolphin. According to the Southwark map of *c* 1542, Bear Alley curved northwards some distance from the street and led to the riverside (Carlin 1996, 39). In January 1551 there was a dispute concerning the lease of a messuage and certain tenements at the sign of 'the Beare at the Bridge Foot' (Loengard 1989, no. 292). By 1614 the Bear and the Dolphin appear to have been amalgamated to form a single property occupied by the Dolphin, in as much as the southernmost of the three properties was then described as adjoining the Bear (Fig 105; Fig 109) (GL, St Barts MS 4383/1, 272). During the 17th century two landlords of the Bear, Abraham Browne and Cornelius Cooke, each issued their

Fig 108 Drawings of the 17th-century tokens issued by Abraham Browne and Cornelivs Cook at the Bear at the Bridge Foot: on the left is a copper halfpenny token, on the observe is a chained bear and the wording 'ABRAHAM BROWNE. AT. YE', on the reverse 'BRIDGFOOT. SOVTHWARK. HIS HALF PENY'; the token on the right is brass or base copper (denomination not stated) with on the obverse a chained bear 'CORNELIVIS. COOK. AT THE', on the reverse 'BEARE. AT THE. BRIDG. FOT. C CA' (Thomson 1827, 385) (scale 1:1)

own tokens (Heal 1931, 329) (Fig 108). On 8 September 1725 all the properties on both the east and west sides of the Bridge Foot were destroyed by fire (CoLRO, BHCJ vol 6, 170, 172).

On 29 March 1757 the London Bridge Passage Committee, appointed under the Act of Parliament of 1756 to improve and widen the bridge (see Chapter 13.2), decided that the Bear tavern should be demolished as part of the widening of the bridge approach (CoLRO, LBC). When the Bear tavern was demolished during 1761, it is claimed that a hoard of Elizabethan (1558–1603) gold and silver coins was discovered under one of its floors (Heal 1931, 329; *Public Advertiser* 1761). The three replacement buildings were erected on the site between the riverwall and Pepper Alley, but no trace of these was found. They were demolished during 1824–6 as part of the scheme to replace the medieval bridge (see Chapter 13.4). Cooke's sketch of the bridge dated December 1826 shows that the site was then vacant (Fig 125).

The property to the south of the Dolphin, referred to in deeds of 1513 and 1554 as formerly belonging to St Thomas's Hospital, came into the possession of the City parish church of St Bartholomew by the Exchange, under the will of Sir George Barnes dated 15 February 1557 (GL, St Barts MS 4384/1, 7–8). In the 1603 churchwardens' accounts it was recorded that the site was occupied by two houses and that it was proposed to rebuild them both to a height of three and a half storeys (GL, St Barts MS 4384/1, 206). It is probable that the remains of the 16th-century stone-built cellar found here represents the easternmost of the two properties on the north side of Pepper Alley (see above). In Wyngaerde's London panorama (*c* 1544) it is clear that both sides of the bridge approach were lined with two- or two and a half-storey high, timber-framed buildings (Colvin and Foister 1996, 26–7) (Fig 109; Fig 113).

Properties on the east side of the Bridge Foot: archaeological evidence

Most of the evidence for the buildings along the eastern side of the Bridge Foot consists of fragmentary and truncated trench-built foundations of mortared chalk rubble, which on stylistic grounds can be dated to the 13th–16th centuries. The northernmost feature was a mortared, chalk rubble, pier base (F29.3) (Fig 106). Nearby was a second fragment of mortared

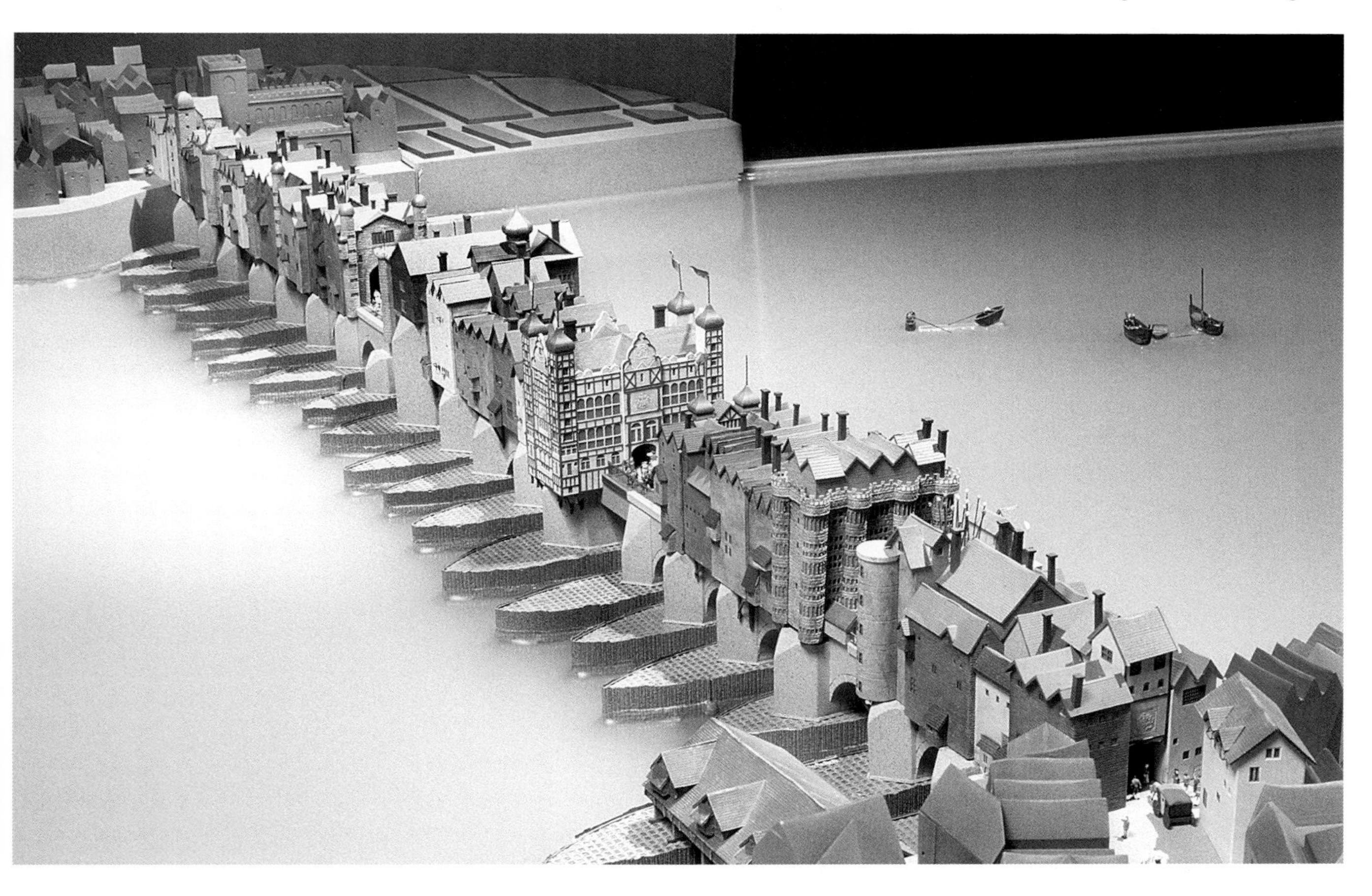

Fig 109 View of the Bridge Foot in c 1600, looking north; part of the model from the Living Bridges *exhibition (Royal Academy; photograph by A Chopping)*

chalk and ragstone rubble masonry, the pitch of this stonework suggesting that it was the basal portion of a relieving arch for a wall aligned north–south (F29.4). This wall probably formed the western edge of a cellared building fronting on to the east side of the bridge approach road, before the 1761 road widening (see Chapter 13.2).

Another fragment of a pre-1761 building was a truncated strip of unmortared, rammed chalk which served as the foundation for an east–west wall (F29.7). Further south there was no realignment of buildings in 1761, for a fragment of mortared chalk wall foundation (F29.6) lay on the same alignment as the external wall of the building standing here in 1821 (Fig 106). Later a small rectangular cellar with mortared chalk rubble walls and a brickearth and mortar floor was added (F29.8: Fig 106). Pottery from within the cellar floor dates to 1270–1500, which suggests that the cellar was built during the 15th century. Later, a mortared ragstone wall foundation was added to the east wall of the cellar, implying that another cellar was added to the adjoining property (F29.9) (Fig 106). This new cellar had a mortar or brickearth floor (F29.10) containing pottery dating to 1480–1500 and two worn Penn floor tiles (see Chapter 14.11). At some time between the late 15th and early 17th century part of the floor was paved with bricks (F30.3) (Fig 107). Probably at the same time, the internal faces of the walls were lined with brickwork, which was robbed out during the late 17th century (F30.4). Finds from the backfill of the robbed out wall foundations included a number of decorated 14th-century Penn floor tiles, two 17th-century tin-glaze floor tiles and a single wall tile (see Chapter 14.11).

To the south of the cellar a brick-lined cesspit was built (F30.5) (Fig 107). In the building to the east of the brick-lined cellar a brick-lined cesspit was constructed during the 15th or 16th century (F30.2). Later, a brick wall was constructed across the space between the cesspit and the cellar, to the west (F30.7); on its north side a brick silt trap, backfilled with coal dust, was later added (F30.8) (Fig 107). During the early 16th century a rubbish pit was dug right up against the existing walls (F30.9); pottery from the backfill of this dates to 1500–50. Half of the pottery sherds found within this pit were locally made Redwares, Tudor Brown ware sherds being the most numerous type (see Chapter 14.5). This pattern is common throughout London ceramic assemblages of 16th-century date, especially during the early part of the century before the impact of the Border ware industry began to make itself felt. There was also a high proportion of imported pottery (33%), including material from the Rhineland, the Low Countries, France and Spain. The standing buildings above these features presumably were destroyed by the fire of 8 September 1725, along with all the other properties around the Bridge Foot (CoLRO, BHCJ vol 6, 170–2).

The eastern side of the Bridge Foot appears to have been speedily developed after the fire of 1725, when a block of brick-walled and floored, cellared buildings was built on the south-east corner of the Bridge Foot (Fig 110); traces of three of these

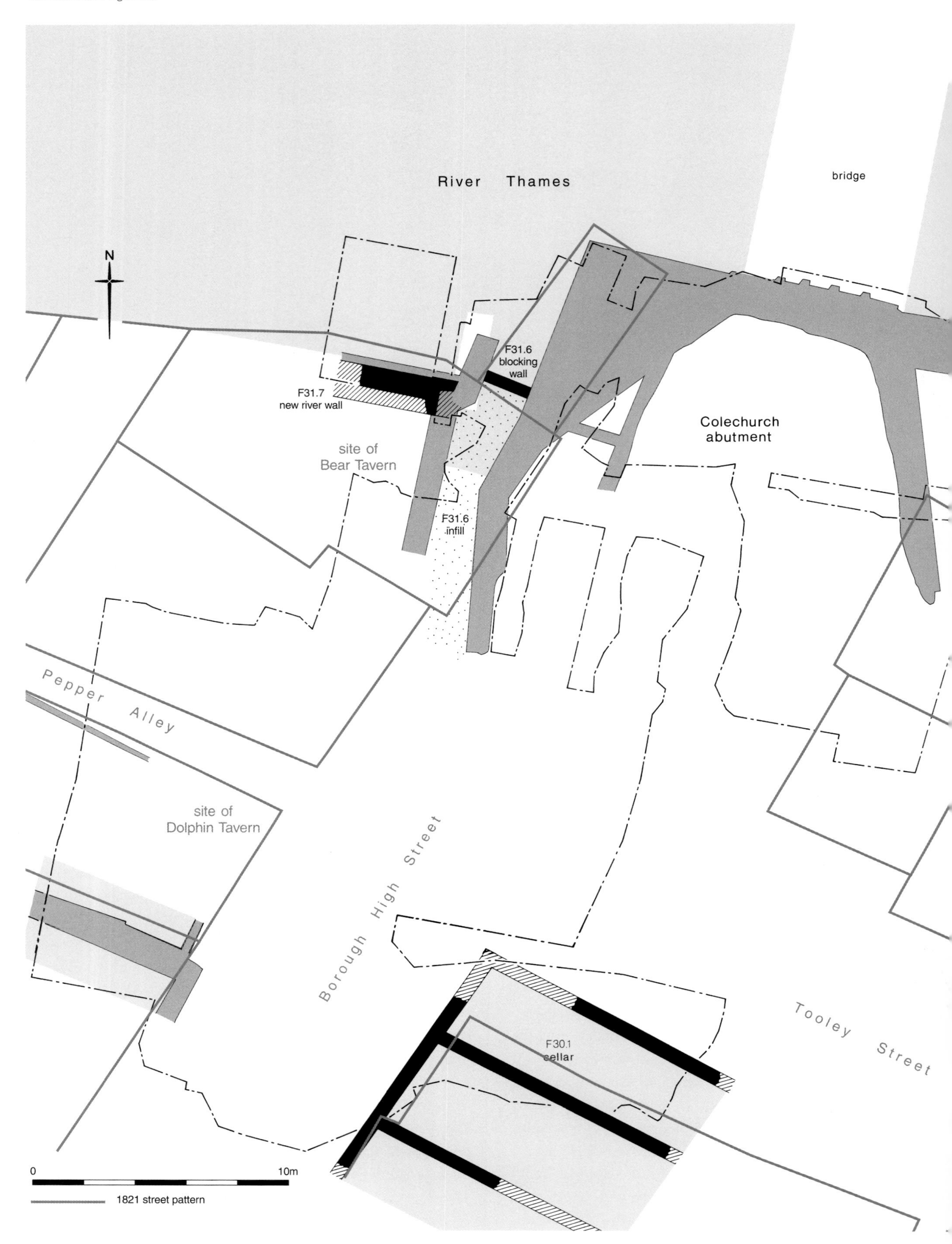

Fig 110 Plan of post-1725 building foundations and other structures found on both sides of the Bridge Foot and on Toppings Wharf; the extents of the streets and properties are taken from the 1821 survey and show the 1761 realignment; by 1747 St Olave's Dock was known as the Tooley River Stairs

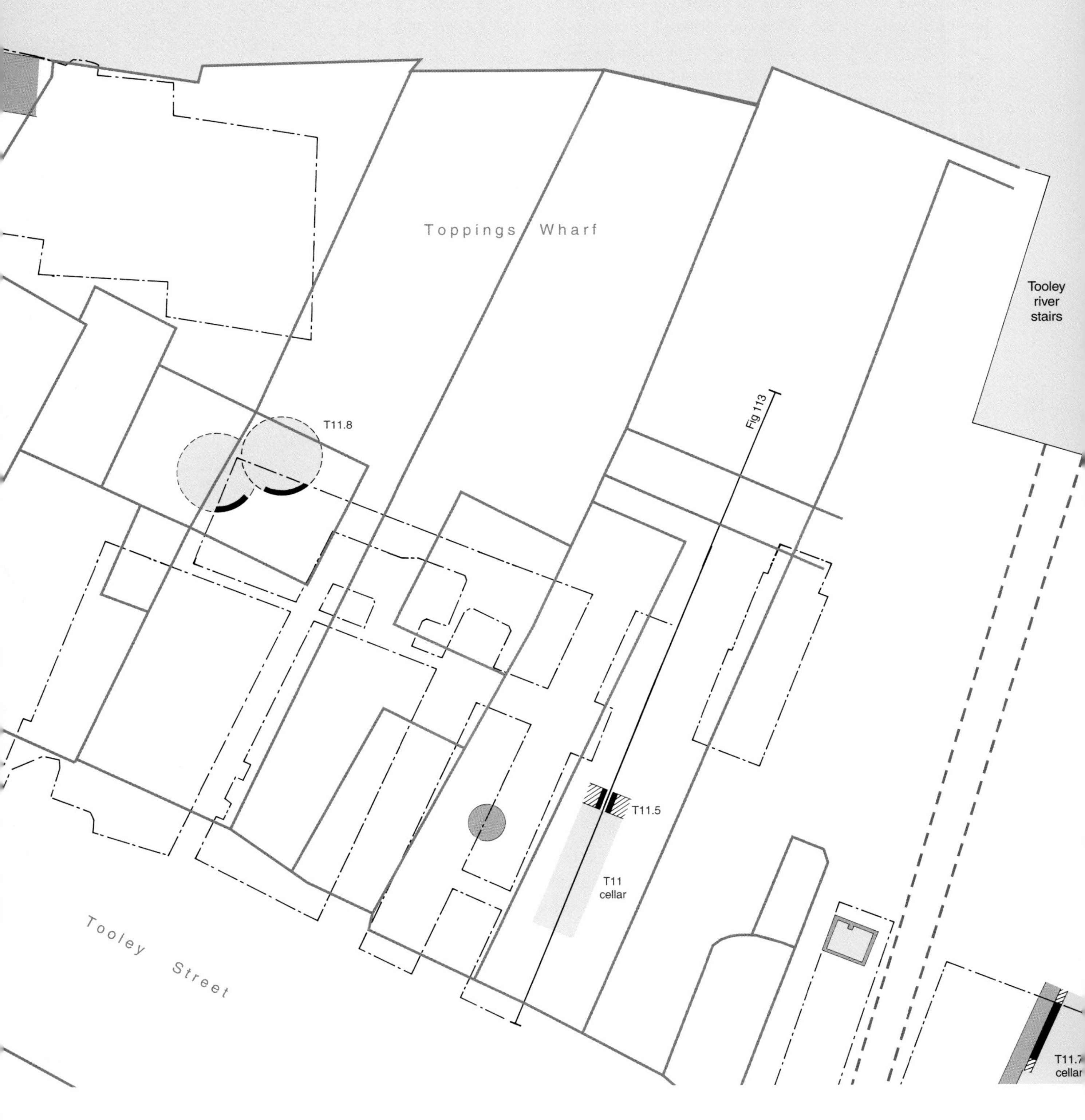
River Thames
Toppings Wharf
Tooley
river
stairs
Fig 113
T11.8
T11.5
T11
cellar
Tooley Street

properties were found (F30.1). In the northernmost building an earlier brick-lined cesspit (F30.2) was backfilled during the late 18th century (Fig 107), probably when it was demolished for the widening of the bridge approach road in 1761. In 1827 there were several four-storey brick buildings of 18th-century date on the east side of the Bridge Foot (Fig 125; Fig 127) and these were still standing in June 1831 (Cooke 1970, fig 6).

The backfills of the various post-medieval features contained an assortment of finds including wine glasses, a copper alloy candle holder, a fragment of tortoiseshell fan, 18th-century halfpennies and a stone 'alley' or marble (see Chapter 14.6). Post-medieval pottery included English Redwares, imported Rhenish stonewares and Chinese porcelain (see Chapter 14.5). Associated animal bones included all the major domesticates: cattle, sheep/goat and pig (see Chapter 14.8). Two unusual finds were the bones of a pheasant (F30.4) and a young calf: the presence of the calf perhaps indicative of the keeping of dairy cattle (Mitchell and Leys 1958, 192).

St Olave's watergate and dock: documentary evidence

A watergate or lane leading northwards from Tooley Street to the river near St Olave's church was in existence by the mid-13th century (Carlin 1983, 368). This suggests the existence here of some kind of river stairs or landing place, possibly at the mouth of a small natural creek or stream (see above), where river craft could moor away from the tidal river channel (Fig 105). A deed of 1353 relating to the property west of St Olave's church describes its western half as consisting entirely of a quay. In 1389 it was described as a tenement with adjacent wharf (Carlin 1983, 368). 'Quay' and 'wharf,' respectively the French and English terms for the same thing, were used interchangeably in medieval London, and they denoted the timber revetments which protected the property from the river (and, by extension, the property itself), more often than they denoted a 'dock' or 'harbour' (Dyson 1981). But the wording of the early documents concerning St Olave's Wharf makes it clear that a dock or landing place of some description was in existence here by the mid-13th century. It has been suggested that the earliest phase of the dock might have consisted of a steep break of slope along the southern side of the later dock (Sheldon 1974, 26–7), which could have been used for beaching boats.

Archaeological evidence confirms the existence of a substantial dock by the late 14th or early 15th century (see below). The earliest explicit reference to a dock occurs in 1539 when an ordinance of Guildable Manor (the name given to the bridgehead manor from the 14th century) forbade the casting of filth into the dock next to the watergate (Carlin 1983, 372). In 1546 it was described as 'Seynt Olaves docke within the Watergate' (Carlin 1996, 30). The St Olave's churchwardens' accounts of 1564–5 contain a detailed record of repair and maintenance work carried out on the dock. The sides were reinforced with elm piles and the 'entrye' was lined with 160ft of 2in-thick boards. A wooden platform was provided for parishioners to do their washing, a privy was built or rebuilt, and the wharf paved with ragstone and gravel (Carlin 1983, 372). There was also a public landing place here, later known as 'St Olave's Stairs' or 'Tooley Stairs'. The earliest representation of the dock is in Wyngaerde's panorama of *c* 1544, which shows a large rectangular inlet surrounded by hard standing. Access to the dock from Tooley Street was by way of a large gatehouse or archway (Colvin and Foister 1996, 31) (Fig 111). By 1716 the southern part of the dock had probably been filled in (Edwards 1974, 4), as by 1747 was the whole dock, except for a tiny inlet or landing place known as 'Tooley Stairs' and reached by a narrow alleyway (Hyde 1981, 13).

St Olave's watergate and dock: archaeological evidence

The 1970–2 excavations and 1984 watching brief revealed various fragments of the late medieval and Tudor dock, the earliest of them being part of the east wall (Fig 106). This was founded on a double line of piles, which supported sillbeams. The eastern or external side of the wall was faced with ashlar blocks, fixed together with iron cramps set in lead, behind which were two phases of late medieval mortared wallcore (T10.3). The probable southern continuation of the east wall of the dock consisted of two phases of adjoining masonry, probably of late 14th- or 15th-century date. One wall was constructed of ragstone and the other of chalk (Sheldon 1974, 26; Orton et al 1974, 75). Set into the east side of the wall were three large horizontal beams, interpreted as supports for a structure on the dockside. Part of the wall was robbed out in the 16th century or later, possibly when the dock was infilled during the 17th century.

The west wall of the Tudor dock was founded on a series of piles set in the base of a construction trench, and on the piles and overlaid by a series of oak sillbeams. The east or external side of the wall was faced with ashlar blocks with a mortared rubble wallcore, and there was evidence of a partial rebuild (T10.7). The positioning of the east and west walls of the Tudor dock shows that it was slightly wedge-shaped and extended for more than 32m north–south, and about 5–6m wide (Fig 106). It appears that the dock walls were of several phases of construction and that the east wall is certainly of pre-16th-century date.

At the southern end of the west side of the dock was a rectangular structure consisting of decayed horizontal beams or baseplates fixed by posts (Fig 106). It is interpreted as the foundation of a riverside quay, which apparently also served as part of the west wall of the Tudor dock. The dumped deposits below the structure contained early 15th-century pottery (T10.1). To the west of the structure, were a number of posts which may have been part of an earlier structure, and inside it were other timbers which may have formed part of a lined drain flowing into the dock (Sheldon 1974, 27). Set within the north-west corner of the structure was an unmortared, oval, chalk rubble foundation, which could have been the

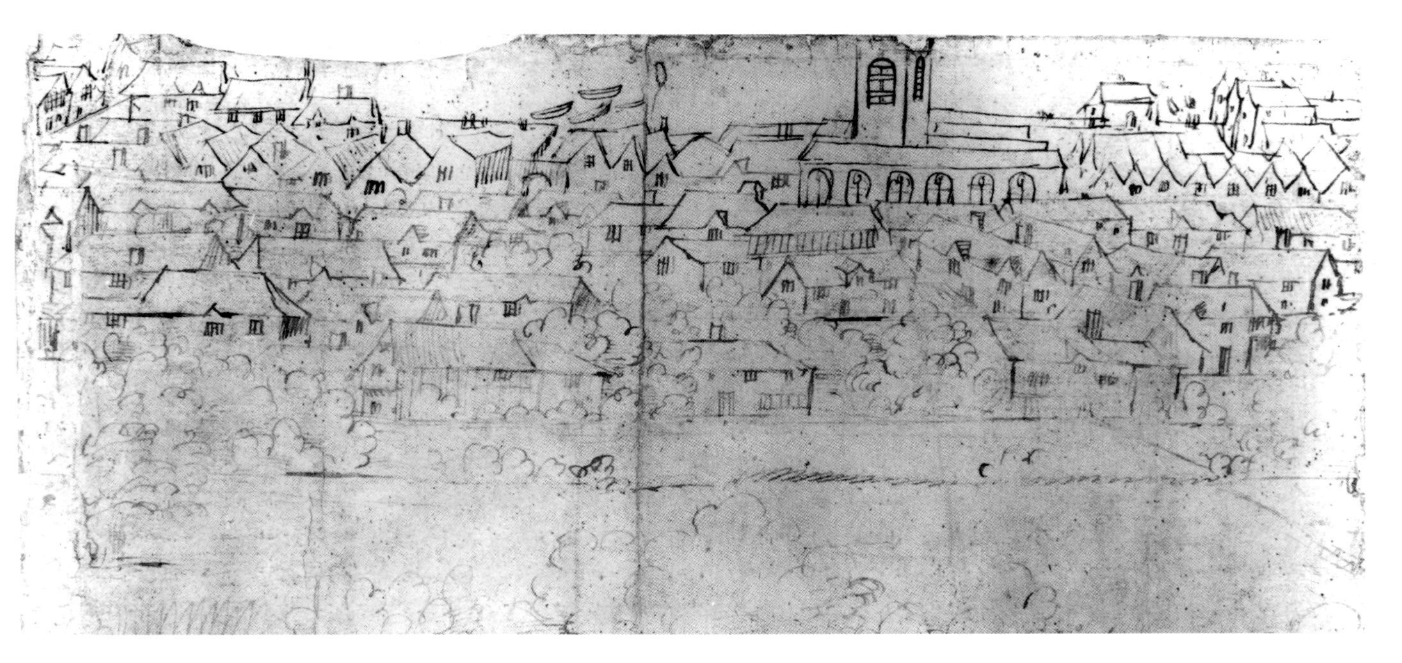

Fig 111 *The portion of Wyngaerde's panorama (c 1544) showing St Olave's church and the adjoining dock, which was entered via an arched gate; on the top left is the Bridge Foot and its surrounding buildings; view looking north (© Ashmolean Museum; reproduced from Colvin and Foister 1996, 28–9)*

foundation for the dockside cranebase seen on Norden's 16th-century view of London Bridge (Sheldon 1974, 26) (Fig 67). According to this view, the cranebase was situated on the extreme northern end of the west wall of the dock and, as it is not shown on Wyngaerde's depiction of the dock (*c* 1544) (Fig 111), it was presumably built during the late 16th century.

To the west of the dock was the western portion of a second timber structure consisting of large decayed horizontal beams or baseplates fixed by posts (T10.2). Above were traces of superstructure described as 'planking' (Sheldon 1974, 27). It is probable that this structure served as part of the west wall of either the late medieval or Tudor dock. To the south of the structure lay the base of a timber-lined drain.

The late medieval and post-medieval buildings at St Olave's Wharf

Unmortared, L-shaped, chalk rubble foundations, found on the west side of the medieval dock, formed part of a large rectangular cellared building with a gravel floor (T9.1) (Fig 106). They are now considered to be of late 14th-century date, as this whole area had been reclaimed by *c* 1350 (see above; Sheldon 1974, 25), although soil dumping still may have continued on the landward side of a presumed new riverwall.

To the west of the rectangular building were a number of small blocks of mortared chalk rubble masonry, interpreted as pier bases and truncated relieving arches (T9.2, .3) (Fig 106). The alignment and spatial pattern of these foundations suggest the existence of several cellared buildings, confirming that a very large area of land had been reclaimed by this date, and suggesting that by the 15th century a riverwall had been constructed, possibly in a similar position to the present one (see Chapter 11.2). Part of a late medieval cellared building, fronting on to Tooley Street, was represented by a dressed chalk rubble wall foundation with a mortared core (T9.5) (Fig 106). Part of this wall was robbed out during the 17th century (Sheldon 1974, 26). By 1325, this area to the west of the dock consisted of a messuage with 14 small shops at 'la Watergate', and the tenements on the site were rebuilt during the late 14th century. In 1555 the site was occupied by the 'Ship' and six tenements (Carlin 1983, 375).

The late medieval cellared buildings on the west side of the dock possibly remained in use until the area was devastated by fire in 1725. Post-medieval features on the west side of the dock consisted of a circular brick-lined cesspit (T11.1) (Fig 107), and two large circular brick-lined features of uncertain function, which were probably linked or interconnected (T11.8) (Fig 110); they were probably of 18th-century date, and backfilled during the 19th century (Sheldon 1974, 28). By 1799 this area was occupied by the premises of Thomas Preston, a lead merchant and shot maker. In 1806 a shot tower was built here (Edwards 1974, 4) (Fig 120; Fig 127). Molten lead was poured from the top of the tower, through a sieve or drop-pan, into a water tank at ground level. Possibly these two large circular features were part of the lead works and were water tanks for quenching hot metal.

The archaeological remains of the post-medieval buildings on the west side of the dock were very fragmentary. The main evidence was a sequence of superimposed cellar floors. The earlier floor, of late 17th-century date (T11.2), was overlaid by rubble and fire debris (T11.3), probably derived from the fire of 1725 (Fig 110). On this rubble was laid a new brick-paved floor (T11.4), containing a George II farthing of 1736. Contemporary with this floor was a fragment of brick-built cellar wall, aligned east–west (T11.5) (Sheldon 1974, 28) (Fig 110; Fig 113).

To the east of the dock during the late medieval period a deep rectangular cellar was constructed out of coursed chalk rubble blocks (T9.4) (Fig 106); a recess in the south wall is interpreted as a window. The building above the cellar was demolished during the 16th century (Sheldon 1974, 26). The most common pottery types from the backfill of the cellar, and the other 16th-century features, were Redware bowls, dishes, jars and dripping pans, some of Dutch manufacture (Orton et al 1974, 76). This assemblage has been reassessed and dated to *c* 1480–1550 (see Chapter 14.5). The cellar (T9.4) would have been part of a property, lying between the east side of the dock and St Olave's church, which was first documented in the mid-13th century when it was held by Isabella de Budely. Later the property was divided into four parts, then re-amalgamated as two (Carlin 1983, 368). In 1358 it consisted of two units, the western one consisting entirely of quay. In 1482–3 the property was conveyed to Magdalen College Oxford and, in 1555, was occupied by 12 tenements and a wharf (Carlin 1983, 368–70).

To the south-east of the dock, on the site of the medieval cellar, a brick- and stone-walled cellar (T11.6) was constructed during the 17th or 18th century to replace the earlier building (T9.4) (Fig 106; Fig 107). Later this cellar was relined with brick walls and the floor repaved (T11.7) and it probably remained in use until the 19th century. Nearby was a large rectangular brick-lined cesspit (T11.1), which was backfilled during the 19th century (Sheldon 1974, 28) (Fig 107). In 1843 all the buildings around the former St Olave's Wharf were destroyed by fire and a new block of warehouses constructed on the site, afterwards named Toppings Wharf after a previous tenant. These warehouses were demolished in 1970 (Edwards 1974, 4).

12.2 The 13th-century river erosion at St Olave's Wharf

During the 13th century the Southwark foreshore around the bridgehead suffered from very serious river erosion. At St Olave's Wharf this resulted in the removal of the northern wall of the 12th-century cellared building (see Chapter 8 and Fig 61) and of any sign of the associated revetments, which must have lain further north. The erosion was also represented by a massive truncation horizon aligned north-west to south-east (T8.1) (Fig 112; Fig 113), its alignment suggesting that St Olave's Dock may already have existed (see below). Perhaps the dock originated as a small natural creek draining northwards into the river, which, greatly enlarged by erosion, was formed into a distinct inlet downstream of the bridge abutment. Alternatively, the alignment of the truncation horizon could have resulted from the bridge abutment upstream acting as a breakwater, and protecting the foreshore immediately downstream from erosion. The highest observation on this truncation horizon was at 1.80m and the lowest at -1.80m OD.

After the erosion ceased, fluvial sands and gravels representing the new foreshore were laid down (T8.2) within the new inlet (Fig 113). Pottery from these deposits is now dated to *c* 1230–70 (see Chapter 14.5), a little earlier than the published date for this material (Sheldon 1974, 25; Orton et al 1974, 66). Above these deposits accumulated a band of blue-grey gleyed fluvial clay 60–150mm thick, its top at 0.03m to -0.52m OD (T8.3). Later, chalk rubble was dumped on the foreshore sands to consolidate the ground surface, and a beech baseplate aligned east–west laid down. This was anchored by stakes (T8.4) and was part of a revetment of unknown design; its upper part was probably washed away by another phase of flooding. Over time the new foreshore was consolidated by a build-up of sands, gravels and silts (Fig 113). These layers appear to be a mixture of fluvial deposits and dumped material deposited on the southern edge of the sloping foreshore to consolidate and reclaim the area (T8.5). Later, there was a period of sustained build-up of fluvial deposits consisting of up to 800mm depth of grey and brown clays (T8.6), containing large lenses or bands of organic material, their top at 1.56m OD. This was followed by the silting up of the inlet with a depth of over 1.2m of silts and peats (T8.7) and containing an *ex situ* oak beam. Finds from these deposits included a lead ampulla of St Thomas (see Chapter 9.4 and Fig 77) and pottery now dated to *c* 1240–70 (Sheldon 1974, 25; see Chapter 14.5). These sediments were sealed by soil dumping (Fig 113), which contained pottery which is now dated to *c* 1270–1350 (see Chapter 14.5). This soil dumping probably occurred during the early 14th century, as part of a reclamation of the area in preparation for the construction of various wall foundations.

The 13th-century river erosion in north Southwark

It is clear from both documentary and archaeological evidence that the area of foreshore around the Southwark bridgehead suffered from serious erosion on several occasions (see Chapter 6.1). The 'Annals' of Bermondsey Abbey record local floods in 1099, 1208, 1230 and 1294 (Luard 1866). The flood of 18 October 1294 is said to have devastated the abbey's estates, which included much of the riverside from Southwark to Rotherhithe, and this implies that any revetments or embankments were swept away or breached by the floodwaters. In 1303 the prior of St Mary Overy petitioned the king concerning the 'continual resistance which without ceasing we attempt against the violence of the river Thames, on whose banks our home is situated' (Taylor 1833, 37) (Fig 114). In 1327 the Bishop of Winchester, as diocesan, received a papal mandate ordering him to relax 60 days of enjoined penance to all those who contributed within three years to the repair of St Olave's church, which had been damaged by the tide beating up against its walls and carrying off bodies from the graveyard (Bliss and Jesse 1895, 256) (Fig 114). Numerous commissions for the repair of stretches of the riverside walls

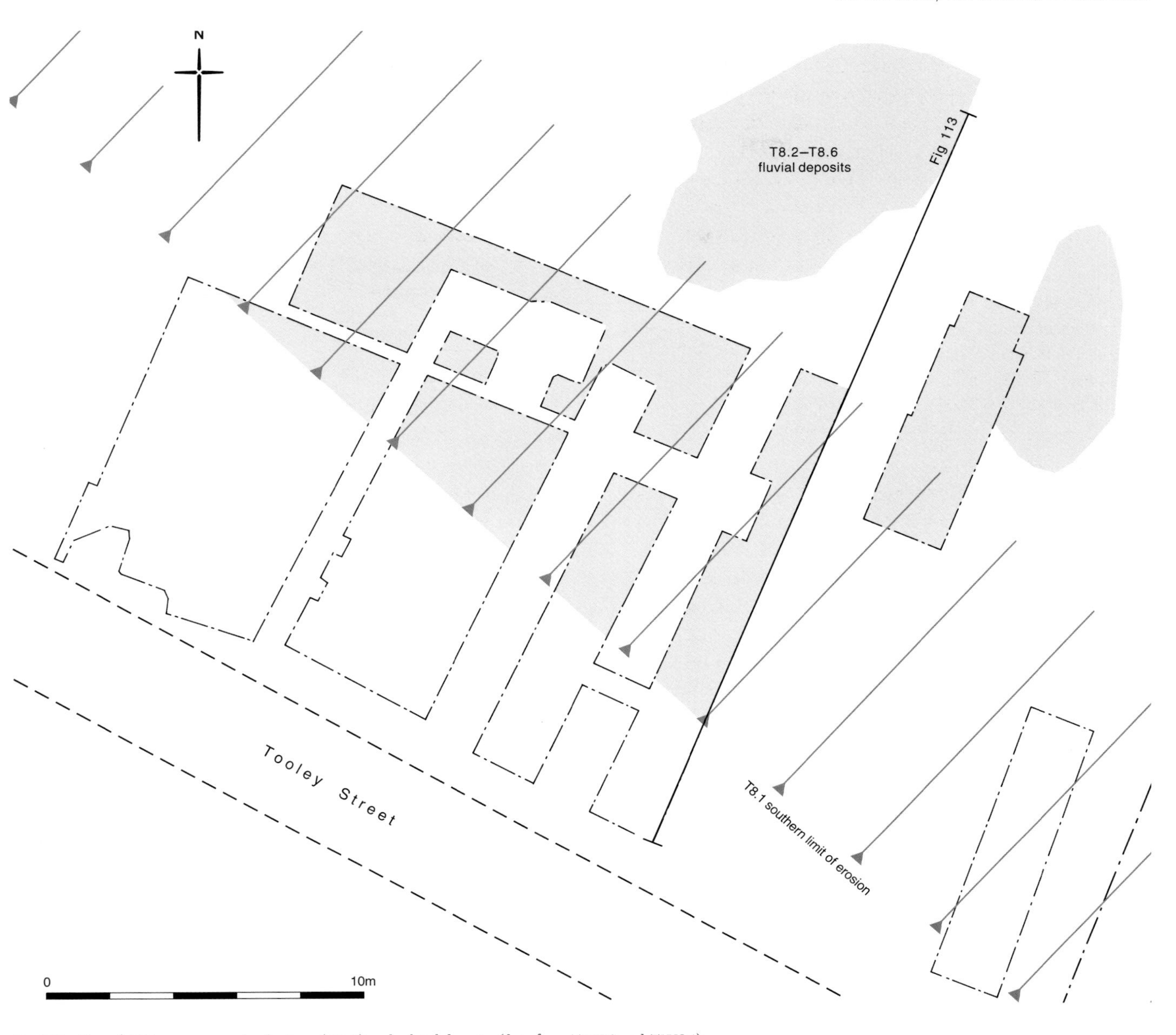

Fig 112 Plan of 13th-century erosion horizon (T8.1) and related deposits (data from TW70 and TW84)

and revetments in the Southwark area were issued – in 1298, 1303, 1309, 1311, 1320 and 1325 – showing that flooding was a recurrent problem (Edwards 1974, 5).

It has been suggested that the riverine erosion at St Olave's Wharf dates to the late 13th century, and so represents part of documented foreshore erosion of c 1294–1327. This, however, amounted to a series of separate floods and erosion phases, rather than the single event identified at St Olave's Wharf (Sheldon 1974, 25). Moreover, it was primarily a general absence of Surrey Whitewares that suggested a late 13th-century date for these deposits, on the assumption that Surrey Whitewares were in common usage only by c 1300 (Orton et al 1974, 66). Since 1974, however, the dating of medieval pottery in London has been refined, and the introduction of Surrey Whitewares in London is now dated to c 1230 at Kingston-upon-Thames and they are found in deposits at Billingsgate dating to c 1250 (Pearce and Vince 1988, 13); an earlier date for this erosion should now be considered.

The pottery from the erosion deposits at Toppings Wharf has now been redated to c 1230–70, so that the observed erosion should perhaps be identified with the documented floods of 1208 or possibly 1230.

Evidence of 13th-century foreshore erosion has been found on a number of sites on the upstream side of the Southwark bridgehead, confirming that it was a widespread phenomenon. Excavations in 1973 at New Hibernia Wharf, upstream of the bridgehead, revealed two phases of 13th-century foreshore erosion (Fig 114), the first phase followed by the deposition of fluvial sands and gravels. This deposition was interrupted by the excavation of two east–west aligned features. The earlier feature was probably a linear ditch dug to provide material for building new revetments (Evans 1973, 100); pottery from silts infilling this ditch dates to the late 13th century. The later feature appeared to be the construction trench for a chalk rubble riverwall, founded on sillbeams. This wall was apparently destroyed by a second phase of erosion, followed

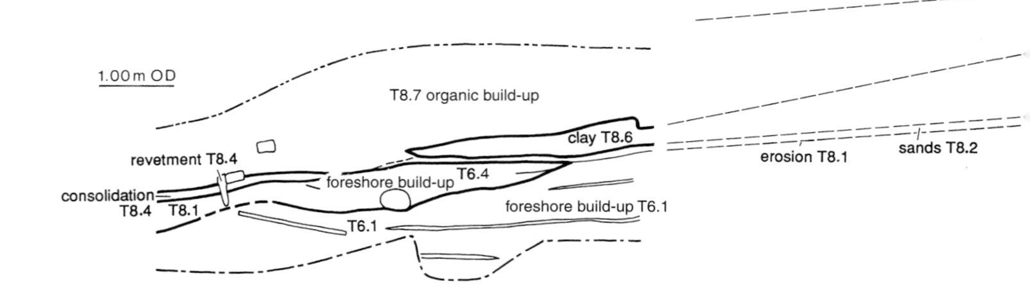

Fig 113 Cross section through the medieval erosion deposits (T8) and later features on Toppings Wharf; see Fig 106 for location of section (redrawn TW70 section A, plus TW84 data)

by the deposition of grey silt and then by gravel representing the new foreshore. All these sediments were later sealed by soil dumping to raise the ground level, as part of a reclamation process that was finally to culminate in the construction of a mortared ragstone rubble riverwall in the 17th century (Evans 1973, 100).

Investigations elsewhere in Southwark reveal a similar pattern. Excavations at Montague Close in 1969–71 revealed evidence of a truncation or erosion horizon, its top at c 1.30m OD, and overlaid by a series of fluvial silts and gravels (Fig 114). There is no precise dating, but the occurrence has been assigned to the 13th century by the presence of a few sherds of Surrey Whiteware (Dawson 1976, 50). Excavations in 1983–4, on the site of the hall of Winchester Palace, revealed that between c 1150 and 1220 there was a phase of foreshore erosion, which, along with subsequent reclamation, must have taken place before the construction of a hall on the site by 1220–1 (Seeley in prep) (Fig 114). The primary deposit above the erosion horizon consisted of gravel, above which was a succession of thin silts and gravels. Later, the ground level was raised by soil dumping, pottery from which, including again Surrey Whitewares, dates to 1150–1350. This suggests an even earlier date of introduction for this type of pottery than previously allowed, further strengthening the case for dating all the archaeologically recorded instances of riverine erosion in Southwark to the first half of the 13th century.

It is thus clear from the evidence of three sites in north Southwark that during the 13th century there were at least two phases of riverine erosion both up- and downstream of the bridge (Fig 114). Of these three observations, only the one at Winchester Palace is closely dated, to 1150–1220, which is interesting as it suggests that the second was later

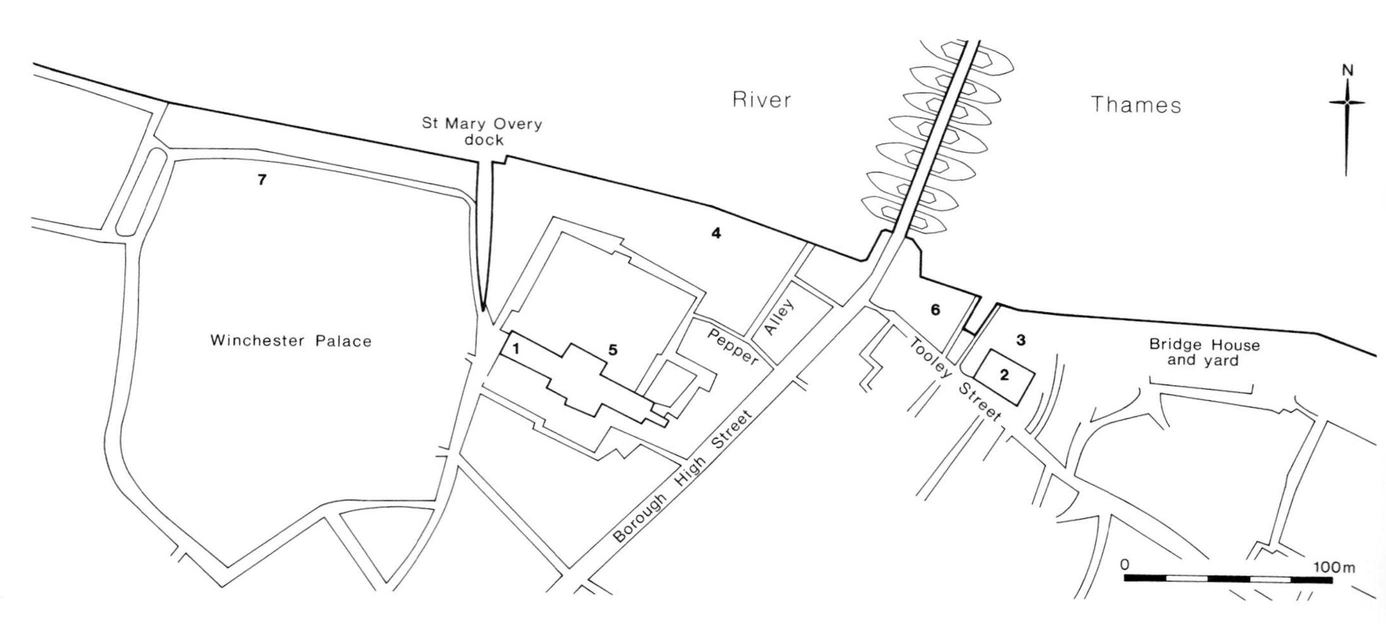

Fig 114 Map of north Southwark showing the 16th-century road network and riverwall, with the sites where riverine erosion occurred during the 13th century; the road network is taken from Carlin (1996, 34)

Key: documented locations of 13th-century erosion: 1 – Priory of St Mary Overy, 2 – St Olave's church, 3 – St Olave's burial ground; archaeological sites where the 13th-century erosion has been located: 4 – New Hibernia Wharf, 5 – Montague Close, 6 – Toppings Wharf, 7 – Winchester Palace

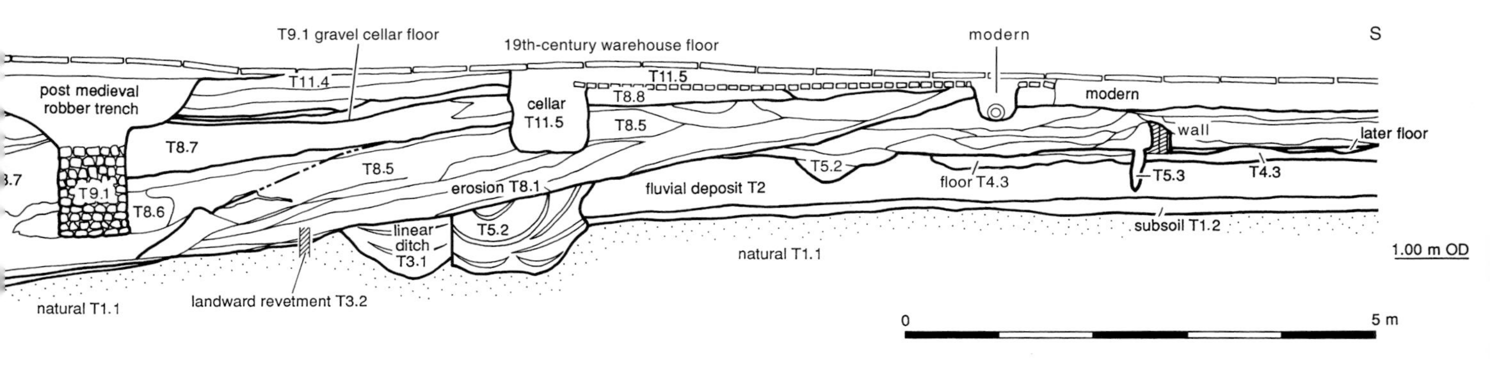

than 1220: possibly the two erosion phases correlate to the documented Southwark floods of 1208 and 1230 (Luard 1866). The original interpretation of the St Olave's Wharf erosion, as the result of the 1294 floods (Sheldon 1974, 25), cannot now be substantiated, as the associated pottery has now been redated to *c* 1230–70 (see Chapter 14.5). The influence of the bridge on the hydrology of the river should also be taken into account, as it effectively acted as a partial dam, increasing the volume of water on the upstream side and accelerating the restricted downstream flow (see Chapter12).

The 13th-century erosion can, thus, be seen as one of several periods of riverine action in north Southwark, beginning shortly before the Roman occupation, when the area of the bridgehead was covered by fluvial clays (see Chapter 2), persisting into the late Saxon period, when the late Roman revetments and part of the settlement itself were swept away (see Chapter 6.1), and on into the late 11th century, when there was considerable erosion of the foreshore on the upstream side of the bridgehead.

13

The last days of old London Bridge

Bruce Watson

13.1 Westminster Bridge

Until the construction of a bridge at Putney in 1729, the nearest upstream crossing from London Bridge, ferries aside, was Kingston Bridge. By the mid-17th century the continual expansion of London and Westminster was generating increasing road traffic and pressure was mounting for a bridge to be constructed in the Westminster area to relieve the congestion. In 1664 the Privy Council examined the case for building a new bridge between Westminster and Lambeth. This proposal was opposed by the City of London authorities, who claimed, falsely, that the present bridge was perfectly adequate and that the building of a new bridge would damage London's trade and so create poverty, cause flooding upstream at Whitehall and hinder navigation (Walker 1979, 33). King Charles II accepted the City's arguments, receiving a £100,000 'loan' from the City in return for opposing the construction of a new bridge (Walker 1979, 34)

Further attempts were made during 1727–36 to build a bridge between Westminster and Lambeth, all of them strongly opposed by vested interests. In 1721, for instance, the House of Commons received a petition from John Pond who rented the toll or wheelage of London Bridge. He claimed that he had paid a large sum for his lease, of which some 20 years remained, and that the construction of a new bridge would ruin his venture (Walker 1979, 45). Despite much opposition, Parliament eventually passed a bill in May 1736 authorising the construction of a new bridge at Westminster. The act stated that no houses were to be erected on the bridge, which was to be funded from the profits of a state lottery (Act 1736). Construction of Westminster Bridge started in 1738, but was not finished until 1750 (Fig 115). Instead of constructing cofferdams (see Glossary), within which the pier foundations could have been constructed, the Swiss engineer Charles Labelye used enormous flat-bottomed barges, some 80ft (24.4m) long and 30ft wide (9.1m) and called 'caissons' (see Glossary). These barges were floated into position with the basal course of stonework already laid, then another three courses of stonework were constructed inside the barges, then they were sunk into dredged voids in the riverbed, the sides detached for reuse. Then the rest of the piers were constructed on top of these foundations (Ruddock 1979, 8, fig 5). There were several flaws in this design. Firstly, there were no piles to strengthen the pier foundations. Secondly, the gravel and sediments that the piers were founded on proved not to be load-bearing. Thirdly, no account was taken of the scouring which undermined some of the piers. During the construction of the bridge, cracking and visible movement of the masonry were noticed and, in September 1744, some of the radial joints on the faces of two arches opened. The fourth pier from the western end suffered badly from subsidence and had to be rebuilt and strengthened with piles (Ruddock 1979, 15–16). After 12 years of construction Westminster Bridge opened to traffic on 18 November 1750.

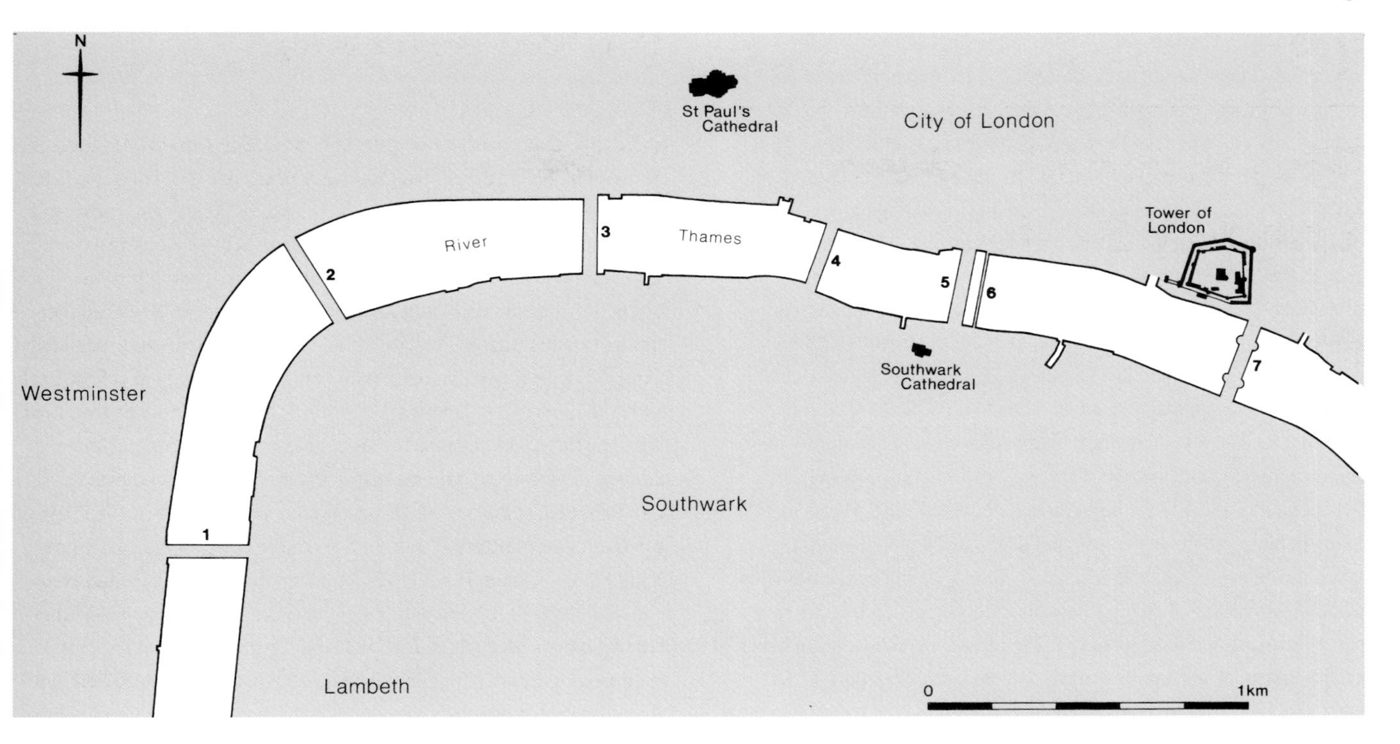

Fig 115 Map of the London area showing the River Thames and the bridges mentioned in the Cities of London and Westminster, plus the Boroughs of Lambeth and Southwark Key: 1 – Westminster Bridge 1738–50; 2 – Waterloo Bridge 1811–17; 3 – Blackfriars Bridge 1760–9; 4 – Southwark Bridge 1814–19; 5 – New London Bridge 1824–31; 6 – Old London Bridge c 1176–1209; 7 – Tower Bridge 1881–94

The new Portland stone bridge had 15 arches and two recesses on each pier for pedestrians to shelter in, 12 of them covered with domes (Fig 116). The bridge was 371.85m long, of which 265.17m (71%) was occupied by arches or voids. The bridge was 13.41m wide, with a central roadway 9.14m wide, flanked by two pavements each 1.82m wide (Walker 1979, 283). The design of the foundations did not prove a long-term success, for by the late 1820s scouring out of the river sediments was causing serious damage to the foundations, and with the completion of a new Westminster Bridge in 1862 the old one was demolished.

The construction of Westminster Bridge brought a shower of criticism upon London Bridge, for the new bridge had a much wider roadway which allowed the easy and fast movement of two-way traffic and its wider arches allowed for safe and easy passage of river traffic, in telling contrast with the narrow and dangerous gullies between the starlings of London Bridge (see below). This criticism led to calls for the replacement and improvement of London Bridge. The restrictive effect of London Bridge on the growth of the port of London is clearly illustrated on Ogilby's survey of London in 1676: downstream of the bridge at Billingsgate are numerous ocean-going vessels, while upstream at Queenhithe there are only small craft and barges (Hyde 1976, 18–19).

Fig 116 Westminster Bridge in 1749, view looking north, as shown in Samuel and Nathaniel Buck's Prospect of London and Westminster (Hyde 1994, pl 40) (MoLPL)

In 1756 a further proposal to rebuild London Bridge was abandoned because of lack of money, it being decided instead to modify the existing bridge (see 13.2 below) and construct a new bridge over the Thames at Blackfriars in 1760–9 (Fig 115). This nine-arch Portland stone bridge was designed by Robert Mylne and was funded by a loan, arranged by the Corporation of London on the understanding that the money was to be repaid by the collection of tolls, although in 1785 the bridge was freed from all tolls (Welch 1894, 135). In 1811–17 another new bridge, designed by John Rennie, was built over the Thames at Waterloo. During 1812–16 a cast-iron bridge was constructed at Vauxhall by John Walker, and in 1814–19 yet another bridge designed by Rennie was constructed at Southwark, between London and Blackfriars Bridges (Fig 115). The Southwark, Vauxhall and Waterloo Bridges were all built by private joint-stock companies as speculative ventures, the intention being to make money out of tolls. In terms of bridge design, the period 1800–35 saw three important developments. Firstly, there was a lengthening of the spans of arches. Secondly, there were changes in the form and architecture. Thirdly, there were great improvements in foundation design, as steam-powered pumps allowed the more efficient use of cofferdams (Ruddock 1979, 175). The construction of pier foundations using sunken barges or caissons was last made extensive use of at Vauxhall Bridge (1812–16), thereafter cofferdams (see Glossary), drained by steam-powered pumps, were the norm.

13.2 Post-medieval modifications to London Bridge

As the roadway of medieval London Bridge was only *c* 3.66m to 4.57m wide (see Chapter 8) it was barely wide enough for two carts or coaches to pass each other. Despite the provision of three pairs of small recesses, from which persons crossing the bridge on foot could view the river and take refuge from the traffic, the narrowness of the road and the volume of traffic was hazardous for pedestrians, who sought to cross in safety by walking behind a vehicle (Hawksmoor 1736, 11). Pennant (1813, 447) described the old (pre-1757) road over London Bridge as 'narrow, darksome and dangerous to passengers from the multitude of carriages'. In spring 1749, after the public rehearsal of Handel's *Music for the royal fireworks* in the Vauxhall Gardens, the crowds which had been attracted by the performance created so much extra traffic that London Bridge was blocked for three hours.

It should be stressed that the roadway of London Bridge was of about average width for a medieval English bridge, the width in the great majority of surviving examples being less than 4.5m wide and those over the Thames at Abingdon, Culham, Newbridge, Radcot and Wallingford all having roadways 3.6–4.5m wide (Harrison 1992, 246). The 1752 survey of the 115 bridges in the West Riding of Yorkshire revealed that only one was wider than 18.5ft (5.63m) wide (including the thickness of the parapets). The great majority of Yorkshire bridges were only 10ft (3.04m) to 14ft (4.26m) wide and, therefore, only suitable for single lane traffic (Ruddock 1979, 27). The general absence of new bridge building in England between 1530 and 1750 suggests that the existing medieval bridges generally sufficed for the volume of road traffic until the economic changes of the 18th century. It was apparently only the massive growth in road traffic, which began during the late 18th century, that triggered the rebuilding or widening of many medieval bridges (Harrison 1992, 257–60). In London, however, the great increase in road traffic would appear to have started during the 17th century, judging by the attempts made from this date onwards to widen the roadway of the medieval bridge.

The first attempt to relieve the traffic congestion on London Bridge was made in 1633, when, after fire had devastated the houses on the northern end of the bridge, it was proposed that the roadway there should be widened from 4.57m to 5.48m on a temporary basis (Chandler 1952, 19). When the northern end of the bridge was refurbished again after the Great Fire of 1666, the roadway there was widened to 6.09m (Home 1931, 244) and the cellars on each side rebuilt (Knight et al 1832, 203). In 1685 the rest of the roadway was widened to 6.09m and, when the houses were rebuilt at the same time, the headroom above the roadway was increased from one to two storeys (to stop high carts crashing into the overhanging buildings) (Seymour 1734, 48). It is probable that, as part of the various road-widening schemes during the 17th century, all of the bridge superstructure was widened by building over part of the starlings. Dodd stated in 1820 that the bridge was originally 24ft (7.32m) wide, with two phases of widening. During the first phase of widening an additional 8ft (2.44m) of masonry was added to the bridge and during the second phase a further 15ft (4.57m) of masonry was added, making a total width of *c* 47ft (14.32m), which fits closely with the width of the bridge in the 1799 survey (Dance 1799). It is clear from various drawings and descriptions of the bridge produced in the early 19th century that most of this earlier widening was rebuilt during the second phase of widening in 1757–62 (see below).

In 1728 the Stonegate was also widened (Seymour 1734, 49). In 1722 a series of tolls was introduced for carts and wagons using the bridge, and the Court of Common Council ordered that, to prevent obstructions on the bridge, three persons were to be appointed to direct approaching traffic to the left side of the roadway (Seymour 1734, 49). This is apparently the first documented observance of the keep left rule, followed in England ever since. Why the left-hand side was chosen is not certain, but the most plausible explanation is the English tradition of mounting a horse from the left and leading a horse on the left side, which makes this half of the road the natural choice for two-way traffic.

In 1736 Hawksmoor suggested that the four arches on either side of the chapel pier should be converted into two much larger spans, of some 24m each, by removing the two central piers of either group (Hawksmoor 1736, 13)

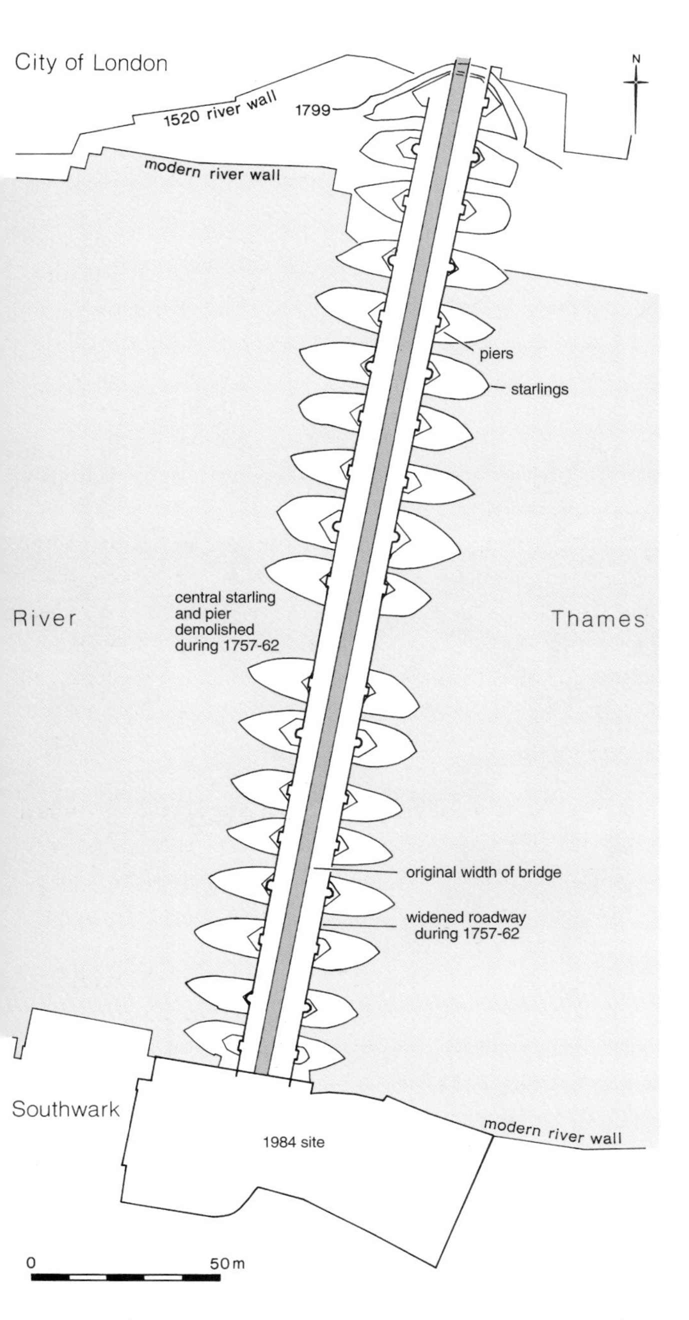

Fig 117 Plan of the widened bridge, showing the narrow medieval portion; data from Dance's 1799 survey

Fig 118 An engraving of the widened bridge by Smith (1814), showing the earlier masonry visible under the arch during the 'Great Frost' of January to February 1814 (MoLPL)

and, with a new bridge costed at £185,950 by George Dance the elder, a modification of the existing bridge was as much as the Bridge House was prepared to consider. In 1756 the decision was taken to go ahead with this option, no doubt influenced by the recent completion of the new bridge at Westminster (Act 1756). It was proposed to remove a single central pier, creating one large central arch, thus improving navigation and water flow (Fig 53; Fig 119), and it was also decided to widen the roadway by demolishing all the buildings on the bridge and widening the piers and arches by extending the existing piers some 4.0m across the starlings (Fig 118; Fig 119). This created a roadway 10.0m wide, flanked by pavements each some 2.1m wide, giving a total width of 14.02m (44ft) (Fig 117). The widened bridge was faced with Portland stone, and each end of the extended piers was decorated with lancet-shaped panels in a mock-Gothic style (Fig 118).

The widened bridge had new baroque-style balustrades, and street lights, and twin alcoves were erected on each pier (Fig 119; Fig 120). Fourteen of these alcoves were covered by semicircular domes (Fig 119), very similar in design to those erected a few years before on Westminster Bridge; four of them still exist in the Greater London area (see 13.5 below). This work was to be paid for partly from an increased scale of tolls for vehicles (hackney coaches exempt) and for boats and barges passing under the bridge (Home 1931, 267–8). During 1761 some of the buildings on the southern Bridge Foot were demolished to widen the approach road to the bridge (see Chapter 12).

These works came under the architectural direction of George Dance the elder, Clerk of Works to the City of London (1743–68), and Robert Taylor (Stroud 1971, 52). Demolition of the buildings on the bridge started in 1757 and finished in 1761. In the period 1758–65 a total of £82,000 was granted by Parliament for the alterations to the bridge (CoLRO, LBC vol 14, 461–542). While this work was in progress traffic crossed over a temporary wooden bridge which was built on the starlings on the west side of the bridge and burnt down on 11 April 1758. It was claimed that the fire was the result of arson and, thereafter, the temporary bridge was guarded by armed men. A second temporary bridge opened on 18 October 1758 (Home 1931, 269; Thomson 1827, 391–2). The new central

Fig 119 *An engraving of the southernmost two arches of the modified bridge at low water and the buildings of the Bridge Foot which fronted on to the river during the 1820s, by E W Cooke (1832), view looking west; notice the alcoves and cutwaters (MoLPL)*

Fig 120 *The London Bridge alcove in the grounds of Guy's Hospital (MoL, R Bartkowiak)*

arch, which required a very elaborate timber former or framework (Fig 121), was completed by July 1759 (Home 1931, 275). The new central arch was 21.46 m wide (Dance 1799). The water was now found to flow much faster through the new central arch than elsewhere, resulting in a local scouring of the riverbed, which in turn weakened the adjoining starlings and piers and caused the masonry to crack. In 1760 rubble from the demolition of the City gates was dumped under the central arch in an attempt to prevent further scouring (Home 1931, 282–3). In 1793–4 the two starlings of the central arch were strengthened by the insertion of a series of nine parallel timber beams to serve as braces, although by 1799 all but two of these timbers had washed away (Dance 1799, surveyor's plan 16). The alterations to the bridge were all completed during 1762. However, the work of cutting arched openings through the ground storey of the tower of the church of St Magnus the Martyr, to allow more space for pedestrians approaching the bridge down the east side of Fish Street Hill, was not completed until 25 June 1763 (Home 1931, 276; Thomson 1827, 405).

These modifications gave the bridge a completely new appearance (Fig 53, Fig 118; Fig 122), giving it (apart from its small arches and starlings) a much more modern aspect and not least ensuring that it continued in service for a further 69 years. The small arches still hindered river navigation, however, generating renewed pressure for complete rebuilding. This

Fig 121 The elevation of the timber former for the central arch of London Bridge, published in 1760 (Soane, drawings vol 19/28)

Fig 122 Modified London Bridge after widening and creation of the great arch, view looking east, by Major George Yates c 1820; notice the shot tower on Topping Wharf (right-hand end of the bridge) (MoLPL)

was compounded by an ever increasing volume of traffic, as people preferred to use London and Blackfriars Bridges, which charged little or no tolls, rather than the new private bridges at Southwark and Waterloo. A traffic census revealed that, during one day in July 1811, 89,640 pedestrians, 2924 carts and drays, 1240 coaches, 769 wagons, 764 horses and 485 gigs and taxed carts passed over London Bridge (Welch 1894, 127).

In 1800 a House of Commons Select Committee resolved that London Bridge should be rebuilt to open up the area between London and Blackfriars Bridges (Fig 115) as part of the port of London. The new bridge was to be built of cast iron, with a large central arch which would stand at least 19.8m above high water mark, to allow ocean-going ships with struck top masts to pass under, and was to be sited upstream from the existing bridge (CoLRO, LBC vol 14, 543–603). The driving force behind these resolutions, which helped seal the fate of the medieval bridge, was the rapid expansion of the port of London, where the value of the import and export trade it handled rose from £10,263,325 in 1700 to £22,992,004 in 1790, an increase of 248% (Report 1796, 267).

13.3 The influence of London Bridge on the hydrology of the Thames

The medieval bridge was more than a feature in the riverscape: it was a major influence on the hydrology of the river, the massive bulk of its starlings and piers so restricting the water flow as to act as a partial dam. The effect of the construction of the Colechurch bridge was apparent upstream at Queenhithe; here the estimated highest tidal level of *c* 1.35m OD at a phase of revetment dating to 1165–70 compares with one of *c* 1.80m OD at a revetment dating to 1181 (Ayre and Wroe-Brown in prep). As this dramatic increase in river level coincided with the construction of the stone bridge (*c* 1176–1209) (see Chapter 8), the two phenomena must be linked. The medieval bridge over the Saône at Lyons, France, also had the effect of a partial dam, and, after its demolition in the early 19th century, the river level dropped so much that a dam was needed to enable upstream navigation to continue (Boyer 1976, 156). The passage of water through the bridge was further restricted by the installation of a number of waterwheels on certain of the northernmost and southernmost starlings, to power corn mills and extract water from the river (see Chapter 9.7); the last of these wheels was only removed in the 1820s.

In 1763 John Smeaton reported that the average height of the tides on the upstream side of the bridge had increased by 0.10m since the water flow was improved by the construction of the new central arch (Smeaton 1763). Rees (1819) noted that when the tide rose above the level of the starlings, some 166.1m (or 58%) of the 286.5m length of the bridge was blocked by its bulk, although, when the level dropped below the level of the starlings, the space through which the water could flow was only 62.2m, or 22% of the length of the bridge, the remaining 78% of the length being effectively blocked by the starlings. The main result of this partial damming was to encourage the build-up of sediments, both up- and downstream of the starlings, while in the narrow channels between them the rush of the ebb tide created deep channels or gullies. It is this continual scouring that explains why the starlings were in need of constant repair (see Chapter 10) and, although the introduction of the wide central arch in 1757–62 improved the water flow, it also created a new bank of sediment a little further downstream. Study of the average level of high water at London Bridge in 1820 revealed that the water level on the upstream (west) side was 7.5in (0.19m) lower than on the downstream (east) side, due to the bulk of the modified bridge acting as partial dam (Giles 1843, 88).

Telford pointed out in 1823 that the removal of the waterworks wheels from the bridge arches in 1822 (see Chapter 9.7) had greatly improved the situation by reducing the average difference or fall in water level between the up-

Fig 123 *An oil painting by Abraham Hondius, 1677, depicting the frozen Thames and medieval London Bridge; view looking west (MoLPL)*

and downstream sides of the bridge from 2.4–3.7m to only 0.9–1.2m. He calculated that the volume of water passing through the bridge between the lowest ebbs and high water of the ordinary (4.3m) tide was 582 million cubic feet and, that at low water, difference in water levels on each side of the bridge was 1.37m. From this he concluded that the removal of London Bridge would have serious consequences for navigation, in that it would allow a greater volume of water to pass faster both up- and downstream. This would have resulted in parts of the riverbed, at such upstream places as Teddington, being nearly dry for several hours during the ebb tide and, to maintain the present depth of water, he contended, dredging would be necessary (Telford 1823, 5). However, there was also the risk that any dredging, in combination with the increased velocity of the water flow, would damage the inadequate foundations of Westminster Bridge, which were founded only on river sediments (see above).

The rapid flow of water between the closely spaced starlings during the ebb and flow of the tides presented a particular hazard to the small craft passing under the bridge in either direction. This involved the dangerous practice of 'shooting the bridge', when the waterman would ship his oars and shoot through, while his passengers either stayed on board or disembarked and rejoined him on the other side. Samuel Pepys in his diaries (1660–2) describes 'shooting the bridge' a number of times, emphasising that the practice was both wet and dangerous, and twice noting that his travelling companions refused to pass under the bridge (Latham and Matthews 1970a, 323; Latham and Matthews 1970b, 59; Latham and Matthews 1970c, 52, 198, 260). Often craft passing under the bridge overturned or struck the starlings. In 1428 the Duke of Norfolk's barge overturned while passing downstream under the bridge (Home 1931, 123). In 1693 15 people were drowned when their boat overturned while passing through the bridge and in 1697 ten people drowned when their boat struck a pier (Quarrell and Mare 1934, 26). In 1710 Conrad von Uffenbach wrote in his journal: 'rowed Londonwards with the tide; and we risked passing through the centre of the bridge, where the eddy and waves were so violent that, when we were under the arch, the water piled up on either side of us to a much greater height than our little ship' (Quarrell and Mare 1934, 26). The hazards of passing under the bridge were immortalised in the proverb 'London Bridge was made for wise men to go over and fools to go under' (Ray 1737, 13, 15).

Another consequence of the bridge's restriction of the water flow was that it encouraged the build-up of ice on the upstream side of the bridge, and the narrow channels between the starlings were easily blocked by ice floes (Fig 123). During severe winter weather the Thames on the upstream side of the bridge could freeze solid for long periods of time and on a number of occasions from 1607 until 1814 'frost fairs' were held on the river (Currie 1996, 4, 60).

13.4 The rebuilding of London Bridge

In 1801 a House of Commons Select Committee had reviewed a number of proposed designs for a new London Bridge, but none was chosen (Report 1801). During the winter of 1813–14, long stretches of the River Thames froze, enabling the last great 'frost fair' to be held on the river during January and February 1814 (Currie 1996, 60–6) (Fig 118). The build-

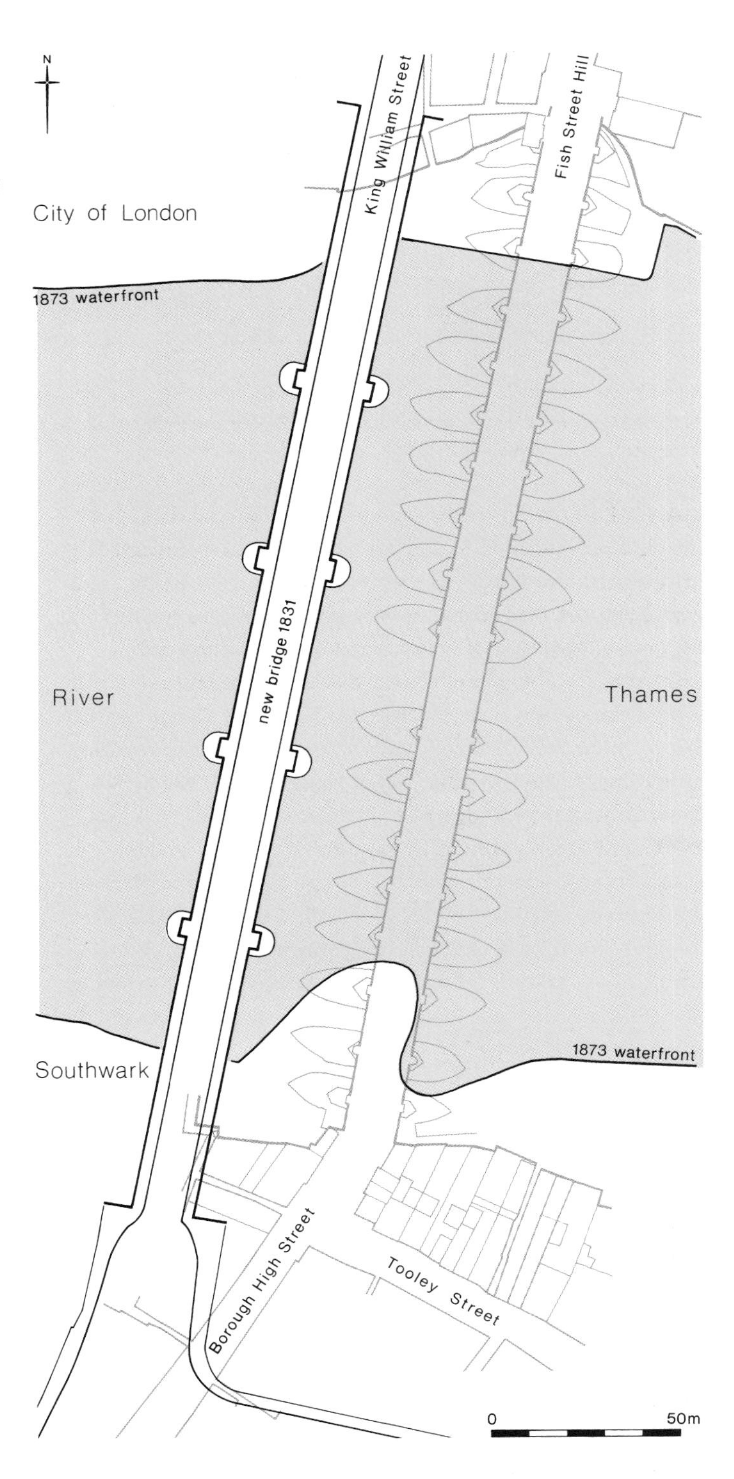

Fig 124 Plan of the new road links and Rennie's bridge, plus the existing medieval bridge and its associated roads

Fig 125 The construction of the new London Bridge in December 1826, view looking north across the former area of the Bridge Foot; notice the buildings on the eastern side of the old bridge approach road and the massive timber cofferdam around the new abutment (Cooke 1970, no. 5)

up of ice on the upstream side of the bridge also damaged its fabric, resulting in fresh proposals to remove another four bridge piers to improve the water flow. In 1820 Ralph Dodd wrote that the bridge was in very poor condition owing to its 'annual battering of craft and ice', and pointed out that there was a 1.20m deep 'chasm' in the starling on the north side of the central arch. In 1821 the House of Commons Select Committee decided that it would be better to rebuild, rather than further modify, the bridge (Home 1931, 291). Designs for a new bridge were called for and, on 12 March 1821, a design submitted by John Rennie for a five-arch granite bridge was selected. The design for the new bridge was actually produced by his elder son George Rennie. After Rennie's death on 4 October 1821, his younger son John, who had assisted his father with the design and construction of both Blackfriars and Waterloo Bridges, took over as chief engineer of the project.

On 23 July 1823 'An Act for the Rebuilding of London Bridge and for the improving and making suitable Approaches thereto' was passed. The new bridge was to be built slightly upstream of the existing one, and the approach roads on both sides were to be completely rebuilt and widened (Act 1823) (Fig 124). On the north bank a completely new road, King William Street, was to be laid out, running from the Royal Exchange to the new bridge. On the south bank the existing bridge approach road, Borough High Street, was to be realigned and widened. At each end of the new bridge massive ashlared abutments were to be constructed to carry the new approach roads (Fig 125; Fig 126). Work on the new bridge started on 15 March 1824 and, on 15 June 1825, the foundation stone was laid in the first cofferdam (Fox 1992, no. 148). The new bridge piers were constructed inside massive timber cofferdams, consisting of two ovals of closely spaced wooden piles, lined with vertical planking, caulked with hemp or oakham and pumped dry by steam engines (Fox 1992, no. 149) (Fig 125).

From 1824 until 1841 there was a massive programme of dredging on the site of the new cofferdams, and later on the site of the old bridge, to remove the build-up of sediments downstream of the starlings and to improve navigation. This dredging revealed many Roman and medieval objects (see Chapters 3 and 8), which were collected by Charles Roach

Fig 126 The east side of the Southwark abutment for the 1820s London Bridge, revealed during the redevelopment of Fennings Wharf in June 1984; view looking west; this abutment was much wider than the original bridge, so was retained during the 1960s replacement of the bridge (MoL)

Fig 127 The southern end of the old London Bridge in March 1827, view looking south, showing one of the piers which was removed in 1826–7 and replaced by a wooden span; in the background is the shot tower, which stood on the site of Toppings Wharf (Cooke 1970, no. 1)

Fig 128 Tinted engraving of the northern end of the new London Bridge, view looking east on the day it opened – 1 August 1831 (MoLPL)

Smith and other antiquarians (Rhodes 1991, 179). In 1826–7 parts of two of the piers and starlings of the medieval bridge were dismantled, as they were obstructing the construction of the new bridge (Knight 1831) (Fig 127). As part of the construction of the new bridge, a new ashlar riverwall was constructed on the east side of the new abutment (F32).

The new London Bridge was opened on 1 August 1831 by William IV, having taken seven years and seven months to build and costing £2,556,170, including the cost of building the new approaches and demolishing the old bridge (Fig 128). In 1831 the younger John Rennie was knighted for directing the construction of the new bridge.

Fig 129 *The demolition of old London Bridge in January 1832, view looking north towards the church of St Magnus (Cooke 1970, no. 4)*

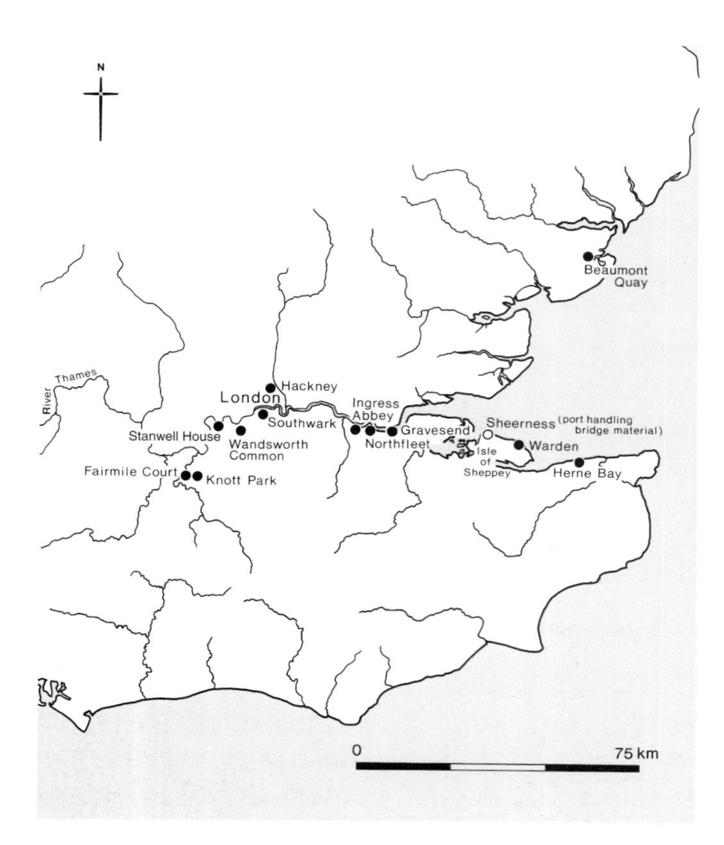

Fig 130 *Map of south-east England, showing the location of the* ex situ *fragments of London Bridge*

In November 1831, some 622 years after medieval London Bridge was completed, its systematic demolition began, and most of the superstructure was removed during 1832 (Fig 129), although the removal of the old starlings below water level, and other tasks, were not finished until mid-1834. The demolition of the old bridge excited a great deal of public interest in the history of the structure (Thomson 1827), and also produced a vast tonnage of dismantled masonry. Much of this stonework ended up in parts of Greater London, Essex, Kent and Surrey, often at places on the shore of the Thames estuary where the material could easily be transported by barge (Fig 130). One voussoir and an iron-shod elm pile from the medieval bridge still remain in the City of London, and are displayed within the Livery Hall of the Company of London Water and Lighter Men.

13.5 Fragments of old London Bridge in south-east England today

Before reviewing known locations of *ex situ* bridge masonry, two points need to be borne in mind. Firstly, almost all the surviving stonework dates from the 1757–62 modification of the bridge (see above). Secondly, the modified bridge was completely refaced with blocks of Portland ashlar during 1757–62, which means that almost all the surviving ashlar dates from the mid-18th century. Despite this, there have been many claims made that such material is actually derived from the medieval bridge.

There are various unsubstantiated claims that numerous fragments of masonry in southern England are derived from medieval London Bridge. Firstly, it has been claimed that the clock tower at Swanage, Dorset, was one of the toll houses on London Bridge. The tower in question never stood on the bridge: it was the Wellington Memorial clock tower and was originally situated near Southwark cathedral (Home 1931, 342). Secondly, it has also been claimed that the chancel of Steep church, Hampshire, is paved with 'York stone from old London Bridge'.[35] Thirdly, a sculptured stone shield, bearing three lions or leopards supported by two angels,[36] which is now serving as the keystone in the tower arch of Merstham church, Surrey, is said to be from London Bridge (Home 1931, 184). Lastly, it has been alleged that the baroque-style, stone balustrade in the grounds of Gilwell Park, in the London Borough of Waltham Forest, is from London Bridge, but examination of the balustrade suggests that this is unlikely.[37] Some cast-iron railings, from the 1757–62 modifications to the bridge, formerly flanked the footpath through St Botolph's church Bishopsgate in the City of London (Home 1930, 302), but they were apparently taken for scrap metal during World War Two.

Heathfield Road, Wandsworth Common

Masonry from the demolished bridge was used in the building of a residence known as 'Stone House' on the edge of Wandsworth Common in south London (Fig 130, 12/18) and, on the demolition of the house *c* 1909, some of the rubble masonry was reused in the facade of the ground floor of 49 Heathfield Road (Fig 131), built on the site of the 'Stone House'. More masonry was used to build the garden wall in front of nos 49 to 73 Heathfield Road (Home 1931, 301). As the geology of the stone in the garden wall and the facade of no. 49 is mainly Kentish ragstone, it is probable that this material was part of the medieval bridge, rather than part of the modified bridge (Home 1931, 339). The Heathfield Road material is the largest remaining group of stonework from the medieval bridge.

Ingress Abbey

Ingress Abbey, is a derelict mock-Tudor country house at Greenhithe, near Dartford in Kent (Fig 130), built in 1832–3 for James Harmer (1777–1853), a lawyer and City alderman (1833–40), and it has been stated that the Portland ashlar facing of the house derived from old London Bridge (Welch 1894, 133) (Fig 132). Surviving documentary evidence makes no mention of any purchases of building stone by Harmer, who, on 15 March 1833, was appointed a member of the London Bridge Committee that was overseeing demolition

Fig 131 The facade of 49 Heathfield Road, Wandsworth Common; view looking south (MoL, R Bartkowiak)

of the old bridge, and so would have been in a good position to purchase materials (Watson 1997b, 14).

During 1831–2 some unspecified 'old materials' from the bridge, perhaps the balustrades and alcoves, were sold off (CoLRO, BHCJ vol 36, 106b), it having already been agreed as part of the demolition contract in 1831 that the London Bridge Committee was to have the option of keeping the ashlar for building new riverwalls or 'embankments', and that some of the rubble was to be used to infill the holes left by the removal of the starlings (CoLRO, BHCJ vol 6, 882, 914). There were also some unauthorised disposals. In January 1832 (long before Harmer's appointment to the committee), there was complaint that four or five barge loads of rubble that should have been used for filling holes in the riverbed had been

Fig 132 Ingress Abbey, Greenhithe, in 1996, view looking east (MoL, R Bartkowiak)

removed to Sheerness and elsewhere, while an unknown quantity had gone by land (CoLRO, BHCJ vol 7, 66). The London Bridge Committee took a very serious view of this loss of material and even considered dismissing Sir John Rennie (acting as a consultant of the demolition of the old bridge) and George Hollinsworth, the Clerk of Works, from the project (ibid, 7, 66). The threat to dismiss such senior staff from the project would imply that a great deal of material had already been removed from the site in an improper manner. Perhaps it was some of this missing stonework that ended up at Ingress Abbey, although after June 1832 any remaining stonework from the bridge not required for embankment works was probably sold off to the demolition contractors, Jolliffe and Banks. It is worth noting that Warden church (see below) near Sheerness was rebuilt in 1836 by the son of Sir Edward Banks, a partner in the contracting firm, using stone from the medieval bridge.

Beaumont Quay

In 1833 a small quay was constructed at Beaumont (Fig 130), Essex, as part of the Walton Havens, to berth sailing barges. The face of the quay was built out of ashlar blocks from the 1757–62 modification of the bridge (Fig 133). The quay and adjoining small warehouse were built by the governors of Guy's Hospital (Booker 1974, 143).[38] According to the records of Guy's Hospital, the stone for the quay was purchased in December 1832 from Jolliffe and Banks, the contractors who were demolishing the old bridge, and the quay itself was built during March 1833 (GLRO, GH Ledger Book G, 102–05).

Next to the warehouse on the quayside are the remains of a brick-built limekiln of a similar date to the quay (RCHME 1996, 7–17). From the mid-19th century, due to the development of the railways, the quay was only used intermittently for handling bulk cargoes and by c 1918 it had completely gone out of use (RCHME 1996, 5).

Gravesend cemetery

On top of one of the tombs in Gravesend cemetery, Kent (Fig 130), was set one of the balusters from the modified bridge (Webster Smith 1956, 35). The baluster no longer survives.[39]

Herne Bay pier

In the early 19th century Herne Bay, Kent (Fig 130), attempted to establish itself as a channel port. To this end a wooden pier 3,615ft (1,101m) long was built in 1831–2 to provide deep-water moorings for passenger ferries. However, the bid failed and the pier had serious structural problems, not least attack by seaworms; it was closed in 1861 and demolished in 1871. The stone-built approach or abutment of the original pier, reused by the two subsequent piers, included two long lengths of stone balustrade from the 1757–62 modification of London Bridge (Home 1931, 301) (Fig 134). The balustrade was destroyed in the great east coast storm of 1953 (Watson 1997b, 16).

Northfleet dockyard

The mock castle-style gatehouse, which was the main entrance to William Pitcher's Northfleet Dockyard, Kent (Fig 130), was allegedly constructed out of ashlar from old London Bridge. The gatehouse was demolished in 1932 (Howe 1964, 155).

Warden church

In 1836 the tower of St James's church at Warden on the Isle of Sheppey in Kent (Fig 130), was rebuilt by Delamark Banks, son of Sir Edward Banks (died 1835), one of the partners in the firm that demolished the medieval bridge. At this date the Banks family had commercial interests in this area of Kent and had developed part of Sheerness town and harbour, where stone from the bridge had been shipped in 1831–2) (Fig 130).

Fig 133 *Beaumont Quay, near Thorpe le Soken, Essex, view looking east; the building in the centre is the warehouse built at the same time (MoL, R Bartkowiak)*

Fig 134 *The second pier at Herne Bay (opened 1873), showing the London Bridge balustrade on the approach to the original pier, view looking north (Herne Bay Record Society)*

The inscription on the Warden church tower stated that all the stone was derived from the medieval bridge.[40] Owing to coastal erosion, the church was demolished in 1876.

An inscribed ashlar block from the bridge was built into the facade of the two-storey brick-built property in Warden village known as 'Bridgestone House', built in 1838 by Delamark Banks for his bailiff (Judge 1988, 606). This house is about to be destroyed by coastal erosion.

The bridge alcoves

Only four of fourteen alcoves added to the bridge in 1757–62 have survived. Three examples survive in the London area and a fourth is in Surrey. Two of the former are to be found in Victoria Park, Hackney, and a third stands in the quadrangle of Guy's Hospital, Southwark (Fig 120; Fig 130). The Guy's alcove was purchased by the hospital in May 1861 for ten guineas, as a garden shelter for convalescing patients.[41] The two alcoves in Victoria Park (opened 1845) were both erected in 1860 and were previously owned by one Benjamin Dixon, presumably the same person who sold the third alcove to Guy's a year later.[42]

Two of the alcoves and a 260ft (79m) length of balustrading from the modified bridge were reassembled as a garden feature in the grounds of Stanwell House, East Sheen in Surrey (Fig 130; Fig 135), built in 1830s (see Chapter 14.13). When Stanwell House was demolished in 1937 and replaced by Courtlands Flats, Sheen Road, all the balustrading and one of the alcoves were demolished. The remaining alcove still survives in the grounds of Courtland Flats.

Fig 135 The London Bridge balustrade on the terrace wall at Stanwell House in 1935, view looking south towards the surviving alcove (Raymond Gill)

13.6 Replacing Rennie's bridge

Sir John Rennie's bridge was not destined to last as long as its predecessor. In 1913 it was widened to accommodate a four-lane roadway, and by the 1960s it had been decided to replace it with a much wider bridge, which would cope better with the ever increasing volume of traffic and pedestrians. During 1967–71 Rennie's bridge was dismantled and replaced by a new three-arch bridge with pre-stressed concrete cantilevers and a six-lane roadway (Fig 136). Dredging on the site of the old bridge in November 1967 revealed more iron-shod elm piles from the starlings of the old bridge (see Chapter 14.6) and other finds including coins, three iron grappling hooks, three axe heads, padlocks, two pewter mugs and four spurs (Marsden 1971, 12). The new bridge was opened by Queen Elizabeth II on 16 March 1973. It is a notable example of continuity that some two thousand years after the first bridge was built here people are still crossing the river in almost the same spot (Fig 136).

Fig 136 The modern bridge, view looking northwards, with the outline of the reconstructed medieval bridge superimposed (see Fig 68)

The ashlar facing from Rennie's bridge was sold to the McCullock Oil Company, who rebuilt the bridge during 1968–71 at Lake Havasu City, Arizona, in the United States of America. The only remaining portion of the Rennie bridge in London is the Portland ashlar faced Southwark approach and abutment (Fig 126).

13.7 The rediscovery of medieval London Bridge

After the medieval bridge was demolished, its northern and southern ends were redeveloped as quayside warehouses. At the northern end, London Bridge Wharf was constructed and, at the southern end, Fennings, Toppings and Sun Wharves on the site of the abutment and Bridge Foot area. The construction of the warehouse basements and deep foundations caused considerable damage to the archaeological deposits, yet the redevelopment also ensured that portions of the medieval bridge escaped demolition in 1831–4. At the northern end of the bridge, the next to last arch and one starling were incorporated within the buildings and roadways at London Bridge Wharf, Adelaide Place (Fig 91; Fig 137), and the arch was rediscovered during the redevelopment of the site in July 1921 (*The Builder* 1921, 103).

Attempts were made to preserve the arch *in situ*, or dismantle and rebuild it at the forthcoming 1924 British Empire Exhibition at Wembley, but these efforts failed because of lack of money and the arch was demolished in October 1922 (*The Builder* 1921, 337. 630, 703; *The Builder* 1922 (vol 122), 67, 103, 104, 251, 313, 347, 436, 898; *The Builder* 1922 (vol 123), 40, 46, 190, 237, 535, 570; Jackson 1971, 114). Fortunately, the arch was drawn by T L Cooke before it was destroyed (Fig 138): the drawing shows that the earliest masonry, interpreted as part of the medieval bridge, was 20ft (6.09m) wide, with a chamfered offset just above the top of the starling. The interior of the arch featured two sets of voussoirs, described as the inner and outer arch, which were braced with three sets of stone ribs and dated 1703. The 1757–62 ashlar extensions to each side of the pier were not fully exposed, but were described as 13ft long (3.96m) with mooring rings at the ends.

Some of the masonry from this arch was saved by the building contractors, McAlpines, and moved to the grounds of Fairmile Court, Cobham, and Knott Park, Oxshott, both in Surrey, and to another house at Nutley in Sussex (Fig 130).[43] Some of the stones were built into the retaining wall of the tennis court at Knott Park, where formerly there was a cement plaque on the wall inscribed: 'These are stones of old London Bridge'. The stones at Knott Park include two large Portland stone blocks with a chamfer at one end, probably corbels from below the balustrade of the modified bridge. In the grounds of Fairmile Court is a garden wall, mostly constructed out of large ashlar blocks of Portland stone, also derived from the modified bridge. Other masonry from the arch was formerly displayed in a rooftop garden at Adelaide House, built on the site of the arch in the 1920s (Home 1931, 306), while three large blocks of Portland stone from the arch are still to be found in the garden of the church of St Magnus the Martyr, all of them part of the facing of the modified bridge (Jackson 1971, 114).

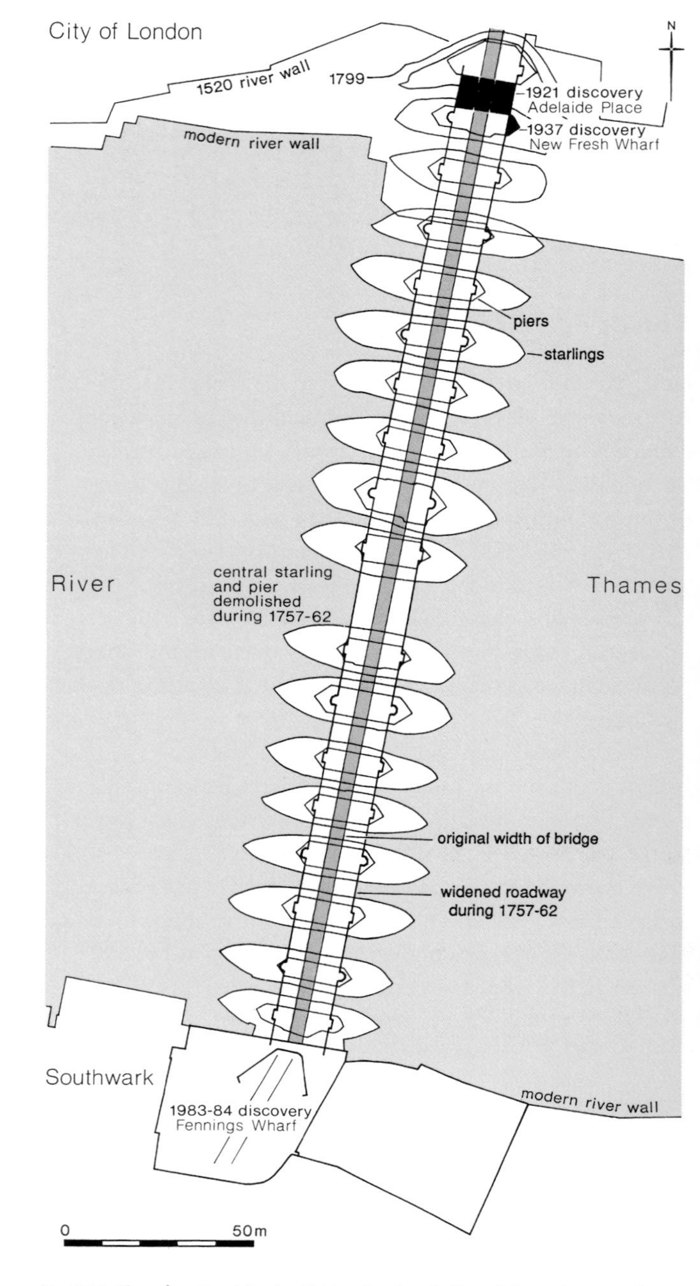

Fig 137 Plan of medieval London Bridge showing the line of the modern waterfront, plus the various portions of the bridge which were rediscovered during 1921, 1937 and 1983–4

Another fragment of the same bridge pier found in 1921 was located in 1937 at Fresh Wharf, just east of Adelaide Place (Fig 137). This consisted of the eastern end of the 1757–62 bridge pier extension, and was recorded by Frank Cottrill, archaeologist at Guildhall Museum (GM site 125: Schofield 1998, 73). This triangular shaped block of masonry was faced with large ashlar blocks, each 900mm long, while the core was mostly constructed from large blocks of ragstone and Reigate rubble. The masonry was founded on a mass of wooden piles. Dredging on the site of the medieval bridge during 1967 revealed a number of ashlar blocks and two portions of the balustrade from the modified 18th-century bridge, plus a number of iron-shod square piles, interpreted

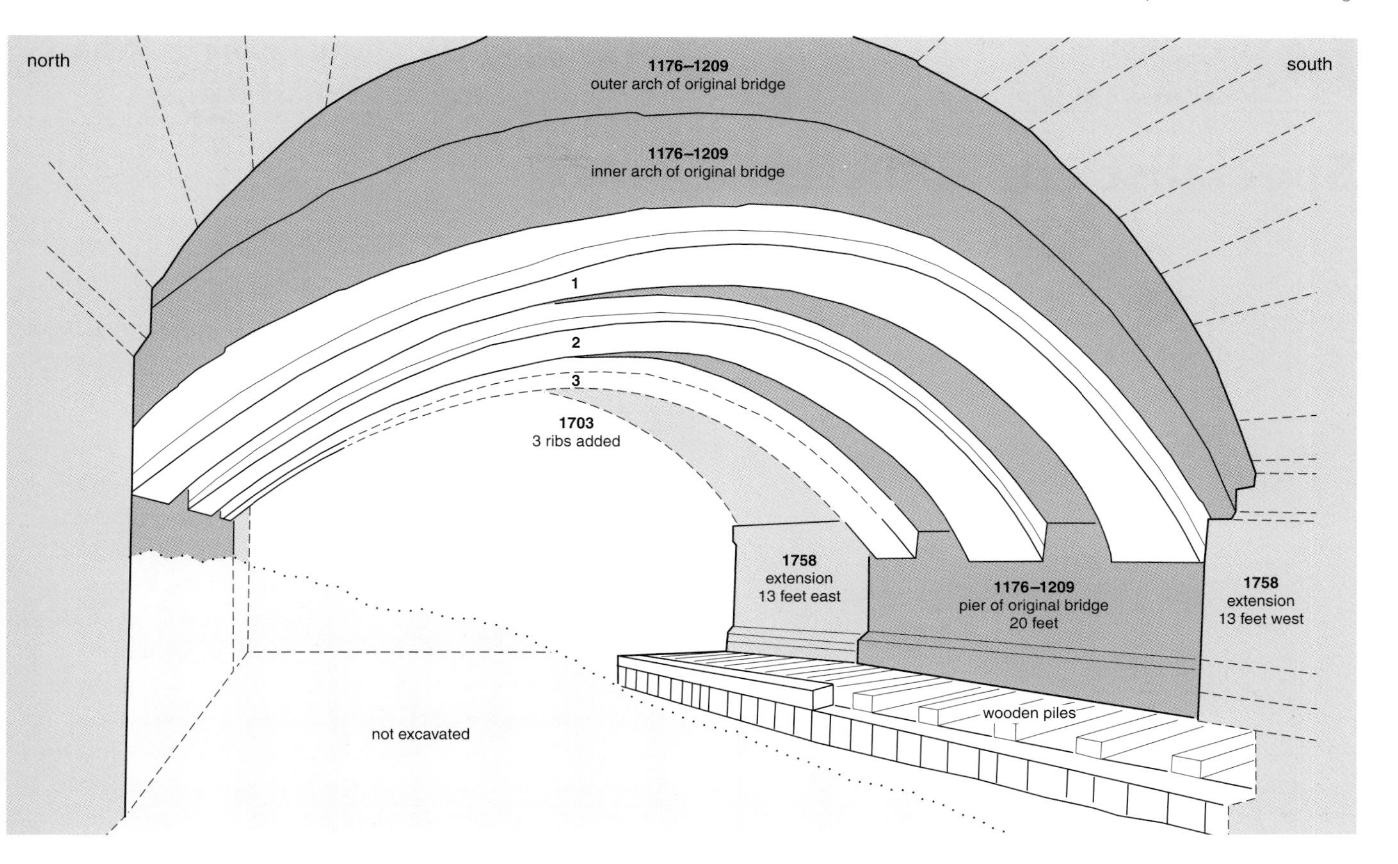

Fig 138 Diagram of the London Bridge arch discovered in 1921 at Adelaide Place; view looking east (after T L Cooke in The Builder *1922, vol 122, 68)*

as part of the foundations of the Rennie bridge (Marsden 1971, 14).

The final fragment of the medieval bridge to be rediscovered was the southern abutment at Fennings Wharf (Fig 52) in 1983, which was subsequently examined and then destroyed in 1984 (see Chapter 1), when the site was redeveloped. Thus London Bridge was 'broken down' for the last time.

14

Specialist appendices

14.1 Summary of the Fennings Wharf and Toppings Wharf dating sequence

Fennings Wharf

Bruce Watson with contributions by Ian Betts (building materials), Robin Symonds and Jo Groves (Roman pottery), Jacqueline Pearce (post-Roman pottery), Louise Rayner (prehistoric pottery) and Ian Tyers (tree-ring dating)

Introduction

The dating evidence outlined here provides the definitive site-wide chronology, and is based on five sources. Firstly, dendrochronology or tree-ring dating (the most important source) has been used to date the waterfront sequence. All dates cited are in accordance with the conventions listed in the group description text (see Chapter 14.2 for further details). Only the latest or relevant dates from each group are included here and it should be stressed that, in the absence of bark edge from timber samples, only an estimated felling date can be provided. All bark, bark edge and sapwood dates are included for the sake of completeness. A complete listing of all sampled timbers is included in the site research archive (Tyers 1993).

Secondly, pottery has been used to date many of the features and structures south of the foreshore (see below and Chapter 14.5). Thirdly, radiocarbon determinations have been used to date the prehistoric ring-ditch; the method of calibration is outlined in Chapter 1.2. Fourthly, in some instances it has been possible to link the archaeological sequence of events, with historically documented events with varying degrees of precision. Lastly, brick size and fabric has enabled approximate date ranges to be assigned to certain post-medieval brick structures (see Chapter 14.11). The few Roman coins all appear to be residual.

Some groups or subgroups lack any independent dating, and in these cases broad dates have been assigned from their relative position in the stratigraphic sequence.

Dates earlier than AD 1000 have the appropriate BC suffix (cal BC = calibrated radiocarbon chronology) or AD prefix.

Key to abbreviations

For an account of conventions used generally in this report see Chapter 1.2. Abbreviations specific to this section are given below.

DENDROCHRONOLOGY

All dates cited are from oak (*Quercus* spp) unless otherwise stated, some beech (*Fagus sylvatica*) dates also being available. For individual tree-ring dates the following code is used:
B: bark present (actual year the timber was felled)
b: possible bark edge
S: sapwood present, the number of sapwood rings

measured being cited
H/S: heart/sapwood boundary present
h/s: possible heart/sapwood boundary present
H: heartwood only

After the sample number an estimated felling date is given where necessary. For sapwood dates an estimated range of between plus ten and 55 years is given for samples with less than ten rings, while for those samples with more than ten sapwood rings the lowest estimated date is the last measured ring, plus one. For heartwood dates an estimated felling date of later than plus ten years is given for English oak (no Baltic oak was found on site) (Hillam et al 1987). The beech heartwood dates are problematic as beech does not have a clear heart/sapwood differentiation, so that a figure of plus one year is given as an estimated felling date for these samples. It must be stressed that the oak heartwood dates can be very misleading, since in some instances carpentry may have removed the last 100 years or more worth of annual growth rings from some samples, while in other samples fewer than 20 rings appear to be missing. In the case of the beech samples the difference between the heart and bark dates from the same phase of activity is often less than five years. During the Roman and medieval periods the oak timbers used in construction were normally green or unseasoned, therefore, their felling date (unless the timber was reused) gives a precise date for the erection of structures. However, tree-ring dating has demonstrated that on some medieval construction projects there was a delay of several years between felling and the actual use of materials due to stockpiling (English Heritage 1998, 12).

FINDS

The following codes are used:
Rpot: Roman pottery;
Mpot: medieval pottery;
Ppot: post-medieval pottery.

For all finds dates the context number [ ], ceramic period and date range is cited. In the few cases where no date range is given the assemblage cannot be precisely dated. Residual material is not included. Only relevant small finds data are included. For selected post-medieval contexts the dates assigned to brick fabrics are included.

Group F1: natural geology

F1.1 Pleistocene sand and gravel.
F1.2 Prehistoric subsoil, pre-1000 cal BC, containing sherds of Late Bronze Age pottery (*c* 1000–700 cal BC) and two fragments of perforated clay slabs of unknown function.

Group F2: ring-ditch

F2.1 Creation of ring-ditch, undated.
F2.2 Primary fill of ring-ditch, one sherd of Bronze Age pottery.
F2.3 Secondary fill of ring-ditch, six radiocarbon dates from charcoal spreads and one sherd of Bronze Age pottery (2000–700 cal BC).
F2.4 Partial recut of east side, two radiocarbon dates from charcoal spreads and two sherds of Bronze Age pottery (2000–700 cal BC).

The full details of the eight radiocarbon determinations appear in Chapter 1.2, Radiocarbon determinations. The close range of the radiocarbon dates confirms that the ring-ditch is of Bronze Age date and was in use as a cremation cemetery between *c* 1900–1600 cal BC.
F2.5 Central feature within the ring-ditch, containing two individual vessels, a cup and a bowl; similar vessels are found in assemblages dating to the Late Bronze/Early Iron Age transition period (*c* 1000–600 cal BC).

The prehistoric pottery from the ring-ditch and the subsoil horizon is not precisely dated due to the lack of decorated sherds, but some sherds have Late Bronze Age characteristics (*c* 1000–700 cal BC). The perforated clay slabs are of Late Bronze Age date.

Group F3: prehistoric features

F3.1 Linear gully, no datable finds.
F3.2 Pit or hollow, one flint-tempered pot sherd and three fragments of Late Bronze Age (*c* 1000–700 cal BC) perforated clay slab.
F3.3 Pit or hollow, no datable finds.
F3.4 Pit, no datable finds.
F3.5 Pit, two flint-tempered pot sherds of Late Bronze Age date (*c* 1000–700 cal BC).

Group F4: fluvial deposits

Contained eight flint-tempered Late Bronze Age pot sherds. The early Roman sherds present in these deposits are considered to be intrusive. Dating to late 1st century BC or early 1st century AD on other sites in north Southwark. At Toppings Wharf (T2) this deposit is of pre-Flavian date.

Group F5: features of uncertain date

There are no datable finds from this group. These features are interpreted as initial Romano-British activity (mid-/late 1st century AD).

Group F6: linear gullies

F6.1 Rpot [1069] AD 40–100.
F6.2 Rpot [1076] AD 40–100.

Part of initial Romano-British activity (mid-/late 1st century).

Group F7: early Roman buildings

F7.1 The earliest context in this subgroup is a layer [1068] of make-up containing Rpot of AD 70–160. The floor layers above it have date ranges of Rpot [1061] AD 150–250 and Rpot [1062] AD 40–100, suggesting an early 2nd-century date for the construction of the building (certainly after AD

70). Gravel make-up [1049] for the final floor contains Rpot AD 70–150, which is all residual.
F7.2 Rpot [1117].
F7.3 Rpot [1114].

The contexts in F7.2 and .3 both contain Romano-British grog- or shell-tempered ware which is not precisely dated. On stratigraphic grounds these two subgroups post-date the initial Romano-British activity and are assigned to the late 1st or early 2nd century AD.

Group F8: Roman structural features

F8.1 None.
F8.2 Rpot [1055] AD 40–80.
F8.3 Rpot [1051] AD 50–160; [1052] AD 50–160; suggesting that this foundation was robbed out during the mid-2nd century. Its construction is not dated.
F8.4 Rpot [2068] AD 80–300; 3rd century.

Group F9: early Roman features

F9.1 Rpot [2168] AD 70–100; one illegible coin SF <46>. This is a very good group of Flavian samian containing a number of profiles and five stamps. The date is based on the presence of South Gaulish Drag (Dragendorff) form 37 bowl of Highgate 'C' ware and of Verulamium region mica-dusted ware.
F9.2 Rpot [2102] AD 50–100.
F9.3 Rpot [2177] AD 180–300; [2179] AD 100–250 (residual finds include [2103] SF <50>, a coin of Nero, AD 64–8). The well was probably constructed during the late 1st and backfilled during the late 3rd century. This group mainly belongs to the first half of the 2nd century, although the date depends on the presence of a jar sherd of Thameside Kent ware which, if intrusive, would give an overall date of AD 120–200.

Group F10: late Roman features

F10.1 Rpot [1056] AD 150–250 (other finds residual); early 3rd century.
F10.2 Rpot [1041] AD 50–160.
F10.3 Rpot [1086] AD 180–300. Tree-ring date [1081] AD 85–122, H estimated felling date after AD 132. 3rd century.
F10.4 Rpot [1064] AD 200–300; [1078] AD 180–300; [1079] AD 180–300. Tree-ring date AD 85–122 H, estimated felling date after AD 132. Probably later 3rd century. This subgroup contains a very unusual amphora, possibly an Almagro type 50 (cf Peacock and Williams 1986, 130–1, class 22). The dating is provided by dishes, cups and a beaker in Central and East Gaulish samian ware; various forms of Black-burnished ware types 1 and 2, and jars in Thameside Kent ware. There is also a roughcast beaker in Colchester colour-coated ware.

There is no significant difference between the end dates for the construction of the well (F10.3) and its infilling (F10.4), suggesting the well went out of use within a short space of time. The tree-ring date confirms that the well was built after AD 132.

F10.5 Rpot [2004] AD 150–250. This subgroup contains a Central Gaulish samian Drag form 45 mortarium, as well an unusual lid-seated, necked jar in Verulamium region ware.
F10.6 None.
F10.7 Rpot [2028] AD 180–300; 3rd century. This subgroup would be dated to AD 70–100 except for the presence of a lattice-decorated jar in Thameside Kent ware.
F10.8 Rpot [2071] AD 250–400; 4th century. This subgroup contains several jars in Alice Holt and Farnham wares.
F10.9 Rpot [2044] AD 200–300; [2062] AD 300–400; 4th century. This subgroup contains Drag form 38 in Trier samian ware, a 'Castor Box' in Nene Valley colour-coated ware and a very large mortarium probably from the Rhineland. Other late material includes a bowl in Oxfordshire colour-coated ware with a white painted decoration, a probable jar in Alice Holt ware, Farnham ware and an unusual moulded samian Drag form 37 bowl which may be from either Colchester or more rarely represented East Gaulish production centres.

Group F11: erosion of the Saxo-Norman foreshore

F11.1 None.
F11.2 None.
F11.3 None.

There are no finds or tree-ring dates for this group. However, as one of the two phases of erosion (F11.3) is earlier than F12 it must be pre-1080 (F12.5 dates to 1081). F11.3 can, therefore, be dated to the early/mid-11th century. The earlier erosion scour, F11.1, is undated, although on stratigraphic grounds it is of 11th-century or earlier date, most of this scour being later removed by another phase of erosion during the early 12th century (F13.1).

Group F12: stabilisation of the foreshore

F12.1 Mpot [2257] 1000–1150.
F12.2 now F14.11.
F12.3 None.
F12.4 Mpot [2260] 1050–1150.
F12.5 Tree-ring dates [2089] {168} 1056 S (two rings), estimated felling date 1064–1109; [2089] {179} 1072 S (20 rings), estimated felling date 1073–1107; [2089] {199} 1054 S (one ring), estimated felling date 1063–1108; [2089] {202} 1072 S (45 rings), estimated felling date 1073–1082; [2089] {207} 1081 B; [2089] {224} 1081 B; [2089] {225} 1079 S b? (39 rings), estimated felling date 1080–1095; [2089] {227} 1081 B; [2089] {240} 1081 B; [2089] {241} 1046 S (11 rings), estimated felling date 1047–1090; [2089] {243} 1081 B. Mpot [2089] 1050–1100; [2267] 1000–1150.
F12.6 now F14.12.
F12.7 now F14.13.
F12.8 now F14.14.
F12.9 Mpot [2110] 1050–1150.
F12.10 Tree-ring dates [2190] {268} 1073 S (24 rings), estimated felling date 1074–1104; [2109] {269} 1080 S

(30 rings), estimated felling date 1081–1105.
F12.11 No finds or tree-ring dates.
F12.12 Tree-ring dates [2088] {188} 1095 B; [2088] {189} 1086 b; [2088] {192} 1076 S (21 rings), estimated felling date 1077–1110; [2088] {235} 1095 B. Mpot [2065] 1050–1150; [2088] 1050–1150.
F12.13 Tree-ring date [2022] {48} 1034 H, estimated felling date after 1044, this timber is reused.
F12.14 None.
F12.15. Mpot [2003] 1050–1150.

This group represents the stabilisation of the foreshore upstream of the bridgehead after a period of erosion (F11). It is now clear that it includes two separate sequences of events which are not linked (according to the tree-ring dates) and the 1994 group sequence has been revised accordingly. F12.3 and F12.4 are earlier than F12.5, which dates to 1081. Subgroups F12.6–.8 are later and unrelated to the rest of the group. The first revetment (F12.1) is dated by pottery to pre-1150. The earliest revetment which can be precisely dated is F12.5 assigned to 1081 by five bark dates from a single context, as well as by pottery dating to 1050–1150. A later addition or repair to this revetment (F12.5) has been dated to 1133 (F14.12). The latest datable revetment in the sequence is F12.12 (overlying F12.5, .10), with two dates of 1095 B. The latest event in this group was a series of dump layers (F12.15) dated to 1050–1150. In summary, this group dates from *c* 1081–1100 and is interpreted as part of the attempt to stabilise the upstream portion of the foreshore, during a documented episode of flood damage in 1097.

Subgroups F12.2 and .6–.8

These subgroups are now reinterpreted as dating after a period of erosion of the Saxo-Norman foreshore *c* 1100–1130, possibly including the documented erosion of 1097, and are now defined as part of F14.
F12.2 now F14.11.
F12.6 now F14.12.
F12.7 now F14.13.
F12.8 now F14.14.

Group F13: revetments and foreshore deposits

F13.1 Phase of erosion may be contemporary with F14.10.
F13.2 Tree-ring dates [2129] {394} 1148 B; boat wale [2124] 1018 felled after 1030 (Tyers 1994, 207).
F13.3 No tree-ring dates.
F13.4 No tree-ring dates. Mpot [2114] 1180–1250.

This group represents a phase of erosion followed by three phases of revetment. The erosion is clearly later than F11.3 and dates to the early 12th century, possibly being contemporary with F14.10. The earliest revetment F13.1 dates to 1148 and included a length of 11th-century boat wale. This date suggests that the erosion of this section of the foreshore still persisted into the early 12th century. The two later revetments (F13.2, .3) produced no tree-ring dates, but Mpot dating is 1180–1230.

Group F14: revetments

F14.1 No tree-ring dates.
F14.2 Tree-ring dates [2243] {873} 1075 S (eight rings), estimated felling date 1077–1122; [2244] 1041 H/S, estimated felling date 1051–1096.
F14.3 Tree-ring date [2240] {850} 1019 H, felled after 1029. This is reused timber.
F14.4 Tree-ring date [2252] {859} 1097 B.
F14.5 Tree-ring dates from boat planking [2157] dating to after 1085 (Tyers 1994, 207) and reused as staves in baseplate [2239] {866} 1083 S (three rings), estimated felling date 1090–1135.
F14.6 None.
F14.7 None.
F14.8 Tree-ring date [2265] {921} 1165 B.
F14.9 Mpot [2140] 1180–1230; [2185] 1000–1150.
F14.10 Period of erosion, may be contemporary with F13.1, predating F14.11.
F14.11 Tree-ring date from collapsed cask [2203], date range 1054–1143 H, felled after 1153. Mpot [2165] 1150–1200.
F14.12 Tree-ring dates [2189] {533} 1133 B; another timber [2191] {539} 1082 S (eight rings) has an estimated felling date of 1084–1129. This revetment is a repair or addition to F12.5 following a period of erosion.
F14.13 Tree-ring date [2184] {535} 1070 H, felled after 1080.
F14.14 No tree-ring dates.

The earliest subgroup (F14.1) consists of a number of *ex situ* timbers or driftwood on the eroded late 11th-century foreshore (F11.3). A collapsed revetment (F14.2) dates to *c* 1077–1100. A later revetment (F14.3) is dated by a reused timber to after 1029. An area of isolated revetment (F14.4) (sealed by F14.9) dates to 1097. A substantial baseplate revetment (F14.5) has an estimated felling date of 1090–1135. This baseplate contained a number of boat planks, dating after 1085, reused as staves (Tyers 1994, 207). Two isolated lengths of revetment F14.6 and F14.7 have no tree-ring dates, but both predate F14.8 and so are earlier than 1165. F14.8 consisted of a number of piles and other *ex situ* timbers, one pile dated to 1165.

Subgroups F14.10–.14 represent a sequence of events which cannot be precisely phased within the group. On the basis of the tree-ring dates F14.10–.14 are considered to be later than the rest of the group. The earliest event is a phase of erosion F14.10 (later than F11.3 and probably contemporary with F13.1) dating to the early 12th century and sealed by F14.11, suggesting that it was finished by *c* 1150. Revetment F12.5 was possibly damaged by this phase of erosion, which would date it to the late 11th century After the erosion ceased, fluvial deposits F14.11 accumulated, containing Mpot 1150–1200 and a collapsed cask (dating after 1153). Some of the material from this subgroup probably belongs to F14.14, which was not properly sealed. After the erosion a log revetment (F14.12) was constructed to protect the remaining foreshore, perhaps serving to repair the erosion damage (F14.10) to the existing log and clay bank (F12.5). There are two tree-ring dates for F14.12: 1133 B and an estimated range of 1085–1129. The next revetment (F14.13) consisted of a north–south baseplate dating after 1080, although on stratigraphic evidence it (F14.13) dates after 1133. The last event (F14.14) is a phase of fluvial build-up or flooding, not directly dated but of early 12th-century date on stratigraphic grounds (predating F15.8).

Group F15: revetments and foreshore deposits

F15.1 Tree-ring date [2237] {814} 929 H, felled after AD 939; probably reused or perhaps all the later rings were removed during the conversion of this timber as this date is 100 years earlier than expected.
F15.2 Tree-ring dates [2236] {820} 1116 S (17 rings), estimated felling date of 1117–54; [2235] {819} 1118 B.
F15.3 Tree-ring date [2234] {849} 1137 B.
F15.4 Mpot [2231] 1050–1100.
F15.5 None.
F15.6 Mpot [2080] 1080–1150.
F15.7 No tree-ring dates.
F15.8 Tree-ring dates [2163] {494} 1118 h/s, estimated date range 1128–1173; [2175] {495} 1133 B.

The earliest structure (F15.1) is a tieback post-dating F12.5 (1081) and so probably of very late 11th-century date. The next revetment (F15.2) dates to 1118. A third revetment (F15.3) consists of scattered posts one of which dates to 1137. This subgroup includes a number of undatable *ex situ* timbers perhaps derived from the destruction of this revetment by flood waters or from the erosion of earlier structures. This phase of destruction was followed by the build-up of fluvial deposits (F15.4) dating to 1050–1100. This deposit was overlaid by a later revetment tieback (F15.5), which in turn was largely destroyed by flood water; its remains were sealed by a further build-up of fluvial deposits (F15.6) dating to 1050–1100. Later two phases of baseplate revetment were built (F15.7, .8), the latter dating to 1133.

Group F16: fluvial deposits

F16.1 Mpot [2152] 1150–1200; [2161] 1180–1250.
F16.2 Tree-ring date [2021] {45} 1126 H, felled after 1136. Mpot [2020] 1150–1250; [2021] 1150–1200; [2150] 1150–1200; [2151] 1180–1250; [2228] 1080–1200.

A series of fluvial deposits post-dating F15.8, of 1133. Finds dating and one tree-ring date confirm that this foreshore build-up is of late 12th-century date and predates the construction of the Colechurch bridge (F26). Pottery from all contexts except one has an end date of 1200.

Group F17: Saxo-Norman occupation around the bridgehead

F17.1 Mpot [1046] 1050–1150; [1037] 1000–1150.
F17.2 None.
F17.3 Mpot [2036] 1000–1150.
F17.4 Mpot [2043] 1050–1150.

Group F18: pre-Colechurch bridge, downstream revetments

F18.1 None.
F18.2 None
F18.3 Tree-ring date [40] {338} 985 H, felled after AD 995.
F18.4 Tree-ring date [143] {995} 1157 h/s, estimated felling date 1167–1212. Mpot [41] 1050–1150.
F18.5 Mpot [410] 1050–1150.

The earliest two revetments (F18.1, .2) are undated, but are earlier than 1167. The third revetment (F18.3) dates to after AD 995. The fourth revetment (F18.4) dates to *c* 1167–1212; it was sealed by deposits (F18.5) containing only residual pottery.

Group F19: the late Anglo-Saxon and Saxo-Norman bridge abutment

F19.1 Tree-ring date [156] boat plank 976 H, felled after AD 986 (Tyers 1994, 207).
F19.2 Tree-ring dates [51] {356} 925 H, felled after AD 935; [51] {358} 939 H, felled after AD 949 – reused timber; [148] {1002} 1046 H after 1056.
F19.3 None.
F19.4 Tree-ring dates [583] {1117} 985 H felled after AD 995; [150] {1003} 977 h/s, estimated date range AD 987–1032.
F19.5 Tree-ring date [584] {1114} 1122 H, felled after 1132.

These timbers predate the mid-12th-century bridge caisson. The earliest feature F19.4, dating to *c* AD 987–1032, is interpreted as the *ex situ* remains of the earliest bridge which was sealed by F19.1. The next structure was an abutment baseplate (F19.2) dating after 1056. The relationship between F19.2 and F19.3 is unknown. On stratigraphic grounds F19.3 is of either late 11th- or early 12th-century date. The latest feature is a cluster of piles (F19.5) dating after 1132.

Group F20: erosion and revetments on the upstream side of the Saxo-Norman causeway

F20.1 None.
F20.2 Mpot [566] 1200–1300.
F20.3 Tree-ring dates [542] {1069} 999 h/s, estimated felling date 1009–1054; [573] {1086} 1055 H, felled after 1066; [574] {1087} 1116 B.
F20.4 Tree-ring dates [576] {1092} 1069 H, felled after 1079; [578] {1090} 1062 H, felled after 1072.
F20.5 Tree-ring dates [523] {1071} 1128 B; [523] {1072} 1108 S (six rings), estimated felling date 1112–57.
F20.6 None.

The erosion of the bridge causeway (F20.1) is dated to

the late 11th century. This is probably part of the same period of erosion as F11.3. The subsequent fluvial deposits are dated to 1200–1300 on the basis of two London ware jug sherds which could be of late 12th-century date. This suggests that intrusive material is present. The earliest revetment (F20.3) dates to 1116, the next (F20.4) to after 1072 and the latest to 1128. These dates confirm that these three revetments were erected over a 12-year span.

Group F21: the construction of the first 12th-century bridge caisson

F21.1 None.
F21.2 Mpot [110] 1150–1200.
F21.3 Tree-ring dates [128] {984} 1022 h/s, estimated felling date 1033–77; [139] {960} 1094 H, felled after 1104; [139] {992} 1148 h/s, estimated felling date 1158–1203.
F21.4 Tree-ring dates [116] {964} 1103 h/s, estimated felling date 1113–58; [118] {974} 1109 H, felled after 1119.
F21.5 None.

The caisson is not precisely dated, but was constructed after 1155 during the period *c* 1158–70. The estimated felling dates confirm that it was built during the late 12th century, and the associated pottery dates to 1150–1200.

Group F22: the second 12th-century bridge caisson

F22.1 Tree-ring dates [150] {1003} 977 h/s, estimated date range AD 987–1032; [569] {1105} 984 h/s, estimated date range AD 994–1039; [569] {1123} 995 h/s, estimated date range 1005–50; [572] {1082} 994 H, felled after AD 995.
F22.2 Tree-ring dates [555] {1060} 1145 S and b (15 rings), estimated felling date 1146–85.
F22.3 Tree-ring dates [120] {976} 964 h/s, estimated felling date AD 974–1019.
F22.4 Tree-ring dates [124] {980} 1141 S (18 rings), estimated felling date 1142–78; [556] {1078} 1130 S (four rings), estimated felling date 1140–85; [567] {1080} 1051 H/S, estimated felling date range 1061–1106.

This caisson is not precisely dated, but was constructed after 1151 according to its tree-ring dates, several timbers having estimated felling dates of *c* 1150–80. However, on stratigraphic grounds it is later than the earlier caisson (F21) and, therefore, must have been built after 1156. This caisson is actually later than F21, though its date range is slightly earlier than that of F21 owing to the quantity of reused timber. As this caisson is earlier than the Colechurch bridge (F26), it must have been built before *c* 1176, so that a date range of *c* 1160–78 can be suggested for its construction. This date range strongly suggests that the caisson was part of the bridge structure repaired by Colechurch in 1163 (Stow 1603, i, 22).

It is probable that a number of the older timbers within the caisson derived from earlier bridge structures, for example [150], [569], [572] and [120] which date from AD 974–1039, making them contemporary with the first Saxon bridge abutment (F19.4).

Group F23: upstream bridgehead revetment

F23.1 Tree-ring dates [543] {1100} 1125 H; [543] {1101} 1100 h/s, estimated date range 1110–55; [543] {1100} 1125 H, felled after 1135; [543] {1103} 1125 H, felled after 1135.
F23.2 None.

This revetment is later than F20.5. It was built after 1135 and has a date range of *c* 1136–55, which makes it contemporary with the first 12th-century caisson (F21). It was still in use when the second 12th-century caisson (F22) was built.

Group F24: preparatory consolidation for the Colechurch bridge

F24.1 None.
F24.2 Mpot [564B] (1150–1200).
F24.3 Tree-ring date [527] {1027} 1042 H, felled after 1052.

This preparatory work for the construction of the Colechurch bridge (begun in *c* 1176) post-dates F22 and so must be later than 1156. Associated finds date to 1150–1200. The date range for this group is estimated as the 1160s or 1170s.

Group F25: erosion of the bridgehead and foreshore build-up

F25.1 None.
F25.2 None.
F25.3 Mpot [500] 1170–1200.
F25.4 Tree-ring date [121] {977} 1145 H, felled after 1155.

The only dating evidence for this group is one tree-ring date of after 1155 and one finds date of 1170–1210. On stratigraphic grounds it dates to the 1170s or 1180s (pre-1187).

Group F26: the construction of the Colechurch bridge abutment

F26.1 Tree-ring dates [87] {973} 1174 H, felled after 1179; [109] {942} 1185 S and b (34 rings), estimated felling date *c* 1186; beech [117] {965} 1187 B; beech [117] {967} 1188 B; beech [117] {973} 1185 H, felled after 1186; [531] {1037} 1183 S and b (14 rings), estimated felling date *c* 1184; beech [533] {1042} 1188 B; beech [533] {1043} 1184 H, felled after 1185; beech [533] {1047} 1186 H, felled after 1187; beech [533] {1051} 1188 B; beech [533] {1053} 1187 B; beech [533] {1055} 1188 B; beech [533] {1056} 1188 B; beech [533] {1108} 1185 H, felled after 1186; beech [533] {1109} 1187 b.
F26.2 None.
F26.3 None.
F26.4 Mpot [35] 1200–1350.
F26.5 Tree-ring dates [30] {296} 1175, H felled after 1185; [36] {297} 1164 H, felled after 1175; [88] {782} 1187 B; [597] {1127} 1173, h/s estimated felling date 1183–1228.

Mpot [7] 1200–1350; [8] 1200–1350.
F26.6 Mpot [10] 1150–1200.

The sills and associated beech piles from the east side of the abutment (F26.1) date to 1187 and 1188, while those from the west side date to 1187. This confirms that the abutment was constructed during 1189 or perhaps 1190. These dates fit within the documented dates of *c* 1176–1209 for the construction of the Colechurch bridge.

Group F27: revetments, riverwall and foreshore build-up contemporary with the Colechurch bridge

F27.1 None.
F27.2 Mpot [580] 1240–1300; [503] 1150–1200; [504] 1200–50.
F27.3 Tree-ring dates beech [39] {290} 1197 B; beech [48] {319} 1187 B; [48] {320} 1175 H, felling date of after 1176; beech [91] {794} 1197 b.
F27.4 Tree-ring date [28] {248} 1201 S (17 rings), estimated felling date of 1202–39. Mpot [20] 1200–1300; [29] 1080–1200.
F27.5 None.
F27.6 None.
F27.7 None.
F27.8 None.
F27.9 Tree-ring date [24C] {82} 1274 B; [24C] {89} 1190 B.
F27.10 None.
F27.11 None.
F27.12 Tree-ring dates beech [24E] {58} 1272 b. Mpot (bay 19) 1270–1500.
F27.13 None.

The earliest post-Colechurch bridge downstream revetment (F27.2) was sealed by fluvial deposits dated to 1230–70. The next downstream revetment (F27.3) dates to 1197. This revetment was obviously built some nine years after the construction of the abutment (F26). F27.3 was sealed by a build-up of material containing a number of timbers (F27.4), one of them with an estimated felling date of 1202–39. A later revetment (F27.5) and foreshore build-up (F27.6) are undated, although on stratigraphic grounds F27.5 predates F28.6, which is interpreted as part of the post-1437 remodelling of the southern abutment.

On the upstream side of the Colechurch bridge, driven into the foreshore (F27.8), were four phases of piles (F27.9–.12). One pile (F27.9) below the riverwall (F27.13) dates to 1190, while one of a series of piles (F27.12) set in front of the riverwall to protect its foundations from erosion dates to *c* 1272. Associated finds date to 1230–70.

Group F28: later additions to the Colechurch bridge

F28.1 Mpot [34] 1450–1550; [79] 1400–1500; [83] 1270–1500; and Ppot 1670–1750 – this must be the result of contamination.
F28.2 Tree-ring date [68] {604} 1352 H, felled after 1362.
F28.3 Tree-ring date [2145] H, samples from this oak beam were dated in 1984 to the mid-15th century (Watson 1997c, 324), but re-analysis has not supported these initial findings and the samples are now considered to be undatable (Tyers 1997a).
F28.4 None.
F28.5 None.
F28.6 Tree-ring dates [95] {798} 1163 H, felled after 1173; [95] {799} 1110 H, felled after 1120. The mortar is of late medieval date. Later additions to this wall (F31.1) are of post-medieval date.
F28.7 None.
F28.8 None.

Finds from the foreshore (F28.1) date to *c* 1450–1500 although there is residual material, in one instance post-medieval, interpreted as contamination probably caused by undetected machine disturbance. The addition to the north-west side of the abutment (F28.2) is dated to after 1364 and the addition to the south-east side of the abutment (F28.3) has no tree-ring dates. These two additions to the original abutment are both interpreted as reinforcements of the bridge after the collapse of 1437 (see Chapter 11). Timbers from below the addition to the east side of the abutment (F28.6) date after 1173 and must be fragments of earlier structures: the oak sillbeams from this addition were undatable. The masonry and mortar used in F28.6 are very similar to that used in F28.3, suggesting that all these additions date from before 1600.

Group F29: late medieval features on the southern approach to the bridge

F29.1 None.
F29.2 None.
F29.3 None.
F29.4 None.
F29.5 Mpot [2070] 1270–1350.
F29.6 None.
F29.7 None.
F29.8 None.
F29.9 None.
F29.10 Mpot [1035] 1270–1500; [1036] 1480–1500.

Most of these fragments of masonry are dated to the 13th–16th centuries on stylistic grounds. One cellar floor (F29.10) is dated to *c* 1500.

Group F30: post-medieval features and structures on the bridge approach

F30.1 Bricks [1001] late 17th–18th century.
F30.2 Bricks late 15th–early 17th century (pre-1666). Ppot [1006] 1720–70; [1009] 1630–1700.
F30.3 Bricks [1025] 15th–16th century.
F30.4 Ppot [1007] 1670–1700.
F30.5 Bricks [1022] late 15th–early 17th century (pre-1666).
F30.6 None.
F30.7 Bricks [1012] late 15th–early 17th century (pre-1666).
F30.8 Bricks [1010] late 15th–early 17th century (pre-1666).

F30.9 Ppot [1013] 1610–40; [1017] 1500–50.
F30.10 None.
F30.11 Bricks [2010] late 15th–early 17th century (pre-1666). Ppot [1212B] 1500–1600.
F30.12 Bricks [2000B] late 15th–early 17th century (pre-1666). Use: Ppot [2002] 1650–1750. Systematic infilling: Ppot [2001] 1690–1750.
F30.13 Bricks [2024] late 15th–early 17th century.
F30.14 Bricks [2092] late 15th–early17th century. Ppot [2091] 1700–1800, contaminated by modern pottery
F30.15 Probably of 16th-century date on stratigraphic grounds. The mortar type is late medieval.
F30.16 Later addition to F30.15 and of post-16th-century date on stratigraphic evidence.
F30.17 Bricks probably of late 15th- to early 17th-century date.
F30.18 Ppot [2039] 1810–1900.
F30.19 None.
F30.20 None.
F30.21 None.
F30.22 None.
F30.23 Modern features.

Many of these features and structures are dated from their brickwork, the associated finds dating being limited. Post-medieval bricks cannot be closely dated by their size or fabric, so broad date ranges are cited. Thus F30.1 is of early 18th-century date on stratigraphic grounds as it post-dates F30.4 and is interpreted as part of the redevelopment after the 1725 fire. The largest finds groups are from the various cess and rubbish pits (F30.2, .9, .14, .18), one well (F30.12), as well as the backfill of some robber trenches (F30.4). The backfill of one rubbish pit (F30.9), which is later than (F30.7), dates to the early 16th century. One cesspit (F30.2) contained five complete early 17th-century vessels, including two Bellarmine jugs.

Group F31: post-medieval additions to the abutment and riverwall

F31.1 None.
F31.2 Ppot [16] 1720–70.
F31.3 None.
F31.4 Ppot [18] 1740–70.
F31.5 None.
F31.6 Coin [59] <61> 18th- to 19th-century halfpenny. Mpot residual 1180–1250.
F31.7 None.
F31.8 None.
F31.9 None.
F31.10 None.

Many of these subgroups are dated to this phase on either stratigraphic or stylistic grounds. The only closely datable subgroups (F31.2, .6) are both of mid-18th-century date.

Group F32: riverwall associated with the Rennie bridge

This riverwall and stairs are dated to 1824–32 as they are believed to be original features of the Rennie bridge.

Toppings Wharf

Bruce Watson, with contributions by Jacqueline Pearce (post-Roman pottery)

Introduction

The southern part of Toppings Wharf (TW70) was excavated during 1970–2 (Sheldon 1974, 8–23), and the northern part of the site was investigated while ground reduction was in progress in August and September 1984 (TW84). To present the two phases of archaeological work as a coherent picture, it was decided to reassess the 1970–2 material and present the data as a single site-wide sequence. As part of this reassessment the published post-Roman ceramics were re-examined and some date ranges revised in the light of recent research (see Chapter 14.5).

Group T1: natural geology

The sub-soil horizon (T1.2) contained flints of prehistoric date and pottery of either Late Bronze Age or Early Iron Age date (*c* 1000–200 cal BC).

Group T2: fluvial deposit

Major marine transgression of either late 1st-century BC or early 1st-century AD date (pre-Roman).

Group T3: early Roman revetment

Revetment ditch infilled during early Flavian period (AD 69–96).

Group T4: Roman pits and alleys

Building 1 (T4.1) was constructed during the Flavian period and remained in use until the early 2nd century. No dating is available for Building 2 (T4.2). Building 3 (T4.3) went out of use during the Trajanic period (AD 98–117). Both phases of construction of Building 4 (T4.4) are probably Flavian (AD 69–96) and Building 5 (T4.5) certainly was. Building 6 (T4.6) was also of Flavian date and went out of use during the Trajanic period. The later additions to Building 4 (T4.7) were of late 1st-century date. The later activity within Building 5 (T4.8) was sealed by dumped material of early 2nd-century date. Three pits (T4.9) were of early 2nd-century date (Sheldon 1974, 14–23).

Group T5: later Roman deposits and features

A timber-lined well (T5.1) was backfilled during the mid- to late 2nd century. One pit and two gullies (T5.2) were of late Roman date, one containing a 4th-century coin. Building 2 was sealed by dumped soil and late Roman dark earth (T5.4). Building 3 was sealed by dumped material of early 2nd-century date. Buildings 5 and 6 were sealed by dumped material during the early 2nd century. These deposits were later sealed by dark earth, containing late Roman pottery (T5.4) (Sheldon 1974, 7, 23).

Group T6: medieval foreshore

T6.1 [43] contained an *ex situ* oak pile dated 1122B. Subsequent activity (T6.2–.7) is undated and revetment timbers were not sampled. The latest event (T6.7) predated the erosion of *c* 1300 (T8.1).

Group T7: medieval pits and masonry building

A pit T7.1 (recorded as FW84, [26]) contained two late Saxon shelly ware cooking pots of *c* AD 900–1050. Two rubbish pits (T7.2) and the 'pre-erosion levels' are now dated to *c* 1140–1220. They were sealed by a masonry building (T7.3), the north wall of which was destroyed by erosion (T8.1).

Group T8: erosion

After a period of erosion (T8.1) during the early or mid-13th century (not *c* 1294–1327 as previously thought: see Chapter 12.2), there was a build-up of erosion deposits (T8.2) containing pottery dating to *c* 1230–70 (see 14.5 below). The later build-up of fluvial and dumped deposits (T8.3, .5) contains pottery dating to *c* 1240–70. No timbers from revetment T8.4 were dated. Pottery from the later silts and dumps (T8.6, .7 and .8) (the post-depositional levels) dates to *c* 1270–1350.

Group T9: late medieval buildings

After the reclamation and stabilisation of the foreshore during the early 14th century, a series of buildings were constructed (T9.1–.5), which probably continued in use until the 16th century. The backfill of the cellared building (T9.4) contained 16th-century pottery (Edwards 1974, 5).

Group T10: Topping's Dock

The dock was dated by associated pottery to the Tudor period. The Tudor features have a date range of *c* 1480–1550 (see Chapter 14.5).

Group T11: post-medieval features

Two brick-lined cesspits (T11.1), along with two large oval pits, are probably of either 17th- or 18th-century date and were both backfilled during the 19th century. The cellared building (T11.2) is of late 17th-century date; it was apparently destroyed by the 1725 fire and sealed by fire debris containing pottery and clay pipes of early 18th-century date (T11.3) (Edwards 1974, 5). Another cellar was built after the fire (T11.4) and its floor contained a farthing of 1736 (Sheldon 1974, 28). A fragment of brick cellar wall (T11.5) is probably contemporary with the later cellar floor. To the east of the dock during the 17th or the early 18th century was constructed another cellar (T11.6); associated with its construction was a halfpenny of William III (1699–1701). This cellar was built over an earlier one (T9.4). At some later date (perhaps after 1725) this cellar was refloored and lined with brickwork (T11.7). Two large circular brick-lined tanks (T11.8) are possibly part of the late 18th-century lead works (established by 1799).

14.2 Tree-ring analysis of the Roman and medieval timbers from medieval London Bridge and its environs

Ian Tyers

Introduction

Tree-ring analysis of the many samples of timbers excavated at Fennings Wharf and Toppings Wharf during 1984 has been carried out piecemeal, and for a variety of purposes, during the extended period between their excavation and the compilation of this report. Most of the analytical work was completed by the end of 1993 (Tyers 1993), since when an updated archive incorporating more recent work, including changes to the interpretative stratigraphic framework that resulted from post-excavation analyses, has been produced (Tyers 1997a). The following report is a summary of all the previous work and includes a re-evaluation and synthesis of the results.

Not all the 795 samples are detailed here, but they are included in the dendrochronological site archive, copies of which are held by the Museum of London and the University of Sheffield Dendrochronology Laboratory (ARCUS project 257; Tyers 1997a). A single figure shows the summary date of the principal phases for which dating was obtained (Fig 139). Additionally, there are diagrams showing the dated position of every dated sample from the site (Fig 140; Fig 141; Fig 142). The estimates of felling dates in the absence of bark or bark edge are based on the sapwood estimate of 10–55 rings (see Hillam et al 1987 for further details). The ring width data from all the measured samples are stored in the Sheffield Dendrochronology Laboratory where they can be consulted.

General descriptions of dendrochronology are given in Baillie (1982), Schweingruber (1988) and English Heritage (1998). Tree-ring dating is theoretically simple but in practice rather complex. Timbers of unknown date are correlated with reference to tree-ring chronologies that have been built up starting with present-day trees and extending backwards in time using a reliable and repeatable methodology (Fig 7). Trees in northern temperate areas grow only during the spring and summer periods. Each tree-ring is formed during the growing season and is marked by the cessation of growth at leaf-fall in autumn. Each tree-ring is formed immediately

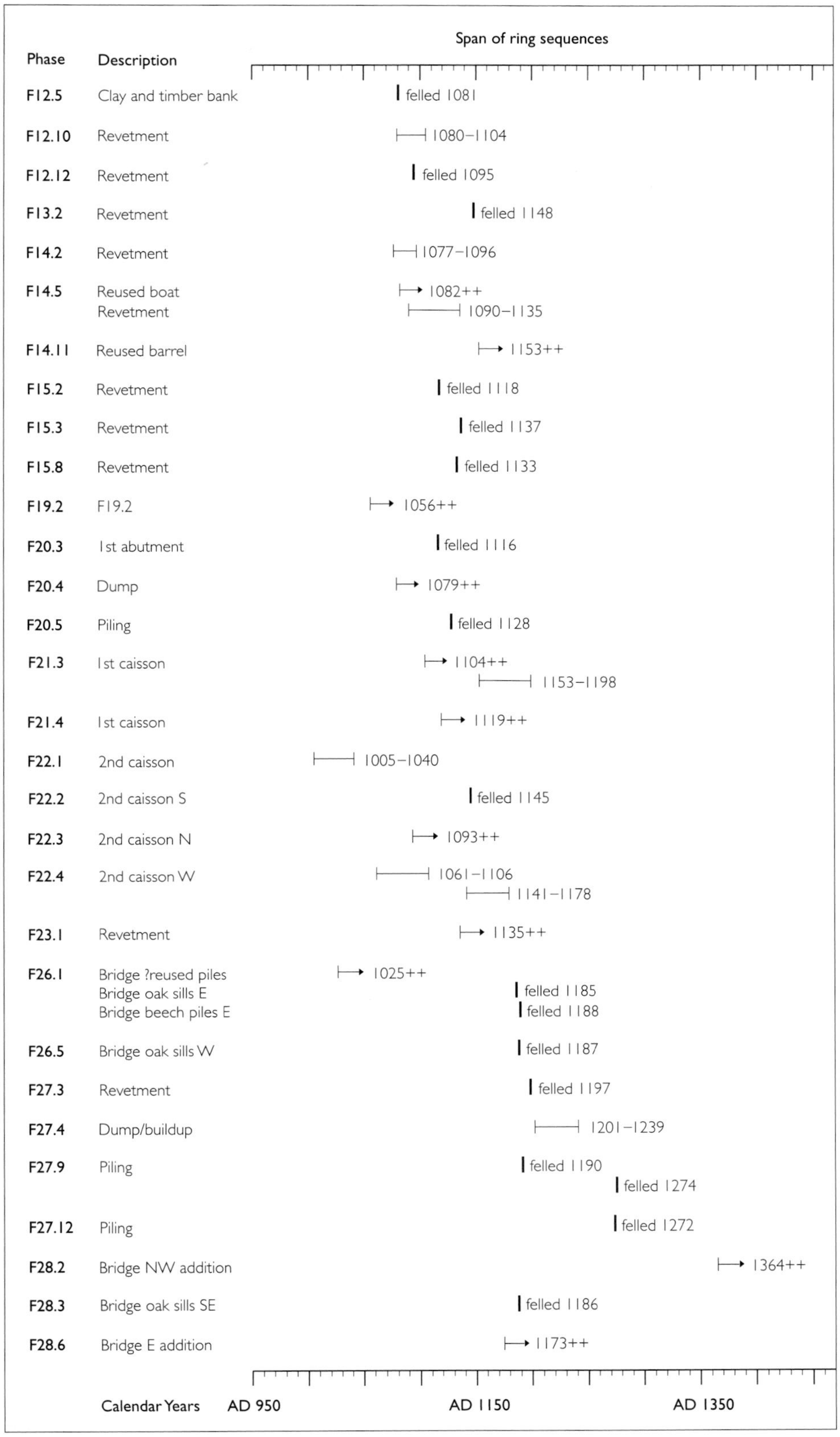

Phase	Description	Span of ring sequences
F12.5	Clay and timber bank	felled 1081
F12.10	Revetment	1080–1104
F12.12	Revetment	felled 1095
F13.2	Revetment	felled 1148
F14.2	Revetment	1077–1096
F14.5	Reused boat	1082++
	Revetment	1090–1135
F14.11	Reused barrel	1153++
F15.2	Revetment	felled 1118
F15.3	Revetment	felled 1137
F15.8	Revetment	felled 1133
F19.2	F19.2	1056++
F20.3	1st abutment	felled 1116
F20.4	Dump	1079++
F20.5	Piling	felled 1128
F21.3	1st caisson	1104++
		1153–1198
F21.4	1st caisson	1119++
F22.1	2nd caisson	1005–1040
F22.2	2nd caisson S	felled 1145
F22.3	2nd caisson N	1093++
F22.4	2nd caisson W	1061–1106
		1141–1178
F23.1	Revetment	1135++
F26.1	Bridge ?reused piles	1025++
	Bridge oak sills E	felled 1185
	Bridge beech piles E	felled 1188
F26.5	Bridge oak sills W	felled 1187
F27.3	Revetment	felled 1197
F27.4	Dump/buildup	1201–1239
F27.9	Piling	felled 1190
		felled 1274
F27.12	Piling	felled 1272
F28.2	Bridge NW addition	1364++
F28.3	Bridge oak sills SE	felled 1186
F28.6	Bridge E addition	1173++

Calendar Years AD 950 AD 1150 AD 1350

Fig 139 Summary of dating evidence for phases with more than a single dated timber

under the bark. Thus the most recent rings are on the outside of a trunk, whilst the oldest rings are at the centre of a trunk. The amount individual trees grow each year is dependent upon a number of external influences. One of the major influences on most trees in most years is the climate of that year.

The Fennings Wharf and Toppings Wharf timbers

The timbers were sampled by the author during excavations on the site of London Bridge in May to December 1984. Work on the 791 samples from the Fennings Wharf excavations and the four samples from Toppings Wharf began with the spot-dating of a small group in the same year. The rest of the material was

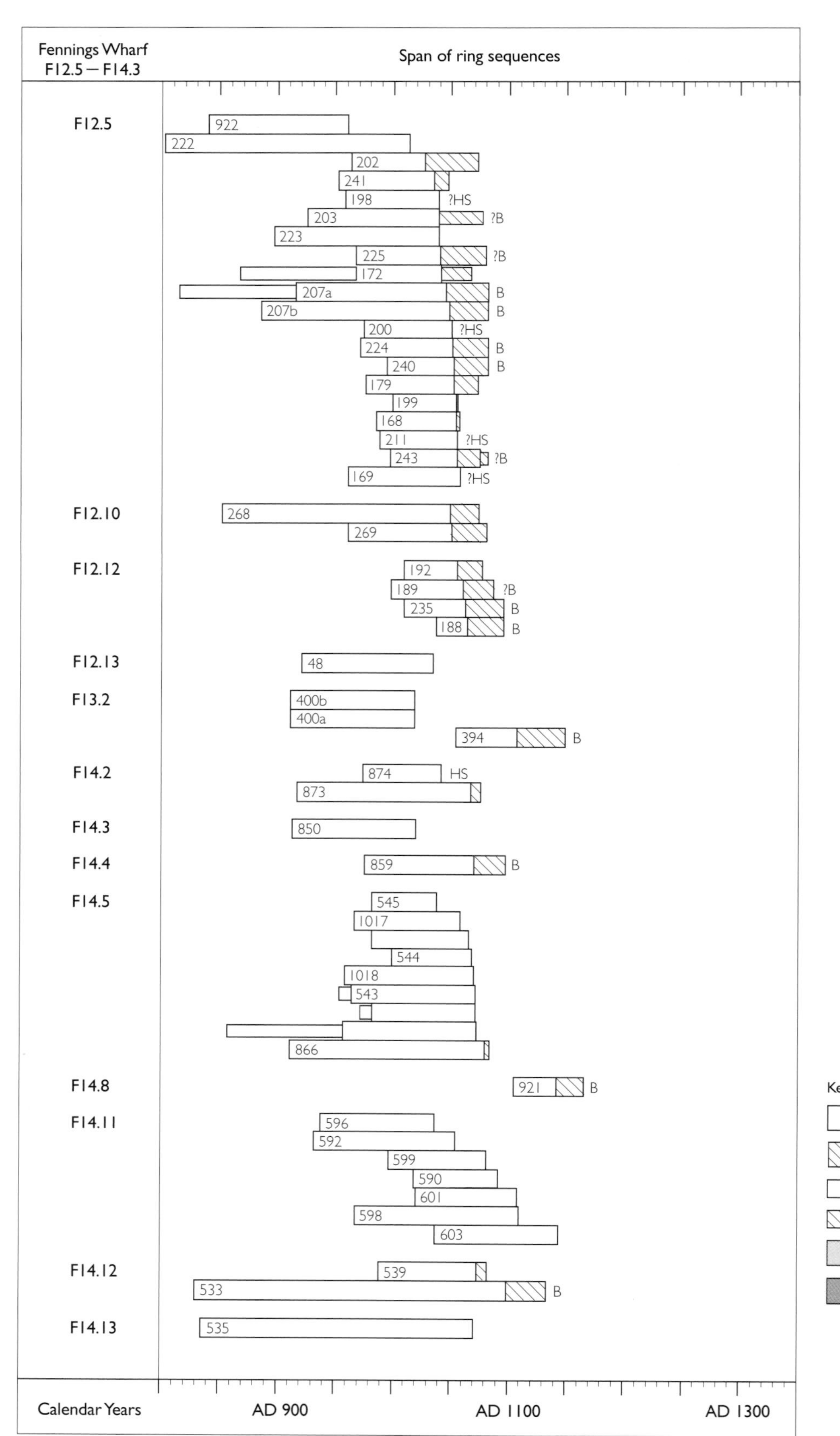

Fig 140 Fennings Wharf: tree-ring dated samples (numbered on date spans) from groups F12.5 through to F14.13

Key
- oak heartwood
- oak sapwood
- unmeasured oak heartwood
- unmeasured oak sapwood
- beech wood
- elm wood
- HS heartwood/sapwood boundary
- B bark boundary

catalogued, sorted by species and stored. The reused boat timbers were analysed in 1990 (Tyers 1994) and, in late 1990 and early 1991, further samples were analysed during a multi-site research project aimed at assessing the viability of beech and elm timbers for tree-ring analysis (Tyers in prep). The bulk of the samples were assessed, prioritised and analysed in 1993. The residue of analytical work and the compilation of the final archive report and the publication text were completed in 1997.

The sampling programme originally adopted for this site

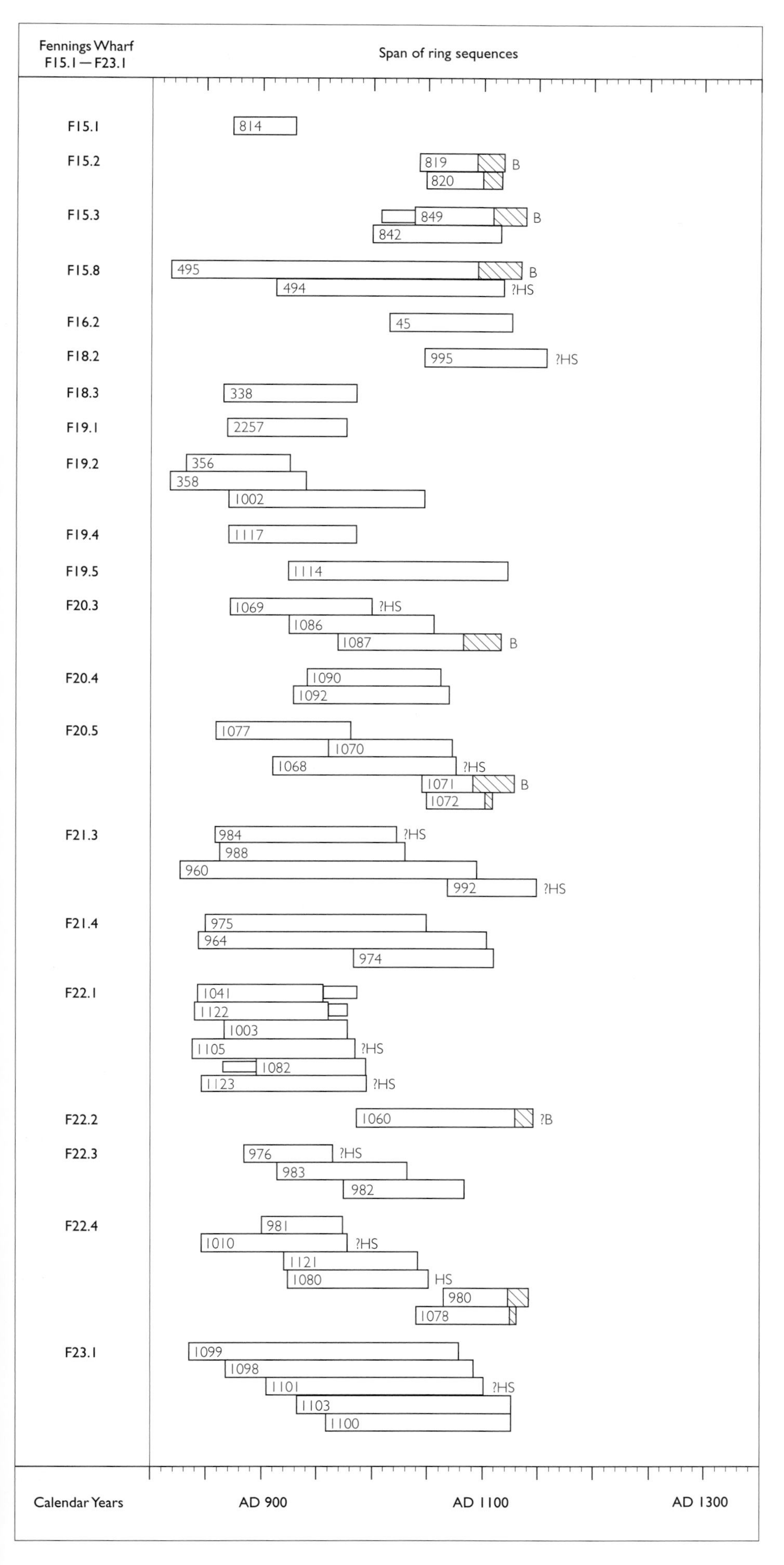

Fig 141 Fennings Wharf: tree-ring dated samples (numbered on date spans) from groups F15.1 through to F23.1; see key on Fig 140

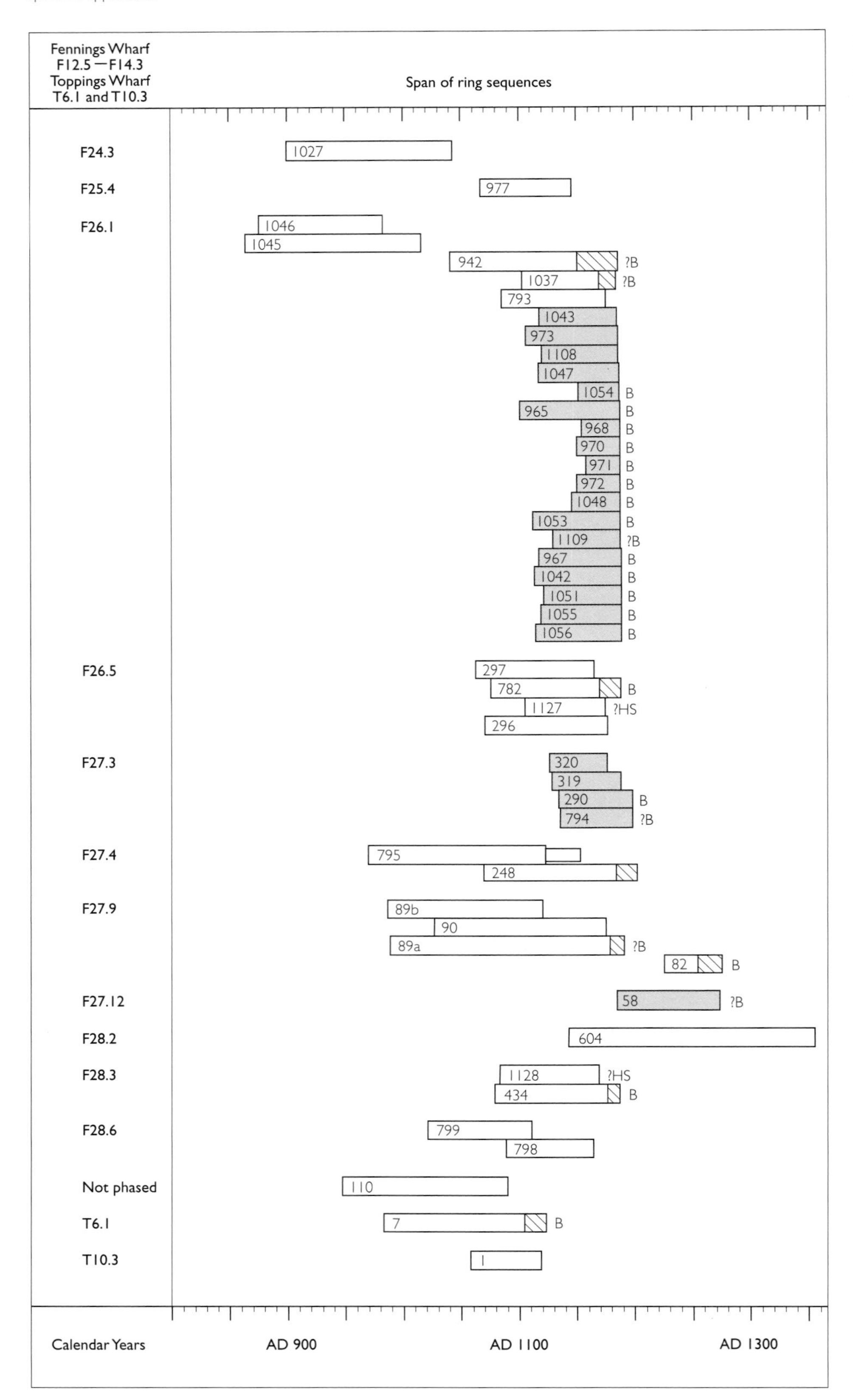

Fig 142 Fennings Wharf: tree-ring dated samples (numbered on date spans) from groups F24.3 through to F28.6 and Toppings Wharf groups T6.1 and T10.3; see key on Fig 140

evolved from experience of tree-ring analysis of material from a handful of sites in Southwark in the early 1980s. These convincingly demonstrated that a near-complete sampling strategy was the only way of maximising the potential of tree-ring dating for archaeologists on complex, deeply stratified, multi-period urban sites. There was little idea at the start of the London Bridge excavations that so many timbers would be involved, let alone of the quantity that was to amass as a result of applying the same strategy to sites in London throughout the building boom of the later 1980s and very early 1990s. With hindsight it could be argued that the strategy was over-ambitious, especially in view of the resources available, but the results have been highly beneficial for later archaeological sites in and around London. They include the creation of unparalleled sequences of dated structures and finds that have made possible a detailed comparison of the development of different parts of the settlement of London, new opportunities to develop typologies based on reliably dated material and the opportunity to reassess whole categories of widely applied finds typologies. At the time of the Fennings Wharf excavations, however, the sampling strategy was designed to meet two key considerations. Firstly, it was impossible to tell during the excavation which samples were likely to be of value in interpreting the site. Civil engineering constraints in particular meant that the site could never be seen as a whole and sampled selectively at that stage (Watson and Tyers 1995). Secondly, any overall improvement of the available reference data with the creation of new data sets would greatly enhance the dating potential not only of this site but of others in the future. As it happens, one result of the delay between excavation and analysis is that this second, although not secondary, aim of consolidating and strengthening the regional reference chronologies has become almost redundant. It is difficult to remember just how weak the London tree-ring sequence was in this era (eg Morgan and Schofield 1978), compared with the present situation. Since the excavations at Fennings Wharf vast quantities of samples have been excavated in London and much has been analysed to create an extremely well replicated reference chronology covering the periods *c* 250 BC–AD 300 and *c* AD 400–1700. The Billingsgate material analysed between 1985 and 1988 still constitutes the largest single site collection of medieval London tree-ring reference data (Hillam 1992).

The Roman timbers from Fennings Wharf consist of two well linings (Tyers 1993, table 1). One well was lined with silver fir (*Abies alba*) staves from a reused barrel (F9.2) and the other with oak planks (F10.4). Four of these planks were dated, the latest producing a heartwood date of AD 122. The 774 samples from post-Roman timbers excavated from Fennings Wharf were divided into 18 major phases (F12 to F31, there being no timbers from F17 or F29) and 77 sub-phases; 16 timbers could not be assigned phases. The four Toppings Wharf timbers are derived from three phases, one timber being unassigned to a phase. Altogether 273 samples from 267 timbers were measured from the Fennings Wharf site, with three timbers from Toppings Wharf. These comprise 206 oak, 43 elm, and 27 beech timbers. The rest of the material was unsuitable for tree-ring analysis for a variety of reasons: some samples were of unsuitable species, others contained too few rings, some again had anatomical peculiarities that prevented reliable measurement and a few had unfortunately disintegrated during their long period in storage. For 422 of these otherwise unsuitable timbers the details of species, ring counts, presence of pith and sapwood or bark, the average growth rates, the cross section, and the cross-sectional dimensions have been recorded, while for the remaining 80 samples only the wood types were identified. These records are listed in the archive report (Tyers 1997a, table 1). The less common wood types were identified using a variety of reference slides and appropriate identification keys (eg Schweingruber 1982; Wilson and White 1986; Schweingruber 1990) and using high magnification microscopes.

A total of 132 oak samples and 23 beech samples were successfully dated by tree-ring analysis. These provide a variety of dates from the 11th to 14th centuries for 47 sub-phases from 18 of the site phases (Fig 139), and there is one dated but unphased timber. The majority of the dated timbers are from the pre-bridge (F12–23: Fig 140; Fig 141) or bridge construction phases (F24–7: Fig 142). This distribution results from the more favourable preservation conditions of the deepest timbers. There were disappointingly few oak samples of any great utility from the post-bridge structural sequence (F28–31: Fig 142), and no results of great significance were forthcoming from this material. The two dated samples from Toppings Wharf material have similarly minor significance (Fig 142). The discussion of the tree-ring results from these sites concentrates on four aspects of general interest: the Saxo-Norman bridge, the date of construction of the Colechurch bridge abutment, the analyses of beech timbers from the site and the analyses of elm timbers from the site. Details of individual dates of chronological significance are listed in Chapter 14.1.

The Saxo-Norman timber bridges

The most interesting group of timbers excavated at Fennings Wharf are the fragmentary remains of earlier bridge structures found underneath the later 12th-century abutment. The groundworks for the Colechurch bridge obliterated almost all of these structures, and what little was seen of them appears to have been greatly disturbed by these works. Relevant groups are the clay and timber bank F12.5 (see Chapter 6.3), the Saxo-Norman abutment F19.1–.5 (see Chapter 5.5), the first 12th-century caisson F21.2–.5, and the second 12th-century caisson F22.1–.4 (see Chapter 7.3). The tree-ring dates of these groups produced an intriguing assortment of results.

The revetment or bank F12.5 incorporates many timbers, mostly lengths of split larger trees. The analysis identified ten of these as derived from only one tree, whilst two other pairs of fragments derived from the same trees were also identified. The group appears to have been felled in 1081–2 and this structure, therefore, provides one good fixed reference point for the development of the foreshore installations in this area.

The Saxo-Norman abutment group (F19) is more complex. Apart from the pile clusters (F19.5), none of the rest of the oaks retained much sapwood. This deficiency may indicate either heavy erosion, or reuse, or post-constructional disturbance of the timbers. Of some concern to the validity of the results is the presence of a piece of reused boat (F19.1), made from timber felled after AD 986 (Marsden 1994, 156–8). The latest ring was present on baseplate {1002} (F19.2) and this indicates that felling must have occurred sometime after

1056. A curious feature is that the rest of the timbers in this group all have late 10th-century end-dates, although only heartwood is present on any of them.

The first caisson group (F21) is equally problematic, nearly all the timbers in F21.3 or F21.4 being found to derive from a single, obviously large, tree. Several problems arise here: firstly this group was noticeable for its absence of sapwood, with the same suggestion of erosion, disturbance, or reuse; secondly, several timbers assigned to these groups are also associated with others in the intermingled second caisson, for example a same-tree link between {974} (F21.4) and {982} (F22.3). The second caisson group (F22) is again dominated by samples with no sapwood, the dated ones include many with late 10th-century end-dates. The dating of much of F22 suggests that it could include many timbers apparently earlier than those in the structure it overlies.

To summarise, the widespread presence of heartwood-only samples, many with later 10th- or mid-11th-century dates, throughout F22, F21 and F19, and the identification of several inter-phase linkages of individual trees perhaps indicate that many of these features are simply reconstructions of earlier structures, almost entirely using timbers from these or other such structures. Instead it is the very much smaller number of later timbers which, it is hoped, indicate the real dates of these phases. From F19.5 there is a pile {1114} dated after 1132, F21.3 includes {992} with a felling date range of 1158–1203, while F22.2 has one timber probably felled in 1145 {1060}, and F22.4 includes two timbers {980} and {1078} derived from the same tree felled in 1141–78.

The date of the Colechurch bridge abutment

There are two principal components to the samples from the bridge abutment. Firstly, there are six samples from the large oak sillbeams which provide the footings for the eastern (F26.1) and western (F26.5) pier walls of the abutment. A further pair of samples from the south-eastern area were interpreted as disturbed original sills (F28.3). Secondly, there are 27 recovered beech piles (F26.1) directly overlaid by the eastern sills.

The recorded dimensions of the eight sampled oaks are shown in Table 1; all are logs, converted into beams, scarf-jointed and pegged together. Note that although in no case were the entire lengths observed *in situ*, one beam was recorded in some detail *ex situ*. This timber must have been obtained from a very large mature oak tree (Table 1; Fig 143).

These planks are half trees (box halved), cut lengthways. They mostly had a single squared face, which was the exposed face, whilst the more ragged wavy edges were always towards the inner and bottom surfaces. All eight of these timbers were suitable for analysis and all were successfully dated. The timbers were not especially long lived despite their size and they provide an interesting set of results. Firstly, two pairs of timbers were found to be derived from single trees: the two samples from F28.3 and, more surprisingly perhaps, sample {297} from the western sills derived from the same tree as

Table 1 Recorded dimensions of eight sampled oaks

Area	Context/sample	Width x thickness x length (m)
Eastern sills F26.1	[109] {942}	0.63 x 0.22 x 2.50+
	[531] {1037}	0.60 x 0.16 x 3.50+
Western sills F26.5	[30] {296}	0.48 x 0.15 x 2.78+
	[36] {297}	0.56 x 0.26 x 4.84+
	[88] {782}	0.60 x 0.22 x 6.35
	[597] {1127}	0.58 x 0.23 x 2.48+
SE area F28.3	[598] {1128}	0.70 x 0.21 x 2.65+
	[2139] {434}	0.76 x 0.22 x 3.10+

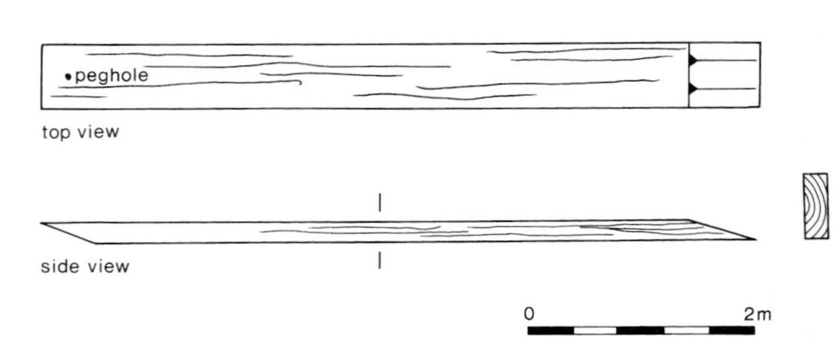

Fig 143 Drawing of the complete sillbeam (F26.5) from the Colechurch bridge abutment

sample {942} from the eastern. This linkage provides a good example of how tree-ring analysis assists the interpretation of the stratification, for engineering constraints resulted in these two areas being excavated many weeks apart, with no continuous section available. The results helped to confirm that the two structures were part of a single feature. The felling dates of these oaks appear to be spread over a couple of years: 1186 for {434}, 1187 for {782} and possibly 1185 for {942}. Such a variation is a relatively unusual finding from archaeological sampling, especially as there is little benefit to be gained by seasoning timbers intended for such positions. The variation in felling dates perhaps indicates that the construction programme required the stockpiling of such timbers for the sillbeams. The 19 piers and two abutments presumably required more than 100 large trees for the sills alone, which may have proved difficult to obtain, suitable specimens being instead stockpiled whenever they were available. The generally high average growth rates observed amongst these samples may suggest that these trees derived from relatively open areas, perhaps parkland or hedgerows. Analysis of timbers from Ware Priory, Hertfordshire, produced felling dates ranging from 1391 to 1416, showing that material had been stored or stockpiled for some years before use (Howard et al 1997).

The beech piles from under the eastern sills were whole round sections of trunks ranging between 240 and 340mm in diameter. Most were found to be complete to the bark edge, although difficulties of sampling in the best locations prevented that being the case with all the samples. Twenty of the samples proved suitable for analysis, 18 of them being matched together to form a single coherent group of data, of which 14 were complete to either bark or probable bark edge. When the composite London beech chronology was absolutely dated (see next section), these piles were found to be felled in the winter of 1187 and during the spring or early summer of 1188. Several timbers may be derived from single trees, although the reliability of same-tree identifications for beech is not yet proven. The felling dates provided by the beech (latest 1188) confirm a construction date of 1189 or perhaps 1190 for the southern abutment, as the relatively small size of the beech timbers used for the piling argues against the need for any sort of long-term stockpiling of this type of material.

The beech analyses

Tree-ring analyses in the British Isles and parts of Europe have concentrated almost exclusively on oak (*Quercus* spp), probably because, at the time of the initial development of the technique, oak provided the most abundant and viable samples. The technique is not, however, exclusively dependent on oak. Almost all tree species are probably useful; nearly all North American and Scandinavian tree-ring dating, for example, concentrates on conifer species (Schweingruber 1993).

Beech (*Fagus sylvatica*) has never been recovered from Roman structures on London sites, but several features of 11th- to 13th-century date have been excavated where beech was a major component. The reasons for this apparently short-term favouring of a fairly unsuitable building timber are not entirely understood, although one possibility is increasing shortages and rising costs, resulting in a diversification into other types of timber, a hypothesis also supported by the increasing importance of elm. An equally plausible hypothesis is that the features concerned were of types for which alternatives to oak were preferred.

Although beech timbers had probably been noticed on several earlier sites, Fennings was the first one where they were sampled. At the time they were simply stored, awaiting further examples; by 1988 there were about 150 beech samples with sufficient rings from nine London excavations, all thought to be Saxo-Norman or later. There were four research aims for the analysis of this material.

1) Could multiple radii from single samples be matched together?
2) Could multiple samples from single features be matched together?
3) If the second aim was realised, could different site groups be matched together?
4) If the third aim was realised, could the resultant multi-site chronology(s) be dated absolutely?

In 1990 a pilot study, using the archaeological samples and a small group of modern trees, concluded that the first two aims were possible. Of great significance for the usefulness of beech in archaeological work was the observation that beech does not have differentiated sapwood and heartwood and that groups of samples generally yield excellent sets of replicated felling dates for each analysed group. This is at least partly due to the general usage of this timber either in the round, or as simply split billets, rather than squared timbers, as with oak.

Stage 3 yielded two different groups of data: a rather longer, slower grown group of chronologies which matched each other extremely well and a rather shorter, faster grown group of data which also matched each other extremely well. The two groups did not appear to be contemporary and for each group there were a number of oak felling dates from associated contexts, which gave some indication of the approximate date of each composite sequence. The final objective was to produce absolute datings for these composite sequences. At the time, the present writer did not know of any other contemporaneous beech reference sequences and there was no opportunity to build a continuous sequence from the present day with the samples. Instead it was hoped that the beech sequences could be cross-matched directly to the medieval oak sequence, oak-to-beech cross-matching having been observed during the analysis of samples from living beech trees.

Stage 4 of the initial pilot survey did manage to match the chronologies constructed at that stage to the local oak sequences. Since these matches suggested dates for the beech sequences that were compatible with the dates indicated by oak samples from the same or related features, there was considerable cause for optimism (Fig 142). Subsequently, between 1990 and 1997, nine further excavated groups have been analysed, including samples from Winchester and South Wales. These have consolidated the two sub-chronologies, and linked the two groups (Fig 144). It has also been found that groups of contemporaneous beech data from excavations in France, Belgium and Germany form a coherent group of reference sequences consistent with the London material. Thus the analysis of archaeological beech from London and elsewhere in the UK has resulted in a reference sequence that, although only 460 years in total length, appears to cover the entire period of major use for the timber. The absence of differentiated heartwood and sapwood has the potential to produce dating results of unexpectedly high precision.

The excavated beech samples from Fennings Wharf comprise a large number of piles from directly below the bridge abutment (F26.1), discussed in the previous section, and a few other items from slightly later features (F27.3, F27.12) which are of relatively minor interest.

The archaeological sampling of waterlogged beech shows that under such conditions it becomes a much weaker timber than oak. Experience shows that whereas waterlogged oak from medieval excavations can often only be cut with special chainsaw blades, which are continually sharpened, medieval beech can usually be sampled with a bread knife! In this light any use of beech in any waterlogged position comes as something of a surprise. Clearly, the builders of the various beech structures that have been excavated in London and elsewhere were unaware how soft the material became, or had no alternative timbers to hand, or perhaps only intended the structures to be short-lived. The fact that beech piles were used under the bridge abutment (F26.1) might suggest that they were unaware of its long-term weakness, since there is no other indication that the bridge construction programme

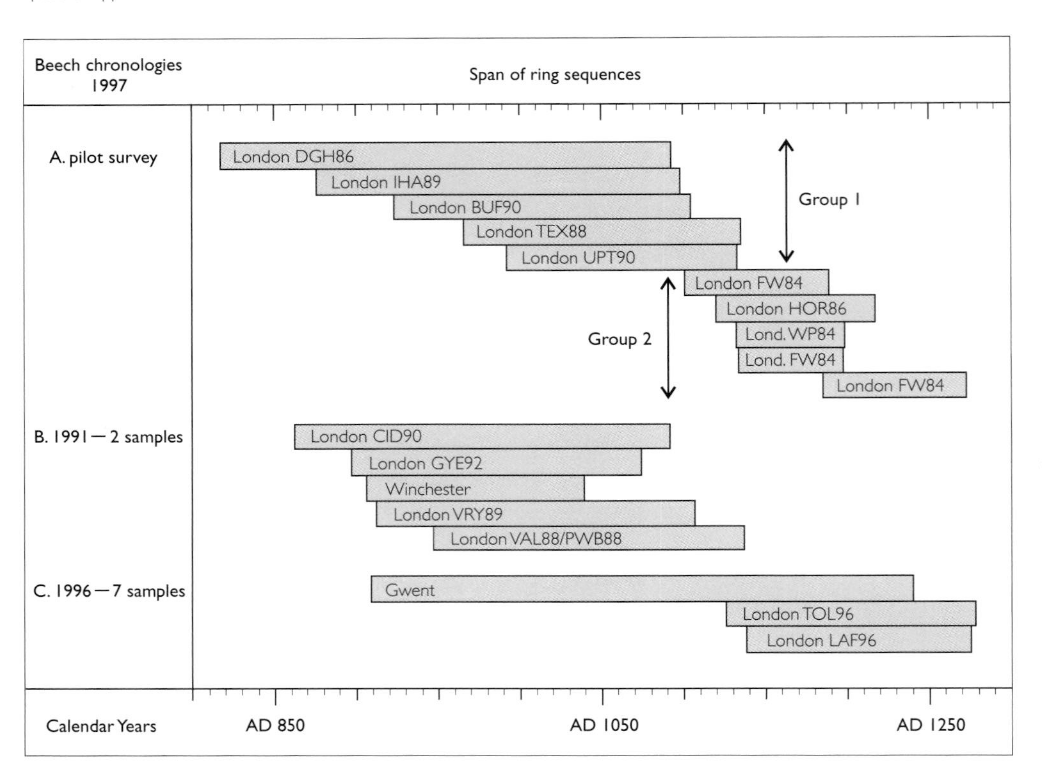

Fig 144 The current state of the archaeological beech reference chronologies from England and Wales; the initial pilot survey produced two groups of chronologies, which subsequent work from other excavations consolidated; more recent work has joined the two groups and extended the chronology slightly

was short of timber. Indeed, apart from this bridge abutment (1188), several important structures in the London area have now been dated from integral beech timbers, notably the pier starlings at Kingston bridge (1212 and 1216, Tyers 1991; see Potter 1991), which included beech piles, and parts of both Henry III and Edward I's building works at the Tower of London (1240, 1275 and 1276; Tyers 1997b; Impey and Keevill 1998, 11–19), which incorporated large numbers of them. This implies that the use of beech was 'normal' and that the material was believed 'suitable'. The use of beech baseplates in F27.3 of *c* 1197 and the occurrence of a single dated beech in F27.12 of *c* 1272 cannot of themselves indicate whether these structures were intended to be temporary, or something more substantial.

The elm analyses

In contrast with beech, elm (*Ulmus* spp) piles had been observed and even sampled from a number of excavations in London before the Fennings excavations (see eg Brett 1978). Elm piles are often observed, apparently preferentially used, in certain types of contexts on the waterfront structures of London, perhaps reflecting a cultural assumption that elm was better able to withstand the continuous wetting and drying caused by the tidal Thames. This assumption was still evident into at least the 19th century, when elm was often used for waterwheel boards and hollowed elms were widely used for waterpipes throughout the post-medieval period.

The use of elm for tree-ring analysis requires that the same four steps outlined for the beech have to be achieved. Elm is significantly more widespread, both on archaeological sites and in standing buildings, than any other non-oak species except post-medieval softwoods and the opportunity of dealing with elm from these excavations was of more than passing interest.

The Fennings Wharf elms were mostly derived from the pile groups forming the protective bridge starlings. A total of 215 samples were obtained, 198 of them from just six structures (F27.4, F27.11, F28.2, F28.3, F28.5 and F28.6) which were the principal exposures of the bridge starlings. Nine of the other samples came from the engineering spoil heaps and were probably derived from these areas. A total of 195 of the sampled elms were roundwood, complete to the bark edge. One problem with some of the data was the risk of intrusive samples, as it is thought probable that some of the area covered by F27 was open foreshore. For instance, F27.11 consisted of elm piles from both below and in front of riverwall F27.13 and this subgroup could have been added to by repiling as part of a maintenance programme for the riverwall. However, this was not the case with F28, which was sealed by the enlarged 15th-century abutment, but it was possible that this area of foreshore was partly piled at an earlier date to protect the Colechurch abutment. Another possible problem is the fact that, during the 15th century, elm is known to have been purchased from many places in south-east England, from trees that had doubtless grown in different soil types and geographical settings, for example hedgerows or woodland, circumstances which may have affected their growth ring pattern (see Chapter 10).

Comparing the complete roundwood samples from the main excavated groups highlights a few systematic differences between the parts of the starlings. Table 2 suggests that F27.11 and F28.3 appear to include a smaller average size of timber, derived from younger trees.

Table 2 Comparisons of complete roundwood samples of elm from the main excavated groups

Group	No. of roundwood samples	Average rings	Average diameter (mm)	Average growth rate mm/year
F27.11	14	20	166	4.44
F28.2	54	47	307	3.61
F28.3	46	29	224	4.06
F28.5	30	41	310	4.06
F28.6	24	48	374	4.03

Table 3 Comparison of oak and beech assemblages

Species	No. of samples	Average rings (and range)	Average growth rate (mm/year) (and range)	Average size (mm) (and range)
Oak	493	70.6 (8–316)	2.38 (0.69–8.57)	182 (40–760)
Beech	45	51.2 (27–88)	2.54 (1.03–6.0)	234 (110–360)
Elm	215	41.5 (13–155)	3.88 (1.43–8.75)	283 (85–530)

A comparison with the entire oak and beech assemblages shows the elms as a group are characterised by younger trees with significantly faster average growth rates and larger samples (Table 3).

There is also a tendency for the elm sequences to include periods of abrupt growth reduction or enhancement (Fig 145), a characteristic that in oak samples is usually taken as indicative of anthropogenic influences. This feature, together with the presence of tangential banding in the summer-wood in elm (Fig 146), makes the material very difficult to measure reliably since the differentiation of narrow rings and latewood bands is not easy.

A total of 43 elms from Fennings Wharf were suitable for measurement (out of a total 215 sampled) and these include 37 from the starling groups F28.2, F28.5 and F28.6. Almost no convincing evidence of inter-sample matching was obtained, although a pair of samples probably derived from a single tree were identified from F28.6 ({801} and {803}). This pairing result is hardly surprising considering that the observed lengths of the piles (*c* 1.5 to 2.1m long) suggests that several could have been obtained from any individual tree. A considerable amount of effort was expended obtaining and then analysing these samples, but no absolute dating was achieved for any sequence, which is disappointing. The Fennings Wharf elm group is the largest group of elms examined by tree-ring analysis in England thus far, but the results from these samples indicate that many elms are of extremely marginal utility for tree-ring analysis.

On a more positive note, perhaps groups of elm samples from a single phase of a rural or provincial structure are less likely than metropolitan samples to be derived from widely scattered sources. Analysis of such rural or provincial groups, where some indication of the likely felling date has already been determined from oak samples, may yield significantly better results than the Fennings material. The suggestion that the study of elm samples from provincial sites could be more useful has already been borne out by the reported dating of a post-medieval elm timber from Upwich saltpit, Droitwich, Worcestershire, by cross-matching it with the local oak sequence (Groves and Hillam 1997).

The purchase of elm trees in large numbers is documented in the Bridge House accounts, together with place of purchase (Fig 85). But nothing is recorded in the accounts concerning the intended use of this material. For instance, in May 1382 the following transaction was recorded: 'Item paid for 100 elms bought of Matthew Langrich at 22d each, £9 3s 4d' (Harding and Wright 1995, 22). It is probable that most of these elms were used as piles to repair the starlings (see Chapter 10). However, it is also clear from the records that not all the elm purchased was used for this purpose, as in 1420–1 the accounts mention paying for 'felling and quartering elms' (Harding and Wright 1995, 94), while in the accounts for 1460–1 men were paid for hewing and squaring elms (ibid, 142), presumably for use as beams.

Fig 145 Typical examples of the type of sequences measured from a beech, elm and oak sample, showing that the elm has both periods of remarkably faster growth and abrupt changes in the growth rate; these features are probably preventing the material from being successfully dated at present

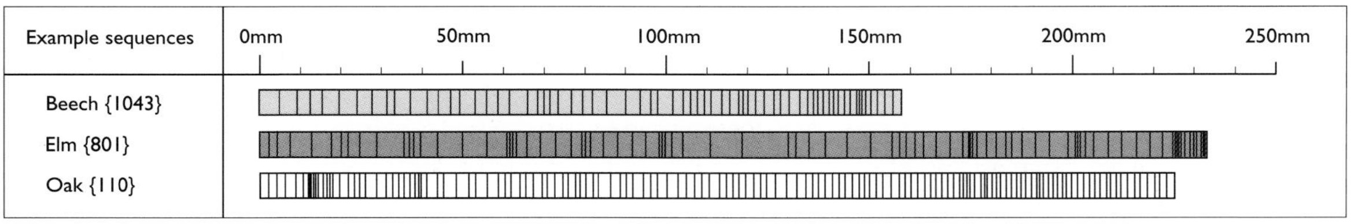

Example sequences	0mm	50mm	100mm	150mm	200mm	250mm
Beech {1043}						
Elm {801}						
Oak {110}						

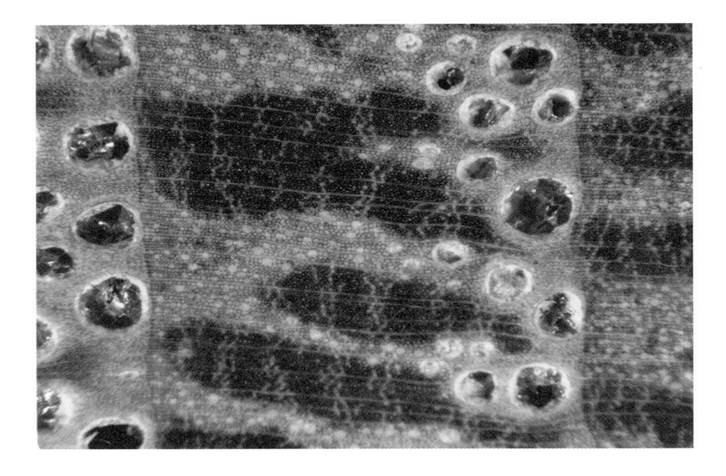

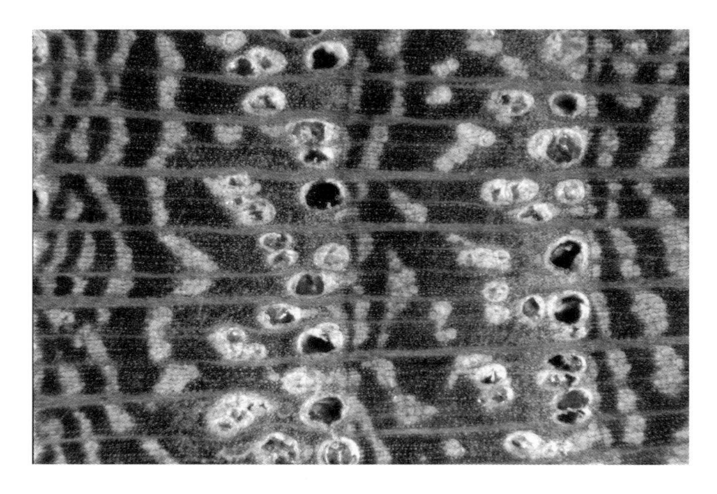

Fig 146 The anatomical differences between oak, beech and elm; all examples show growth from right to left; (top) oak, clearly defined growth rings; (middle) beech, less well defined; (bottom) elm, the latewood banding can make identification of narrow rings difficult (magnification c × 40) (I Tyers)

Elm sillbeams were used in part of the 15th-century bridge abutment (F28.7) and an upstream breakwater featured an elm sillbeam (F28.5). The latter sillbeam was 3.86m long, 1.19m wide and 140mm thick, fashioned from a thin section of tangentially faced log that had been obtained from a very large elm tree. Sadly, none of the samples from these structural elm timbers was analysed owing to deterioration and loss while in storage. A very interesting aspect of the Bridge House purchase records of elm trees is the wide geographical spread of the 30 documented locations (Fig 85), which could be interpreted as evidence that the maintenance of the bridge required so much timber that the Bridge House officials were obliged to purchase material from far and near.

Elm is not normally regarded as a forest tree and it most frequently occurs (or occurred) in hedgerows. The inability of the samples to provide much in the way of cross-correlations between individual piles could indicate that the trees used there originated in hedgerows. Poor levels of inter-correlation appear to be the normal experience of tree-ring analyses of oak timbers thought to derive from hedgerows, it being disputed, for example, whether many of the jointed crucks from Devon buildings were obtained from hedgerow trees (C Groves, pers comm) and the dating of these has proved troublesome. The continued analysis of the Devon cruck buildings and modern hedgerow trees may help determine whether hedgerow trees generally have low correlations with contemporary material, or whether the problem is worse with elms.

14.3 Aspects of the woodwork from medieval London Bridge and its associated waterfronts

Damian M Goodburn

Introduction

The hurried nature of the 1984 archaeological investigation of Fennings and Toppings Wharves resulted in a less than adequate field record of the excavated timbers, none of which were retained for study except for a number of reused boat timbers which have already been published (Marsden 1994, 156–9). In terms of timber studies the most significant aspect of the London Bridge assemblage is that it was the first for which an extensive timber sampling policy for both tree-ring dating and species identification was conducted (see Chapter 14.2; Watson and Tyers 1995, 61). This report is intended only as an introduction to the site research archive and as a summary of the aspects of the structural woodwork relevant to this publication. The complete archive report is lodged at the Museum of London.

Changes and innovations in the structural woodwork of London's medieval waterfronts

The well-preserved sequences of medieval timber structures investigated in London and Southwark offer a large assemblage of well-dated, structural woodwork, which allows the range of joints, techniques of log conversion, types of tools used and other aspects of design, including innovations, to be studied in detail (Brigham et al 1992; Goodburn 1995). The vast majority of these timbers have come from the numerous waterfront excavations within the medieval City (Milne and Milne 1982, 14–38; Milne and Milne 1992).

Saxo-Norman (AD 900–1100) structural woodwork in London was typified by three distinctive techniques as follows.

1) The use of a small range of simple joints in the prehistoric woodworking tradition, such as laft joints (edge notches), lap joints, tongue and groove jointing, through-splayed scarf joints, scarf joints and sockets (see Glossary and Brigham et al 1992). These joints were mainly cut with axes and adzes. The pegged mortice and tenon joint was not used in London during this period.
2) There was a lack of accurate or complete prefabrication of framed timbers and a predominant use of building techniques involving either earth-fast timbers, or laftwork or stavework construction (see Glossary) at this period. There were two phases of early or mid-12th-century bridge caisson of laftwork construction.
3) The most common timber conversion technique was cleaving (controlled splitting) of logs, both radially and tangentially. The cleft timbers were then hewn, which did not result in truly straight, square or standardised components. Timbers were not sawn long or across the grain in London at this period.

It is apparent that between *c* 1180 and 1200 there were a number of important changes in the practice and technology of structural woodwork in London (Brigham et al 1992; Goodburn 1995). The four most important innovations were as follows.

1) The manufacture of accurate, prefabricated components used to make 'timber-framed' structures, such as buildings and revetments.
2) The use of the mortice and tenon joint (see Glossary) with neatly cut shoulders and locked with pegs. One example of this type of framing work found at Fennings Wharf dates to 1197 and another to *c* 1167–1212 (discussed later).
3) The production of relatively well-squared and consistently sized timbers.
4) The sawing of timbers lengthwise initially to produce planking and later for framing timbers. One 15th-century elm plank used as a sillbeam was 1.19m wide (F28.5) and a large ripsaw would have been required to cope with the depth of cut needed to produce planks of this width.

These technical innovations could be interpreted as meeting the needs of the vast number of large construction projects taking place in late 12th-century England, including the construction of the first stone bridge over the River Thames (see Chapter 8).

Discussion of selective structural timbers from Fennings Wharf

The following comments are intended only to supplement the information in the narrative text, not to provide detailed descriptions of the material. All the dating evidence is outlined above (14.1).

The early 12th-century bulwark revetment (F20.5)

The technique of bulwark (see Glossary) construction was used in London during the 11th and 12th centuries for constructing both walls (Goodburn 1995, 51) and revetments (see Glossary) (Milne and Milne 1992, 60–1). One of the series of revetments upstream of the bridge was of this construction (F20.5: Fig 34), the best preserved portion of this consisting of two squared posts, 260 by 270mm and 320mm square, set 3.05m apart in stone-packed post pits. Cut into opposite faces of each post were two vertical grooves (*c* 60–80mm wide and 100mm deep), housing horizontal, radially cleft and hewn oak planks laid on edge, three of them still *in situ* (Fig 36). The three planks were *c* 3.15m long, 160–190mm wide and 55–90mm thick. The ends of all the planks had been roughly bevelled with an axe to help fit them into the grooves (Fig 36). No frontbraces such as those found at the 12th-century bulwark revetment at Thames Exchange (Milne and Milne 1992, 60–1) were found at Fennings Wharf, an absence probably due to their having been robbed out for reuse. From tidal levels it can be estimated that the Fennings Wharf bulwark revetment originally stood *c* 3.0m high.

The early 12th-century stave-work revetment (F23.1)

The revetment consisted of two squared, flat-bottomed posts (240mm and 250mm) set in post pits 4.36m apart (Fig 39). Cut into opposite faces of the southern post were two vertical grooves (each 90–95mm wide and 60–65mm deep); there was only one groove in the northern post, which marked the end of the revetment. The grooves were intended to retain the adjoining staves (see Glossary). Set in between the posts was a large baseplate made from a roughly squared, hewn, oak log (Fig 39). It was 4.29m long, 240mm wide and 270mm high. There was a continuous groove in the upper face of the baseplate, intended to retain staves (thin, radially cleft, oak planking), the basal portions of which remained *in situ*. Some of the staves were too thin to fit properly and had had to be secured with timber wedges. The varied thicknesses (45–75mm) and widths (190–360mm) of the eight *in situ* staves suggest that a number of them may have been reused timbers. One of the staves contained a small, neatly cut diamond-shaped hole, which might have been a ventilation hole if this timber was originally a stave in a partition wall of a building (Fig 39). The baseplate was certainly reused, as cut into its east side were two sloping, pegged lap joints and in the upper face two rectangular recesses, probably fragments of mortice type joints. Three spoon-auger holes in the east face of the baseplate may be relict features from the marking-out associated with the previous use of the timber. The southern (landward) continuation of this revetment probably consisted of horizontal, edge-laid planking, all of which had been robbed out.

The late 12th-century baseplate revetments

Two of the truncated revetments on the downstream side of the bridge represent early examples of prefabricated carpentry and the use of mortice and tenon joints (discussed earlier). The first revetment consisted only of a squared, beech log felled in 1197, set on cross-bearers and retained by roundwood stakes (F27.3: Fig 55). In the upper face of the baseplate was a series of rectangular mortices intended to hold the vertical elements of the superstructure. The second revetment consisted of a squared oak log baseplate over 3.12m long and felled *c* 1167–1212 (F18.4: Fig 55). Cut into its upper face were a series of rectangular mortices intended to hold verticals, all of which would have been secured by oak pegs: a number of these remained *in situ* (Fig 59). These two revetments are described in Chapter 8.

The Saxo-Norman bridge abutment (F19.2)

Timber {148} is interpreted as the baseplate for the northern side of a bridge abutment (see Chapter 7). The dimensions of this beam, over 6.31m long (both ends were damaged during lifting and before recording) and roughly 400mm square, show that it was part of a substantial structure. It had been produced from a large, straight oak log (Fig 42). At each end of the upper face of the beam was a through-socket for holding stakes to anchor it. Adjoining these were a pair of shallower sockets (not cut through the whole beam), presumably intended to hold vertical timbers (perhaps end posts for the superstructure). The intervening space in the upper face of the beam contained a continuous vertical groove (*c* 90mm wide and deep), which would originally have retained staves. As the rest of the structure was not found, reconstruction of its plan or superstructure is uncertain; possibly the rest of it (the landward portion) was less substantial and was washed away or salvaged for reuse. The design of the baseplate is similar to that of one found in a mid-12th-century stave-built revetment at Billingsgate (Milne and Milne 1992, 28).

The Saxo-Norman bridge abutment (F19. 3)

Timber {147} was over 4.25m long (both ends were broken during lifting) and rectangular in cross section, 360mm wide and 120mm thick and hewn to this shape from half an oak log (Fig 43). Cut into the upper face were a series of at least four irregularly spaced through-sockets (440mm to 880mm apart), several of which contained roundwood stakes intended to anchor the timber. This timber was obviously intended as a baseplate, but its interpretation is hindered by the fact that the rest of the structure was missing. In plan it closely resembled socketed baseplates or top plates from various 11th- or 12th-century London buildings found at Billingsgate (Brigham 1992b, 86–7). This could mean either that the timber was reused as part of the foundations of a bridge abutment, or that the same techniques were used in building a substantial timber house and a bridge abutment during the Saxo-Norman period.

The 12th-century bridge caissons

The first phase of caisson (F21) contained a number of timbers which, from relict joints and other features, appear to be reused (Fig 44) (see Chapter 7). The second phase of caisson F22 was well preserved (Figs 45–9), except for the missing east side. A general description of the caisson appears in Chapter 7.3. The three essential elements of the structure were as follows.

1) The thick baseplates were laft-jointed (¼ thickness notches) and fixed at the corners by square stakes set in through-sockets (Fig 48). All the baseplates were rectangular in cross section, made from cleft and hewn half logs and anchored by square oak stakes set in through-sockets in the upper face. The baseplate on the south side was 430mm wide and 150mm thick in cross section, and contained five irregularly spaced through-sockets. The baseplate on the north side was 290mm wide and 130mm thick in cross section and contained three such sockets. The baseplate on the west side was 350mm wide and 150mm thick in cross section. All the baseplates were founded on cross-bearers or reused timbers.
2) Traces of the superstructure remained *in situ* (Fig 49). It was represented on the south side by fragments of two rectangular, horizontal oak timbers made from quartered logs and set on edge. The lower of these two timbers {554} was laft-jointed at the west end; like the baseplates, it was 190mm high and 130mm wide. The fragmentary upper timber {553} was 170mm high and 140mm wide.
3) The interior of the caisson was filled with chalk rubble, set in a clay/silt matrix (F22.6).

The original height of the caisson from the top of the baseplate to the underside of the roadway or decking is estimated from contemporary tidal data (see Chapter 3.5 and Chapter 6.5) to have stood at least 3.0m high (Fig 49). That height was easily attainable in laftwork (see Glossary) and indeed there are a number of contemporary Russian bridge piers or caissons of laftwork construction standing several metres high (Opolovnikov and Opolovnikova 1989, 78). The weight of the rubble used to infill the caissons would have created a very strong lateral pressure, which would have resulted in the sides of the caisson bulging outwards, unless they were internally or externally braced. In Russian bridge piers or caissons, interlocking notched cross-ties are used to brace the structure (Opolovnikov and Opolovnikova 1989, 80).

The possible design of the rest of the bridge structure is discussed in Chapter 7.4: it should be noted that any reconstruction is conjectural as there is no archaeological evidence for its design. At the edges of the river, where the water was shallow, some of the bridge roadway was possibly supported by laftwork caissons, which could have been constructed at low tide and served as piers. Some form of bracing was probably used to strengthen the underside of the bridge roadway and prevent it flexing. It would have been difficult to fit diagonal braces or struts to the laftwork caissons, unless it had a substantial middle rail which could also have served as part of the external bracing.

14.4 The Roman pottery from Fennings Wharf

Robin P Symonds and Jo Groves

Introduction

This report summarises the site archive (Symonds and Groves 1997), which contains all the spot-dates and includes full details of the fabric and form codes, as used in Davis et al (1994) and Symonds and Tomber (1991), with minor revisions. Significant spot-dates and other related details are incorporated in Chapter 14.1 above.

Summary of the Roman pottery from Fennings Wharf

The Roman pottery from this site covers the full range from the early Roman period, the earliest context being dated AD 40–80, to the latest possible Roman date of AD 350–400. The date ranges shown are wholly typical of sites in Southwark and the City of London: if all the dates of all the contexts from all recent excavations in those areas could be entered on a single graph in *terminus post quem–terminus ante quem* order, its shape would be almost identical to the one for Fennings Wharf.

When the date ranges are ordered by stratigraphic groups, as shown in Fig 147, they seem to be rather less neatly defined. However, the solid line which crosses the centre of the graph divides purely Roman groups/contexts (above) from groups/contexts which contain post-Roman pottery and in which the Roman pottery is, therefore, residual. It is, therefore, only in the top part of the graph that a chronological sequence might be expected to be visible. In fact, given that the quantities of pottery are not especially large, the image in the top part of Fig 147 is more or less predictable. The image in the lower part is virtually random, given the indeterminable nature of the post-Roman disturbance of Roman features on the site.

Along with the dating, the pottery types present are also wholly typical of pottery assemblages from Southwark and the City. The types which date the contexts are all listed above in the dating summary and all of the few unusual types present are also mentioned.

There are some interesting assemblages, although these are mainly of interest because they are typical of their periods. For example, it is not at all surprising that F9.1, a rubbish pit [2167], with finds in [2168] and [2169], is the largest assemblage of Roman pottery: pit groups of the Flavian period are very often the largest assemblages on London sites.

The well, F9.3, is somewhat more problematic, since it contains a jar of Thameside Kent ware (AD 180–300). If this is intrusive, the overall date of the group becomes mid- to late 2nd century. In fact this is a relatively small amount of material for a well group and it is, therefore, not as securely dated as some other well groups from Southwark.

The cesspit, F10.1, is more homogeneous in composition, with a good sample of types which belong happily to the late 2nd or the first half of the 3rd century. Although there is some earlier material, it forms a relatively small proportion of the assemblage.

The well backfill, F10.4, is somewhat later than F10.1, but has a similar degree of homogeneity. The pit, F10.9, is probably the latest undisturbed Roman feature. It contains some residual early Roman pottery, but most of the material belongs convincingly to the 4th century. There are some types represented in the post-Roman contexts which are even later, that is dated to the second half of the 4th century, but none of these can be associated with Roman features.

14.5 Medieval and later pottery from excavations at Fennings Wharf with a reassessment of the Toppings Wharf material

Jacqueline Pearce

Fennings Wharf – introduction

The following is intended as an introduction to the post-Roman ceramic archive report (Pearce 1997). The pottery from Fennings Wharf was initially spot-dated to provide a chronological ceramic sequence for the site, based on the presence and absence of diagnostic fabrics and forms, as well as other significant dating features, such as styles of decoration. A date range was assigned for each context, giving a *terminus post quem* derived from the start date of the latest type(s) in contemporary use. The dating of London's medieval pottery is based principally on the massive groups of material dumped behind successive timber revetments on the northern foreshore of the Thames (Vince and Jenner 1991; Vince 1991b; Pearce et al 1985; Pearce and Vince 1988). These dates remain substantially unaltered by subsequent work in London and consequently form the basis for the ceramic chronology proposed for Fennings Wharf, according well with other dating evidence from the site (see Chapter 14.1 for details).

All fabrics (medieval and later) have been recorded using standard codes (generally of four letters) in both the MoLAS computerised record and in the paper archive; these codes are listed in Table 4 and are used throughout this report. Form names are given when there is sufficient evidence for a reasonable identification; where this is uncertain, sherds are recorded as 'miscellaneous' (MISC).

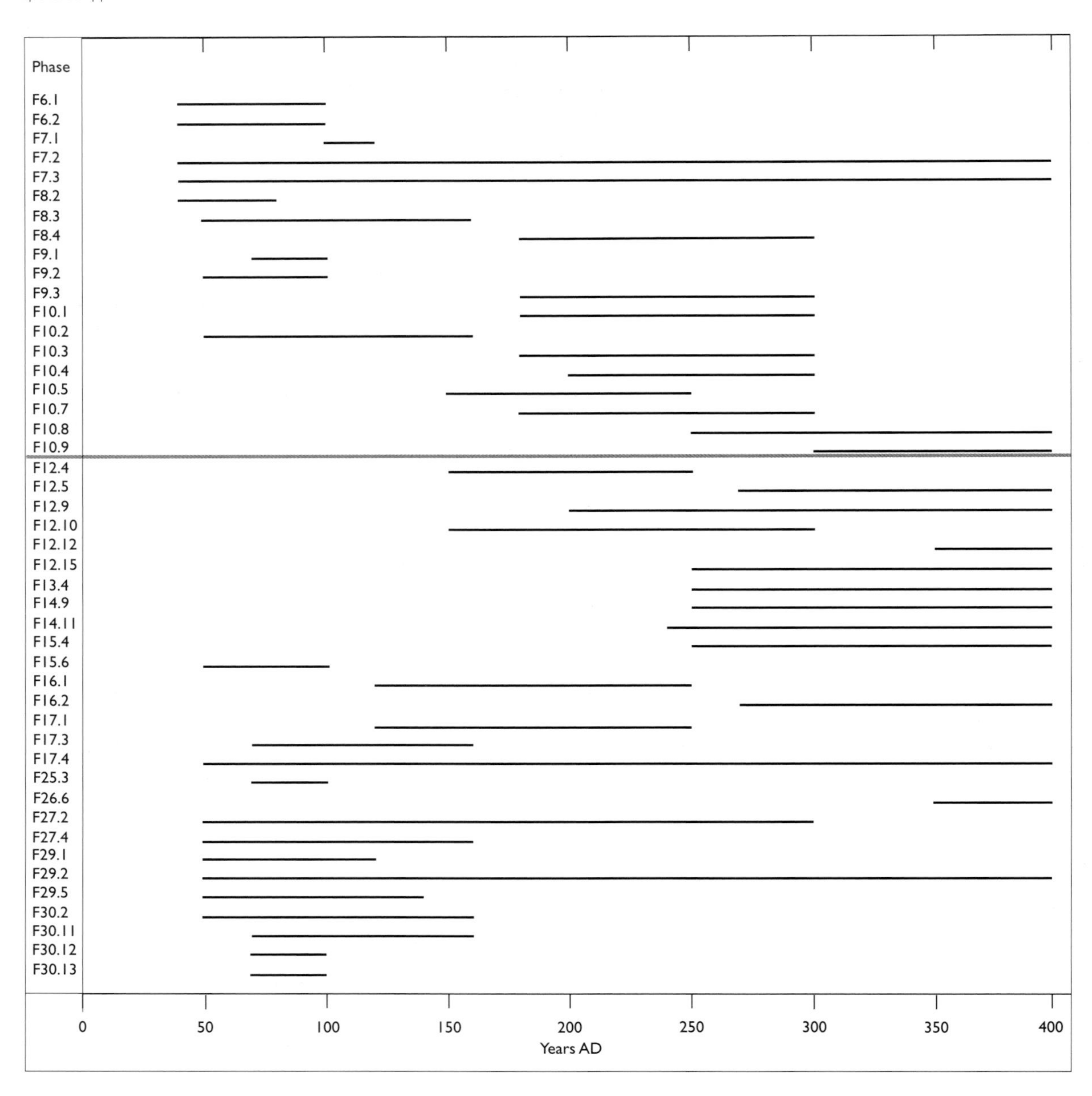

Fig 147 Date ranges for the Roman pottery from Fennings Wharf ordered by stratigraphic groups

The distribution of pottery across the site is summarised in Table 5. For full details of the analysis of this assemblage see the site research archive (Pearce 1997). As part of this report the published medieval and Tudor pottery from the 1970–2 Toppings Wharf excavation (Orton et al 1974) has been reassessed.

The Saxo-Norman and medieval pottery

The pottery is discussed more or less in chronological sequence by successive stratigraphic groups (see Chapter 14.1 for other dating evidence).

Group F12: stabilisation of the foreshore

The bulk of the pottery has been dated to *c* 1050–1150, although a date in the second half of the 11th century seems most likely, since no 12th-century fabrics or forms were recovered (eg London-type wares).

The pottery from Group F12 is dominated by handmade coarsewares (74% by ENV), in a range of fabrics and forms typical of late 11th-century assemblages from the City and Southwark. By far the most common fabric in the whole group is EMSS (29 ENV, or 36%), generally dated to *c* 1000–1150 in London (Vince and Jenner 1991, 63). It occurs with EMSH, ESUR and EMCH, all of which were used in London from *c* 1055 onwards (15%, 11% and 2% respectively). Apart from a single bowl with incised decoration in ESUR from [2089], cooking pots are the only form found in all fabrics.

There are very small quantities of non-local pottery: a single sherd of THET, probably part of a storage jar, from [2260], and one sherd from a STAM pitcher from [2110]. Continental imports are more numerous, consisting of REDP (10%) and ANDE (14%), principally spouted pitchers. There is a relatively

Table 4 MoLAS fabric codes and date ranges for the medieval and later pottery from FW84

Code	Common name	From	To
Medieval			
ANDE	Andenne-type ware	1050	1200
BLGR	Blue-grey or Paffrath ware	1050	1200
CBW	Coarse Border ware	1270	1500
CHEA	Cheam whiteware	1350	1500
DUTR	Dutch red earthenware	1350	1650
EMCH	Early medieval chalky ware	1050	1150
EMFL	Early medieval flinty-tempered ware	970	1100
EMGR	Early medieval grog-tempered ware	1050	1150
EMS	Early medieval sandy ware	970	1100
EMSH	Early medieval shelly ware	1050	1150
EMSS	Early medieval sand and shell ware	1000	1150
ESUR	Early Surrey ware	1050	1150
KING	Kingston-type ware	1230	1400
LCALC	Calcareous London-type ware	1080	1200
LCOAR	Coarse London-type ware	1080	1200
LLSL	Late London slipware	1400	1500
LMHG	Late medieval Hertfordshire glazed ware	1350	1450
LOGR	Local greyware	1050	1150
LOND	London-type ware	1080	1350
LSS	Late Saxon shelly ware	900	1050
MG	Mill Green ware	1270	1350
MISC	Miscellaneous unidentified wares	900	1500
NFM	North French monochrome ware	1170	1300
NFRE	Miscellaneous North French unglazed wares	900	1200
REDP	Red-painted or Pingsdorf-type ware	900	1250
ROUE	Early Rouen ware	1170	1300
SHER	South Herts/Limpsfield greyware	1140	1300
SIEG	Siegburg stoneware	1300	1500
SPOA	Miscellaneous Spanish amphorae	1200	1500
SPOW	Miscellaneous medieval Spanish wares	1250	1500
SSW	Shelly-sandy ware	1140	1220
STAM	Stamford-type ware	1050	1150
THET	Ipswich/Thetford-type ware	1000	1150
THWH	White Thetford-type ware	970	1150
VALE	Early Valencian lustreware	1380	1450
Post-medieval			
BEAU	Beauvais sgraffito ware	1500	1630
BORD	Surrey/Hampshire Border ware	1550	1700
BORDB	Brown-glazed Border ware	1620	1700
BORDG	Green-glazed Border ware	1550	1700
BORDY	Yellow-glazed Border ware	1550	1700
CHPO	Chinese export porcelain	1580	1900
DUTR	Dutch red earthenware	1350	1650
DUTSL	Dutch slip-coated ware	1500	1650
EBORD	Early Border ware	1480	1550
FREC	Frechen stoneware	1550	1700
ISAB	Isabella polychrome tin-glaze ware	1500	1550
LONS	London stoneware	1670	1900
METS	Metropolitan slipware	1630	1700
OLIV	Spanish olive jars	1550	1750
PMBR	Post-medieval bichrome redware	1480	1600
PMFR	Post-medieval fine redware	1580	1700
PMSR	Post-medieval slipped redware (Guys ware)	1480	1660
PMSRG	Post-medieval green-glazed slipped redware	1480	1660
PMSRY	Post-medieval yellow-glazed slipped redware	1480	1660
PMR	Post-medieval redware	1580	1900
PMRE	Early post-medieval redware (Tudor brown)	1480	1600
RAER	Raeren stoneware	1480	1550
RBOR	Red Border ware	1580	1800
SWSG	Staffordshire white salt-glaze stoneware	1720	1780
TGW	English tin-glaze ware	1570	1800
WEST	Westerwald stoneware	1590	1800

high incidence of vessels represented by more than one sherd in the pottery from the two revetments (F12.5 and F12.12), amounting to almost 50% by minimum vessel count from F12.5. This suggests that the material may have been discarded as contemporary rubbish, rather than redeposited.

Group F13: revetments and foreshore deposits

Medieval pottery was found in context [2114] which formed part of the silt/peat accumulation sealing the third revetment, post-dating revetment F13.2 which is dated to 1148. It can be dated to the second half of the 12th century and possibly as late as *c* 1180+. In addition to ESUR, EMSS and EMSH cooking pots, probably residual, there is a single cooking pot rim in SSW, which reached the peak of its use in London *c* 1140–1220 (Vince 1991b, 271; Blackmore and Vince 1994, 48, 69–70). There are also sherds from jugs or storage jars in SHER, which was in use *c* 1140–1300, and from jugs in LOND, LCOAR and LCALC. With one exception, all London-type jugs in this context are of early rounded form, with clear or copper-stained lead glaze, typical of the second half of the 12th century (Pearce et al 1985, 127–9). However, there is also part of a rod handle with green glaze over a white slip, more characteristic of 13th-century LOND baluster jugs in the North French or Highly Decorated styles than of 12th-century jugs (Pearce et al 1985, 26–7). If the sherd is not intrusive, therefore, a date towards the end of the 12th century seems most likely. Imports are represented by a single sherd of ANDE with rouletted decoration.

Group F14: revetments

A relatively low proportion of vessels in this group are represented by more than one sherd (14%). The large number of residual Roman sherds (24%) in context [2165] indicates a high level of redeposition, which is not unexpected in an unsealed fluvial deposit such as this, and the late 12th-century pottery and collapsed cask in this subgroup are likely to be intrusive. The bulk of the pottery (69%) consists of late 11th- to early 12th-century local wheelthrown coarsewares, with later SSW and LCOAR cooking pots, and LOND and LCOAR early rounded jugs.

A sequence of revetments to the north of the clay and log bank (F12.1) was buried by an accumulation of silt/peat (F14.9), dated by the pottery to the late 12th/early 13th century, the latest material in the group. More than half the pottery (58%) consists of early medieval handmade fabrics: EMS, EMFL, EMSH, EMSS, EMCH, LOGR and ESUR, as well as imported ANDE and REDP. Among these are the rim and short tubular handle of a socketed bowl in EMS, a form better known in EMSS (eg Vince and Jenner 1991, fig 2.39, no. 87); part of a bowl in EMS, sooted externally (ibid, fig 2.34, nos 57–8); and the everted rim, with fingertip impressions around the top, of a cooking pot in EMSH, sooted both inside and out. In addition, there are sherds from 17 different vessels in London-type wares (LOND, LCALC and LCOAR).

Table 5 Distribution of pottery across the site by stratified group in estimated number of vessels (ENV) and sherd count (SC)

Group	Subgroup	Contexts	ENV	SC
F12	F12.1; F12.4; F12.5; F12.8; F12.9; F12.12; F12.15	[2257]; [2260]; [2267]; [2089](M); [2152A]; [2110]; [2063]; [2003]	81	131
F13	F13.4	[2114]	22	29
F14	F14.9; F14.11	[2140](M); [2158]; [2165](M)	80	102
F15	F15.4; F15.6	[2231]; [2080]	20	29
F16	F16.1; F16.2	[2161]; [2152]; [2020]; [2021]; [2151]; [2228]	64	79
F17	F17.1; F17.3; F17.4	[1037]; [1038]; [1046]; [2043]; [2036](L)	13	171
F18	F18.5	[41]	1	1
F20	F20.2	[566]	2	2
F21	F21.2	[110](M)	4	33
F24	F24.2	[564B]	4	8
F25	F25.3	[500](M)	54	69
F26	F26.4; F26.5; F26.6	[35]; [7]; [8]; [10]	14	21
F27	F27.2; F27.4; F27.12	[503]; [504]; [580]; [20]; [29]; [24E]	-	56
F28	F28.1	[34]; [79](M); [83](M)	56	95
F29	F29.5; F29.10	[2070]; [1035]; [1036]	5	6
F30	F30.2; F30.4; F30.9; F30.11; F30.12; F30.14; F30.18	[1006](M); [1009](M); [1007](VL); [1013]; [1017](L); [2102]; [2001]; [2002]; [2091]; [2030]	102	499
F31	F31.12; F31.3	[16]; [18]	5	32
Total			**583**	**1379**

Context size is indicated by the letters M (medium), L (large) and VL (very large) after the context number; all other contexts are small

These include two cooking pots and a spouted pitcher in LCOAR, and several undecorated early rounded jugs. All these London-type wares can be dated to the late 12th century, but there are two sherds from different vessels which may be later.

Group F15: revetments and foreshore deposits

The pottery consists predominantly of early medieval handmade wares – ESUR, EMSH, LOGR, EMSS – current *c* 1050–1150, together with sherds from five ANDE spouted pitchers and three REDP pitchers, including part of the shoulder of a vessel with parallel horizontal rows of painted 'commas'. A single rim sherd from a ?pitcher in THWH from [2080] is the only example of this ware recovered from the site (cf Vince and Jenner 1991, fig 2.98, nos 222–3). THWH is not common in London and its source has yet to be located, but sherds occur in contexts dated from the late 9th to the late 12th centuries (ibid, 95). No other fabrics or forms introduced after *c* 1080 are present.

Group F16: fluvial deposits

This group consists of a series of waterlaid deposits which accumulated during the late 12th century, starting before work began on the construction of the Colechurch bridge (*c* 1176). All but one context have been spot-dated to *c* 1150 or later.

Sherds from six SSW cooking pots come from [2161] and [2151]; and part of a base from [2152] has splashes of glaze, a feature which suggests that SSW may have been produced by the same potteries which made local glazed wares (Vince 1991b, 271). A cooking pot rim in SHER from [2151] dates from the same period or later. Handmade local coarsewares dating to *c* 1050–1150 (14% of all pottery from group F16) include a sherd from a large vessel in EMSS with vertical, thumbed, applied strips and part of a circular hole, which may have formed a lug handle for a bowl (cf Vince and Jenner 1991, fig 2.39, no. 88).

Continental imports comprise 12% of all pottery in this group and include, apart from ANDE and REDP, a single sherd from a BLGR ladle [2151], imported from the middle Rhine valley and not at all common at the site, and part of a ?pitcher in NFRE with a broad, thumbed, applied vertical strip, from [2021], the only occurrence of this ware on the site. Sherds of NFRE are found in London as early as the 10th century, although they are more common in mid- to late 11th-century contexts (ibid, 108; fig 2.115, nos 276, 286). A small sherd of STAM from [2228] is another uncommon find at the site.

London-type wares are the most common pottery in group F16 (52%). Undecorated early rounded jugs are the main form, although there are sherds from three cooking pots and a bowl in LCOAR from [2151] (see Pearce et al 1985, fig 67, nos 337, 341; fig 72, no. 388). Only three small decorated sherds were recovered, two with applied white slip pads or discs under a green glaze (from [2020] and [2161]) and one with a pattern of small finger-impressed dimples (from [2151]). Both kinds of decoration are known on early rounded jugs, although they are not so common as white slip decoration (ibid, fig 18, nos 31–2; fig 25, no. 50; fig 22, no. 46; fig 18, nos 30–32).

The latest datable LOND comes from [2151]: two sherds from a jug decorated in the Rouen style and one green-glazed sherd with a vertical applied strip typical of the North French style. Both styles were current during the first half of the 13th century; they may have been first produced as early as *c* 1170 and certainly by *c* 1190 (Vince 1991b, 268). A small green-glazed sherd of LCALC from [2021] has part of an applied strip and may have come from a jug with North French style decoration. This is of interest, since LCALC went out of use in London at the end of the 12th century.

Group F17: Saxo-Norman occupation around the bridgehead

The main fabric in this group is EMSS (85%), mostly cooking pots. This includes sherds from five different vessels in rubbish pit [2036], one of them almost complete but very fragmented (149 sherds). Two of the rims have continuous fingertip impressions around the top (cf Vince and Jenner 1991, fig 2.39, no. 81). There are also sherds from a spouted pitcher in EMSS, an uncommon form in this fabric, from [1038] (ibid, fig 2.41, no. 91). The only other fabrics in group F17 are a single sherd of EMGR from [1046] and the rim of a bowl in EMSH from [2043], both datable to *c* 1050–1150. There is no pottery introduced after *c* 1080 (eg LOND, LCOAR), so the most likely date for the group is *c* 1050–1100.

Group F18: pre-Colechurch bridge downstream revetments

A single sherd from a cooking pot in EMSH was found in subgroup F18.5 [41], which dates the context to *c* 1050–1150.

Group F20: erosion and revetments on the upstream side of the Saxo-Norman bridge causeway

The only pottery in this group consists of two jug sherds in London-type ware from [566], part of the foreshore build-up. One is green-glazed LCALC, decorated with ring-and-dot stamps on applied discs or pads of clay around the neck, typical of the 13th-century Rouen and Highly Decorated styles (eg Pearce et al 1985, fig 26, no. 59; fig 55). However, LCALC appears not to have been made after *c* 1200, although ring-and-dot stamps are not a common feature of 12th-century London-type decoration. Perhaps the sherd is best seen as coming from a vessel made at the turn of the 12th/13th centuries, at a time when new styles of decoration were coming into fashion and when LCALC was nearing the end of its life.

The other sherd from [566] is in LOND, decorated with horizontal rows of long, nicked impressions under a white slip and green glaze. These have the appearance of pellets, but are not applied. They were probably made by drawing a tool or fingertip downwards in the soft clay, and recall late 12th-century finger-impressed decoration (Pearce et al 1985, fig 17, no. 28).

Group F21: construction of the first 11th-century bridge caisson

Sherds from four vessels were found in context [110]. Apart from a single sherd of EMSH, all are late 12th-century LOND early rounded jugs. One has traces of white slip decoration in the early style and the other two are green-glazed with no decoration. Thirty sherds have survived from one of these jugs, clearly discarded whole or freshly broken, allowing a complete profile to be reconstructed. The vessel has a collar rim, a lightly rilled neck and a strap handle with regular thumb impressions down each side (cf Pearce et al 1985, fig 18, no. 34), but is unusual in having a slightly recessed base, a feature more usually associated with North French or Rouen style baluster jugs (eg ibid, fig 25, nos 51–3); convex or sagging bases are the norm for early rounded jugs. A further unusual feature is a small circular hole, neatly drilled through the base near the circumference after firing. Since it would have prevented the jug from holding liquid, its purpose is uncertain – unless the vessel had been adapted as a water sprinkler for plants (eg Landsberg 1995, 93).

Group F24: preparatory consolidation for the Colechurch bridge

Sherds from four vessels were found in context [564B] which is dated to the late 12th century by the presence of SSW and SHER cooking pots and the base of a LOND early rounded jug. There are, in addition, five sherds from a Spanish amphora or jar (SPOA) – a relatively thin-walled vessel with calcareous inclusions in a pink fabric, white-slipped externally. The body is rounded or globular, but too little remains for the complete form to be reconstructed. Sherds of a medieval Spanish amphora were found in an early 13th-century context at Seal House, London (Vince 1982, 138–9; Vince 1985, 48), but none has been recorded from earlier contexts.

Group F25: erosion of the bridgehead and foreshore build-up

London-type wares predominate (63%), but there is also a relatively high proportion of SSW (26%). Other local fabrics are limited to odd sherds of SHER, residual EMFL and EMS. Non-local wares are represented by a single sherd of STAM and the complete handle of a ladle in BLGR.

The London-type wares are almost evenly divided between LOND and LCOAR, with sherds from six different jugs in LCALC, two of which have decoration painted in red slip under a clear glaze in the early style. This parallels the use of white slip on LOND and LCOAR, and is more common on LCALC jugs, with their paler fabric (eg Pearce et al 1985, fig 17, nos 25–7). There is also a sherd from a jug in LCALC, with a mottled green glaze and vertical applied strips, typical of the North French style, suggesting a date in the last quarter of the 12th century. The group has been dated to the 1160s or 1170s on stratigraphic grounds. NFM and ROUE have been found in small quantities in London in contexts of *c* 1170/80 (Vince 1985, 47–8) and this appears to be an early example of the style.

The majority of the London-type wares from [500] consist of undecorated early rounded jugs. There are, however, single examples of a jug with pellet decoration; a green-glazed jug with incised decoration, probably in a trellis pattern (cf, for example, Pearce et al 1985, fig 18, no. 32); and a strap handle.

Group F26: construction of the Colechurch bridge abutment

A felling date of 1187–8 was obtained for the sills from the east and west sides of the abutment, suggesting that construction took place in either 1189 or 1190, while documentary evidence dates the construction of the Colechurch bridge to *c* 1176–1209. Four sherds from three LOND jugs, one with early style decoration, were found in the ashlar wall and core of the Colechurch abutment (F26.5) and probably date to the late 12th century.

A more unusual find came from the ragstone rubble beneath one of the sills on the west side (F26.4): half a LOND dripping dish packed full of concreted shells and other food remains such as nuts and animal bone. The vessel is of the usual elongated oval shape, slab-built and heavily knife-trimmed externally, with a pinched pouring lip at the surviving end and probably originally at the other end as well. It is likely that there was a single handle in the middle of one side, although this is now missing (Pearce et al 1985, 43; fig 70, no. 371). The vessel is heavily burnt and sooted from its use as a receptacle to collect the juices from spit-roasted meat. Dripping dishes are not common finds at any date. They appear to have been made in LOND from *c* 1210 until the mid-14th century (ibid, 43, 19–20). The Fennings Wharf example probably dates to the early years of the 13th century.

Sixteen sherds of predominantly late 12th-/early 13th-century pottery came from the infill of the abutment (F26.6). They include SHER and SSW cooking pots and LCOAR and LOND jugs, one of which has red and white slip decoration, possibly in the Rouen style. There is also a small sherd of LCALC and a sherd of REDP from an unidentified form with straight-sided cut-outs, made before firing, and running diagonally across the throwing marks.

Group F27: revetments, riverwall and foreshore build-up contemporary with the Colechurch bridge

London-type wares constitute the bulk of the pottery in this group (66%), with early rounded jugs the most common form. Part of a rod handle with deeply stabbed herringbone decoration from [580] may come from an early rounded jug or a tripod pitcher (cf Pearce et al 1985, fig 22, no. 45; fig 21, no. 43) datable to the late 12th century. Sherds from two different early rounded jugs, from [303] and [504], are decorated with alternate red and white slip diagonal stripes (cf ibid, fig 17, no. 26; fig 20). There are also examples of late 12th-century pellet decoration in the same contexts. However, an early 13th-century date may be indicated for subgroup F27.2 by the presence of sherds of LOND with North French style and Rouen style decoration, if these are not intrusive. The angular everted rim of a small vessel in LOND from [504], with glossy dark green glaze inside and out, may have come from a pipkin, several of which have been found in waterfront contexts dated after *c* 1210 (ibid, fig 68, nos 349–60, especially no. 355). The extent and quality of glaze cover is unusual for this form and a different function cannot be ruled out. The rim and short tubular handle from a cooking vessel, sooted externally, again from [504], may also have come from a pipkin, although ladle-type handles are the more usual form (ibid, 42; fig 68, no. 363).

The most unusual find from subgroup 27.2 is the modelled head and part of the body of a human figure in LOND. The head is crudely formed and luted to the body. The nose is pinched out from the clay of the face, the eyes are stabbed and the mouth incised. The type of vessel to which this head was applied cannot be identified with certainty from the fragment of attached vessel wall; this sherd has a thick green glaze inside and out. It may be compared with a large vessel found at Billingsgate Lorry Park and termed a 'vase' or jar (Pearce et al 1985, 46, fig 76, no. 417), which may have had anthropomorphic decoration. A more likely alternative is that it came from a large and elaborately decorated chafing dish, a form known in LOND from *c* 1210 including an early example with modelled human heads around the rim (ibid, 44, fig 73, nos 400, 404). Chafing dishes were made throughout the 13th century, although excavated finds are rare. Stylistically, the Fennings Wharf sherd has much in common with anthropomorphic decoration on mid-13th-century jugs (ibid, 30, 132–3; fig 56) and it is on this basis that context [580] has been dated to *c* 1240–70.

Group F28: late additions to the Colechurch bridge

The pottery recovered from this group dates largely to the 15th century, with a latest date of *c* 1450–1500 from context [34]. There is a relatively low proportion of residual 13th- to early 14th-century material (21%). These include a SHER cooking pot, a jug and part of a dripping dish in LOND and a number of KING jugs, one with polychrome decoration, the rest undecorated.

CBW is the most common fabric (41%). Several different forms are represented, including: jugs; part of a cistern with red-painted decoration; cooking pots with either flat-topped or bifid, lid-seated rims, introduced *c* 1380; bowls and a ?lobed cup. One unusual bowl from [79] may have been a distilling base. The vessel is glazed inside and has slight traces of a yellow-tinged deposit. Externally, it is sooted right up to the rim and the sides are thoroughly knife-trimmed and smoothed. It has a bifid or lid-seated rim and a flat base, and may well have been used as part of a distilling unit, probably with an alembic of Moorhouse's type 2 (Moorhouse 1972, 104, 11–15, fig 31).

Other pottery includes sherds from a cooking pot and a pipkin in LLSL, in use from *c* 1400; part of a dripping dish in LMHG from [79]; and sherds from four CHEA jugs. There are also a few interesting Continental imports. A SIEG drinking bowl or lid was found in [34], dated to the late 15th/early 16th century (Hurst et al 1986, fig 88, no. 257). Part of a tall baluster-shaped jug in SIEG, possibly a *Jacobakanne*, comes from the same context and dates to the late 14th/15th century (ibid, 180, fig 88, no. 263). The base of a DUTR frying pan was found in [79]. DUTR became more common in London from the mid-14th century onwards, although it is present in late 13th-century contexts (Blackmore 1994, 37). The remaining imports, also in [79], come from Spain and consist of the rim of a large jar in a fine red micaceous fabric with a pale buff slip (SPOA); and part of a small flanged bowl or dish in VALE, with blue interlace around the rim and faint traces of lustre externally (cf Hurst et al 1986, fig 17, dated *c* 1375–1425). The earliest occurrence of VALE from London dates to the late 14th century and it became more common during the 15th century (Blackmore 1994, 39).

Group F29: late medieval features on the southern approach to the bridge

An area of flooring (F29.10) in the cellared building (F29.8) included a sherd of CBW, which could date from any time between *c* 1270 and 1500; and from the floor make-up [1036] there are two sherds from a cup in EBORD. Made at Farnborough Hill, and probably other sites in the area *c* 1480–1550, this may be seen as the direct precursor of the Border ware industry of the late 16th to 17th centuries.

Group F30: post-medieval features/structures on the bridge approach

The earliest pottery from the group comes from the fill of rubbish pit [1016] (F30.9); the large context [1017] is closely datable to the first half of the 16th century and includes sherds from 20 vessels, eight of which are represented by several sherds each. Four of these are substantially complete, although fragmented. There is nothing which can be dated with certainty later than *c* 1550. However, there is a high proportion of imported pottery from various sources (40%). These include two cooking vessels in DUTSL. The first is a small cauldron and the second is a skillet.

Other imports in context [1017] come from Spain, France and the Rhineland. Part of a dish in Isabella Polychrome (ISAB) is decorated in mauve and blue and can be dated to *c* 1500–50 (Hurst et al 1986, 54–7; figs 24–5; pl 11). ISAB was not regularly traded by the Spanish, but shipped by them as their everyday pottery. The rim and small horizontal lug are all that remains of a bowl in Beauvais single sgraffito, a type made throughout the 16th century (ibid, fig 52, nos 162–3). Sherds from four RAER drinking jugs of classic form are among the most common imports into London at this period, datable to *c* 1475–1550.

The local redwares, which comprise 60% of the pottery in this group, are principally PMRE (40% of the context), together with PMBR and clear-glazed PMSR. These include substantial remains of a deep carinated bowl of skillet form in PMSR, with a curved ladle handle springing from the body at the mid-point of the profile and thumbed feet around the base. The influence of 16th-century Low Countries redwares is further reflected in the vessel's profile and base form. The vessel is slipped and glazed inside only. Kingston-type redwares provide further parallels for another PMSRY vessel from [1017]: a wide bowl or 'pancheon' (Hurst et al 1986, fig 3, nos 11–14) with a flanged rim and pulled 'feet' around the base, in Dutch fashion.

Apart from one small sherd of PMBR, all remaining vessels are in PMRE, the principal London area redware of the 16th century: sherds from two unglazed jugs or pitchers; part of a cooking vessel, possibly a cauldron; and bowls. The rim of an unglazed, deep bowl has a single circular hole through the flange, made before firing. Similar perforated PMRE bowls are known from numerous sites in London, but their purpose is uncertain. The context also includes a complete large wide bowl in PMRE (35 sherds), similar in form to the PMSR bowl or pancheon described above. The latest pottery from the F30.9 pit is a large almost complete dish or charger in TGW from [1013]. It is decorated in dark blue with stylised, foliate scrolls in Orton's group A (Orton 1988, 321; cf fig 136, no. 1363), dated in London *c* 1612–40/50.

Most of the remaining contexts from group F30 have been dated to the mid-17th to early 18th centuries. Among these, context [1009] is of particular interest. It comes from the backfill of a brick-lined cesspit [1005] in subgroup F30.2, which contained pottery made and used from at least the mid-17th century [1009] until the mid-18th century [1006]. The earlier context includes five complete vessels and part of one other. There are two complete FREC Bartmänner or 'Bellarmines', one with a highly stylised, devolved face medallion on the body and one with a ten-petalled rosette and a face mask, closely comparable to an example from Rotterdam dated *c* 1650–75 (Hurst et al 1986, pl 44 right). There is also a chamber pot in RBOR (cf Pearce 1992, 32–4; cf fig 40, no. 324) and part of a mug or caudle cup, also in RBOR. Two PMFR bowls complete the pottery from context [1009]: a large deep, handled form and a smaller similar bowl without handles. Both are glazed inside and out. PMFR, largely from kilns in Essex, was used in London from *c* 1580 to *c* 1700 (Orton 1988, 298).

The other context from the same cesspit [1006] is also medium-sized, although the pottery is much more fragmentary, with no complete vessels. It also includes a number of 17th-century fabrics and forms – a sherd each of METS, FREC and BORDG – as well as wares which continued to be made into the 18th century, such as PMR, RBOR and TGW. These include the rim of a flanged dish in RBOR with trailed slip decoration, which is much less common in the City and Southwark than METS; and sherds from two TGW dishes, one very abraded, and the other decorated in Orton's group D, dated in London to *c* 1640–80 (Orton 1988, 327).

Small quantities of late 17th-/early 18th-century pottery come from the backfill of a brick-lined well (F30.14): FREC Bartmänner; a bowl or porringer in RBOR; a PMR bowl; and a PMFR flanged dish with incised curvilinear decoration.

Three small sherds of TGW come from a dish in [2001] with decoration in Orton's group D (Orton 1988, 327); a mug or caudle cup with blue, scroll-like decoration; and a vessel with at least one elaborate scrolled handle terminal [2002]. The sherd is decorated in blue over a thick, glossy white glaze and probably comes from a posset pot, a form which was most popular in the second half of the 17th century, dying out in the 18th century. Tightly curled handle terminals were made from the late 1660s and, from what little remains of the decoration on this sherd, a date in the 1690s is suggested (cf Lipski and Archer 1984, 200; nos 927, 1546).

The largest single context from the site by sherd count ([1007], F30.4: 231 sherds) comes from the backfill of a robber trench in the cellar F30.3. The group includes two more or less complete but smashed PMR cisterns (135 sherds). Both are very large and heavy, and have horizontal loop handles, a broad band of closely spaced thumbing around the neck and a plain bunghole just above the base. Both are glazed inside and can be compared with storage jars without bungholes made in Woolwich fabric E2, dated *c* 1660–80 (Pryor and Blockley 1978, 61–372, fig 15, fig 16, no. 82). Other fabrics and forms in [1007] are those common in London during the 17th century: sherds from a bowl, jug and cooking pot in PMR; three FREC Bartmänner; part of a WEST chamber pot; two Staffordshire butterpots; and tripod pipkins in BORDY and RBOR. There is also an almost complete flanged dish in RBOR (29 sherds) and six sherds from a BORDB porringer.

A number of sherds are worthy of note: part of a dish in PMSRY with sgraffito decoration, and probably of 16th-century date; a Spanish olive jar; and the rim of a dish in biscuit TGW with raised bosses and a serrated edge. Comparable dishes were made at Pickleherring Wharf, which started production in 1618, and a large Adam and Eve charger from this factory, with a similar rim, is dated 1635 (Britton 1987, pl D; see also Garner and Archer 1972, pl 5, dated 1637). There are, in addition, sherds from six other vessels in TGW, including a plate with a flat base and a bowl, both decorated with 'Chinamen among grasses', a short-lived style based on late Ming designs and datable to *c* 1670–90 (Garner and Archer 1972, 17–8, 25; Orton 1988, 327, group F). Part of the base of a large dish or charger has decoration in Orton's group D, probably a tulip design executed in blue, yellow and green, with lead glaze externally, and datable to *c* 1650–80 (ibid, fig 134, no 1354; Garner and Archer 1972, pl 16B).

Group F31: post-medieval additions to the abutment and the riverwall

Contexts [16] and [18] are both dated to the mid-18th century by the presence of SWSG, including several sherds from a small teabowl with 'scratch blue' decoration from [18] and the lower body and base of a tankard from [16]. The base of an undecorated, straight-sided TGW jar from [18], is typical of long-lived 18th-century forms. Context [16] also includes 12 sherds from the rim and nipple end of two PMR sugar-cone moulds.

Table 6 Distribution of Saxo-Norman/early medieval pottery from groups F12–25, pre-Colechurch bridge construction, by estimated number of vessels (ENV)

Pottery code	F12	F13	F14	F15	F16	F17	F18	F20	F21	F24	F25
LSS	-	-	-	-	2	-	-	-	-	-	-
EMS	5	1	4	2	2	-	-	-	-	-	1
EMFL	2	-	1	-	-	-	-	-	-	-	1
EMSS	20	1	7	1	2	11	-	-	-	-	-
EMSH	12	4	13	3	4	1	1	-	-	-	-
EMGR	-	-	1	-	-	-	-	-	-	-	-
ESUR	9	1	9	2	5	-	-	-	-	-	-
LOGR	-	-	5	1	-	1	-	-	-	-	-
EMCH	2	-	2	-	-	-	-	-	-	-	-
LOND	-	5	10	1	20	-	-	2	3	1	15
LCOAR	-	2	13	-	10	-	-	-	-	-	13
LCALC	-	1	2	-	3	-	-	-	-	-	6
SSW	-	1	2	-	6	-	-	-	-	1	14
SHER	-	4	-	-	1	-	-	-	-	-	13
STAM	1	-	-	-	1	-	-	-	-	-	6
THET	1	-	-	-	-	-	-	-	-	1	14
THWH	-	-	-	1	-	-	-	-	-	1	2
REDP	8	1	5	3	3	-	-	-	-	-	-
ANDE	11	1	6	5	3	-	-	-	-	-	-
NFRE	-	-	-	-	1	-	-	-	-	-	-
BLGR	-	-	-	-	1	-	-	-	-	-	1
SPOW	-	-	-	-	-	-	-	-	-	1	-
MISC	1	-	-	1	-	-	-	-	-	-	-
Total	**81**	**22**	**80**	**20**	**64**	**13**	**1**	**2**	**4**	**4**	**54**
Terminus post quem	1050	1180	1180	1080	1180	1050	1050	1180	1150	1150	1170

Discussion

By comparison with the massive London medieval foreshore revetment dumps, the overall quantity of pottery recovered from Fennings Wharf is relatively small (1379 sherds, or a minimum of 583 vessels), much of it from small contexts. Accordingly basic quantification was carried out during spot-dating by means of a sherd count (SC) and minimum vessel count (estimated number of vessels or ENV), more appropriate to the quantity of material recovered. The absence of large revetment dumps at Fennings Wharf imposes certain limitations on the quantified data.

The Saxo-Norman/early medieval sequence (groups F12–25) c 1000–1200

Groups F12–F25 represent activity around the bridgehead before the recorded construction of the Colechurch bridge (c 1176–1209): the build-up and erosion of waterlaid deposits, timber revetments and earlier bridge structures, as well as limited evidence for occupation. A total of 785 sherds, or 345 ENVs, was recovered from these groups, although F18, F21 and F24 are too small to be statistically viable. The distribution of pottery by fabric within each group is given in Table 6.

Groups F12, F15 and F17–18 have all been dated to the late 11th/early 12th century by the presence of large quantities of early medieval handmade wares current c 1050–1150 (more than 45%) and the absence of locally made, wheelthrown glazed pottery in use from the mid-12th century (SSW, SHER, and early style jugs in London-type wares). Tree-ring felling dates of 1081–95 have been obtained for group F12 and 1118–37 for group F15, which agrees well with the ceramic dating. Out of a minimum of 81 vessels, there are no London-type wares in the earlier group F12 and only one vessel in F15 (5% of the group), a clear-glazed jug or spouted pitcher. Although small quantities of local glazed pottery were found at Billingsgate dated to c 1055–85, it is not surprising that groups F12 and F15 should be dominated by handmade wares (73% and 45% respectively), since these provided the mainstay of the pottery used in London until the ceramic industries of the region were transformed by the spread of wheelthrowing and glazing technology during the 12th century. EMSS is by far the most common ware in group F12 (48% of all handmade pottery), more than twice as common as EMSH (21%) and nearly three times as common as ESUR (15%), although in subsequent groups (F14–16) EMSS is noticeably less frequent than EMSH, while the frequency of ESUR fluctuates (see Table 6). Group F17 is exceptional, consisting entirely of handmade wares, including the remains of five smashed EMSS cooking pots (85% of the group). At Billingsgate, EMSS and ESUR are the most common wares in groups dating to c 1055–1108 (Vince 1991b, fig 3), making an interesting comparison with the Fennings Wharf pottery, particularly with group F12, the earliest in the Saxo-Norman sequence. EMSS, EMSH and ESUR are clearly shown as the major wares used during the late 11th and early 12th centuries, although the reasons for the relative abundance of EMSH and the variable proportions of ESUR are difficult to explain except by accidents of survival. The overall quantity of pottery is too small to argue for different patterns of distribution north and south of the Thames on the evidence of this site alone. Handmade wares constitute the bulk of the pottery found at Billingsgate from c 1055–1108, decreasing in quantity in subsequent deposits. At Fennings Wharf, handmade wares account for between 23% and 53% of the pottery from groups F13, F14 and F16, all of which are dated to the second half of the 12th century, but only 3% of group F25, which is comprised mainly of wheelthrown coarsewares (30%) and local glazed wares (63%), and which immediately predates the construction of the Colechurch bridge during the last quarter of the century. In all these groups, London-type wares constitute more than 30% of the pottery, and tree-ring dates range from 1148–1165, by which time handmade wares were going out of use in London in the face of competition from superior technology and new forms (Table 7).

The highest proportion of wheelthrown coarsewares (SSW and SHER) occurs in groups F13 and F25 (23% and 30% respectively) and in smaller quantities in the other main groups dated to the second half of the 12th century. They are not present in any of the groups dated to the late 11th or early 12th century. This corresponds with their main period of use in the City, although SSW is present in small quantities at Billingsgate in deposits of c 1085 and later (Vince 1991, 271).

Imported wares from the Continent form a relatively high proportion of the pottery from the late 11th- and early 12th-century groups at Fennings Wharf (24% of group F12 and 40% of F15), and from 9% – 14% of the late 12th-century groups F13, F14 and F16. In all groups, the main wares are ANDE from the Meuse Valley and REDP from the Rhineland, in roughly similar proportions, although ANDE is more common in F12 and F25. This presents an interesting contrast with the pottery from the waterfront site at Bull Wharf, Upper Thames Street (UPT90), where BLGR was a relatively common import (Pearce in prep). Other rare Continental imports at Fennings Wharf are single examples of NFRE and SPOW in groups F16 and F24, both dated to the second half of the 12th century. The latter is an extremely unusual find at this date, but NFRE occurs at Billingsgate in deposits dated c 1108, as well as other French wares (North French glazed ware, Rouen ware, and Normandy

Table 7 Relationship of ceramic sources in ENV (%) by stratigraphic groups, including more than 20 vessels from the Saxo-Norman/early medieval period (groups F12–25)

Ceramic sources	F12	F13	F14	F15	F16	F25
Early medieval handmade wares	73	32	53	45	23	3
Wheelthrown coarsewares	-	23	2	-	11	30
Local glazed wares	-	36	31	5	52	63
Non-local wares	2	-	-	5	2	2
Continental imports	24	9	14	40	12	2
Miscellaneous	-	1	-	-	5	-
Terminus post quem	1050	1180	1180	1080	1180	1170

Gritty and Glazed wares), none of which was found at Fennings Wharf.

At the end of the 11th century, the cooking pot was the main form produced in all the handmade industries which supplied London (see Table 8), constituting 70% of all pottery from group F12. Bowls, generally used for cooking, are rare at Fennings Wharf at any time in the 11th and 12th centuries, but do occur both in handmade fabrics (ESUR, EMSH) and local glazed wares (LCOAR, LOND). With the exception of two examples in EMS and LSS, residual in F15 and F16, all spouted pitchers came from the Stamford area or were imported from the Continent and are either decorated or glazed or both (REDP, ANDE and STAM). Together, these form a relatively high proportion of the late 11th-century pottery from group F12 (25%), decreasing in frequency in those groups in which London-type jugs are found in significant numbers (eg F16 and F25, where jugs constitute 41% and 56% respectively of all forms and spouted pitchers only 12% and 4% – see Table 8). The use of glaze and of a better-fired, less porous fabric, as well as the potential for decoration offered by well-made, wheelthrown vessels, are important features of forms which were to be used at the table for serving liquids as well as for storage.

Table 8 Relative proportions of forms in groups F12–25, in ENV (%)

Form	F12	F14	F16	F25
Cooking pot	70	55	42	35
Bowl	1	4	2	-
Jug	-	27	41	56
Spouted pitcher	25	14	12	4
Other	4	-	3	5

Later medieval pottery (groups F26–9) c 1200–1500/50

The total quantity of pottery recovered from the groups associated with the construction of the Colechurch bridge and later medieval activity around the bridgehead and approach is far smaller than the sum of the pottery recovered from the late 11th- to 12th-century groups, predating the stone bridge construction. In all, 194 sherds or 131 ENV come from groups F26–9. All contexts are small, except for two of medium size in F28, and there is less dating evidence from other sources, such as tree-ring dating, to complement the ceramic chronology. Consequently, these 13th- to 15th-century groups do not lend themselves to the same level of analysis as the earlier groups, although they do conform to the broad patterns of usage and supply demonstrated by large waterfront sites in the City.

Group F26, associated with the construction of the Colechurch bridge (1176–1209), yielded little pottery (21 sherds) and has been given a broad 13th-century date range, probably before *c* 1230 when KING was first used in London. This fits well with documentary evidence for the date of work on the bridge. Group F27 is considered to be contemporaneous and includes late 12th-century contexts (F27.4 and F27.12). The latest pottery in the group is a sherd of CBW in F27.12, which also yielded a tree-ring date of 1272b, coinciding closely with the spot-date of 1270+.

The dominance of London-type ware is clearly seen in all subgroups (59–70%), production of the fine LOND fabric having almost completely displaced the coarse glazed fabric (LCOAR) by the start of the 13th century (64% of F27.2 as opposed to 6% LCOAR). Wheelthrown coarsewares are also present in relatively high proportions in the larger subgroups F27.2 and F27.4, and SHER is more numerous in both than SSW, which had a comparatively short lifespan during the second half of the 12th century, representing the tail end of shell-tempered pottery production in the London area. Imports from other parts of England and from the Continent are of limited occurrence, consisting of small quantities of REDP, STAM and THET only. However, although French wares became the most common imported pottery in the City during the 13th and early 14th centuries (Vince 1985, 47–8), no French pottery of this date was found at Fennings Wharf.

Jugs and cooking pots are the main forms in the late 12th- to 13th-century groups (F26, F27). Cooking pots (28% of all forms) are made principally in SHER and SSW, although there are some examples in LOND. There is also a single pipkin, a specialised form developed in the early 13th century (Pearce et al 1985, 42), as the range of vessels used for cooking began to expand from the simple cooking pot. Jugs are the most common form produced by the London-type ware industry and this is reflected by their abundance in group F27 (61%), including both Rouen and North French style vessels designed to look attractive as well as being functional and intended for serving rather than for storage.

Post-medieval pottery (groups F30–1) c 1500+

The post-medieval features associated with the brick-built cellared building on the approach to the bridge yielded a total of 499 sherds, or a minimum of 101 vessels, and included the largest single context on the site ([1007]: 231 sherds or 35 ENV). One other large context [1017] can be closely dated to the first half of the 16th century and represents the most significant group of pottery of this date found at Fennings Wharf. It comes from a rubbish pit ([1016]: subgroup F30.9), dated at the latest to the first half of the 17th century by a sherd of TGW. Local redwares (PMRE, PMBR and PMSR) constitute the bulk of the pottery (50%), with PMRE the most numerous (33%). This pattern is common throughout the London area during the 16th century, especially before *c* 1550, when the impact of the Border ware industry began to make itself felt. After a hiatus in the life of the London medieval redware industries during the late 14th century, there was a revival of redware production after *c* 1400 (LLON and LLSL), although the scale of production was more limited than in the 13th and early 14th centuries, facing serious competition from the dominant whiteware industries of the Surrey-Hampshire borders (CBW). Coarse Border ware production came to an end at Farnborough Hill at the beginning of the 16th century, to be replaced by specialised, fine green-glazed tablewares on a much smaller scale (EBORD), and the London redware industries revived and took a firm hold on the local market. Known kilns at Kingston, Cheam and Woolwich made a wide and varied range of vessels suitable for storage, cooking, serving and other household needs. It was not until the end of the 16th century, when the revitalised Border ware industry began to assert its position in the London market once more, that the dominance of local redwares was challenged. This meant that the major ceramic suppliers to the capital were obliged to specialise in order to retain their share of the market. Consequently, local redwares focused on heavy-duty cooking vessels (cauldrons, pipkins), bowls and storage vessels, and plain jugs or pitchers, to which they were well suited, while Border wares, with their attractive buff-coloured fabric and green, yellow or brown glazes, produced a far wider range of forms suitable for all manner of household uses.

The pottery from Fennings Wharf group F30.9 context [1017] comes from the period before the introduction of Border wares and hence is dominated by redwares and includes all the most common forms made at this date – jugs and pitchers, bowls, dishes, cauldrons and skillets. There is also a high proportion of imported pottery (33% of [1017]) from a number of different sources: the Rhineland, Low Countries, France and Spain (RAER, DUTSL, BEAU and ISAB). This is a marked concentration of imported wares and may be the result of several different factors: the waterfront location with its proximity to boats unloading goods from various countries which regularly traded with London and contact with their crews and the personal equipment they brought with them (for example, the ISAB bowl, an item not usually traded, but used on board ship); the presence of immigrant communities in the area (eg Flemish exiles escaping religious persecution), using pottery brought with them when they first settled in London, or retaining contacts with their homeland; or a degree of prosperity in the families which generated the rubbish discarded in the pit, enabling them to purchase attractive imported items such as the Beauvais sgraffito dish. Any or all of these factors could have played a part in forming the make-up of the pit's contents.

The largest group of pottery from the site comes from context [1007] (subgroup F30.4), from the robbing-out of brickwork in the cellar, F30.3. It clearly demonstrates something of the diversity and range of pottery available to Londoners during the later 17th century. Local redwares were still the most common source of everyday domestic pottery (31%, principally PMR), a picture which is mirrored at other contemporaneous sites in the City (eg Aldgate, cf Orton and Pearce 1984). However, Border wares are rather under-represented compared with other sites, while TGW accounts for a relatively high proportion of the pottery from this period (23%). This consists of various tablewares in styles current during the mid- to late 17th century, which would obviously have been acquired for their decorative effect, attractive appearance and fashionable appeal, to adorn the house rather than for everyday use. By the end of the 17th century, the quantity and variety of Continental imports into London was diminishing, with the main source centred on the Rhineland until the Fulham stoneware factory of John Dwight began to supply comparable pottery to the capital during the last quarter of the 17th century.

The latest group of reasonable size from Fennings Wharf comes from the backfill of a cesspit (context [1006], F30.2) and includes fabrics and forms datable to the middle of the 18th century, in addition to earlier 17th-century pottery. Red Border ware is far more common than whiteware, a reflection of the phasing out of whiteware production at the kiln sites in the early years of the 18th century, in favour of a more limited range of utilitarian redwares – particularly bowls, dishes and porringers. The proportion of RBOR from F30.2 is significantly higher than local redwares (23% as opposed to 5%), although at other sites in London at this period PMR is generally the main coarseware. The subsequent history of the red Border ware industry has yet to be traced, both in the source area and in London – one of the chief consumers during the 17th century – although it appears to have continued to supply the capital throughout the 18th century.

The Toppings Wharf excavation 1970–2: a reassessment of the pottery

When Clive Orton wrote up the pottery from Toppings Wharf, published in 1974, he used a simple alpha-numeric code to refer to the various fabrics recovered, dividing them into broadly based groups based on major inclusions and colour (Orton et al 1974). It is probably safe to say that at this date the study of London's pottery was in a formative state. Much of Orton's dating, particularly of London-area coarsewares, depended on the Northolt sequence published by Hurst (1961). Since then considerable advances have been made in the identification of fabrics and sources and in the establishment of a refined ceramic chronology derived largely from the London waterfront sequence, allowing a much more closely focused picture to be built up of London's medieval pottery (discussed earlier). In his 1988 report on the pottery from 199 Borough High Street and other sites in Southwark, Orton adopted the common names established by the DUA and made broad comparisons with the most closely corresponding Southwark fabric codes (Orton 1988). However, he pointed out that 'it should be remembered that the correspondence is not always exact, so that without re-examination of the pottery already published, it is not possible to make completely accurate comparisons' (ibid, 295). Since this is hardly practicable in the case of the Toppings Wharf material, the published pottery report has been reassessed, using Orton's later comparisons and the broad fabric descriptions and illustrations as a guide, equating these with current terminology and dating. The main alpha-numeric codes devised by Orton for the Toppings Wharf pottery are listed below with their current equivalents and date range.

Medieval

Fabric A2: Shelly-sandy ware (SSW), *c* 1140–1220
Fabric C1 and 'part of C2': Coarse London-type ware (LCOAR), *c* 1080–200
Fabric C2 (excluding the above element): South Herts greyware (SHER), *c* 1140–1300
Fabric D: London-type ware (LOND), *c* 1080–1350 (this code includes Mill Green ware (MG), *c* 1270–1350; the fabric is virtually impossible to distinguish from the written descriptions, but is visually distinctive
Fabric E: Surrey whitewares, divided into:
E1: 'Tudor green ware' (TUDG), *c* 1380–1500 (this code also includes other very fine whitewares, such as Saintonge ware (SAIN), *c* 1250–1500)
E2: 'this seems to correspond with' Kingston-type ware (KING), *c* 1230–1400
E3: Coarse Border ware (CBW), *c* 1270–1500 (common from *c* 1350)
E3d: Cheam whiteware (CHEA), *c* 1350–1500

Post-medieval

Fabric E4: Beauvais sgraffito ware (BEAU), *c* 1500–1630
Fabric J: Border ware (whiteware only) (BORD),*c* 1550–1700
Fabric I1: Tudor brown ware (TUDB, now PMRE), *c* 1480–1600; and Dutch red earthenwares (DUTR), *c* 1300–1650, including Dutch slip-decorated redware (DUTSD), *c* 1400–1520
Fabric I4: Guys Hospital ware (GUYS), *c* 1480–1660; and Dutch slip-coated redwares (DUTSL), *c* 1500–1650
Fabric K1: Post-medieval redwares (PMR), *c* 1580–1700+ and Post-medieval fine redware (PMFR), *c* 1580–1700
Fabric K2: Post-medieval black-glazed redware (PMBL), *c* 1580–1700
Fabric F: Stonewares, divided into:
F1: Raeren stoneware (RAER), *c* 1480–1550
F2: Siegburg and early Cologne stonewares (SIEG), *c* 1300–1500 (ie medieval)and (KOLS), *c* 1500–80
F3: Frechen stoneware (FREC), *c* 1550–1700
F4: other
Fabric L: Cistercian ware (CSTN), *c* 1500–1600
Fabric M: miscellaneous medieval and post-medieval fabrics
Other fabrics listed in the 1974 report are not so easy to compare with current common names and are not mentioned in the 1988 report. These are as follows.
Fabric A1: described as St Neots-type (NEOT), *c* 970–1100; however, this fabric is not common in London and the description allows the possibility that other shelly wares might be included, for example Early medieval shelly ware (EMSH), *c* 1050–1150, or Late Saxon shelly ware (LSS), *c* 900–1050. Without seeing the pottery it is impossible to discern which of these was recovered.
Fabric B: described as 'hard grey gritty'. This could mean Local greyware (LOGR), *c* 1050–1150, and include SHER (see above), but again it is impossible to be certain.

A further problem lies in the very broad base for the definition of major fabric groups which include more than one common name (eg LOND and MG; SHER and unglazed LCOAR; TUDG and other fine whitewares). Accurate identification of these individual fabrics is fundamental to the establishment of a reliable chronology and cannot be taken for granted in the absence of the pottery. The potential for considerable confusion resides in the definitions given for the various Surrey whitewares and in the lumping together of local post-medieval redwares (PMRE) and Dutch redwares. The only way in which these problems can be resolved is for the pottery to be re-examined.

The Toppings Wharf pottery report is now reconsidered in chronological sequence and redated where appropriate, as far as is possible within the parameters outlined above.

The pre-erosion levels

SSW predominates, and there are sherds of SHER (if this has not been confused with the visually similar and earlier LOGR), as well as fabrics A1 and B. There are no glazed London-type wares or Surrey whitewares (although two sherds in fabric C have patches of glaze and may be LCOAR). Orton dated these levels to the 12th or possibly early 13th century. On the basis of the fabrics listed a spot-date of *c* 1140–1220 would be assigned using current chronology, although the paucity of LOND/LCOAR might suggest a latest date nearer the middle of the 12th century and, not surprisingly, a wide chronological range of fabrics could well be included.

The erosion levels

These features include a higher proportion of LOND, both glazed and unglazed; two sherds have vertical applied strips, one probably rouletted, and can be equated with North French style decoration (NFR), dated *c* 1180–1270, but predominantly early 13th-century. A third sherd has 'scaly' decoration which could mean NFR, Highly Decorated style (1240–1350) or 12th-century pellet decoration. There are also sherds of SSW and SHER and one small sherd of possible KING. The sample was very small (19 sherds) and could easily be contaminated. Nevertheless, Orton's date of 'about or just before 1300' (Orton et al 1974, 71) seems too late, and the material would now be spot-dated to *c* 1230–70 on the basis of the sherds listed, not accounting for the possibility of later intrusion.

The deposition levels

These levels contained the earliest large groups of pottery. LOND appears to comprise slightly more than 50%, with 25% SHER (and LCOAR), 15% shelly wares (mostly SSW) and 4% Surrey whitewares, probably KING. The group C fabrics include both glazed (rare) and unglazed LCOAR. Within the London-type wares, various decorative styles are described, including NFR, Highly Decorated and Rouen styles (ROU, c 1180–1250). Slipped and glazed vessels with combed sgraffito decoration are also listed and probably refer to MG jugs. None of the Surrey whiteware sherds are decorated.

In the discussion which follows, Orton cites various parallels and gives the then accepted date ranges for the various fabrics. These can be revised by reference to the list given above. He concludes that the features contain a wide range of pottery dating mainly from the 12th and 13th centuries, and a latest date 'a little, if at all, after 1300' because of the scarcity of Surrey whitewares (Orton et al 1974, 70–1). The first part of this conclusion agrees with current dating and the pottery would now be spot-dated to the second half of the 13th century (c 1240–70). There is little to suggest the presence of types introduced after c 1300.

The post-deposition levels

These levels show a decline in 12th-century shelly wares, SHER and LOND, compared with a large rise in Surrey whitewares including CBW as well as KING. The number of sherds of CBW is not great compared with KING, but problems of identification cast doubts on the accuracy of these proportions and may affect the date. A small number of whiteware vessels are decorated and from the description these are probably mostly KING Highly Decorated and North French style, but there is also a jug with red-painted decoration which could be CHEA or CBW or possibly mid-14th-century or earlier polychrome KING. There is also at least one sherd of Saintonge ware (SAIN), introduced after c 1250.

Orton commented on the presence of much residual 12th-/early 13th-century pottery. He noted that LOND continued into the first half of the 14th century, which agrees with current dating, but placed the introduction of CBW at c 1350, whereas it is now known only to have been used in London in small quantities c 1270; by 1350 it was the major pottery type in use in the capital. Orton felt that the evidence suggested 'no finer dating than 14th or 15th century for these layers'. However, the text appears to indicate that a date of c 1270–1350 is more likely, although there could be odd sherds as late as c 1350. If the layers were later (ie mid- to late 14th century) a greater quantity of CBW and CHEA would be expected, including diagnostic forms typical of this date, which are notably absent. The foundation trench for wall 2 and the layers above seem later, according to Orton. The former he dated to the late 14th/early 15th century, although it is hard to see how he differentiated this from other features/layers; the latter is probably late 15th/16th century, which appears quite possible (Orton et al 1974, 71–5).

The Tudor features

There appears to be a relatively high proportion of CBW and other medieval Surrey whitewares; CBW and CHEA continued to be made and used in London until c 1500. There are also quite a number of residual earlier medieval sherds (LOND, SSW and so forth). However, these features are dominated by fabric I, which includes PMRE, DUTR and DUTSD. There are also small quantities of TUDG (which could include Early Border ware – EBORD, dated c 1480–1550), Beauvais sgraffito ware and Rouen stoneware. Together, these fabrics suggest a date range of c 1480–1550 for the Tudor features; this early 16th-century date is based on the apparent absence of any fabric known to have been introduced after c 1550, for example Border ware and Frechen stoneware in particular (Orton et al 1974, 76–87).

The range of 16th-century fabrics and forms looks interesting and would probably repay closer re-examination for other projects (ie redwares). The differentiation of Dutch and local products, if possible, would be highly desirable. However, this would have no effect on the date. There is no reason to question any of the dating given by Orton in his discussion of these features.

14.6 The accessioned finds from Fennings Wharf

Angela Wardle

Introduction

The following summary discusses the range of excavated finds by general period and refers to specific items, mostly stratified objects, details of which can be found in the catalogue. Items referred to in the text, but not catalogued, are residual and are described in the finds archive which is available on application to the Museum of London (Wardle 1996). The first part of the study examines objects lost or discarded in the area of the bridge and includes a summary of items found during earlier work, for example the dredging operations in 1967 (Marsden 1971, 12–14). The second part describes structural items all thought to be part of the bridge itself.

The general assemblage

Roman features were identified in groups F5, F7, F9 and F10 but there are relatively few Roman finds from the site (26 accessions) and of these only 12, seven of them small fragments of glass, come from Roman contexts; all other objects of Roman date are residual. There are no finds from the early Roman buildings (F7), contemporary with those further east on Toppings Wharf, where it is assumed that commercial, manufacturing and domestic activities were carried out (Sheldon 1974, 12–13), but it is likely that the Fennings Wharf buildings were of a similar nature. At Toppings Wharf few tools were identified, the assemblage consisting chiefly of personal possessions and domestic fittings. The Roman material from Fennings Wharf, including the residual items, is of the same general character. Among the few stratified items was a coin, an as of Nero (AD 64–8) from a well (F9.3), one of the features which attests Roman activity in the area of the bridgehead during the late 1st century. A 3rd-century well contained two items of interest in addition to small fragments of bottle glass, namely a bone tool, possibly a scoop (Fig 148) and a bone hairpin of common Roman type. Metalwork was not well preserved in these levels and, with the exception of two coins (one illegible), nothing was identifiable.

The Roman material from later levels comprised three coins, a Claudian copy dated AD 45–65 and two 4th-century issues (one of Constantine, AD 319–20), a damaged copper-alloy earring, a basalt lava (Niedermendig) quernstone fragment and several fragments of glass bottles and vessels. Identifiable forms included part of a pillar moulded bowl (Isings form 3) of mid- to late 1st-century date and the handle of an aryballos (Isings form 61) dating from the late 1st/2nd century. Several of the residual Roman artefacts came from foreshore deposits and the very small number of finds in this area is likely to be due largely to the absence of any Roman waterfronts, which appear to have been destroyed in truncation of the deposits caused by periodic flooding. The residual finds from these deposits, which now might be all thought to constitute the only evidence for any later Roman activity on the foreshore, comprise fragments of Roman glass, including the 1st- and 2nd-century forms mentioned above, one bottle fragment that could date from the 3rd or 4th century, a copper-alloy two-strand cable earring and an illegible 4th-century coin.

The excavations produced several medieval finds, covering a broad date range from the Saxo-Norman period to the 15th century. Two skates (Fig 153), both from the late 11th-century foreshore in subgroups F12.1 and F12.15 are typical of those found in 11th- and 12th-century deposits in London, while an elaborately decorated bone handle from the same area (Fig 152) is probably of post-medieval date.

A siltstone spindle whorl (Fig 155) of a type found in both Saxo-Norman and medieval contexts comes from one of a series of revetments predating the Colechurch bridge (F15.3), and fragments of domestic basalt lava rotary querns were found in foreshore contexts, one (from F25.3) in a context which produced a large ceramic group but few finds.

The only personal item of particular interest, from a deposit upstream of the Colechurch bridge abutment, was a gold finger ring set with an almandine garnet (Fig 149), very similar in style to several found on the north side of the River Thames in deposits dating to the second half of the 14th century (Egan and Pritchard 1991, 326–7). The medieval foreshore deposits also produced pins, fragments of wire and unidentified copper alloy and corroded ironwork. Tools are represented only by two schist hones (Fig 154). A rumbler bell (Fig 150) from a late medieval cellar floor (F29.10) is a type used both as a dress accessory or on harness, although the size of this example makes the latter function more probable. The medieval assemblage, therefore, contains a few examples of items seen elsewhere in London in far greater quantity. The comparative lack of finds in the area of the bridge might again be explained by the limited nature of the archaeological investigation.

There are a few finds of post-medieval date, including a coin of George II/III from a brick-lined well (F30.12); other post-medieval coins are unstratified. A group of 18th-century finds came from various post-medieval features on the bridge approach and included a complete candleholder (Fig 156), an alley or stone toy marble (Fig 157), part of a tortoiseshell fan, several glass wine cups and clay tobacco pipes, comprising a common urban assemblage.

Some of the diverse range of medieval and post-medieval finds dredged from the site of the medieval bridge during 1824–41 and 1967 (see Chapter 9.1) include some personal items comparable to those from the excavations. A stirrup-shaped finger ring (ER1279A <8>),[44] the hoop rising to a pointed bezel with a socket for a cabochon, is of a type known between 1150 and 1450 (Egan and Pritchard 1991, 326, no. 1609) and a lace chape <9> is similar to one from Swan Lane (ibid, no. 1443). Certain objects may have a more specific association with the bridge, namely those connected with transportation and perhaps lost from the bridge itself. Three horseshoes are medieval types and four rowel spurs date from the 14th and 15th centuries (Ellis 1995, 139, no. 336; 144–6, nos 352–3, 355). Objects associated with the structure of the bridge itself are discussed below.

Catalogue

This comprises stratified items and other objects of intrinsic interest.

Roman

COINS (COPPER ALLOY)

[2103] <50> F9.3
Nero, as, RIC 319, AD 64–8.

[2168] <46> F9.1
Illegible, plated copy, AE, 19.5mm, 1st–4th century.

BONE

Hairpin [1064] <114> F10.4 well (2nd/3rd century) (not illustrated)
Plain hairpin with conical head, Crummy type 1 (1983); length 113mm.

Tool [1064] <207> F10.3 well (3rd-century to late Roman features) (Fig 148)
Tool made from a cattle scapula; the side edges and apex are smoothed and the working end has a worn blunt bevelled edge. There is a single hole for suspension near the apex. The upper side appears to be the more concave of the two and the object fits naturally into the right hand when held this way up. Length 153mm.

There are several parallels for this type of object, with two further examples from London, one with incised graffito from

Billingsgate Buildings (Chapman 1980, 93, no. 490, fig 53) and a more recently excavated find from 64–66 Cheapside (CED89; [226] <163>). Another comes from Newstead (Curle 1911, 314 and pl 83.2); several are noted by MacGregor (1975, 179) and the existence of similarly shaped implements in iron is noted by Chapman (1980, 93). It is possible that it was used as a scoop, an interpretation favoured by MacGregor (1975, 179), who suggests a domestic function (perhaps for flour) rather than industrial use. However, the worn bevelled edge might suggest that this example was used for smoothing or spreading a soft substance.

Fig 148 *Bone tool or scoop (length 153mm)*

STONE

Quern [2062] <156> F10.9 later Roman features (not illustrated)
Small fragment lava quern, all faces broken. Mayen (Niedermendig) lava.

Medieval

GOLD

Finger ring [79] <66> F28.1 foreshore (Fig 149)
Fine gold wire hoop, with circular section; hoop soldered to an irregularly shaped transverse semicircular bezel set with an almandine garnet cabochon. The hoop is now detached from the bezel on one side.
External diameter 23.5mm; internal diameter 20mm; dimensions of bezel 9 × 6.5mm.

The ring is very similar in style to five gold finger rings recovered from waterfront deposits dating to the second half of the 14th century on the north side of the River Thames (Egan and Pritchard 1991, 326–7, nos 1610–14). These have extremely simple settings with a variety of precious and semi-precious gemstones, including three almandine garnets, which were both fashionable and inexpensive. Two finger rings in the Waterton collection (Victoria and Albert Museum) have similarly plain settings, reflecting the natural outline of the cabochon, as in the present example (Oman 1930, 67, nos 256–7, pl xi).

Fig 149 *Gold finger ring (external diameter 23.5mm)*

COPPER ALLOY

Bell [1035] <31> F29.10 late medieval features (Fig 150)
Lower half of rumbler bell of a type used on occasion as a dress accessory and also on harness or on the collars of pet animals (Egan and Pritchard 1991, 336–7). The present incomplete example is similar in form to no. 1666 from Trig Lane (ibid, 339) and was made in two halves from sheet metal which was hammered in a form. A suspension loop was inserted through a hole in the upper half and an aperture and round sound holes, which can be seen here, were cut n the lower part. An iron pea was inserted into the body and the two halves were soldered together, using a lead/tin solder. It is very common for such bells to split, as here, at this weak point.
Diameter 35mm.

Fig 150 *View of bottom of copper alloy rumbler bell (diameter 35mm)*

Pin [79] <64> F28.1 foreshore (1380–1450) (Fig 151)
Solid spherical head and very fine-gauge wire shank; similar to types found in 14th-century contexts in London (Egan and Pritchard 1991, 302).
Length 56mm.

Fig 151 *Copper alloy pin (length 56mm)*

BONE

Handle? [2003] <4> F12.15 foreshore (Fig 152)
Elaborately decorated implement made from a large cow metatarsal which has been roughly squared to make four faces. On each of the two sides, occupying the central part, an animal has been carved, stylised, but detailed, the hair shown as short wavy lines. is not clear what beasts are represented; they appear somewhat ursine, but the length of their tails might indicate that they are baboons or similar. The upper surface is plain and the lower has crudely carved transverse and linear grooves. A hole 9mm in diameter has been drilled through the upper and lower faces at approximately the central point. There are wear marks (small chips) around the upper hole and a depression around the lower hole that might indicate that a ?metal plate or washer was fastened to it. Both ends

Fig 152 Decorated bone handle (length 102mm)

of the bone have been worked; the narrower end has claw-like grooving and the other is recessed and presumably held an iron tang, but there is no trace of any metal remaining.

The object is so far unparalleled but the general style of the carving and the size of the bone would place it in the post-medieval period (perhaps modern). It may well be an import (?eastern) and is likely to be from a large elaborate knife or possibly a corkscrew. Other finds from this foreshore group date from the 12th century but context [2003] underlay 19th-century warehouse foundations and contamination is highly probable.
Length 102mm; width 30mm; width of socket 17mm.

Skate [2003]<167> group F12.15 foreshore (Fig 153)
Flat pointed toe and no holes at either end. The toe has been very roughly shaped and the posterior surface only slightly flattened. The anterior surface, which has not been flattened, is slightly polished but shows little, if any, sign of wear. Although unpierced skates exist (Pritchard 1991, 266, nos 255, 259), the absence of holes at either end, the poor craftsmanship and lack of wear might suggest that the object is unfinished; the bone is splitting just below the posterior surface and the object may have been discarded. Horse metacarpal.
Length 215mm.

Skates have been found on several sites in London (MacGregor 1976, 65; West 1982, 303) and the examples from 11th- and 12th-century deposits have recently been discussed by Pritchard (1991, 209 fig 3.91; 266, nos 253–9).

The bones used for skates are typically long bones, horse or cattle metatapodia or radii. The bones are shaped at the toe (distal end) and common features include holes drilled through the toe and heel, used to hold thongs for binding the skate to the foot. Transverse cuts along the upper (posterior) surface made to improve the foothold are recorded on some examples, but not in this case. The different methods of fastening skates to the feet are reviewed by MacGregor (1976, 61).

Fig 153 Bone skate (length 215mm)

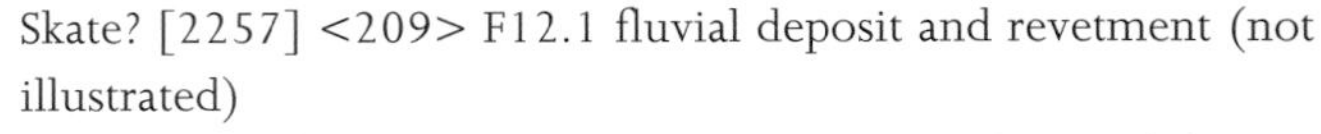

Skate? [2257] <209> F12.1 fluvial deposit and revetment (not illustrated)
Skate? made from cattle metatarsus. The toe is damaged, but there is trace of a vertical hole. The upper (posterior) surface has been flattened at each end; the anterior surface is only slightly worn.
Length 177mm.

STONE

Hone [2140] <158> F14.9 foreshore (1180–1230) (Fig 154)
Rectangular section; one side is very smooth with wear marks and one edge is ridged; the other faces are abraded. Mica-quartz-schist – Norwegian ragstone.
Surviving length 84mm; width (average) 46mm; thickness 25mm.

Fig 154 Hone stone (surviving length 84mm)

Hone [2151] <155> F16.2 foreshore (1180–1230) (not illustrated)
Fine-grained grey stone ??phyllite. Rectangular section, the stone broken at both ends; the upper surface has been worn down at both sides. Two small holes have been drilled in the underside.
Surviving length 80mm; width 12mm; thickness 9mm.

Spindle whorl [2231] <79> F15.3 foreshore (Fig 155)
Light grey fine-grained calcareous siltstone/mudstone, lathe-turned, biconical in shape but with a flat base; turning marks on the upper surface.
Diameter 37mm; height 17.5mm; diameter of hole 8.3 mm; weight 25gm.

The material was commonly used for spindle whorls found

Fig 155 Stone spindle whorl (diameter 37mm)

in London of both Saxo-Norman and medieval date (Pritchard 1991, 165, 259, nos 170–4; Egan 1998, 790–802). Pritchard notes that other examples made of this material have a wide distribution and although the source of the stone has not been located, production of such spindle whorls may have formed part of a specialised industry. There is no evidence for local manufacture. Two spindle whorls, apparently similar, but said to be ceramic, came from Toppings Wharf, in contexts of late medieval date (Schwab in Sheldon 1974, 103, nos 14–5).

Quern [500] <123> F25.3 (not illustrated)
Edge fragment of a rotary quern, made from basalt lava imported from the Mayen area. The underside is flat and worn and the upper surface, which is parallel to it, is abraded but shows a series of tool marks running tangentially across the surface. The material from which early medieval querns was made is that used in London from the Roman period and where only small fragments survive it is not always possible to distinguish between Roman and later examples. In this case, however, the tooling, seen also on other upperstones from London, is a diagnostic feature (Pritchard 1991, 162). The size of such querns varies considerably and although larger than many stones found in 10th- to 12th-century contexts, one from a waterfront dump at St Magnus is of even greater diameter at *c* 1.5m (ibid 258, no.152).
Diameter *c* 600mm; thickness 37mm.

Quern [2089] <206> F12.5 (not illustrated)
Edge fragment of a rotary quern made of basalt lava, similar to the above, but with very deep tool marks on the upper surface.
Approximate diameter 600mm; thickness 37mm.

Post-medieval

COPPER ALLOY

Candle holder [1009] <28> F30.2 fill of cesspit (Fig 156)
Chamber or bedroom candlestick, cast in brass with a one-piece pan and handle. The socket is cast and peened over beneath the pan. The elongated strip handle, which terminates in a large flat ring, is an early feature and the object is of a type known in the early part of the 18th century (*c* 1710–30) (Caspall 1987, 138 and fig 298).
Height 68mm; diameter of pan 96mm; length of handle 101mm.

TORTOISESHELL

Fan [1007] <180> F30.4 cellar (not illustrated)
Fan stick, undecorated; broken at narrower end and at lower end above rivet hole. The overall dimensions and simple design, very similar to examples from Aldgate (Marschner in Grew 1984, 111), suggests a date in the late 17th or early 18th century, consistent with the date of the deposit.
Length 190mm

Fig 156 Copper alloy candle holder (height 68mm)

STONE

Alley or toy marble [1006] <149> F30.2 (Fig 157)
This is of identical size and shape to the examples from Aldgate (Grew 1984, 118) and there are numerous others from London.
Diameter 15mm.

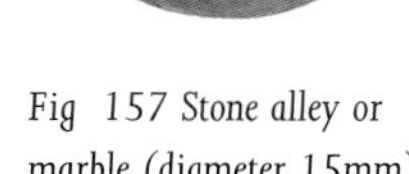

Fig 157 Stone alley or marble (diameter 15mm)

Structural ironwork from the bridge

CRAMPS

During excavation of the ashlar masonry used for the 12th-century Colechurch bridge abutment (F26.5) two tie bars or cramps (see Glossary) were found, used in the construction of the wall facing. Similar ties also came from an ashlar facing which was part of a 14th-century addition to the bridge (Fig 89).

[7] <20> F26.5 construction of Colechurch bridge abutment (Fig 89)
Cramp (tie bar) for ashlar masonry. Stout bar with arms at each end set at right angles. One arm retains a collar of lead into which the bar was set. In construction, the lead was presumably poured in molten and has set around the bar.
Length of cramp 265mm; width 23mm; thickness 13mm; length of arm 43mm.

[8] <5> F26.5 (not illustrated)
Tie bar for ashlar masonry, as above, but the lead casing is now detached.
Length 260mm; width 23mm.

[55] <55> F28.2 later additions to the Colechurch bridge (not illustrated)
Cramp (tie bar) for ashlar masonry of bridge. Stout bar with arm at right angles at each end, with lead fixing.
Length of cramp 255mm; width 25mm; thickness 9mm.

[55] <56> F28.2 later additions to the Colechurch bridge (Fig 89)
Cramp or tie bar for ashlar masonry. Stout bar with arms at both ends (one is now bent at an angle). The remaining arm is encased in lead, which has run down half the length of the bar and has a large flange at the end.
Length (of both bar and lead) 310mm; width of bar 24mm.

PILE SHOES

The 1984 excavations also produced nine iron pile tips or shoes, which served as bindings or reinforcements for the wooden piles used in the construction of the bridge (Fig 90). They are all of similar shape and manufacture, each with a solid point in the form of an inverted pyramid from which rise four broad straps which were nailed to the timber. In several instances part of the wooden pile has survived. The excavated pile shoes fall into two size ranges, a larger one with a maximum length of about 390mm, the other smaller at 250mm, but this does not necessarily imply that smaller bindings were used for smaller timbers.

The pile shoes are directly comparable with those found during dredging of the Thames in 1967, which are apparently from the sixth and seventh starlings from the southern end of the medieval bridge (Marsden 1971, 13–14). Two of these piles ER 1279 <62> and <63> are included in the catalogue. The dredged examples are both of the larger size. Marsden notes that two types of pile were found, the 'older' cut from tree trunks and the more recent, square in section and only superficially discoloured. Both types were fitted with iron points. There is apparently no wood identification for these piles. The wood in the recently excavated iron sheaths includes both circular and square-sectioned timbers, although none of the former approach the size quoted by Marsden (45–60cm). The iron-shod elm piles associated with the 15th-century enlargement of the abutment (F28) were all oval or circular (diameter *c* 230–360 mm) lengths of tree trunk, with sharpened ends (see Chapter 11.2; Fig 93; Fig 94).

The three examples from F25.4 come from the earliest stratigraphical group to contain pile shoes and all are of the larger size. Substantial amounts of wood (oak) remain in the sockets of [131] <89> and [154] <88>, circular-sectioned and square-sectioned timbers respectively. The corrosion on the three pile shoes is similar and quite unlike that found elsewhere, suggesting similar burial conditions. Pile [131] (pile shoe <89>) has been identified as one of several used in preparatory work for the construction of the Colechurch bridge abutment (sealed by the infill of the abutment) and an (untipped) pile in the same group [121] {977} has a felling date after 1155. The other iron-shod piles ascribed to this group were not found *in situ*, but as noted above their forms and the extent of corrosion were similar to <89>. On stratigraphic grounds the date of this group is *c* 1170.

The remaining six pile shoes come from F28, the later additions to the Colechurch bridge abutment (Fig 87; Fig 90). Four of these, one from the foreshore (F28.1) and three from F28.2 are of the smaller size, with shorter arms (maximum 260mm). Other finds from the foreshore upstream of the abutment are dated to the late 14th/15th century, (although there is a little post-medieval contamination in context [83]). Pile shoe [34] <21> is not *in situ*. Group F28.2 represents additions to the north-west side of the abutment, dated after 1362 (on date of sillbeam [64]). The final examples from group F28.5, the 15th-century upstream starling, are of the larger size, similar in their proportions to those from the earlier group (F25) but without the extensive corrosion. The wood from piles [73] and [75] has been identified as elm.

The general shape of pile shoes was, therefore, apparently uniform from the 12th to the 15th century. Variations in size, perhaps due to the production of different batches, seem to accord with different structural elements (although only three basic groups are represented).

The practice of reinforcing bridge timbers in this way has been demonstrated from the Roman period. Iron shoes were used on two piles from a Roman bridge at Aldwincle, Northamptonshire, where during a reconstruction the piles were rammed into the collapsed timbers of an earlier phase and additional strength and penetration was necessary (Jackson and Ambrose 1976, 44, 55, fig 5). The form of the Roman shoes differs from the medieval London Bridge examples in that only two arms were set on a tapering base, but the application and method of nailing the straps to the timbers is similar.

Rigold (1975, 52) traces the development of the late medieval wooden bridge to the mastery of deep and accurate piledriving achieved by the late 13th century, when there are increasingly frequent references to a piledriver in English royal accounts (although there are no specific references to pile shoes in this article).

Medieval accounts sometimes refer to iron pile shoes by the weight of the iron purchased. The Fennings Wharf shoes were, therefore, weighed to see if there was any correlation with the conjectured weights of pile shoes thus recorded. There is an obvious difficulty in obtaining accurate weights for such artefacts as the metal was frequently highly corroded resulting in a loss of metal and, therefore, weight. On the other hand, where the timber survived it was impossible to separate it from the iron shoe which resulted in an overestimation. Analysis of the weights obtained, however, shows that the smaller 15th-century pile shoes are consistently *c* 3 to 3½ lb in weight. The lightest example no. <26> is exceptionally corroded. It is interesting that this corresponds with a conjectured weight of about 3lb for pile shoes held in store at London Bridge in 1350 (Salzman 1952, 86), each of which was valued at 4d. The Rochester Bridge accounts for 1438–9 mention the purchase of '50 pile shoes which weigh 101lb' (Becker 1930, 76), an average weight of 2lb each. It is more difficult to obtain an accurate weight for the pile shoes from the later groups, owing to the greater amounts of preserved timber, and the adherence of concretion in some cases, but making allowances for the extra weight of the wood, the shoes weigh in the region of 9lb. Heavier pile shoes than those mentioned in the London Bridge accounts are recorded in the Rochester Bridge accounts. In 1438–9 iron weighing 402lb was purchased for 28 pile shoes, an average weight of 14.4lb each (Becker 1930, 76). The figure of 14.4lb per shoe, suggested by the purchase of 402lb of iron for 28 shoes in 1438–9, seems very high by contrast, but as can be seen from the evidence of the excavated London Bridge shoes, the weight of these items can vary considerably.

In each description the maximum length of the object is given, being the length from the base to the top of the longest strap. In all examples there is some variation in the length of the four straps but this is not all thought to be of significance. The maximum width at the top of the object is the distance between the opposite straps, given as an approximate dimension for the timber which they held. The wood identification is given where available.

[34] <21> F28.1 (Fig 90)
Complete. Pyramidal base with four straps attached, each with a shaped, pointed terminal and two nail holes for attachment to the timber pile. Two incomplete nails survive. Traces of wood remain in the socket.
Maximum length 260mm; length of base (pyramidal terminal) *c* 100mm; maximum width of foot *c* 90mm; maximum width at top 150mm; average width of each strap *c* 40mm.

[33] <24> F28.2 (not illustrated)
One arm missing, the others bent and distorted, but the object is of similar size and form to <25>. The form of the socket above the base is square and there are traces of wood (elm?) in the socket.
Maximum length 250mm; length of base (pyramidal terminal) *c* 80mm; maximum width of foot *c* 60mm; maximum width at top distorted; average width of each strap *c* 50mm; length of nail 70mm; weight 3.75lb (no wood).

[33] <25> F28.2 (not illustrated)
One arm missing, but the remaining arms, each with two nail holes, are complete and the form is clear. A complete nail survives *in situ* and there are traces of wood (elm ?) on the inner faces of the straps.
Maximum length 240mm; length of base (pyramidal terminal) *c* 80mm; maximum width of foot *c* 60mm; maximum width at top 170mm; average width of each strap *c* 50mm; length of nail 70mm; weight 3.5lb (no wood).

[33] <26> F28.2 (not illustrated)
Complete; the four straps are very corroded and distorted (squashed), but of similar form to <25>, with traces of wood (elm ?) adhering to the base.
Maximum length 240mm; length of base (pyramidal terminal) *c* 90mm; maximum width of foot *c* 60mm; maximum width at top distorted; average width of each strap *c* 50mm; weight 2lb (no wood).

[73] <76> F28.5 (Fig 90)
Complete; pyramidal base and four well preserved straps, each with three nail holes and several nails remaining *in situ*. The nails are all of the same form with a rectangular-sectioned shank and square head. Desiccated wood (elm) has survived in the socket.
Maximum length 390mm; length of base (pyramidal terminal) *c* 130mm; maximum width of foot *c* 70mm; maximum width at top 190mm; average width of each strap *c* 50mm; length of nail 70mm; weight 9lb (little wood).

[75] <77> F28.5 (Fig 87)
Complete; solid pyramidal base. The elm pile is well preserved and projects beyond the ends of the four straps, where it has been sawn off. Two nail holes are visible on each strap, but corrosion may conceal a third.
Maximum length (excluding timber) 390mm; length of base (pyramidal terminal) *c* 120mm; maximum width of foot *c* 70mm; maximum width at top 180mm; average width of each strap *c* 50mm; weight10.5lb (includes timber).

[154] <88> F25.4 (not illustrated)
Poorly preserved but of similar form, character and dimensions to other examples. Corrosion has encased half the timber as in <89>, but on two sides fragmentary straps can be seen, one with two ?circular nail holes. The rectangular timber (oak, undatable) pile is well preserved and has been sawn at the top.
Maximum length (excluding timber) 390mm; length of base (pyramidal terminal) *c* 80mm; maximum width of foot *c* 50mm; maximum width at top 150mm; average width of each strap *c* 50mm; weight 10.5lb (includes timber).

[131] <89> F25.4 (not illustrated)
Pile sheath, apparently of similar form to the other examples but only the solid base can be seen, as the iron of the straps has corroded severely, forming a solid cone around the surviving timber. Similar corrosion can be seen on <88> above, where the individual straps can also be distinguished. The surviving timber (oak, undatable) is circular in section and projects about 100mm above the top of the binding.
Maximum length (excluding timber) 390mm; maximum width at top 180mm; corrosion has obscured all other details but the general dimensions are similar to <77>; weight 11.5lb (includes timber).

[151] <87> F25.4 (not illustrated)
Incomplete; corroded and flaking but of similar form to the other examples. The straps are closer together at the top, securing a smaller timber, although the overall length is the same as larger examples. Wood survives within the strapping (not identified).
Maximum surviving length 330mm; width of foot (approx) 80mm; maximum width at top 80mm.

PILES FROM 1967 DREDGING

ER1279 <62> (not illustrated)
Identical in form to the excavated examples with a solid

pyramidal base and four heavy straps with nail holes for attachment to the timber. Traces of wood remain.
Maximum length 320mm; length of base (pyramidal terminal) *c* 90mm; maximum width of foot *c* 50mm; average width of each strap *c* 40mm.

ER1279A <63> (not illustrated)
Similar in form, slightly larger, but less well preserved.
Maximum length 360mm; length of base (pyramidal terminal) *c* 150mm; maximum width of foot *c* 70mm; maximum width at top 190mm; average width of each strap *c* 40mm; length of nail 70mm.

14.7 Two wooden statues from the site of medieval London Bridge

John Cherry

Introduction

Two wooden statues were dredged up from the Thames during the period 1824–41, on the site of medieval London Bridge. The first was a standing figure of a monk, possibly St Benedict, and the second a seated figure of God the Father, personified as a king or emperor (Fig 80; Fig 81). Both statues would originally have been gilded and painted, but no trace of the original decoration survives on either. The original context of both these statues is not known, but it is quite possible that both of them were part of the fittings of the chapel of St Thomas the Martyr, which stood on the large central bridge pier. It is documented that the chapel did contain a number of statues, which are discussed in Chapter 9.4, but to date it has not been possible to identify these two particular statues within the Bridge House records. Perhaps the two statues were thrown into the river during the mid-16th century by iconoclasts. However, it is possible that one of the statues was deliberately hidden within one of the starlings, perhaps in an attempt to avoid destruction (discussed later). The aim of this study is to reconsider the evidence for the identification and date of these two statues.

Charles Roach Smith described both statues as Flemish and he may well have been correct (Smith 1854, 111–12). There are not many free-standing, sculptured figures for comparison and the paucity of the material has already been noted (Gardner 1951). The style of the figure of the monk, with rounded face and straight falling robes, may be compared with the figures on the chapter stalls at Henry VII's chapel (built 1500–12) at Westminster Abbey (RCHM 1924, pls 125–7). The absence of the accounts for this chapel means that we do not know who carved the stalls, but it has been suggested that there is a Flemish influence in the chapel sculpture. Some other examples of late medieval wood carving in England are all thought to be Flemish. The wooden screens in the churches of Lullingstone in Kent and Lavenham in Suffolk may be from either Flanders or the Low Countries (Howard and Crossley 1917, 23). Flemish influence has been identified in the choir stalls at Windsor Castle (Hope 1913, 429–30; James 1933). Although by the early 16th century artists from the Low Countries had ousted indigenous workers in the production of the finest stained glass and manuscripts, it is probable that in carpentry and woodworking the position was different and more complex (Lindley 1995, 29).

Both statues probably date from the second half of the 15th or the beginning of the 16th century. However, further research is needed on late medieval wood statues to formulate more precise opinions concerning both the date of the two figures and the nationality of their carvers.

The standing figure of the monk

In 1854 this figure was in the possession of Charles Roach Smith, who described it as a figure of a Benedictine monk, probably of Flemish manufacture and 15th-century date (Smith 1854, 111). It was acquired by the British Museum in 1856 as part of the Roach Smith collection of London antiquities (BM, MLA 1856, 7–11, 1500).

The wooden figure is carved out of a single block of oak, it is mounted upon a plinth, that is surrounded by three additional moulded pieces of oak with mitred corner joints (Fig 80). The front and back of the figure is well preserved, but the forepart of both arms has been damaged. There is a plugged hole in the back of the figure, which may have been used to secure it to a wall.

The figure is tonsured. The square face is well modelled and the hair on the fringe of the tonsure is indicated by delicate curly lines. At the junction of hood and head is a small hole that may have served for the addition of a halo. His right arm is raised, possibly to hold a crosier, but the left arm curves inwards, possibly to hold a book or point to the cavity in the chest. The figure is wearing three garments. There is an undergarment only visible at the slit in the neck, an overall scapular (monastic habit) and a hood. The scapular has rectangular front and rear portions of uniform width; it falls to the feet and has long sleeves, which hang down from the arms. The figure has an elaborate hood (*caputium*), pointed at the back with folds on either side of the face. It is jointed underneath the neck. In the centre of the forehead is a knob (possibly for the attachment of an inscription). The statue is 480mm high. On stylistic grounds the figure is dated to *c* 1480–1520.

The hole in the breast of the monk was probably to conceal a relic or the host. Statues of Christ sometimes had an opening in the breast for the consecrated host to be placed at Easter. An example was inventoried amongst the treasures of the shrine of St Hugh at Lincoln in 1536 – 'an image of our saviour silver and gilt standing on six lions, void in breast for sacrament for Easter day' (Aston 1988, 230). An actual example of a statue with a cavity in the breast is a suffering Christ (*Christus Elend*) in Brunswick Cathedral. It dates to the end of the 15th century or c 1500. The hole in the Brunswick Christ was closed by green glass and a host was placed there on Gründonnerstag or Maundy Thursday (Westfehling 1982, 87–90). At Durham on Good Friday according to the *Rites of Durham* there was a 'picture of our Saviour Christ, in whose breast they did enclose with great reverence the most holy and blessed sacrament of the altar' (*Rites of Durham*, 12, section V, The quire, the passion). No other examples of a monk with such a cavity appear to have been noticed.

The tonsure and dress indicate that the figure was a monk and possibly a monastic saint. If there was a halo, it is possible that it could have been a representation of St Benedict (c 480–c 550), the author of the monastic rule which bears his name. He was not a popular saint and was most often represented in Benedictine monasteries. Therefore, a chapel on London Bridge would seem to be an unusual place to find a statue of him. However, in the will of John Hatfield (1363) money was left to maintain a chantry in the nearby parish church of St Benedict's (or Benet's), Gracechurch Street, which by 1392–3 had passed by default to London Bridge chapel (see Chapter 9.5). This important chantry could possibly explain the link between the saint and the bridge chapel. Rushforth (1936, 138) notes that there seem to have been very few representations of St Benedict in England. For instance, in the stained glass at Malvern Abbey and Durham Cathedral St Benedict is only represented once in each place. The iconography of St Benedict has been discussed by Dubler (1966); there are also two exhibition catalogues: Schweizerisches Landesmuseum (1980–1) and Certosa di Firenze (1982).

However, there are a number of representations of St Benedict which offer parallels for the London Bridge statue. St Benedict is shown in an 11th-century psalter, where he is represented as a seated tonsured saint. He is in monastic dress, with an elaborately folded hood, secured with a clasp. He holds a crosier in his right hand and with his left gestures towards a copy of his rule. Around his brow is a scroll with the words *HONOR DIE*, which may provide a parallel for the attachment on the forehead of the London Bridge figure (BL, Arundel MS 133, fo 133).[45] It is illustrated in *The Benedictines in Britain* (St Augustine's Ramsgate 1989, pl 3). St Benedict is also shown on the Langdale rosary, now in the Victoria and Albert Museum. On the rosary the saint is wearing a chasuble and holding a crosier (Maclagan and Oman, 1936, 9, pl iv, bead 42a). He is also shown on the painting, formerly in a rood screen and now in the church of Great Plumstead, Norfolk. Here he is shown in monastic dress with a hood and long sleeves. He holds a scroll in his left hand and crosier in his right (St Augustine's Ramsgate 1989, 85).

St Benedict is also shown in the print in the middle of the wooden mazer (silver-mounted drinking bowl) that Robert Pecham gave to the frater of the Benedictine priory at Rochester, which is now in the British Museum. It is hallmarked London 1532. On the print a tonsured, standing St Benedict is holding a book and a crosier between two plants (which it has been suggested are meant to be herb bennet). The identification of the figure is confirmed by the addition of his name: 'S. Benit' (Jackson 1911, 627; Read and Tonnochy 1928, 2). St Benedict is sometimes shown bearded and sometimes clean shaven. For instance, in the painting by Hans Melinc (1487) in the Uffizi, Florence, the saint is clean shaven.

The seated figure of God the Father

The second statue was in the possession of R L Jones before 1854, by which date it was the property of John and William Dent of Sudeley Castle, Gloucestershire, where it can be seen today by visitors in the 'Stone Drawing Room'. In the castle archives there is an (undated) photograph of the statue, stating that it was 'given to John and William Dent by Richard Lambert Jones Esq. chairman of the London Bridge Committee. By some it is supposed to represent Peter of Colechurch, Architect of Old London Bridge in 1205 – from the grandeur of the head and the globe at the feet it is all thought by others to represent the Creator. It was found in a small oratory within one of the piers beneath the watermark' (information from Jean Bray, Sudeley Castle archivist). This information suggests that the statue may have been hidden within one of the starlings during the Reformation, not thrown into the river.

At some time before 1854 two plaster casts of the statue were made by the sculptor Samuel Nixon (1803–54). One cast was presented to the Society of Antiquaries by Henry Stothard in 1847 (Anon 1847); and in 1854 the second one was part of the Charles Roach Smith collection and was acquired by the British Museum in 1856 (MLA 1856, 7–1,1500) (Smith 1854, 11–12). Both casts are now missing.

The only previously published image of the statue shows a seated figure with no base or hands (Anon 1847). It was described as 'God the Father personified as the Pope'. Today, the hands of the figure have been replaced: the present right hand holds an orb and a new wooden base has been added (Fig 81). The figure is carved out of a single piece of oak and set on a modern plinth. He is seated, wearing the robes of a priest (including a scarf with fringed ends) and at his feet is a globe (representing the world), divided into three segments by raised bands. The figure is bearded with projecting eyes and nose. The front of the figure is worn and eroded and the rear of the figure is concave. He is wearing a closed or German crown (not the papal triple crown), decorated with 12 fleur-de-lis. It is suggested that the figure originally held in his hands an orb and sceptre or a representation of Christ on the cross, surmounted by a dove (symbolising the Holy Spirit) – making it a representation of the Trinity. The figure is 590mm high (excluding the modern plinth) and has a front width of 220mm. On stylistic grounds the figure is dated to c 1450–1500, as the folds in the drapery compare very closely with those in some late 15th-century alabaster carvings.

The restoration of the figure gives the impression that it was all thought to be a statue of God the Father alone. However, although the figure of Christ and the dove of the Holy Spirit are missing, it is a possibility that the statue originally represented the Trinity, but if it did there is no evidence to indicate where the cross was placed in front of the legs. This suggests that it was probably God the Father alone, possibly in judgement, rather than the Trinity.

In the Turin-Milan Hours illuminated manuscript (*c* 1440s) there is a painting by a follower of Jan Van Eyck, of God the Father enthroned between Christ and the Virgin Mary (Paris, Musée du Louvre, Cabinet des Dessins, X3107). The figure of God the Father is seated, bearded, wearing a closed crown and holding an orb and sceptre. In the upper central panel of the Ghent Cathedral altarpiece is a portrait of God the Father. It was painted in 1432 by Hubert and Jan Van Eyck. The figure of God the Father is seated, wearing red pontifical robes and a papal triple crown; in one hand he holds a crystal sceptre and with the other he bestows a blessing (Dierick 1970, pl 26).

14.8 The animal bones from Fennings Wharf

Kevin Rielly

Introduction

The common denominator in the various period assemblages summarised below is the overriding abundance of the major mammalian domesticates. Clearly the great majority of these bones represent the remains of food waste dumps, a large proportion of which appear to be *in situ*, judging by the quantity of concentrated bone collections throughout these deposits. It is, therefore, no surprise to find that the non-edible species, such as horse, dog and cat, are either absent or poorly represented. The mixed nature of these dumps lends no insight into the organisation of the various post-mortem products, apart perhaps from a noticeable under-representation of horncores, particularly of cattle. Possibly these were removed for resale to hornworkers. In addition there is slight evidence, based on butchery marks and the poor representation of phalanges, for the removal of skins to the tanner. However, the absence of the latter could just as easily be related to recovery bias.

Roman

It is argued below that the good representation of pig, particularly within the early Roman deposits, is likely to represent reality. Such a high proportion of pig bones is relatively rare amongst similarly dated Roman sites and a close comparison can be made with the adjacent site of Winchester Palace (Rielly in prep b), in respect of both the 1st- to 2nd-century and 3rd- to 4th-century bone assemblages. The good representation of this species in the Roman period may be interpreted as representing a highly Romanised diet, possibly related to high status. High counts of pig bones have been found at numerous villa sites in Italy and also at the 1st-century villa at Fishbourne (King 1984, 201). In terms of the number of the individual Romano-British animals represented in the faunal assemblage from the 1970–2 Toppings Wharf excavations, cattle (49) were the most frequent species, followed by pigs (32), with sheep/goats (16) as the third most numerous species (Rixson 1974, 111).

The presence of game species, particularly deer, may also be taken as an indicator of status, either through the availability of time to hunt or the possession of the wherewithal to pay for luxury food items. It can perhaps be suggested that a large range of food items was available in the London area during this period. The 1970–2 Toppings Wharf excavations revealed bones of deer and part of the ulna of a hare (Rixson 1974, 111).

A notable aspect of the late Roman assemblage is the high proportion of fish and particularly of eel, herring and smelt. The quantities of fish found here and also at other Roman sites in the City strongly suggest the intensive exploitation of particular seasonal estuarine and riverine fisheries. For example, large quantities of smelt were found at 1–7 St Thomas's St, Southwark (Jones 1978), and of herring and sprat at Peninsular House (Bateman and Locker 1982). This last assemblage was interpreted as the waste from a processing centre, probably involved in the production of *garum* (fish paste).

Saxo-Norman and medieval

In comparison with the Roman period, there is a marked decline in the abundance of pig to similar levels in each of the later periods. In general this follows the representation of the species found at a variety of contemporary London sites. Comparisons can also be found for the breeding of pigs within or adjacent to the city (the piglet found in one of the Saxo-Norman fluvial deposits), with especially good evidence from the mid-11th-century site at Westminster Abbey (Davis et al in prep). The local breeding or at least keeping of pigs would appear to have been fairly common within the medieval city (Hammond 1995, 41).

The fallow deer, crane and goshawk are clear indicators of high status. It is perhaps significant that the deer bones were found in 12th- rather than 11th-century deposits; although a small number of hunting parks had been set up within the earlier century, at least one of these preceding the Conquest, a far greater number of parks had been enclosed by the end of the 12th century (Rackham, O, 1986, 123). However, fallow deer bones have been identified in London from earlier deposits, for example at Westminster Abbey (Davis et al in prep). It is of some interest to find a hunting bird as well as one of the highly prized falconry prey species, for example the crane (Huff 1997, 8 and see Description of the bones, below).

It is to be wondered where such high-status creatures could have been kept or eaten. If these dumps represent local deposition events, as is perhaps likely, there must have been one or more high-status household in the vicinity. Examples of such were found at Winchester Palace, in the course of erection in the 12th century and following at least one substantial wooden building dated to the Saxo-Norman period (Seeley in prep).

Fish are very poorly represented among these periods. To a certain extent, as suggested below (see Description of the bones), their absence is certainly related to the limited number of soil samples (from which almost all of the fish bones were obtained) and poor recovery on site (there was no bulk sieving of deposits). It is notable that the only fish found in these deposits were relatively large fragments of cod and whiting. It is inconceivable that their absence is related to poor exploitation, as it is well documented that the medieval period saw a dramatic increase in the use of fish, possibly the result of a stricter adherence to the prescribed diet for the various Christian fast days and holy days (Wilson 1973, 25).

Post-medieval

A deposit dated to this period provided the only example of a young calf. The presence of such a young individual is not without precedents in sites of this period. Several calf bones were discovered at the nearby excavation at Battlebridge Lane (Rielly in prep a), and at the Royal Navy Victualling Yard at Tower Hill (West 1995). West suggested that these animals may have resulted from the keeping of dairy cows in the back gardens of nearby tenement buildings (the culling of young calves being a necessary part of the exploitation process). It is conceivable that dairy cows may also have been kept within this part of Southwark.

The high-status component of the post-medieval assemblage is the pheasant bone. Records from the Tudor period show that the value of a pheasant was about four times that of a rabbit, which, weight for weight, would make this bird as costly as swan (Rackham, O, 1986, 50). It was mentioned that this bird had suffered and survived a broken leg. That the bone was allowed to mend is perhaps unusual for a wild bird, where any physiological weakness would have reduced the chances of survival. One possibility is that it may have lived in managed surroundings, such as a hunting park, where all or most threats to either the game birds or animals were eliminated. Gamekeeping involving pheasants dates from the late 18th century, and involved the employment of professional gamekeepers whose remit was to provide enough pheasants for the shoot and to keep down predators (ibid, 50–1).

Method of study

The bones were recorded using a system devised by the Environmental Section at MoLAS (based on that described by Rackham, D, 1986) and were entered on ORACLE. Both the method and the bone archive are available at MoLAS (Rielly 1998). Bones were identified to species wherever possible and otherwise to size-based categories (see Table 9). The latter categories include ribs, vertebrae (excluding atlas, axis and sacrum) and indeterminate fragments. The majority of the fish bones were identified by Alison Locker and the remainder by Alan Pipe. An important feature of the analysis is the information provided by the sieved assemblages from the soil samples. This information is used in a complementary fashion with the

Table 9 Species representation (hand collected; total fragment count)

Type	Early Roman	Late Roman	Saxo-Norman	Medieval	Post-medieval
Cattle	12	33	50	61	26
Sheep/goat	6	23	50	59	20
Pig	30	18	9	16	5
Horse	-	-	-	-	1
Fallow deer	-	-	-	2	-
Roe deer	-	2	-	-	-
Cat	-	-	1	1	-
Chicken	10	17	5	8	1
Goose	-	2	3	1	-
Mallard	3	-	1	-	1
Goshawk	-	-	1	-	-
Pheasant	-	-	-	-	1
Crane	-	-	-	1	-
Teal	1	-	-	-	-
Cod	-	1	1	-	-
Plaice/flounder	-	2	-	-	-
Cattle size	27	20	34	98	37
Sheep size	17	13	54	73	10
Chicken size	-	1	-	-	-
Goose size	-	-	2	-	-
Total	**106**	**131**	**211**	**320**	**102**

hand recovered data, essentially reducing the biases inherent within the latter recovery method; that is, providing a full range of species and skeletal parts, so increasing the species count as well as testing the representation of the larger species.

Description of the bones by phase

In all periods the hand collected assemblages are dominated by the three major mammalian domesticates, cattle, sheep/goat and pig (Table 9; Table 11). These totals could also include the bones categorised as cattle and sheep size, given that the majority of the identified bones in this size range belong to these three domesticates. Sheep/goat is likely to be represented principally by sheep, as a proportion of bones were clearly identified to this species in each period. Only one goat bone was identified, a horncore from the Saxo-Norman foreshore deposits.

Early Roman

The great majority of the bones dated to this phase were recovered from the backfill of the pit (F9.1), the well (F9.3) providing 12 bones and the floor deposit just four fragments. Relatively few bones were added to the well assemblage by the sample taken from this feature. A significant feature of the pit assemblage is the dominance of pig bones. Here it should be pointed out that the quantities of bones involved are rather small, which precludes any definite conclusions related to species distribution. However, in defence of the noted pig abundance it should also be mentioned that with hand recovery there is generally a bias towards the larger bone fragments, essentially towards cattle. The significance of a high count of pig bones during this period is discussed in the last section of this report (see above, Introduction).

Other species present include both domestic (chicken and possibly duck) and wild birds (teal). Each of these is relatively poorly represented, as would be expected from a hand recovered assemblage. In addition a few fishbones were recovered from the well sample, none of which unfortunately could be identified to species.

Late Roman

Most of the hand collected bones were produced by the backfill (F10.4) of the F10.3 timber-lined well and one out of the three rubbish pits, which had probably been used for the disposal of cess, as well as household waste (F10.9). These assemblages include a large proportion of pig bones, although not as substantial as that shown by the earlier period. The well deposit produced the only roe deer found at this site. Although the cesspit provided very few hand collected bones, the sample produced a large quantity of small fragments, almost entirely composed of fish bones. In addition this pit revealed a number of amphibian bones (all of those shown in Table 10). The presence of these animals clearly demonstrates that this pit may have been open to the elements for a considerable period. Large quantities of fish bones were also discovered within the pit (F10.9). The great majority of the fish bones found in these deposits were identified as eel and herring. These were from relatively small fish, for example the estimated lengths for the eels range from 280mm to 320mm (after Libois et al 1987). The presence of bones from such small fish may indicate either the exploitation of young individuals, for example whitebait (see Introduction), and/or the incidence of decayed faeces in

Table 10 Species representation (sieved; total fragment count)

Type	Early Roman	Late Roman	Saxo-Norman	Medieval	Post-medieval
Cattle	-	-	-	1	-
Pig	-	3	-	-	-
Chicken	-	8	-	-	-
Ray	-	1	-	-	-
Eel	-	201	-	-	-
Herring	-	33	-	-	-
Sprat	-	4	-	-	-
Smelt	-	12	-	-	-
Dace	-	1	-	-	-
Small cyprinid	-	4	-	-	-
Cod	-	1	-	1	-
Large gadoid	-	12	-	-	-
Whiting	-	1	1	-	-
Plaice/flounder	-	2	-	-	-
Flatfish	-	1	-	-	-
Amphibian	-	12	-	-	-
Cattle size	6	8	-	-	-
Sheep size	2	10	2	-	40
Indeterminate mammal	20	-	-	-	-
Chicken size	1	2	-	-	-
Indeterminate fish	10	-	-	-	-
Total	**39**	**316**	**3**	**2**	**40**

these deposits, assuming that their size precluded the necessity for de-boning. It is noticeable that the cesspit fish assemblage was composed solely of species within this size range, while the pit (F10.9) additionally provided a number of larger fish, that is plaice/flounder as well as large gadoid (possibly cod or whiting). Overall, there is a wide range of fish, all of which are likely to have been captured either in the Thames (the cyprinids and eel) or the Thames estuary (the remaining species, see Table 9 and Table 10).

There would appear to be a good representation of chicken. However, most of these bones belonged to a partially articulated cockerel found in one of the pit fills. In addition most of the chicken bones found in the sieved assemblage actually belong to an articulated foot (tarsometatarsus and seven phalanges) from another pit.

Saxo-Norman

All but the latest of these deposits are extremely well dated, the earlier dumps/fluvial deposits dating to the end of the 11th century, while the latest (F12.15) ranges in date from the mid-11th to mid-12th centuries. Most of the bones (178 out of 211 fragments) are from the former deposits, with particular concentrations within Groups F12.2 and F12.5. As stated above, the assemblages from these deposits did produce some gnawed and/or eroded fragments, possibly suggesting that the bones from this period may be less well preserved in comparison with other periods.

The proportion of pig has again declined, while the abundance of cattle and sheep/goat are approximately similar. Otherwise the hand recovered assemblages produced a reasonable range of domestic and wild species, including a small number of fish bones. The sample (from one of the earlier deposits, F12.9) produced very few bones, although it did provide one additional species, whiting (see Table 10). It could perhaps be envisaged that certain smaller bones may not have survived, due to various fragmentation biases, as mentioned above. However, it should be realised that the hand recovered assemblage from this deposit was also very small.

Of especial interest is the presence of goshawk. This undoubtedly represents a high-status find, these birds being trained during the medieval period for the use of the rich (Grant 1988, 180).

Medieval

The bones were obtained from another series of well dated dumps/fluvial deposits. Reasonably sized concentrations of bones were found in each of the three groups analysed, that is groups F14, 15 and 16. These provided roughly similar results to the Saxo-Norman deposits, regarding the major domesticate representations, the range of species and the relatively poor quantity of bones from the sample.

Further similarities include the incidence of high-status items, fallow deer and crane. The former species was undoubtedly imported to the City from one of the newly established deer parks. It has been shown that the Norman gentry were very keen on hunting and that fallow deer were introduced to this country to satisfy this need (Rackham, O 1986, 123; Grant 1988, 166). Crane is mentioned in various historical texts as a luxury item, intended for use solely at high-status feasts (Hammond 1995, 130). Otherwise it was linked, with two or possibly three other birds (heron, curlew and bittern), with hawking. The capture of this bird would have resulted in the falconer gaining a special honour from the king (Huff 1997, 8).

Post-medieval

The great majority of the bones were well dated to the 17th/18th centuries, the exception being the brick-lined cesspit (F30.17) which contained pottery up to the 19th century. This feature provided just two out of the total of 102 hand recovered bones analysed from this period. Out of a total of seven features looked at, only one provided a reasonable quantity of bones: the rubbish pit (F30.9) from which were recovered 55 bone fragments. The species representation tends to follow the pattern described for Saxo-Norman and medieval, although no fish species were discovered (hand or sieved collections) and the birds tend to be relatively poorly represented. There is one interesting find, the pheasant bone, which was recovered from the brick soakaway (F30.14). The single bone, a tarsometatarsus representing a female bird (no spur), had undergone extensive remodelling following a relatively clean fracture. This species can be classed as a luxury item, being either a catch of the hunting fraternity or an expensive food item purchased from one of the City markets.

Use of animals

Major domesticates

The major post-mortem product of these domesticates, cattle, sheep/goat and pig, is meat. Indeed, it can be assumed that most, if not all, of the major domesticate bones represent the waste from animals which have been eaten. There is also some evidence for the use of the other post-mortem products: horn, skins and bone.

The clearest indication of a meat use is the presence of cut marks, several of which were found on each species (including cattle and sheep size) in each group/area assemblage. These cuts could be interpreted as conforming to a sequence of butchery events (shown in all areas), the major meat part of which involves dressing, halving, sectioning and jointing the carcass, followed by de-fleshing and culminating, in cattle, in marrow extraction (shown by split long bones). No obvious differences in these processes between areas were seen, the only noticeable differences being between species where the cattle bones inevitably showed a relatively greater proportion of heavy butchery (ie use of the cleaver). In addition the evidence points to the production of cattle 'stew bones', (shown by excessive chopping of bones into short lengths, beyond the needs of jointing), from which the meat was likely to have been removed.

Table 11 Representation of major domesticates (hand collected)

Period	Cattle (%)	Sheep/goat (%)	Pig (%)	Total no. all fragments
No. of fragments				
Early Roman	25.0	12.5	62.5	48
Late Roman	44.6	31.1	24.3	74
Saxo-Norman	45.9	45.9	8.2	109
Medieval	44.8	43.4	11.8	136
Post-medieval	51.0	39.2	9.8	51
Period	**Cattle (%)**	**Sheep/goat (%)**	**Pig (%)**	**Total weight (g)**
Weight				
Early Roman	34.2	7.4	58.4	801.7
Late Roman	73.9	12.8	13.3	3602.8
Saxo-Norman	67.5	24.7	7.8	4568.1
Medieval	68.2	21.0	10.8	4392.5
Post-medieval	67.8	21.9	10.3	1686.2

Table 12 Age of cattle, sheep/goat and pig: epiphysis fusion (hand collected)

Type	Age groups	Early Roman	Late Roman	Saxo-Norman	Medieval	Post-medieval
Cattle	1	2/-	7/-	5/-	7/-	4/3
	2	1/-	1/-	3/4	3/1	1/1
	3	-/2	-/3	6/1	1/1	-/3
Sheep/goat	1	1/-	1/-	10/1	8/3	8/-
	2	1/1	10/4	6/4	6/-	1/1
	3	-/1	1/2	3/6	4/4	3/3
Pig	1	1/-	3/1	2/1	3/-	-/-
	2	3/5	-/1	1/1	2/-	-/-
	3	-/3	-/4	-/3	-/-	-/3

Age groups: 1 – juvenile (using scapula P, humerus D, radius P and pelvis acetabulum); 2 – immature (using tibia D and metapodial D); 3 – adult (using humerus P, radius D, femur P and D and tibia P)
Numbers in each group (total fragment count): fused/unfused

Taking weight as a rough indication of the proportion of meat supplied (see Table 11) it is clear that cattle provided for the greater part of the meat demand in all but the early Roman period, where cattle comes second to pig.

There is butchery evidence for the removal of horns and skins, which perhaps suggests the use of these products. Incidences of chops through the base of the horncore were found in skull fragments from late Roman (sheep/goat), Saxo-Norman (cattle and sheep/goat) and medieval (sheep/goat), while lower limb bones showing typical skinning cuts were found in late Roman (cattle and sheep/goat) and Saxo-Norman (cattle). Overall relatively few phalanges (all three major domesticates) and horncores (cattle) were found. This could suggest the removal of skins to the tanner and horncores to the hornworker. However, the absence of phalanges could also be related to the inherent bias of hand recovery against the retrieval of the smaller skeletal parts. There are only two incidences of the use of this category of bone for reasons other than culinary: a bone scoop made from a cattle scapula from a later Roman well and a roughly worked and smoothed cattle metatarsus from a Saxo-Norman deposit, possibly representing an unfinished bone skate.

In general the assemblages of all three species show a mix of skeletal parts. Clearly each group is represented by the waste from most, if not all, the stages of the butchery processes relating to these species.

Evidence for the organisation of the meat supply can sometimes be provided by the analysis of the age data, the evidence available here is minimal and precludes any detailed analysis. Epiphysis fusion provided the major body of evidence (see Table 12). Few ageable mandibles were recovered. Essentially the interpretation of this data follows the premise that juvenile, immature and early mature individuals will have been bred for their meat, while older individuals, later mature and adult, will have been bred primarily for some ante-mortem product, for example wool or milk. Juvenile and immature refer to the unfused figures shown in age groups 1 and 2 respectively (see Table 12), while mature and adult refer to those fused in age groups 2 and 3 respectively. In terms of teeth wear, immature/mature refers to the third molar, whether not in wear/in wear. Early and late mature can be seen from the teeth.

For the major mammalian domesticates, the data tentatively suggest the following.

Cattle were generally culled prior to adulthood (with the exception of the Saxo-Norman period), Saxo-Norman and post-medieval being represented by a substantial proportion of immature and juvenile individuals respectively. A post-medieval deposit also produced the partial remains of a young calf.

Table 13 Size of cattle and sheep/goat (all): measurements

Species	Phase	Bone	Dim	No.	Range	Mean
Cattle	late Roman	m'carpus	GL	1	185.0	
	Saxo-Norman			1	176.2	
		m'tarsus		1	205.7	
	medieval	m'carpus		1	174.6	
		m'tarsus		1	205.7	
	early Roman	m'carpus	Bd	1	50.3	
	late Roman			1	52.1	
	Saxo-Norman			3	47.2–47.9	47.5
	medieval			1	46.4	
Sheep/goat	late Roman	radius	GL	1	154.0	
		m'carpus		5	109.7–124.2	117.8
		m'tarsus		3	119.2–136.2	129.9
	Saxo-Norman	radius		2	135.6–150.3	142.9
		m'carpus		2	113.2–120.2	116.7
		m'tarsus		1	115.0	
	medieval	radius		1	140.0	
		m'carpus		2	118.0–122.2	120.1
		m'tarsus		1	127.0	
	post-medieval	humerus		1	147.3	
		femur		1	179.0	
	late Roman	m'carpus	Bd	5	22.7–26.0	24.4
	Saxo-Norman			2	22.0–23.7	22.8
	medieval			2	23.8–24.2	24.0
	early Roman	tibia	Bd	1	22.3	
	Saxo-Norman			4	22.4–25.7	24.3
	medieval			2	24.7–25.2	24.9

Dim: Dimension; No.: Number of measurable bones; GL: Greatest length; Bd: Breadth of distal end
All measurements in millimetres, dimensions following von den Driesch (1976)

Table 14 Size of cattle and sheep/goat (all): withers height (mm)

Species	Period	No.	Range	Mean
Cattle	late Roman	1	1132	
	Saxo-Norman	2	1078–1121	1099
	medieval	2	1068–1095	1081
Sheep	late Roman	9	536–619	594
	Saxo-Norman	5	519–601	558
	medieval	4	560–591	573
	post-medieval	2	621–628	624

Heights calculated using multiplication factors taken from von den Driesch and Boessneck (1974)

Most of the sheep/goat survived to become mature. The teeth evidence for Saxo-Norman and medieval deposits regarding immature/mature is 2/7 and 0/4 respectively. A large proportion of the examples found in Saxo-Norman, medieval and post-medieval contexts survived to full adulthood. Immature individuals are best represented in the late Roman and Saxo-Norman periods.

Pig is represented, with one exception, solely by early mature and younger individuals throughout. The exception, from the Saxo-Norman period, belongs to a very juvenile individual, probably a piglet.

All three species were clearly bred, to a greater or lesser degree, for their meat in all groups. Unusually for cattle, where secondary products are often of greater importance (on the evidence from a variety of sites in London), this seems to have been the case in each phase/area except the Saxo-Norman. The value of sheep/goat meat was probably highest in the late Roman and Saxo-Norman periods, while the pig ageing results are to be expected, since the major products of this animal tend to be post-mortem. It can be assumed, with the almost total lack of very young individuals, that the meat animals were imported into this area. Such young animals are likely to represent infant mortalities, thus providing good evidence for local breeding of stock animals. The evidence is limited to two individuals, a post-medieval calf and a Saxo-Norman piglet.

Other species

All the remaining species, excluding horse, cat and goshawk, are likely to have contributed to the diet. The interpretation of the excluded species as non-edible is based partly on the absence of cut marks. However, only one of the other 'edible' species shows such marks, a Saxo-Norman chicken bone. Horse, cat and goshawk are more likely to have been used primarily for work purposes and afterwards discarded, or else, in the case of horse and cat, skinned.

The age range of the 'edible' species provides some clues to possible exploitation patterns. One of the two medieval fallow deer bones represents an immature individual. It is to be wondered whether this represents a hunt specifically for a younger animal, or a selective purchase from a London meat market. Most of the chicken bones in each period represent adult birds. This mix of young birds amongst a predominance of old boilers clearly demonstrates the importance of egg production.

Chicken, as well as several of the fish species, are represented by a wide range of skeletal parts. It can be assumed that these food items were bought whole and processed by a member of the household. Alternatively the chickens may have been kept rather than purchased.

The late Roman fish species testify to the variety of Roman fishing methods, involving seasonal fisheries in the Thames catching eel, sprats and smelt, and then a number of marine fisheries probably using netting for whiting and herring, trapping or line fishing for flatfish and rays, and line fishing for cod. Marine fishing had become more advanced by the later periods. The cod and whiting identified in Saxo-Norman deposits are likely to be part of a 'white fishery' in the North Sea.

Size and type/breed

Most of the sheep horncores are relatively small, although one large Saxo-Norman ram horn was found. It can be assumed that the smaller cores represent females. Similar sheep horncores have been found at numerous other historical period sites in London. None of the cattle horncores was sufficiently complete to warrant a detailed description. However, it can be said that none appears to represent particularly large cores, as eg the longhorn types found at various medieval and, especially, post-medieval sites in London (Armitage 1982).

The size of the cattle and sheep (Table 13; Table 14) easily fits within the range of corresponding measurements from several Roman through to medieval sites in London. Essentially these animals were relatively small, the cattle resembling the Jersey in size and the sheep, the Soay. From the above evidence, it can be assumed that, in sheep, both sexes were horned.

14.9 The plant remains from Fennings Wharf

John Giorgi

Introduction

During the excavation 26 soil samples were collected for the potential recovery of biological data. Sample size was generally small, with an average volume of *c* four litres a sample. The size of each sample is shown in Table 15 and Table 16. The samples were processed in a flotation machine using sieve sizes of 0.25mm and 1.00mm for the recovery of the flot and residue respectively. After processing, all flots which contained organic material were stored in industrial methylated spirits (IMS), while the remaining flots together with the residues were dried and sorted for biological and artefactual material.

The samples were initially assessed to establish the range, frequency and diversity of the various forms of biological

Table 15 Number of soil samples from each group and feature type

Date	Group	Feature type	No. of samples	No. of samples analysed
Natural	1	-	1	0
Prehistoric	2	ring-ditch fills	5	0
		pit fills	-	0
Early Roman	9	well fills	2	2
Late Roman	10	cesspit fills	3	3
		rubbish pitfill	1	1
Saxo-Norman	11	foreshore deposit	2	1
	12	fills/f'shore deposit	2	1
	13	fills/f'shore deposit	2	2
	17	pit fill	2	2
	18	foreshore deposit	1	1
Medieval	26	layer	1	1
	27	foreshore deposit	3	3
Post-medieval	30	pit fill	1	0
Total			**26**	**17**

remains (Davis 1993). On the basis of the assessment results, 17 of the 26 samples were selected for further analysis. Full details will be found in the archive report (Giorgi 1997). Table 15 shows the number of samples from each group and feature type, and those samples that were included in the post-assessment analysis.

The Roman period

Plant remains were analysed from six samples belonging to the fills of three features: a well (F9.2), cesspit (F10.1) and rubbish pit (F10.9). Most of the botanical remains were recovered from samples taken from the cesspit and rubbish pit (see Table 16).

Cereals

These were mainly represented by charred cereal grains, although a large quantity of bran fragments, preserved by waterlogging, was recovered from the fill of the F10.9 rubbish pit. Most of the charred cereal grains were found in the backfill of the (F10.1) late Roman cesspit. Cereals included wheat (*Triticum* spp), rye (*Secale cereale*) and barley (*Hordeum sativum*). Free-threshing wheat, probably bread/club wheat (*T aestivum* sl) was the best represented wheat grain although spelt wheat (*T spelta*) was also represented by a single glume base in the fill of the F9.2 early Roman well. Six oat grains (*Avena* spp) were recovered, although it was not possible to establish whether they were from a cultivated or wild species. The majority of cereal grains, however, could not be identified to species. The cereals grains were probably accidentally burnt while being dried before storage or hardened for milling into flour.

Charred cereal grains are frequently found on Roman sites in the City and Southwark, the most common species being spelt wheat and barley; free-threshing wheat is sometimes found, while oat is seldom recovered and usually in small quantities, which suggests that it may represent weeds of the other cereals. Two recent excavations in north Southwark have produced extensive deposits of Romano-British charred grain:

Table 16 Plant remains from Fennings Wharf – prehistoric and Roman

Species	Common name	Habitat	Subgroup and sample numbers					
			F9.2 {217}	F9.3 {499}	F10.1 {440}	F10.1 {366}	F10.1 {365}	F10.9 {49}
Charred remains								
Triticum spelta	spelt glume base	FI	1					
T aestivum L sl	bread/club wheat	FI				10		
Triticum spp	wheat	FI		2		8		
Secale cereale	rye	FI				12	2	
cf *S cereale*	rye	FI				4		
Triticum/Secale spp	wheat/rye	FI				15	2	
Hordeum sativum	barley	FI		1		12		
Avena spp	oat	AFI				3	2	
Avena sp	oat, floret	AFI					1	
Cerealia	indet cereal	FI		5	4	134	4	
Agrostemma githago L	corn cockle	AB				1		
Anthemis cotula L	stinking mayweed	ABGH	+					
Lolium sp	rye-grass	BI		1				
Bromus sp	brome	ABD				1		
Poaceae indet	-	ABCDEFHI				3		
Indeterminate	-	-				4		
Indeterminate	wood fragments	-	+++	++	+++	+	++	++
Mineralised remains								
Vitis vinifera L	grape	FI			+			
Pyrus/Malus spp	pear/apple	CFI			+			
cf *Pyrus/Malus* spp	pear/apple	CFI					+	
Rosaceae indet	-	-			++		++	
Polygonum cf *lapathifolium*	pale persicaria	ABE			+			
Sambucus nigra L	elder	BCFGH			+			
Indeterminate	-	-			++		+	
Waterlogged remains								
Cerealia	indet cereal, bran	FI						++++
Ranunculus cf *flammula*	lesser spearwort	EG						+
R sceleratus L	celery-leaved crowfoot	E		+				
Papaver somniferum L	opium poppy	BGHI					++	+
Papaver spp	poppy	ABGHI	+					
Brassica/Sinapis spp	-	ABFGHI				+		+
Silene spp	campion/catchfly	ABCDF						+
Agrostemma githago L	corn cockle	AB						++++
Chenopodium spp	goosefoot etc	ABCDFH		+		++	+	++
Atriplex spp	oraches	ABFGH						++
Rubus fruticosus/idaeus	blackberry/raspberry	CFGH			+++		+++	++++
Apium spp	-	EFI						++
Torilis spp	hedge-parsley	ACD						+
Umbelliferae indet	-	-						+
Polygonum persicaria L	persicaria	ABEH						+
Rumex acetosella agg	sheep's sorrel	AD	+			+		+
Rumex spp	dock	ABCDEFG				+		
Urtica urens L	small nettle	AB						+
Ficus carica L	fig	FGI			+			
Corylus avellana L	hazel	CF		+				
Hyoscyamus niger L	henbane	BDG		+				+
Euphrasia/Odontites spp	euphrasia/red bartsia	ABCDE			+			
Prunella vulgaris L	self-heal	BCDG						+
Sambucus nigra L	elder	BCFGH		+	++		+	+
Chrysanthemum segetum L	corn marigold	AHI				+		
Lapsana communis L	nipplewort	BCF						++
Alisma spp	water-plantain	E	+		+			+++
Juncus spp	rush	ADEH	++	++	+++	++	++	++++
Eleocharis palustris/uniglumis	spike-rush	E	+		+		+	++
Carex spp	sedges	CDEH				+		+
Indeterminate	-	-		+	+			+
Indeterminate	vegetative fragments	-						++++
Indeterminate	wood fragments	-	++					++
Bryophyta indet	moss	-						+

Key: A – weeds of cultivated land; B – weeds of waste places and disturbed ground; C – plants of wood, scrub and hedgerows; D – grassland plants; E – plants of damp or marshy land; F – edible wild plants; G – medicinal plants; H – wild plants with other economic uses; I – cultivated plants

Frequency: + = 1–10; ++ = 11–50; +++ = 51–250; ++++ = >250 plant items hulled barley at the Courage Brewery site (Davis 1995) and wheat and barley at the Borough High Street site (Gray-Rees 1996). Present archaeobotanical evidence from the rest of Roman Britain shows that spelt and barley are usually the best represented cereal crops, with some bread wheat and emmer (*T dicoccum*), the latter not found at Fennings Wharf (Greig 1991, 309).

The recovery of rye grains is interesting, as no definite identifications of this grain have previously been made from London sites, although some grains were classified as wheat/rye in two large grain deposits of wheat and barley from Pudding Lane and the Forum (Straker 1984, 326). Outside London, rye grains have been identified at a small number of Romano-British sites, for instance, in a military warehouse in Coney Street, York (Williams 1979) and in a granary deposit at St Albans (Renfrew 1985, 21). The cereal grains may have been used for bread, porridge, gruel and cakes (Wilson 1991, 234). While wheat was probably used exclusively for human food, barley may also have been used for brewing and animal fodder, particularly for horses.

Fruits

Fruits were represented by mineralised and waterlogged fruit stones and seeds, and were mainly found in the F10.1 cesspit samples. The fruits included fig (*Ficus carica*), grape (*Vitis vinifera*), apple/pear (*Malus/Pyrus* sp), blackberry/raspberry (*Rubus fruticosus/idaeus*), elder (*Sambucus nigra*) and hazel (*Corylus avellana*). A large number of the mineralised remains could not be identified, owing to the absence of surface detail, although their morphology was similar to fruits and seeds of Rosaceae, and may belong to apple/pear or *Prunus* species such as plum/bullace (*Prunus domestica* sl) and cherry (*Prunus avium/cerasus*).

All these fruits are common finds on Roman sites in London; for example, fig, grape, blackberry/raspberry, apple/pear, elder and hazel were among the fruits identified at three Roman waterfront sites in Southwark: Calverts Buildings (Pearson and Giorgi 1992), Winchester Palace (Giorgi in prep a) and Courage's Brewery (Davis 1995). Other Southwark sites have also produced evidence of some of these fruits (eg Tyers 1988b).

The fruits may have been used as food, either fresh or dried, for the preparation of drinks and in cooking. Grape seeds are frequent finds on Roman sites in London; the fruits may either have been imported in dried form as raisins, currants, or sultanas, or possibly cultivated in this country (Wilson 1991, 325). Viticulture in Roman Britain has been suggested on the basis of grape skins from Gloucester and vinerods at Boxmoor villa, Hertfordshire (Renfrew 1985, 24), while recent excavations at Wollaston, in the Nene Valley, Northamptonshire, have revealed possible evidence for an extensive Romano-British vineyard (Meadow 1996). Grapes would have been used for wine for drinking and cooking, for vinegar for use in sauces, for dressings, for preserving fruits and vegetables, and as food. Wine was also imported from Spain and south-west France (Renfrew 1985, 24). Figs may also have been imported as dried fruit, although they could have been grown in sheltered areas in southern Britain (Wilson 1991, 325). The apples and pears represented in the samples may have been cultivated in orchards in Roman Britain (Renfrew 1985, 25) or collected from the wild while in season along with elderberries and blackberry/raspberries.

Other potentially used plants

A number of other plants represented in the samples may have been collected and used for food or other purposes. For instance, the seeds of the *Brassica/Sinapis* group which includes cabbage, swede and rape were recovered, although it is impossible to establish whether these are from cultivated species or their wild relatives. Indeed it is often very difficult or impossible, particularly with mixed urban assemblages, to distinguish the residues of used plants from weeds. For instance, opium poppy (*Papaver somniferum*), which was well represented in the cesspit sample, was used as a culinary herb in Roman Britain, while the seeds were used to garnish bread (Wilson 1991, 327). The seeds of this plant were also found in samples from several sites in the Roman city (eg Willcox 1977). Its use as a food plant in Roman Britain is illustrated by the recovery of large quantities of charred opium poppy seeds in a shop at Colchester, which was destroyed in the Boudican rebellion (Renfrew 1985, 23). However, opium poppy is also a high seed-producing plant and a characteristic arable weed.

A number of the wetland and grassland plants represented in the assemblages, for example the rushes (*Juncus* spp) and sedges (*Carex* spp), may have been collected and used as flooring materials, for example, or to dampen down the smells of cess/rubbish pits. Rushes and sedges are particularly well represented at other Southwark sites, although this is probably a reflection of the local environment and of the fact that both are high seed-producing plants. Charcoal and wood fragments in most of the samples suggest the collection of wood, presumably for building purposes and fuel.

The environment

The use of archaeobotanical remains from urban sites in providing information on the character of the local environment is limited by the difficulty of separating wild plants growing *in situ* from imported food plants and weeds, and the fact that many plants may grow in a range of habitats. Also some wetland or disturbed ground plants are high seed-producers and may be over-represented in the samples, while grassland plants tend to be low seed-producers and may, therefore, be under-represented. Nevertheless, the following broad habitat groups were represented by the plant remains from the Roman samples.

Wetland plants

These were represented by aquatic and bankside/marshland species, for example water-plantain (*Alisma* sp), celery-leaved crowfoot (*Ranunculus sceleratus*), spike-rush (*Eleocharis palustris/uniglumis*), rushes, sedges, lesser spearwort (*Ranunculus* cf *flammula*), and *Apium* sp. This reflects the close proximity of the site to the river and the marshy character of north Southwark, some of these plants possibly being collected and used as flooring or roofing material. A similar range of wetland plants was found on other low-lying Roman sites in north Southwark (Davis 1995).

Disturbed ground plants

Many of the species in this group grow in a range of disturbed ground habitats, such as waste places, gardens and arable fields, for example small nettle (*Urtica urens*), goosefoot/oraches (*Chenopodium/Atriplex* spp), corncockle (*Agrostemma githago*), corn marigold (*Chrysanthemum segetum*) and opium poppy. Some of the species, for example elder, are characteristic of nitrogenous waste ground, such as farmyards and rubbish tips, while some of the charred seeds, for example corncockle, stinking mayweed (*Anthemis cotula*), are common arable weeds and may represent the residues of crop-processing activities.

Grassland plants

Several species are characteristic of open grassland habitats, such as sheep's sorrel (*Rumex acetosella*), henbane (*Hyoscyamus niger*) and self-heal (*Prunella vulgaris*), although these plants also grow as waste ground weeds.

Scrub/hedgerow/woodland plants

These include species such as elder, blackberry/raspberry and hazel, although the recovery of these seeds may simply represent the residues of imported plant foods.

Saxo-Norman/medieval period (1050–1250)

The samples from the Saxo-Norman and early medieval foreshore samples (groups 12, 13, 18, 26 and 27) produced the richest waterlogged plant assemblages (see Table 17).

Cereals were represented mainly by charred grains in nine of the 11 samples, while a small number of rachis fragments and culm nodes were also recovered. The findings were similar to those of the Roman period, bread/club wheat, barley, rye and oats being present, although oat grains were recovered in much larger quantities, suggesting that it was probably cultivated rather than a weed. A large number of the cereal grains, however, could not be identified.

A similar range of cereal types is often found in samples from Saxo-Norman and early medieval sites from the City and Southwark, although seldom in large quantities. For example, bread/club wheat, rye, barley and oat grains were recovered from 9th- to 12th-century samples from several sites in the City (Jones et al 1991).

Oat grains have also been found in 12th- to 13th-century deposits at the Fleet Valley (Giorgi 1993) and in medieval deposits at the waterfront sites of Winchester Palace and Battle Bridge Lane in north Southwark (Giorgi in prep a). Bread/club wheat, rye, barley and oat are also the four main cereals found in samples from sites in southern Britain throughout the medieval period, albeit in variable quantities (Greig 1991, 321).

In the medieval period all these cereals were grown in the London region, wheat being the most important crop and wheat and rye being the main bread grains (Campbell et al 1993, 24). Wheat bread was eaten by the affluent classes while rye, which produces a dark and heavy bread, was generally eaten by the less well-off. Maslin bread, made from wheat and rye flour, and bread made from mixed grain types were consumed in some quantity by the less well-off households in London (Campbell et al 1993, 26). A mix of these cereals was often used in pottage (Wilson 1991). An important use of barley may have been for brewing, although occasionally wheat and oats were also used for malting in medieval London (Campbell et al 1993, 25). Barley and oat grains would have been used as animal feed and stabling for livestock.

Legumes

A number of charred seeds of pulses were found in five samples but could not be identified to species level owing to their poor preservation. Most of the seeds were classified as vetch/tare/vetchling (*Vicia/Lathyrus* spp) although several seeds were more rounded and similar to pea, being included in the category vetch/tare/vetchling/pea (*Vicia/Lathyrus/Pisum* spp). These seeds may represent both wild and cultivated species. Archaeobotanical finds of legumes on medieval sites in London are rare, although documentary evidence indicates that pulses (peas, beans, vetches) were extensively grown around London during this period (Campbell et al 1993, 134); they were an important food resource for both humans, being used in the cheapest bread (Black 1985, 6) and in pottage, and for animal fodder.

Fruits

These were represented mainly by waterlogged fruit stones and fruit seeds, and included grape, blackberry/raspberry, elder, hazel and *Prunus* stones, including peach (*Prunus persica*) in a 13th-century context. A small number of mineralised apple/pear seeds were found in one sample. Fruit seeds are frequent finds on medieval sites in London and Southwark, for example at the waterfront Southwark sites of Winchester Palace and Battle Bridge Lane (Giorgi in prep a).

Grapes were cultivated in Britain in the early medieval period (Wilson 1991, 331); for instance, there are records of vineyards on the Winchester Palace estate in Southwark in the 13th and early 14th century (Carlin in prep). Figs may have been grown but produce only immature seeds, so that the fig seeds at Fennings Wharf probably represent imported fruit. Indeed, both figs and grapes (raisins and currants) were imported from southern Europe as dried fruit in the medieval period (Cobb 1990). Home grown grapes were used mainly

Table 17 Plant remains from Fennings Wharf – Saxo-Norman and medieval (for key see Table 16)

Species	**Common name**	**Habitat**	**Subgroup and sample numbers**										
			F11.2 {387}	F11.3 {272}	F13.1 {389}	F13.1 {353}	F17.3 {361}	F17.3 {34}	F18.4 {288}	F26.6 {30}	F27.4 {36}	F27.4 {37}	F27.4 {287}
Charred remains													
Triticum aestivum L sl	bread/club wheat	FI								21	1		6
T cf *aestivum* sl	bread/club wheat	FI				1		1	1				10
cf *T aestivum* sl	bread/club wheat	FI									3		
Triticum spp	wheat	FI				1				10	6	1	17
Triticum spp	wheat, rachis	FI								4			1
Secale cereale L	rye	FI								2	12		20
S cereale L	rye rachis	FI								2			
cf *S cereale*	rye	FI						1			13		7
Triticum/Secale spp	wheat/rye	FI								2	2		10
Hordeum sativum L	barley	FI				2	1		1	1	6		9
H sativum L	barley, rachis	FI								1	1		3
Hordeum/Triticum spp	barley or wheat	FI				1		1					13
Avena spp	oat	AFI			8	1	1	1	1	3	5		3
Avena spp	oat, floret	AFI			3								2
cf *Avena* spp	oat	AFI			1	1				4	2		8
Cerealia	indet cereal	FI			4	2		4		119	66	1	168
Cerealia	indet cereal, culm node	FI							1	1	5		14
Cerealia	indet cereal, rachis	FI								5			3
Agrostemma githago L	corn cockle	AB											3
Spergula arvensis L	corn spurrey	ADF								2			
Medicago/Trifolium sp	medick/clover	ABDI											1
Vicia/Lathyrus spp	vetch/tare/vetchling	ACDEFI						1			4		13
Vicia/Lathyrus/Pisum sp	vetch/tare/vetchling/pea	ACDEFI								1			1
Fabacea indet	-	-				1		3		4			8
cf *Polygonum lapathifolium*	pale persicaria	ABE								1			
Polygonum spp	-	ABCDEFG									1		
Rumex acetosella agg	sheep's sorrel	AD											1
Rumex spp	docks	ABCDEFG								7	2		2
Lithospermum arvense L	corn gromwell	A											1
Euphrasia/Odontites spp	euphrasia/red bartsia	ABCDE						1					
Galium spp	bedstraw	ABCDE											1
Anthemis cotula L	stinking mayweed	ABGH								365	11	1	16
Chrysanthemum segetum L	corn marigold	AHI								1			
Centaurea cf *cyanus*	cornflower	ABGH											1
C nigra	lesser knapweed	BDG											1
Centaurea spp	knapweed/thistle	ABDGH						1		1	1		
Carex spp	sedges	CDEH								2			
Lolium/Festuca spp	rye-grass/fescue	BCD								2			
Bromus spp	bromes	ABD								2	1		
Poaceae indet	-	ABCDEFHI				1				9	2		8
Poaceae indet	grass, culm node	ABCDEFHI											12
Indeterminate	-	-						2		5	2		2
Indeterminate	wood fragments	-	+	+	++	+	++	++++		+++	+	++	+++
Mineralised remains													
Pyrus/Malus spp	pear/apple	CFI					+						
Rosaceae indet	-	-					+						
Poaceae indet	-	ABCDEFHI					+						
Indeterminate	-	-					+						
Waterlogged remains													
Ranunculus acris/repens/bulbosus	buttercups	ABCDEG			++	++++			+++		++		++
R sardous Crantz	hairy buttercup	ABE	+	++	+++	+++			++		++		++
R flammula L	lesser spearwort	EG		+	+	+					++	+	+
R sceleratus L	celery-leaved crowfoot	E	+		+					+			+
Ranunculus spp	-	ABCDEG	+	++	+++	++++			++++		+++	+	++
Papaver somniferum L	opium poppy	BGHI				+							+
Papaver spp	poppy	ABGHI		+	+						+		+++
Fumaria sp	fumitory	ABC									+		
Brassica nigra (L) Koch	black mustard	BFHI			+	++							
Brassica/Sinapis spp	-	ABFGHI		+++	++	+++			+		++		++
Raphanus raphanistrum L	wild radish/charlock	A		++	++				+		+		+
Thlaspi arvense L	field penny-cress	AB		+	+				+				
Silene spp	campions/catchflies	ABCDF		++	++	++			+		++		+++
Agrostemma githago L	corn cockle	AB		++	++	+++			+		+		+++
cf *A githago*	corn cockle	AB	+										
Stellaria media gp	chickweeds	ABCDE		+	+	++			+		++		++
Stellaria graminea type	lesser stitchwort	CD									+		+

(Table 17 cont)

Species	Common name	Habitat	Subgroup and sample numbers										
			F11.2 {387}	F11.3 {272}	F13.1 {389}	F13.1 {353}	F17.3 {361}	F17.3 {34}	F18.4 {288}	F26.6 {30}	F27.4 {36}	F27.4 {37}	F27.4 {287}
Stellaria spp	chickweed/stitchwort	ABCDEG											+++
Caryophyllaceae indet	-	-									+		
Chenopodium spp	goosefoot etc	ABCDFH		+++	++++	+++		+	+		++++		+++
Atriplex spp	oraches	ABFGH		+++	+++	++			+		++		++
Chenopodium/Atriplex spp	goosefoots/oraches	ABFGH	+									+	
Linum usitatissimum L	cultivated flax	HI		+	++	++			+				
cf *L usitatissimum*	cultivated flax	HI									+		
L catharticum L	purging flax	D			+								
Linum sp	flax	ADHI			+								
Vitis vinifera L	grape	FI										+	+
Rubus fruticosus/idaeus	blackberry/raspberry	CFGH			+	++	+++		+	+	+		++
Potentilla spp	cinquefoil/tormentil	BCDEFGH										+	
Potentilla/Fragaria spp	-	BCDEFGH			+	+					+		
Prunus persica (L) Batsch	peach	FI									+		
Prunus spp	-	CFGI							+				
Aethusa cynapium L	fool's parsley	A		+									+
Conium maculatum L	hemlock	CEG									+		
cf *Conium maculatum*	hemlock	CEG							+				
cf *Apium* spp	-	EFI				+							
Torilis spp	hedge-parsley	ACD			+	+							
Umbelliferae indet	-	-							+				
Bryonia dioica Jacq	bryony	CG		+	++	+			+				+
Euphorbia helioscopia L	sun spurge	AGI				+							
Polygonum aviculare L	knotgrass	ABG		++	+++	++			++		+++		+++
P persicaria L	persicaria	ABEH		+	++	++					+		+
P lapathifolium L	pale persicaria	ABE		++	+++	+++			++		+		++
P hydropiper/mite	water-pepper	E		+	++	++							+
Fallopia convolvulus(L) A Love	black bindweed	ABF		+	++	++							
Polygonum spp	-	ABCDEFG			+	++							
Rumex acetosella agg	sheep's sorrel	AD		+	++	++			+		++		++
Rumex spp	docks	ABCDEFG		+++	+++	++++			+		+++		+++
Urtica urens L	small nettle	AB									+		
U dioica L	stinging nettle	BCDEFGH	+						+		+		++
Cannabis sativa L	hemp	BGHI		+		+			+		+		+
cf *C sativa* L	hemp	BGHI			+								
Corylus avellana L	hazel	CF		++	+	++			++		+		++
Hyoscyamus niger L	henbane	BDG				+					+	+	
Solanum nigrum L	black nightshade	BF									+		+
Lycopus europaeus L	gipsy-wort	EH	+	++	++	++					+		
Prunella vulgaris L	self-heal	BCDG		++							+		
Labiatae indet	-	-											++
Sambucus nigra L	elder	BCFGH		+	++	++			++	++	++		++
Bidens tripartita L	tripartite bur-marigold	E		++	++	+							
Anthemis cotula L	stinking mayweed	ABGH		++	+++	+++			+		++		+++
Chrysanthemum segetum L	corn marigold	AHI		+++	+++	++++			++		++		++
Carduus/Cirsium spp	thistles	ABDEG			+++	++			+		+		++
Centaurea cyanus L	cornflower	ABGH		++	+++								
Centaurea spp	knapweed/thistle	ABDGH				+					+		
Lapsana communis L	nipplewort	BCF		++	+	+							
cf *L communis*	nipplewort	BCF										+	
Leontodon spp	hawkbit	BDF			+								+
Sonchus arvensis L	field milk-thistle	ADE		+	+	+							
Sonchus oleraceus L	milk-/sow-thistle	AB			+								
cf *S oleraceus*	milk-/sow-thistle	AB		+									
Sonchus asper (L) Hill	spiny milk-/sow-thistle	AB		++++	+++	++							+
Sonchus spp	milk-/sow-thistle	ABE		++	+								
Alisma spp	water-plantain	E									++		
Alismataceae indet	-	E								+			
Juncus spp	rush	ADEH	++	+++	+++	++	+++		+++	+++	++++	++	++++
Eleocharis palustris/uniglumis	spike-rush	E	+	+	+	+++			+	+	++	+	++++
Carex spp	sedges	CDEH		++	++	++++	+	+	++		+++	++	++
Poaceae indet	-	ABCDEFHI		++	+++	+++			++			+	++
Indeterminate	-	-	+	++	+	++			+		+	+	++
Indeterminate	leaves	-			+								+
Indeterminate	stems	-		+++	++++	++++			+++		++		++
Indeterminate	vegetative fragments	-	+	++	+++	+++			++		+	+	++
Indeterminate	wood fragments	-	+	+	++	++			++				+++
Bryophyta indet	moss	-			+++	++			++++		++		+++

for verjuice (grape juice made from unripe grapes and used for pickling and cooking) rather than for wine.

Apples were one of the most widely cultivated fruits in the medieval period (Greig 1988) and could have been used for making cider and verjuice. Peach stones are infrequent finds on medieval sites in London, although one was found in a 13th-century context from the waterfront dumps at Billingsgate. Peach trees were growing in the royal gardens at Westminster during the 13th century (Wilson 1991, 331).

According to the documentary evidence, fresh fruit was rarely eaten in the medieval period, being considered unhealthy; instead it was mixed with other foods (eg cereals) and cooked as pottage (Wilson 1991), or made into preservatives, such as jams and jellies.

The other fruits – elder, blackberry/raspberry and hazel – represented in the samples may have all been growing wild as part of woodland/hedgerow/scrub environments and gathered while in season.

Other potential plant foods

Seeds of the *Brassica/Sinapis* group, which includes cabbage, rape and so forth, were found in varying quantities in six samples; like the Roman samples, these may represent the residues of food plants or simply weeds. A number of the seeds in two samples, however, were identified as black mustard (*Brassica nigra*). Black and white mustard seeds were ground together to produce mustard flour (Mabey 1972). The leaves of some of the wild plants represented in the samples, for example the docks (*Rumex* spp) and goosefoots/oraches (*Chenopodium/Atriplex* spp), may have been used in pottage or eaten as green vegetables, although they are also common disturbed and waste ground weeds.

Medicinal plants

According to the medieval herbals many plants represented in the samples were exploited for their medicinal properties, although it is difficult to establish whether or not the plants in question were ever used or simply grew as weeds on the site. For example, opium poppy, henbane (*Hyoscyamus niger*) and black nightshade (*Solanum nigrum*) are all potential medicinal plants, but were only represented by low numbers of seeds in mixed assemblages.

Industrial plants

Two plants used in the textile industry were represented by seeds of flax (*Linum usitatissimum*) and hemp (*Cannabis sativa*) in a number of the samples. Flax seeds have previously been found in medieval deposits at the Royal Mint (Giorgi in prep b), in 12th- to 13th-century deposits at the Fleet Valley (Giorgi 1993) and in large quantities from medieval samples at Winchester Palace, while both hemp and flax seeds were found at Battle Bridge Lane, Southwark (Giorgi in prep a).

Flax and hemp were grown in gardens and orchards in the medieval period (Greig 1988, 114), while hemp was also a field crop (Greig 1988, 122). The garden cultivation of hemp at Winchester Palace, immediately west of Fennings Wharf, is mentioned in the pipe rolls of 1246–7 (Carlin 1983, 106). The fibres of both plants were put to use: flax for canvas, cordage and cloth, and hemp for rope and sacking. Flax seeds were also eaten in bread and stews, and oil extracted from both plants for cooking and lighting (Greig 1988, 122). However, as both plants were only found in relatively low quantities, they may simply represent escapes from cultivation.

The environment

A much greater seed abundance and species diversity was present in the Saxo-Norman and medieval samples, plants from a wide range of habitats being represented. However, the same problems outlined earlier affect the interpretation of these samples. The range of habitats that the plants represent are as follows.

Wetland plants

These are well represented in these samples by a range of aquatic and bankside/marshland species, for example water-plantain, celery-leaved crowfoot, spike-rush, rushes, sedges, lesser spearwort, tripartite bur-marigold (*Bidens tripartita*), water pepper (*Polygonum hydropiper*), gipsy-wort (*Lycopus europaeus*). With a similar range to the Roman samples, the presence of these species reflects the location of the site, although some of the wetland plants, for example the sedges, could have been collected and used.

Plants of disturbed ground

This includes a large number of the wild species represented in the assemblages, both as waterlogged and charred material. Many of these species can grow as weeds of both cultivated ground and waste places. Some of the species which are particularly well represented, for example stinking mayweed, corn cockle, corn marigold and cornflower, are frequently found as weeds of cereal crops in medieval assemblages. It could be argued that the charred seeds of disturbed ground plants are more likely to represent by-products of imported cereals than simply wasteground weeds. Large numbers of charred stinking mayweed seeds were found.

Some of the arable weeds in the samples can provide information on aspects of crop husbandry. For instance, corn spurrey (*Spergula arvensis*) is usually found on sandy cultivated ground and corn marigold is a weed of acid arable soils; both of these weeds may be associated with the cultivation of rye, which is also often grown on sandy soils. Stinking mayweed is associated with heavy soils and could have grown as a weed of bread wheat. Other weeds of waste places and disturbed ground that were well represented in Saxo-Norman samples were chickweeds (*Stellaria media*), campions/catchflies (*Silene* spp), goosefoots/oraches, docks, thistles (*Carduus/Cirsium* spp) and various species of *Polygonum* and *Sonchus*.

Woodland/scrub/hedgerow plants

Some of the plants in this habitat group could also appear in the previous category as weeds of disturbed ground; for instance, elder, which grows in hedges, woods, shrubberies, waste and rough ground, especially on manured soils (Stace 1991). Other plants in this category include brambles (*Rubus* spp), hazel and bryony (*Bryonia dioica*). The fruits of some of these plants, for example hazel, brambles and elder, could have been exploited, and so their presence may not be an indication of local vegetation but the residue of food consumption.

Grassland plants

These are represented by a small range of plants characteristic of grassland/meadow environments and include self-heal, hawkbit (*Leontodon* sp), 'buttercups' (*Ranunculus acris/repens/bulbosus*), sedges and rushes, and large numbers of seeds belonging to various grasses.

14.10 The bones of Peter of Colechurch

Jan Conheeney

A wooden casket on display at the Museum of London (accession no. NN 17445) reputedly contains part of the skeletal remains of Peter of Colechurch (died 1205), the builder of London Bridge (see Chapters 7 and 8). As the bones have never been examined before it was decided to undertake this work to verify or refute this claim. Consequently, the purpose of this examination was firstly to establish whether the bone fragments were human and secondly to see whether there were any characteristics consistent with their belonging to the skeleton of Peter of Colechurch. It is often difficult for the untrained observer to identify small fragments of bone as either animal or human. Contemporary documentary evidence suggests that these bones may not all be genuine (see Chapter 9.4).

The bones are housed in a heavy glass container, the lid of which has been glued on, but the glue had degenerated and the seal was broken. On the outside of the container, in the centre of the lid, is a small, inscribed brass plaque which reads: *Remains of Peter the Engineer of* OLD LONDON BRIDGE *who died in* 1205. The container fits closely inside the wooden casket. Black and gold ribbon is fixed around the inside of the joints of the walls and the joints in the base of the box are lined with strips of black velvet. This lining may have been intended to cushion the inner glass container or perhaps marks a further attempt to seal the remains inside the casket, in addition to gluing the lid of the inner box. The casket is of oak, and a second brass plaque, again in the centre of the lid, recounts that it was made from a bridge pile removed during demolition work on Colechurch's bridge.

Inside the glass container were five small fragments and several unidentifiable 'crumbs' of bone less than a few millimetres in size. Only one of the identifiable fragments came from a human skeleton, the remainder deriving from animals.

The fragment of human bone was from the proximal end of a humerus (upper arm, shoulder end). Most of the articular surface was present, attached to a sliver of the shaft from the medial aspect of the long bone. The manner in which the bone had broken made it difficult to assign it to a particular side of the body, but it was probably from a left arm. Measurements of the vertical diameter of the humeral head can be indicative of the likely sex of a skeleton, although in this case the head was incomplete, due to post-mortem damage to the edges of the articular surface. The measurement, excluding the damaged sections, was 45.4mm, a minimum value since the diameter of the undamaged head would have been greater. This is interesting as Bass (1987, 150–1) quotes research work which demonstrates that humeral heads measuring more than 47mm in this dimension are generally from adult males. The indication would, therefore, be that the humerus was probably male, but this conclusion should be treated with some caution as assigning sex on the basis of a single characteristic is not reliable: in this case the individual could have been a very robust female. Reliability is increased by utilising as many sexing characteristics as are available, not only dimensions which reflect merely the relative size differences of male and female skeletons, but also the more definitive differences in the pelvis. Clearly this was not possible in this examination.

One of the four remaining bone fragments was from a goose, as probably were two others, probably all from the same goose; the fourth was from a cow-sized animal. The bone positively identified as goose was a fragment of the proximal end of a right humerus. It was an adult goose as the epiphysis was fully fused. The two probable goose fragments were of the right shape and thickness to be midshaft fragments of the same humerus, but could not be conclusively matched. The cow-sized fragment was a 64mm long piece of one of the squarer-sectioned ribs rather than one of the more flattened ones. It probably actually was from a cow, but as there is no way to distinguish it from the rib of a small horse, closer identification is perhaps best left open. The fragment had a cut or butchery mark at one end.

In conclusion, of the five fragments of bone examined only one is human (part of an adult humerus), the others being from a goose and a cow. Therefore, it seems probable that the claim made on the casket label is fraudulent.

14.11 The building materials and geological samples from Fennings Wharf

Ian M Betts

Introduction

The building material from Fennings Wharf includes both ceramic brick and tile and a variety of stone of Roman, medieval and post-medieval date. All the ceramic building material has been recorded by fabric and form using standard Museum of London recording sheets and fabric codes. Samples of these fabrics are held at the Museum and the types found at Fennings Wharf are fully described in the site research archive (Betts 1997).

Roman ceramic building material

1st- to mid-2nd-century origins of ceramic building materials

Most of the ceramic building material found at Fennings Wharf was made with clays containing varying amounts of quartz, which when fired turn various shades of red. The tiles come from a number of tile kilns situated within 30km of London (Betts 1987, 27–8), most of them apparently situated along Watling Street between London and St Albans. It is not possible to state in most cases at which particular kiln site the tiles were made, as most tileries used very similar clays (fabric group 2815). One exception is the products from the tile kiln at Radlett in Hertfordshire, which are made from a clay containing numerous small black iron oxide inclusions (fabric type 3023). A Roman roofing tile made from this clay was found in a post-Roman deposit (F12.5) at Fennings Wharf.

Another early source of tile was the tilery situated in the Eccles area of north-west Kent. This produced distinctive yellow and white tiles (fabric 2454), a small number of which have been found at Fennings Wharf.

Mid-2nd-century and later ceramic building materials

By the mid-2nd century many, although perhaps not all, of the kilns situated close to London seem to have fallen out of use (Betts and Foot 1994, 33–4). Instead, London was obtaining tiles from an increasingly diverse range of sources, only a few of which can be identified. At Fennings Wharf there were a small number of roofing tiles in a distinctive calcareous fabric (type 2453) which date from the mid-2nd to the 3rd century (F10.9; F12.5; F17.1; F30.15). The location of the tilery supplying these calcareous tiles has still to be established, although it is very unlikely to have been in the London area (Betts and Foot 1994, 32–3).

Types of tile and other Roman building material

BRICK

Fabric: group 2815; types 2454, 3018, 3070

Roman bricks were made to more or less standard sizes based on the Roman foot (*pes*) which equals 296mm. The Fennings Wharf examples are all too fragmentary to show which types of bricks are present. One brick (F10.9) has the remains of an animal print on its upper surface.

ROOFING TILE

Fabric: group 2815; types 2453, 2454, 3023, 3238

The roofing tile present is either flanged tegula or curved imbrex tiles. A number of roofing tiles (fabric group 2815) were found associated with the remains of a sequence of early Roman buildings (F7).

The most important roofing tile, which is an unstratified find, is an imbrex stamped PPBRILON (Betts 1995, 226). This was made at an official local government-run tilery which is believed to have been situated at Brockley Hill in Hertfordshire. The lettering is interpreted as: p(*rocuratores*) p(*rovinciae*) Bri(*tanniae*) Lon(*dini*): 'The procurators of the province of Britain at London'. The stamp die type present can be identified as die 2A (Frere and Tomlin 1993, 2485.2; formerly MoL type 6). Recent work has confirmed that procuratorial stamped tiles were used in the construction of London's major public buildings, which were constructed during the period AD 70/80 to 120/125 (Betts 1995, 222–4).

CAVITY WALLING

Fabric: group 2815

Included under this heading are box-flue tiles and a probable fragment of wall tile. The purpose of these tiles was to create a cavity through which heated air from the hypocaust below floor level could circulate up through the building. The possible source of these tiles might have been the Roman masonry hypocausted building, which was found during 1975 in Tooley Street – south of St Olaf House (Fig 2) (Graham 1988, 49).

Two box-flues came from an undated pit (F5.3) and one of them has scored keying. In London this is normally a feature of box-flue tiles of 1st- or early 2nd-century date. The other box-flues, one of which is combed, all came from post-Roman contexts.

The probable wall tile was also found in a post-Roman context. This tile has lattice scored keying on its sanded base and would originally have had a thin rectangular shape, with small notches situated in the long edges near each corner. Wall tiles were set vertically, parallel to the wall surface, to create a cavity for the movement of hot air. The tile was separated from the wall by circular-shaped clay objects known as spacer bobbins. The tiles were held in place by iron cramps, or nails, which fitted through the gaps left by the notches and passed through the centre of the bobbins into the solid wall behind. The keyed surface of the tile was fixed facing into the room to allow the attachment of wall plaster.

The presence of examples found elsewhere in London with procuratorial stamps suggests that wall tiles may have been made exclusively for use in London's public buildings during the late 1st- to early 2nd-century (Betts 1995, 214).

CERAMIC TESSERAE

Fabric: group 2815; types 3060, pottery
There are 30 individual ceramic tesserae, of which 24 came from late Roman features (F10.1; F10.3). This figure includes three tesserae made from pottery sherds. Their presence suggests that a building with a tessellated floor existed locally: possibly these tesserae were derived from the nearby hypocausted building in Tooley Street (Graham 1988, 49). However, this is only conjecture as the nature of the flooring in this building is unknown (ibid, 49).

DAUB

The only daub from a sealed Roman context came from a rubbish pit (F10.9) associated with pottery of 4th-century date.

Roman stone building material

Only three stones of definite Roman date were recovered. The first was a fragment of brown ferruginous sandstone found in the same 4th-century rubbish pit (F10.9) as the daub. This stone was used wherever cut and moulded stone was necessary, such as the plinth of the city wall, as well as for decorative effect; its brown colour contrasting with the red tile and grey Kentish rag used to construct most of Roman London's masonry buildings.

The two others are blocks of fine-grained laminated sandstone found in the backfill of a 3rd-century well (F10.4). The location of the quarry producing these stones has not been identified, although, as with the ferruginous sandstone, a source in south-east England seems most probable.

Medieval ceramic building material

Types of floor tile

DECORATED PENN TILE

Fabric types: 1810, 3076
Penn in Buckinghamshire was the location of one of the most successful commercial medieval tileries known in Britain, its products were widely distributed over the area of the Chilterns and the Middle Thames Valley. Large quantities of Penn floor tiles were used in London from the 1350s until the 1380s. The four examples from Fennings Wharf are believed to date to this period.

The Penn tiles all came from a backfill of a robbed out wall of a post-medieval cellar (F30.4), so there is no indication as to the building in which they were originally used. These tiles are decorated with three separate designs, all of which have been published by Eames (1980), these being designs E2037, E2234 and two examples of E2839 (the letter 'E' denotes an Eames design number). The tiles, one of which is complete, measure 111–118mm square by 18–24mm in thickness.

PLAIN GLAZED FLEMISH TILE

Fabric types: calcium carbonate varieties 2497, 2504; slightly silty variety 3082
Flemish tiles can usually be identified by the presence of nail holes in the top surface coupled with their distinctive fabric types. Plain glazed Flemish floor tiles are extremely common in London parish churches and religious houses (Betts 1994, 134).

The earliest examples from Fennings Wharf were found in the floor make-up of a cellar (F29.10), associated with pottery dated 1480–1500. They must have been of some antiquity before they were discarded, as both tiles are extremely worn, with no glaze remaining on the upper surface. The tiles measure around 118mm square and have four round (1.5mm diameter) nail holes, one in each corner. The floor tiles from Group F29.10 are in a calcium carbonate fabric type, as are the plain glazed Flemish floor tiles from F30.2 and F30.4. These tiles have either a yellow, green or brown glaze. The tiles, some of which are reused and incomplete, all measure between 30–32mm in thickness.

The single tile in a slightly silty fabric was found within foreshore deposits (F28.1). This is probably Flemish, although many tiles found elsewhere in London in the same fabric (type 3082) seem to lack nail holes. The tile, which cannot be dated with any accuracy, has a thickness of 24mm and a mottled yellow and brown glaze.

PLAIN UNGLAZED FLEMISH TILE

Fabric: 2318
By at least the mid-17th century, Flemish tiles imported into London were no longer glazed. The two examples from Fennings Wharf have the same distinctive silty fabric types discussed above. One example found on the foreshore (F27.4) is quite thick (40mm), indicating it is part of a tile larger in size than the earlier glazed examples. However, the other came from the same late 17th-century robber trench (F30.4) as some of the glazed floor tiles (see above).

Types of roofing tile

FLANGED, CURVED AND SHOULDERED PEG TILE

Fabric type: 2273
The earliest roofing tile found at Fennings Wharf comprised flanged tile, curved tile and shouldered peg tile, most of which was found dumped in late 12th- to early 13th-century foreshore deposits (F13.1, F16.1, F16.2, F25.3, F27.4). Flanged tiles and curved tiles were used together in the same manner as Roman tegulae and imbrices. The evidence from the Cheapside area indicates that all these early tile types first appeared in London sometime during the period 1135 to 1150 (Betts 1990, 221).

PEG ROOFING TILE

Fabric types: mainly 2271, 2276; one 3204

A small quantity of peg roofing tile was found in a number of medieval and post-medieval contexts (F25, F27–F31). Such tiles represent the standard type of ceramic roof covering used in London from the late 12th century until the widespread introduction of pantiles after the Great Fire of 1666. There seems little doubt that the vast majority of the peg tiles found at Fennings Wharf were made in tileworks situated close to the capital, most of which seem to have been situated to the east of London (Cherry 1991, 194; McDonnell 1978, 114). However, one peg tile (F30.11) is in an unusual slightly silty fabric (type 3204), perhaps derived from a more distant kiln source than the other tiles.

The earlier peg tiles found at Fennings Wharf and elsewhere in London are frequently, although by no means always, glazed. This can be in the form of a splash lead glaze or, more rarely, a more uniform covering. The glaze was restricted to the bottom third of the upper surface in the area which would have been exposed to weathering. See Table 18 for peg tile sizes.

Table 18 Peg tile sizes

Group	Fabric	Length (mm)	Breadth (mm)	Thickness (mm)
F30.11	2276	266	150–157	13
F30.12	2276	-	151–162	12

PANTILE

Fabric: 3202

A fragment of pantile roofing was recovered from the fill of a brick-lined well (F30.12). This is probably contemporary in date with the associated pottery dated 1690–1750. The tile may have been made locally or have been imported from Holland.

RIDGE TILE

Fabric types: 2271, 2276

Curved ridge tiles set along the ridge of the roof were used in conjunction with both peg tile and pantile. A few small fragments were found in medieval (F13.1, F28.1) and post-medieval contexts (F30.4). Their fabric types are the same as that of the peg tiles, which would indicate that both were made together at the same tileries. As with peg tile, only the earlier examples have glaze present.

Brick

YELLOW BRICK

Fabric type: 3031

Of the two fragments of yellow brick, one came from a 15th-century foreshore deposit (F28.1), whilst the other came from a brick wall foundation (F30.11). The latter measuring 226 × 105 × 45–50mm in size.

Yellow bricks of this type first appear in London during the second half of the 14th century and continue in use during the 15th century. It is not certain whether these bricks were made in the London area or were brought in from elsewhere. Similar bricks in Essex are interpreted as Flemish imports (Ryan 1996, 32).

RED BRICK

Fabric types: 3032, 3033, 3034

It is only in the mid- to late 15th century that brick began to be widely used in London. These were locally made red bricks (Schofield 1984, 129). By the 17th century there were a number of centres involved in brick manufacture, such as Islington, Spitalfields, Moorfields and the parish of St Giles in the Fields.

Red brick was used extensively in various post-medieval structures on the bridge approach (F30) (see Table 19).

It is not possible to give precise dates to most of the red brick present. Those in fabric types 3032 and 3034 are almost certainly post-Great Fire, but those in fabric 3033 could be of any date between the late 15th and 17th centuries. All that can be said is that the thinner bricks in fabric 3033 are probably earlier in date than the thicker examples.

ABNORMALLY LARGE BRICKS

Fabric types: 3044, 3242

Two bricks stand out from the rest by virtue of their unusually large size. The complete example is much larger than any other brick found so far during excavations in London or Southwark.

Table 19 Red brick in post-medieval structures on the bridge approach (F30)

Group	Structure	Fabric	Length (mm)	Breadth (mm)	Thickness (mm)
F30.1	cellar walls/floor	3032	227	98	65
		3033	217–227	97–114	52–64
		3034	220	95	60
F30.2	cesspit	3033	220–228	104–107	60
F30.3	cellar floor	3033	222–227	100–104	50–53
F30.7	wall foundation	3033	215–220	110–112	50
F30.8	silt trap	3033	215–227	105–107	58–56
F30.11	wall foundation	3033	220–227	105–117	50–58
F30.12	well	3033	227–230	107–110	55–59
F30.13	wall foundation	3033	222–227	102–107	55
F30.14	soakaway/cesspit	3033	217–222	100–108	52–58

The complete example, which measures 319 × 147–8 × 75mm is made from a yellowish clay with red bands (fabric type 3044). The other brick, which has a thickness of 70mm, is made from a distinctly sandy reddish-orange clay (fabric 3242). There is no indication what these large bricks were used for: one came from a medieval masonry foundation (F29.4) and the other one (fabric type 3044) came from the backfill of a post-medieval brick-lined well (F30.12).

Tin-glazed tile

FLOOR TILE

Fabric types: 3067, 3067 (near 3086)
Two decorated tin-glaze floor tiles were recovered from the same robber trench backfill (F30. 4) as the medieval Penn floor tiles discussed earlier.

The earlier tile has a geometric and foliate design in blue, green, yellow and brown on a white background. This tile may be a Dutch import; Pluis (1997, 212: A.01.03.54) illustrates a Dutch tile with the same design dated 1560–1600. Alternatively, it may have been made in London at Aldgate, where production of tin-glaze pottery and tile by Flemish craftsmen began in 1571.

The second, slightly later tin-glaze floor tile has a so-called medallion design, which comprises a landscape with what appears to be the back half of a dog set within a multi-circular border. The tile, which shows no signs of wear, is painted in blue, green and brown on a white background

Tiles of medallion design with either birds and animals in their centre were produced at the tin-glaze tile factory at Pickleherring Quay in Southwark from *c* 1618 to the mid-17th century (Hume 1977, 55, fig I, nos 1–3; Britton 1987, 35–6; 172–3, figs 186–191). Some of these are painted in the same colour scheme as the Fennings Wharf example, which would suggest that the Pickleherring factory is the source of the tile.

WALL TILE

Fabric: 3078
A small fragment of blue on white tin-glaze tile (thickness 8mm) was found associated with the floor tiles discussed above. The tile shows part of a landscape within a Dutch style octagonal border. Van Dam (1991, 113, figs 135–6) illustrates a number of similar Dutch examples dated 1750–1800/10.

Non-ceramic building material

Colechurch bridge and riverside wall

The majority of medieval and later stone comprises samples collected from the ashlar facing and rubble core of Colechurch bridge (F26) and from the associated riverwall. The documentary evidence for the purchase of stone in the Bridge House accounts is discussed in Chapter 9.

KENTISH RAG

This is a hard grey sandy limestone from the Cretaceous Lower Greensand. Large quantities of Kentish rag were brought into medieval London from quarries around Maidstone and from Aylesford to the north and Boughton to the south (Salzman 1952, 128). Kentish rag was widely used in the mid-15th-century additions to the bridge abutment (see below).

HASSOCK

Hassock, a medium grained sandstone found interbedded with Kentish rag, must also have been shipped into London, as this stone in mentioned in the Bridge House accounts for 1463–64. This stone is nothing like as hard as Kentish rag, so was probably only used as rubble infill.

WHITE QUARTZOSE SANDSTONE

According to Dr Worssam (pers comm) this white and pale grey coloured sandstone 'matches some samples of sarsen stone' and would be from the London Basin, probably the 'North Downs, Bagshot area, or Chilterns'.

WEALDEN 'MARBLE'

Wealden 'marble' is the term used to describe a limestone characterised by the presence of large freshwater snail shells of the genus *Viviparus* (formerly known as *Paludina*). The 'marble' is given different names in different parts of the Low Weald where it is found, such as Bethersden marble, Petworth marble or Sussex marble (Tatton-Brown 1991, 361–2). All are similar in appearance, so it is not possible to give an exact source for the London Bridge examples.

PURBECK MARBLE

This is a grey limestone from the Isle of Purbeck, Dorset, readily identifiable because it is composed of closely-packed small freshwater snail shells.

PURBECK STONE

A whitish or grey coloured crystalline shelly limestone with a 'featherbed' texture, also from the Isle of Purbeck. Some of the London Bridge examples are unusual in having a green or mottled green and red tinge.

REIGATE STONE

This is a fine-grained grey or green sandstone from the Cretaceous Upper Greensand. According to documentary evidence, from the 13th century onwards, this type of stone was being quarried in the Reigate and Merstham area of Surrey (Salzman 1952, 129).

PORTLAND STONE

This white oolitic limestone, derived both from the Isle of Portland and from the Isle of Purbeck in Dorset, was widely used in London after the Great Fire of 1666. In 1350 Portland stone was included in the inventory of the Bridge House stores (Welch 1894, 259).

Construction of the bridge abutment (F26)

The construction of the Colechurch bridge abutment is dated to *c* 1188–1189.

1) East side – basal coarse of ashlar (F26.3) and west side foundations (F26.4)
   These comprised a white quartzose sandstone.
2) Standing masonry (F26.5)
   The ashlar blocks on the west wall comprise both Purbeck marble and Purbeck stone (some of which is an unusual green colour) together with Wealden 'marble', Kentish rag and white quartzose sandstone. The same white colour sandstone was also found forming the east side.

Later additions to the bridge abutment (F28)

The majority of stones used in the later, mid-15th-century additions to the bridge abutment and the riverside wall comprise Kentish rag limestone. The south-west addition to the bridge abutment contains the Purbeck stone with the same unusual green tinge as that found on certain stones from the initial construction of the bridge. This would suggest that some of the stones used in the additions to the bridge abutment were reused during alterations to the original bridge.

1) Additions to north-east side (F28.2)
   These additions comprise Kentish rag, both Purbeck marble and Purbeck stone and a white oolitic limestone, almost certainly Portland stone.
2) Additions to south-west side (F28.3)
   These comprised Kentish rag and Purbeck stone.
3) Riverside wall (F28.4); addition or enlargement of wall (F28.5) and probable riverwall build against east side of abutment (F28.6)
   These were all constructed using only Kentish rag.
4) North side of abutment (F28.7 and F28.8)
   Kentish rag was used as both rubble fill (F28.7) and for the standing masonry (F28.8).

Post-medieval additions to abutment and riverwall (F31)

The presence of Wealden 'marble', used in the 12th-century bridge abutment, suggests that this stone was reused in the post-medieval period, when the bridge was rebuilt. This reuse would account for the presence of fragments of Reigate stone moulding in the north-west abutment. There would have been no need from fresh supplies of either stone when superior Portland stone was available.

1) Additions to eastern riverside wall (F31.1)
   These comprised Portland stone.
2) Additions to north-west corner of abutment (F31.2)
   These comprised Wealden 'marble'.
3) Further addition to north-west abutment (F31.3)
   These again comprised Wealden 'marble' together with fragments of Reigate stone moulding.

14.12 Saxo-Norman London bridge and Southwark – the saga evidence reconsidered

Jan Ragnar Hagland

London bridge in the Olaf sagas

Snorri Sturluson (1178/9–1241) wrote two versions of the Olaf saga. The earlier one – the great saga of St Olaf (*Olaf Helga*) – was written *c* 1220 and is a biography of the king (Johnsen and Helgason 1941). The later Olaf saga was written before *c* 1235, and formed part of the *Heimskringla,* a history of Norway which contains biographies of all the Norwegian kings up to Magnus Erlingson (*c* 1162–84) (Adalbjarnarson 1945). There is also a third text, the legendary saga of St Olaf (*Olafs saga hins helga*) (Johnsen 1922): this belongs to the religious legendary tradition about St Olaf and is believed to be based on the same literary sources as the Snorri sagas (Seip 1929, 11).

The purpose of this contribution is to reconsider the Old Norse wording in the Olaf sagas which describes the structure or design of the 11th-century London bridge. The standard English translation of chapter 12 of *Heimskringla* states: 'King Olaf and the Northmen's fleet with him, rowed quite up under the bridge, laid their cables around the piles which supported it, and then rowed off with all the ships as hard as they could down the stream. The piles were thus shaken in the bottom and were loosened under the bridge' (Laing 1964, 124). Study of the two original texts (which are identical) shows that the key portion of the text should read: 'and under the bridge were staves [stafir, ie vertical timber poles or pillars] and those [they] stood down on the shallow [parts] in the river'. In the St Olaf sagas this passage reads: 'Vndir brygionum voro stafir. oc stoðo þeir niðr grvnn í anni' (Johnsen and Helgason 1941, 43).[46]

The wording of this passage in the legendary saga of St Olaf is slightly different: 'And the bridges [plural] were [constructed] thus that they stood out in the River Thames, and [there] were poles [stolpar] below [undir] down in the river which held up the bridge [singular]'. 'En sva varo hattaðar bryggiurnar alat þær ftoðo ut a ana tæms oc varo stolpar undir ì ana niðr þæir er upp hellòo bryggiunar' (Johnsen 1922, 10).

In the standard English translation of chapter 11 of *Heimskringla* it reads: 'on the other side of the river (from London) is a great trading place, which is called Southwark (*Súðvirki*). There the Danes had raised a great work, dug large ditches, within had built a bulwark of stone, timber, and turf, where they had stationed a strong army ...' (Laing 1964, 123). This passage corresponds to chapter 23 in the saga of St Olaf (*Olaf Helga*) (Johnsen and Helgason 1941, 42). The passage about Southwark has no parallel in the legendary saga of St Olaf, it only occurs in the Snorri Sturluson's sagas of Olaf. It should be noted that the defences of Southwark are also mentioned in a scaldic stanza, by Olaf's (Icelandic) court scald Sigvatr Þordarson. The stanza belongs to the so-called *Vikingarvisur* (stanza 6), dated by Finnur Jónsson (1912) to 1014–15. Snorri himself in general explicitly puts much emphasis on scaldic poems as historical sources. Thus it is interesting that he quotes Sigvatr's stanza as evidence for Olaf's London experience. This particular stanza is also quoted in the legendary saga of Olaf and the *Fagrskinna* version of Olaf's saga, a Norwegian text included in a collection of the kings' sagas – independent of Snorri Sturluson's work. None of these other texts expands on the information contained in the stanza.

Passage from chapter 23 of the great saga of St Olaf

'avðrom megin arinnar er mikit cauptun er heitir Svðvirci. þar haufðv Danir mikinn umbunat grafit dici stor. oc setto fyr innan vegg[47] með viðum oc grioti oc torfi. oc hofðv þar lið mikit. Aþalraðr konungr let veita þar at socn micla. en Danir vavrðv. oc fecc Aðalraðr konungr þar ecki at gert. Bryggior voro þar ifir ana milli borgarinnar oc Svðvirkis. sva breiðar at aca matti vaugnom a vixl. a bryggionom voro vigi gor beði castalar oc borðþac forstreymis sva at toc upp fire miðian mann' (Johnsen and Helgason 1941, 42–3).

Translation: On the other side of the river is a great trading place which is called 'Suðvirki'. There the Danes had, with great care [mikinn umbunat], dug large ditches [and made a wall and a road on the inside][48] of wood, stone and turf, and had a great army [lið mikit] there. King Ethelred attacked the place [þar] heavily, but the Danes defended [it] and King Ethelred did not accomplish anything. There were bridges [plural, but this is not unusual for 'a bridge'] over the river between the city [borgarinnar] and 'Suðvirki', so wide that it was possible for [two] carriages to bypass [each other]. On the bridge[s] downstream [forstreymis] were built fortifications [vigi], strongholds [castalar; Latin *castellum*] as well as parapets [borðþac] up to a level over a man's waist [sva at toc upp fire miðian mann, literally 'so that it reached over a man's waist'].

Stanza 6 of the Vikingarvisur by Sigvat Þorðarson, the scald

Ret er at socn hin setta
snarr þælngill bauð Ænglum
at þar er Olafr sotte
uggsó Lunduna bryggiur
sværð bitu volsk en varðu
vikingar þa dyki
atte sumt i sletto
Suðrvirki lið buðir

Translation: it is correct that the sixth battle took place where Olafr attacked London bridge[s]. The brave warrior offered the Angles a fight, the foreign or French swords bit, but the Vikings defended the ditches; part of the army had booths [buðir] on the plain of 'Suðrvirki'.

According to Finnur Jónsson in his edition of *Fagrskinna* (1902–3, 141), this text agrees very well with the versions of the same stanza given in the Olaf sagas.

14.13 The *ex situ* London Bridge material from Stanwell House, Richmond, Surrey

Raymond Gill

At Marshgate, East Sheen, during the 18th century a property later known as Kenyon House was constructed within a ten-acre plot on the south side of the road from Putney to Richmond (Fig 130). The property extended from the entrance to East Sheen Common to Queen's Road. In 1830 it was purchased by the Hon Heneage Legge, a younger son of the third earl of Dartmouth. A senior gentleman usher to the Queen Victoria and one of the commissioners of the Board of Customs, he had previously lived at Putney. A few years later he decided to demolish the old house and construct a new one on the elevated ground to the south. This rebuilding was presumably finished by 1837, as the rating assessment was increased that year. The new property was known as Stanwell House. The ancient barony of Stanwell had passed to another branch of the Legge family, but had become extinct in 1820.

The Tithe Apportionment map of 1839 shows the layout of the new house. Between the east front of the house and the paddock a 260ft long terrace wall was constructed. It was topped by stone balustrading added to London Bridge during the modifications of 1757–62 (see Chapter 13.2; Fig 119; Fig 135). At each end of the terrace one of the discarded alcoves from the bridge was set up to serve as a summer house or shelter. Some stone blocks from the bridge were reused in the pillars flanking the entrance gates.

Legge died in 1844, aged 57, but his widow continued to live there. Eventually, in 1868, Legge's trustees sold the estate to Sir Henry Watson Parker, who lived there until his death in 1881. When Sir Henry's widow died, the property passed to his niece Emma Jane Vaughan Arbuckle for her life, provided that she made it her permanent home. After her death in 1934 the property passed to her son Major Charles Lionel Vaughan Arbuckle, who died in 1935. The property was then put up for sale by auction on 25 June 1935, when it was described as 'ripe for immediate development' and was sold for £33,400. The estate was resold by auction for £36,500 on 14 July 1936. In 1937 the house was demolished and the building of the new flats and garages largely completed by 1938. During this redevelopment the entrance gates, the balustrading on the terrace and the northern alcove were destroyed, but the alcove at the southern end of the terrace was left undisturbed and stands today among the trees along the southern boundary of the estate. The only known record of the balustrading is a series of photographs of the house and grounds taken in 1935 (Fig 135). The site is now occupied by Courtlands Flats, Sheen Road, in the London Borough of Richmond (Fig 130).

NOTES

1 The present is internationally defined as 1950.

2 HAR-3931, 263080 BP.

3 HAR-3925, 301070 BP; HAR-3926, 277080 BP; HAR-3927, 257080 BP.

4 Beta-68571, 298060 BP, 1390–1000 cal BC; Beta-70405, 297060 BP, 1380–1000 cal BC.

5 Beta-70881, 296080 BP.

6 The radiocarbon dates from Oxford were all obtained from charcoal *Quercus* sp sapwood.
Subgroup F2.3: FW84 {546a}, OxA-8765, 334545 BP, 1750–1520 cal BC; FW84 {546b}, OxA-8766, 342540 BP, 1880–1640 cal BC; FW84 {547a}, OxA-8767, 342040 BP, 1880–1630 cal BC; FW84 {547b}, OxA-8768, 349040 BP, 1930–1690 cal BC; FW84 {573a}, OxA-8769, 343045 BP, 1890–1630 cal BC; FW84 {573b}, OxA-8770, 354540 BP, 2030–1750 cal BC.
Subgroup F2.4: FW84 {435a}, OxA-8763, 336040 BP, 1750–1520 cal BC; FW84 {435b}, OxA-8764, 340045 BP, 1880–1610 cal BC.
All these dates were calibrated using the data set of Pearson and Stuiver (1986) and the maximum intercept method of Stuiver and Reimer (1986) so that they conform with the rest of the text and other dates cited in this publication (see Chapter 1.2, Radiocarbon determinations).

7 It should be noted that the radiocarbon dating programme for the ring-ditch is not complete yet and that the high-precision dates from Queen's University, Belfast, are not yet available. The dating of the ring-ditch and the analysis of its cremations and finds will be fully discussed in Sidell et al (in prep).

8 It has distinct points of similarity with another early river embankment and ditch excavated on the north bank at Dowgate Hill House (DGH86) in 1986 (Brigham 1990b, 129–31). This was constructed with the local clay in the later 1st or early 2nd century; it was revetted, and the ditch on the landward side was replaced several times, eventually silting on each occasion until the whole installation was replaced in the late 2nd or early 3rd century. This approach was used in open areas where there was no existing riverbank. Where there was a riverbank, as at Regis House (Brigham et al 1996b) and Billingsgate Buildings (Jones 1980), revetments were constructed against the bank. Examples of both types were found at Pudding Lane and Peninsular House (Milne 1985).

9 Several further examples of this type of structure have been found at Newgate Street (GPO75), in the City of London, excavated between 1975 and 1977. These were comparatively small, being c 6.5m in diameter, as well as slightly earlier, almost certainly pre-AD 60 (Perring and Roskams 1991, 3–6, figs 4 and 7). There were no similar industrial finds, although two (Buildings D and E) were identified as possible ancillary structures, the earliest (Building A) apparently residential. Excavations at Winchester Palace (WP83) in Southwark, 1983–4, revealed a late 1st-century structure (B6), probably of circular plan with an internal diameter of 7.0m, although this may have been a tower (Yule in prep). A circular building excavated at Courage's Brewery (B1 in area F) had a diameter of c 10.0m (Cowan in prep).

10 A river vessel found at New Guy's House in 1958 (Marsden 1965, 118–31; 1994, 97–104) had been abandoned further up the same channel, indicating that boats were capable of penetrating some distance inland from the Thames.

11 The Vauxhall causeway bridge has been radiocarbon dated to 1870–1530 cal BC (Beta-122970, 3380± 40 BP) and 1630–1310 cal BC (Beta-122969, 3180± 70 BP) (Haughey 1999, 18; calibration by Peter Marshall).

12 Excavations at the Eton rowing lake, Buckinghamshire, during 1995 revealed within an infilled Thames meander two phases of causeway bridge or deep-water jetty (as the piles did not apparently extend the full width of the channel (Allen and Welsh 1996, 125). The earlier phase of the structure has been dated to 1420–1210 cal BC (BM-3020, 3050± 40 BP) and 1530–1310 cal BC (BM-3022, 3150± 40 BP), and the later phase to 770–400 cal BC (BM-3023, 2450± 50 BP) and 770–390 cal BC (BM-3021, 2420± 50 BP).

13 That is in the form of relatively small irregular rectangular facing blocks set in lime mortar, with a mortared rubble core. The technique is often termed '*petit appareil*'.

14 British Museum, Department of Medieval and Later Antiquities catalogue nos: axehead 1856.01701.1426; spearhead 1856.01701.1439; 9th-century spearhead 1838.01052.

15 Æthelred II also known as 'the unready' or *unraed* was deposed in 1013 by Swend, a Scandinavian. After Swend's death in February 1014, his followers chose his son Cnut as king. However, the *witan* chose Æthelred, who in 1014 with the help of Olaf (see Chapter 14.12) recaptured London and regained the throne; he then drove Cnut out of England. In the spring of 1016 Cnut returned to England and was preparing to besiege London when Æthelred died; after his death Cnut was accepted as king.

16 Earl Godwin (died 1053) and his family had been in dispute with King Edward and he was proclaimed an outlaw in 1051; his unopposed return to London in 1052 heralded his reinstatement (McLynn 1998, 83–4). After Edward's death on 5 January 1066, one of Godwin's sons, Harold Godwinson, became king of England.

17 For an explanation of how these numbers are used to define and describe the archaeological sequence on a site-wide basis see Chapter 1.

18 The radiocarbon date originally published by Greensmith and Tucker (1969) and republished by Wilkinson and Murphy (1986) was AD 560–710. This date was incorrectly calibrated using the atmospheric curve; it was recalculated during 1999 by Alex Bayliss and Peter Marshall (pers comm), using a marine R value of 540 BP using the data set of Stuiver et al (1986, fig 10B).

19 *Prunus* is the genus used to describe a number of shrubs and trees bearing stoned fruits including: blackthorn or sloe, cherries, damsons and plums (Bentham and Hooker 1947, 134–5). More precise identification of these timber samples is not possible.

20 *Rederesgate* (1108–48) means 'cattlegate', the first element in the name of this watergate being derived from the Old English *hrýðer* (a horned beast). This place name could mean that this watergate was used to land cattle intended for the meat market at East Cheap (Ekwall 1954, 154–5).

21 The possibility that some of the excavated timbers (F15 and F37) were part of an earlier timber bridge was considered by Huggins (1970, 129), but this idea was rejected in favour of other interpretations, despite the fact that there is documentary evidence for the existence of a timber bridge.

22 The most accurate dimensions of the medieval bridge are those cited by Dance (1799). Dance cites two different figures for the length of the bridge: west side 915ft 10in and east side 905ft exactly. See Home (1931, 340–1) for a summary of these figures and those from earlier surveys.

23 It is possible that the 12th-century bridge at Avignon possessed stone piers and a timber superstructure, and that the surviving masonry arches date from 1345 to 50.

24 The evidence for a cofferdam consisted of one large oak beam F15 described as a soleplate (see Glossary), and above it was a vertical break in the sequence of sediments, interpreted as the remains of decayed planking or sheathing. A length of planking F33 was described as sheathing for a cofferdam; it was supported by two vertical stakes (Huggins 1970, 130–3). It is hard to see how a watertight barrier could been constructed out of a single skin of planking, as the soleplate was laid on 'grey silt' and water would have presumably percolated through it flooding the cofferdam. The tree-ring dates quoted in this report (Appleby 1970) have never been checked against the modern chronology for south-east England (Ian Tyers, pers comm 2000).

25 The width of the 18th-century new central arch was 21.46m; see Chapter 13.2.

26 Exeter Bridge, constructed c 1196–c 1214, possessed St Edmund's church, a chantry chapel, and was partly lined with houses and shops by the 14th century (Brown 1991, 2–3).

27 See the painting 'Ouse Bridge, York' c 1763, by William Marlow, York City Art Gallery (accession no. 176).

28 High Bridge is a single arch stone bridge, spanning the River Witham, Lincoln. The row of timber-framed shops on the western side of the bridge was largely dismantled and re-erected during 1900–1 (Pevsner and Harris 1989, 523), but they probably date from the late 15th or 16th centuries.

29 This woodcut map of London is widely known as the 'Agas' map. However, the map has no connection with the Elizabethan surveyor Ralph Agas (1545–1621). It was produced some time during 1561–70 and may have been published in 1562–3. It appears that much of the detail on this map was taken from a slightly earlier copper plate map dating from c 1553 to 1559 (Fisher 1981, Intro).

30 Many of the 1824–41 finds are part of the Charles Roach Smith collection in the British Museum Department of Medieval and Later Antiquities. Other finds from the area of the bridge are held by the Museum of London's Early Department. Most of the finds from the 1967 dredging were not accessioned into the main MoL collection, but were catalogued under the Guildhall Museum ER (Excavation Register) system as ER1279 A.

31 Layerthorpe postern gate, York, was situated at one end of a medieval bridge spanning the River Foss, not on the actual bridge, therefore, it is not counted as a fortified bridge. The postern was demolished in 1829–30 as part of the preparations for rebuilding the bridge (RCHM 1972, 137).

32 The 'Bastard Fauconberg', Thomas Fauconberg or Nevill, was the Earl of Warwick's illegitimate son (Seward 1997, 61).

33 The central portion of Kingston Bridge was constructed of timber in the late 17th century according to John Aubrey (Walker 1979, 29).

34 A note in the *Gentleman's Magazine* (Anon 1833a, 68), which discussed the claim of Knight to possess some of the bones of Colechurch, stated that Knight had been informed by the author that Colechurch's bones had been discovered during the demolition of the bridge 'but were thrown into the river by the workmen'. By this date Knight was dead and his collection of books and archaeological finds was being sold, so the matter was never resolved.

35 All Saint's Church, Steep, Hants: Anon n d.

36 Since 1198 three lions or leopards *passant gardant* have served as the English royal arms.

37 Gilwell Park, Bury Road, Chingford, is a late 18th-century rebuild of an earlier country house. The balustrade is situated in the grounds on the west side of the house. Comparison of the balusters with the two half balusters on each end of the Guy's Hospital alcove revealed that the Gilwell Park examples are 65mm shorter than the Guy's examples and the Gilwell Park examples also have a slightly different profile. It should be borne in mind that this design of baluster was very popular during the late 18th and early 19th century. Since 1919 Gilwell Park has been the National Scout Training Centre.

38 There is a very badly eroded inscription on the quayside warehouse which reads: 'THIS BUILDING and QUAY; was ERECTED by the GOVERNORS of GUY'S HOSPITAL 1832; [indecipherable] HARRISON ESQ [indecipherable]; The stone used to [unreadable] Quay [unreadable]; of LONDON BRIDGE built about 1176.'

39 Sadly, many of the monuments in Gravesend municipal cemetery have been vandalised in recent years.

40 The inscription on the plaque formerly built into the church tower reads: 'THE TOWER OF THIS CHURCH WAS ERECTED AT THE EXPENSE OF DELAMARK BANKS ESQr; MAGISTRATE FOR THE COUNTY ANNO DOMINI 1836: WITH THE STONE OF OLD LONDON BRIDGE; WHICH WAS BUILT IN THE YEAR 1176; AND TAKEN DOWN IN THE YEAR 1832' (length 840mm, height 60mm, width 335mm). In 1998 the Purbeck marble plaque was in the possession of Mrs Robinson of Bromley, Kent; her father acquired the stone in the 1950s, when he purchased a house at Warden on the Isle of Sheppey. During 1999 Mrs Robinson presented the plaque to the Museum of London (cat ID 99.11, location EWR ST 16.2).

41 Guy's Hospital: Journals 1856–62 (GLRO, H9/GY/D5/14), 329 (8 May 1861): 'Paid to B Dixon esq for Alcove £10 10s'. The inscription on the alcove states that it was originally situated in the 'park' and was moved to its present position in the hospital quadrangle in 1926.

42 The inscription inside the two alcoves in Victoria Park reads: 'This alcove which stood on Old London Bridge was presented to Her Majesty by Benjamin Dixon Esqre J P for the use of the public and was placed here by order of the right Honourable W Cowper first Commissioner of Her Majesty's Works and Public Buildings 1860'.

43 In a letter of 24 December 1975 Sir Edwin McAlpine stated that some of the London Bridge masonry from Adelaide House was collected by Sir Thomas McAlpine and taken to Fairmile Court, Knott Park and other family houses (Watson 1997a).

44 Guildhall Museum Excavation Register, this unpublished ring from the 1967 dredging was not mentioned in Marsden's (1971, 12–14) report.

45 This psalter was produced for Christ Church Cathedral Priory, Canterbury, by Eadvius Basan, after 1012.

46 In the published saga texts some words or parts of words are printed in italics to represent where abbreviations in the original MSS have been expanded; these italics have been omitted from the lengths of text reproduced here to improve clarity.

47 Editor's emendation: the main MS together with the other three MSS have *veg* = road.

48 Alternative reading from main MS: [and built a road behind].

FRENCH AND GERMAN SUMMARIES

Résumé

C'est grâce à l'estuaire de la Tamise qui fournit une voie navigable excellente s'étendant de la Mer du Nord jusqu'au centre de l'Angleterre vers l'Ouest que Londres s'est développée comme port et comme cité. A Londres là où le chenal se rétrécit et la topographie devient favorable, les romains décidèrent de construire un pont en travers de l'estuaire. Les preuves de l'existence de ce pont romain sont en grande partie indirectes mais sont suffisantes pour permettre de proposer une séquence de trois ponts; le dernier aurait été construit après *c* AD 90 et avant *c* AD 120. Le premier et le second pont étaient tous les deux probablement des constructions en bois tandis que le troisième était une construction composite avec des butées et des piles en maçonnerie et une infrastructure en bois. Les pièces de monnaie romaines récupérées lors du dragage de la Tamise en 1824–41 suggèrent que la troisième phase du pont était sans doute terminée avant AD 330.

La date de remplacement du pont pendant la période saxonne est incertaine; ce remplacement n'a probablement pas eu lieu avant que Londres ne soit réoccupée à la fin du 9ème siècle. Les documents indiquent que le pont a été remplacé avant *c* AD 1000. En 1984, des fouilles archéologiques du site de la butée sud ont permis la découverte d'une sequence de structures de pont saxo-normandes en bois, la plus ancienne de celles-ci ayant été construite avec du bois coupé en *c* AD 987–1032. Au moins cinq phases de reconstructions de caissons en bois et de butées semblent avoir pris place entre la fin du 10ème et le début du 11ème siècle et *c* 1160–78, beaucoup de caissons et de butées ayant été emportés par des inondations. Tout au long du 11ème siècle, la tête de pont de Southwark fut soumise à l'érosion et l'on essaya de nombreuses fois de stabiliser le rivage avec des revêtements.

La construction du premier pont en pierre par Peter, le chapelain de St Mary Colechurch, prit place entre *c* 1176 et 1209. La butée sud contenait des bois de construction provenant d'arbres qui avaient été coupés pendant l'année 1187–8, ce qui confirme que la construction eut lieu en 1189 ou peut-être en 1190. Le pont en pierre mesurait 276,09m de long et était soutenu par 19 piles entourées de défenses. Entre les piles, il y avait 19 arches en pierre, l'une d'elle était enjambée par un pont basculant. La largeur d'origine de la route reste inconnue, mais il est probable qu'elle était de 6,09m et se rétrécissait sans doute jusqu'à 3,66m à cause de la présence de bâtiments de chaque coté du pont car nous savons que des gens vivaient sur le pont en 1281. Sur le côté est de la pile centrale, il y avait la chapelle de St Thomas le Martyr. Le pont faisait aussi partie des défenses de la Cité: un corps de garde (Stonegate) ou barbacane était situé au niveau de la seconde pile à partir de l'extrémité sud et le pont basculant enjambait la septième ouverture en partant du sud.

Les ponts du Moyen-Age étaient souvent attaqués par les inondations et la glace mais la maçonnerie du pont de Londres semble n'avoir souffert que deux effondrements catastrophiques à la fin du Moyen Age en 1281–2 et en 1437. Les deux fois, ces effondrements furent précédés d'une période d'abandon partiel sans doute due à des problèmes économiques qui auraient causé la réduction du programme d'entretien habituel. Le pont de Londres avait un vice de construction fondamental: les fondations de ses piles étaient des poutres et des pilotis bas protégés de l'érosion par d'énormes défenses qui, d'après les comptes de Bridge House, avaient constamment besoin d'être réparées. Des 'fissures' sur les arches furent remarquées et notées avant les deux épisodes d'effondrements ce qui suggère que les fondations des piles avaient été minées par l'érosion, probablement à cause du manque d'entretien des défenses.

En 1281–2 de la glace accumulée causa l'écroulement de cinq arches et en Février 1282, le roi donna l'ordre au maire et aux citoyens les plus riches de contribuer au coût de la construction d'un pont en bois temporaire. En Janvier 1437, une semaine après que les dommages aient été notés, le corps de garde (Stonegate) et les deux arches voisins s'écroulèrent. Un pont de bois temporaire fut de nouveau construit en travers de la brêche. L'effondrement de 1437 a encouragé la reconstruction de la partie sud du pont y compris l'agrandissement et la reconstruction partielle de la butée. Les archives de Bridge House montrent que dès le début des années 1460 jusqu'au milieu des années 1490, ce qui restait du pont fut reconstruit arche par arche.

Vers le milieu du 17ème siècle, l'expansion de Londres et de Westminster furent la cause d'un accroissement de la circulation que le pont de Londres avec sa voie étroite ne pouvait accommoder. La voie fut donc élargie pendant le 17ème siècle et en 1722, la conduite à gauche fut introduite pour essayer de contrôler la circulation.

A partir de 1664, de nombreux projets de construction pour un nouveau pont à Westminster furent tous opposés par la Corporation de Londres. Cependant, la construction d'un nouveau pont à Westminter commença en 1738 et fut achevée en 1750. Ce nouveau pont fut la cause d'un grand nombre de critiques contre le pont de Londres. Bien qu'il n'y ait pas eu suffisamment de fonds pour reconstruire le pont de Londres, il fut modifié en 1757–62. Pour l'élargir, on construisit au-dessus des piles et tous les bâtiments qui étaient sur le pont furent détruits pout créer une voie plus large et des trottoirs pour les piétons. Une des piles centrale fut aussi enlevée pour créer une nouvelle arche au centre du pont afin d'améliorer l'écoulement des eaux de la rivière et la navigation.

En 1820, il fut décidé de reconstruire le pont dont la structure était en mauvais état. La construction d'un nouveau pont parallèle à l'ancien commença en 1824; en même temps on réaligna le tracé des routes venant des environs de Bridge Foot à Southwark et on enterra les fondations des bâtiments voisins médiévaux ou plus récents sous le sous-sol d'un nouvel entrepôt. Lorsque le nouveau pont fut achevé en 1831, le vieux pont de Londres fut démoli. Des fragments du pont médiéval aux extrémités nord et sud échappèrent à cette destruction et furent scellés sous les nouvelles fondations. Les restes du pont médiéval de l'extrémité nord furent mis à découvert pendant les périodes de redéveloppement en 1921 et 1937, tandis que les restes situés au sud, soit du côté de Southwark, furent redécouverts en 1983 et fouillés avant le redéveloppement en 1984.

Zusammenfassung

Die Entwicklung Londons als Hafen und Stadt ist allein auf die Nähe der Themsemündung als bestem Ausgangspunkt für den Schiffsverkehr von der Nordsee westwärts bis tief nach Zentral-England hinein zurückzuführen. Bei London, wo sich die Fahrrinne verengt, und die Topographie günstig ist, entschieden die Römer, eine Brücke über den von Ebbe und Flut abhängigen Fluß zu schlagen. Die meist indirekten Hinweise auf die Existenz einer römischen Brücke sind jedoch ausreichend, um eine Folge von drei Brücken nachzuweisen, von denen die letzte ungefähr zwischen 90 und 120 n. Chr. gebaut wurde. Die erste und zweite Brücke waren wahrscheinlich Holzkonstruktionen, während die dritte möglicherweise aus verschiedenen Materialien bestand, das heißt aus gemauerten Widerlagern und Pfeilern mit einem hölzernen Überbau. Von den römischen Münzvorkommen, die 1824–41 bei Baggerarbeiten zutage traten, kann geschlossen werden, daß die dritte Brücke ungefähr nach 330 n. Chr. nicht mehr benutzt wurde.

Wann die Brücke in sächsischer Zeit ersetzt wurde ist ungewiß, aber fand wahrscheinlich nicht vor dem späten 9. Jh. während der Wiederbesiedlung Londons statt. Dokumentarische Quellen deuten darauf hin, daß die Brücke im Jahre 1000 wieder ersetzt worden war. Ausgrabungen am südlichen Brückenpfeiler förderten 1984 eine Folge saxo-normannischer Holzbrückenkonstruktionen zutage, deren früheste Teile aus Holz bestanden, das zwischen 987 und 1032 gefällt worden waren. Zwischen dem späten 10. und frühen 11. Jh. und zwischen 1160–78 scheinen mindestens fünf kurzlebige Senkkästen und Pfeiler gebaut worden zu sein, die meist von der Flut wieder weggeschwemmt wurden. Während des späten 11. Jh. litt der Southwark-Brückenkopf unter Erosion und es wurde wiederholt versucht, ihn durch den Bau von Uferbefestigungen zu stabilisieren.

Peter, Kaplan der St Mary Colechurch, ließ die erste steinerne Brücke zwischen 1176 und 1209 erstellen. Der Südpfeiler enthält Stämme, die während der Jahre 1187/88 gefällt wurden, eine Bestätigung, daß die Bauarbeiten während der Jahre 1189/90 stattfanden. Diese Steinbrücke war 276,09m lang und wurde von 19 Pfeilern mit Strombrechern getragen. Zwischen ihnen spannten 19 Bögen, einer von ihnen in Form einer Zugbrücke. Die ursprüngliche Straßenbreite ist nicht klar, lag aber vermutlich bei 6.09m, die sich auf Grund von Gebäuden auf beiden Seiten sehr wahrscheinlich auf 3.66m verengte. Wir wissen, daß um 1281 Menschen auf der Brücke gelebt haben. Die Kapelle St Thomas der Martyrer befand sich auf der Ostseite des Zentralpfeilers. Die Brücke diente der City auch als Verteidigung: sie hatte ein Torhaus (Stonegate), oder einen Wachturm auf dem zweiten Pfeiler vom Südufer, während die Zugbrücke über den siebenten Zwischenraum von der Südseite führte.

Mittelalterliche Brücken wurden oft das Opfer von Überflutungen und Eisgang. London Bridge scheint aber nur zwei katastrophale Zusammenbrüche während spätmittelalterlicher Zeit, 1281/82 und 1437, erlitten zu haben. Beiden Ereignissen ging ein Zeitraum der

Vernachlässigung voraus, vermutlich ein Zeichen finanzieller Schwierigkeiten, während derer die Instandhaltungsarbeiten reduziert wurden. London Bridge hatte eine grundsätzliche Konstruktionsschwäche: ihre Pfeiler ruhten alle auf Fundamentbalken und kurzen Pfählen, die durch Strombrecher vor Erosion geschützt wurden und, nach Unterlagen des Brückenkontors zu urteilen, laufend Reparaturen erforderten. Es ist belegt, daß jeweils zeitlich vor den beiden Zusammenbrüchen Rissen in den Bögen entdeckt wurden. Dieses weist sehr darauf hin, daß die Pfeilerfundamente, vermutlich auf Grund ungenügender Wartung der Strombrecher, von Erosion unterhöhlt worden waren.

1281/82 brachte Packeis fünf Brückenbögen zum Einsturz, so daß in Februar 1282 der König dem Bürgermeister und den reicheren Bürgern auferlegte, zum Bau einer provisorischen Holzbrücke beizutragen. Im Januar 1437 stürzten, eine Woche, nachdem Schäden an der Brückenstruktur festgestellt worden waren, das Torhaus (Stonegate) und zwei benachbarte Bögen ein. Wieder wurde eine provisorische Holzbrücke über die Einbruchstelle geschlagen. Der 1437er Einsturz zeitigte größere Wiederaufbauarbeiten am Süd-Ende der Brücke, wo das Widerlager vergrößert und teilweise neu erstellt wurde. Aufzeichnungen des Brückenkontors zeigen, daß von den frühen 1460er bis weit in die 1490er Jahre der größte, verbleibende Teil der Brücke Bogen für Bogen neu gebaut wurde.

Um die Mitte des 17. Jh. schufen das Wachstum Londons und Westminsters ein größeres Verkehrsaufkommen als London Bridge mit ihrer engen Fahrbahn verkraften konnte. Während des späten 17. Jh. wurde sie verbreitert und 1722 der Linksverkehr eingeführt, um den Verkehrstrom zu regulieren.

Von 1664 an wurden verschiedene Anläufe gemacht, eine Brücke in Westminster zu bauen, sie wurden jedoch alle von der Corporation of London zu Fall gebracht. Schließendlich wurde jedoch 1738 mit dem Bau einer neuen Brücke in Westminster begonnen und 1750 vollendet. Diese neue Brücke brachte London Bridge eine Flut von Kritik ein. Da für einen Neubau nicht genügend Mittel zur Verfügung standen, wurden von 1752–62 immer wieder Änderungsarbeiten vorgenommen: Die Brücke wurde, teilweise die Strombrecher überragend, verbreitert und alle Gebäude darauf abgerissen, um eine breitere Fahrbahn und einen Fußgängersteig zu schaffen. Das Entfernen einer der zentralen Brückenpfeiler führte zum Bau eines neuen Zentralbogens, wodurch der Wasserfluß und die Navigation verbessert wurden.

1820 wurde entschieden, die Brücke, die sich wieder in schlechtem baulichen Zustand befand, neu zu bauen. Die Bauarbeiten an der neuen Brücke entlang der alten begannen 1824. Die Neutrassierung der Zufahrtsstraßen führte zur Bildung von 'Southwark Bridge Foot' und die Grundmauern der daran liegenden mittelalterlichen und späteren Gebäude fielen dem Neubau eines Lagerhauses zum Opfer. 1831, nach Fertigstellung der neuen Brücke, wurde die alte London Bridge abgerissen. Teile der mittelalterlichen Brücke, sowohl auf der Nord- als auch auf der Südseite entgingen der Zerstörung, wurden aber unter den neuen Fundamenten begraben. Die Reste der mittelalterlichen Brücke auf der Nordseite wurden 1921 und 1937 bei Neubauarbeiten freigelegt, während die Überreste auf der südlichen, der Southwark-Seite, 1983 wiederentdeckt und archäologisch untersucht wurden, bevor 1984 neue Gebäude darüber errichtet wurden.

GLOSSARY

Abutment The landward end of a bridge, open on one side to incorporate a ramp or causeway for the bridge approach road

Baseplate The lowest horizontal member of a timber structure

Bridge Foot This term was used to describe the properties on London Bridge on the south side of the Stonegate and those that adjoined the southern end of the bridge. However, its precise limit is uncertain as it seems to have been used as part of the address of various post-medieval properties lining the approach road to the bridge (Home 1931, 328)

Bridgehead The area of land adjoining either end of a bridge

Bridgemasters or wardens These were the men appointed to run the Bridge House Trust and maintain London Bridge. From the late 13th century the existence of these officers is documented (Harding and Wright 1995, x)

Bulwark A revetment or wooden wall composed of lengths of edge-laid horizontal planking, slotted into vertically grooved posts (Goodburn 1995, 49, 51)

Burh A *burh* is an Old English term denoting a fortified place or defensible place, whether a private residence or, increasingly from the late 9th century, a town (later 'borough'). Many of the latter type were either built or restored under Alfred (AD 871–99) in Wessex and Edward the Elder (AD 899–924) in Mercia, and were manned from the districts (often whole counties in the Midlands) which they protected

Caisson This word has a number of meanings in bridge studies, and in literature on modern bridges it is generally used to describe a watertight chamber which is open at the bottom, like a cofferdam. It is used here in Chapter 4 to describe Roman, timber lap-jointed, bridge pier foundations. In Chapter 7 it is used twice to describe different timber structures. Firstly, it is defined as a rectangular, four-sided, box-like structure serving the same function as a bridge abutment. A bridge caisson could be described as a landward pier (without cutwaters) since it is not an integral part of any approach ramp or causeway. Secondly, it is used to describe a type of Scandinavian laftwork quay foundation (Chapter 7). Thirdly, in Chapter 13 it is used to describe a certain type of open-topped, watertight, flat-bottomed, wooden barge in which the basal portion of the piers of Westminster Bridge were constructed; the barges were sunk (using sluice valves) in dredged voids in the riverbed. The base of the sunken barge then became part of the pier foundations and the sides of it were detached for reuse (Ruddock 1979, 8, 210)

Cofferdam A watertight, wooden structure, enclosing an area of a riverbed in which either a bridge pier or abutment foundations can be constructed. Today such structures are normally constructed using interlocking, ribbed, steel sheeting

Cramp An iron bar used to hold together large blocks of stone, usually by the insertion of the turned-down ends of short bars into holes in both blocks. The two ends of the cramps were normally secured by pouring molten lead into the holes (Ruddock 1979, 120)

Cubitus An ancient measurement equivalent to 0.444m (two spans) and 1.5 *pedes monetales*. It was derived from the length of the forearm with the fingers extended and was commonly used as a timber width in Roman carpentry

Digitus An ancient measurement (pl *digiti*) equivalent to 18.5mm. Four digits make a hand, 16 make a foot (*pes monetalis*). It was derived from the width of a finger

Frontbrace Normally a diagonal brace intended to prevent the front of a riverside revetment from collapsing forward

Laftwork A style of timberwork in which horizontally laid beams, logs or planks are laft-jointed together, one on top of another, log cabin style. This style of timberwork is seen in many vernacular Scandinavian buildings

Mortice and tenon This category of joints consists of a slot (mortice) cut into one timber to accommodate a tongue (tenon) cut into the end of another (Brigham et al 1992, 15). The development of this joint is one of the most important innovations in late medieval English carpentry (see Chapter 14.3)

Palmus An ancient measurement (pl *palmi*) equivalent to 74mm. Three *palmi* make one span, four palmi make a foot (*pes monetalis*). It was derived from the width of the palm of the hand and, therefore, is equivalent to four fingerwidths (*digiti*)

Pes monetalis The Roman foot (pl *pedes monetales*) was equivalent to 0.296m. It was used singly and in multiples in carpentry for both the length and width of timbers. It was derived from the nominal length of the human foot

Pier This word has two related meanings. In stone bridges it is the upstanding masonry (and foundations) from which the arches are sprung (Hopkins 1970, fig 4); a masonry pier may have cutwaters at one end or both. In timber bridges piers are the upstanding foundations (of varying shape and size, and infilled with rubble) that support the beams of the roadway

Pile A sharpened log (which may be iron-shod) driven into position by a piling engine, either a 'gin' or 'ram' (Harding and Wright 1995, xxii). These medieval engines would have consisted of a wooden tripod frame from whose apex a heavy weight – the 'ram' – was suspended by rope. The weight was hauled up to the apex of the frame by a gang of labourers and then allowed to fall on to the top end of the pile, driving it into the ground. On the London Bridge starlings a team of 20–25 men worked the 'great ram'

Regression Phases of sea-level fall relative to the adjoining land mass are termed regressions; this is the opposite of transgressions

Reused timber A timber with features such as empty slots or pegholes which have no apparent function in the context in which it was found (Brigham et al 1992, 14)

Revetment A timber or composite structure erected to prevent erosion of the foreshore by tides and river current

Scarf joints A variety of end-to-end joints intended to make one long timber from two or more shorter lengths by joining their ends (Milne and Milne 1982, 111)

illbeam A horizontal beam on which a masonry wall was founded

Socket A rectangular timber slot cut into the upper face of baseplates, beams, and so on, to accommodate upright timbers. Often a socket was cut completely through the timber. Sockets cannot be described as mortices as there is no evidence for the use of tenons in this type of joint

Soleplate The lowest horizontal member of a wooden bridge trestle (Rigold 1975, 75–6)

Span An ancient measurement equivalent to 0.22m. It was derived from the width of two hands with the fingers slightly spread. It appears in Roman carpentry

Starling A low, boat-shaped foundation structure at water level around a bridge pier, composed of lines of piles into which rubble was dumped to produce a solid platform, the top of which was strengthened by a network of horizontal timbers (Knight 1831, 118–19). Its function was to prevent erosion of the pier foundations

Staves The vertical elements of a wall or revetment formed of contiguous, vertical planks, normally set in the upper face of a baseplate

Tieback A horizontal brace retaining the front of a riverside revetment (see above) from the landward or rear side to counteract the weight of the landfill material (Brigham et al 1992, 15)

Transgressions Phases of sea level increase relative to the adjoining land mass are termed transgressions, this is the opposite of regressions

Trestle A timber framework for supporting the roadway of a bridge. The basal portion of this framework normally consisted of one or two soleplates, into the upper face of which were slotted two or more vertical timbers. On top of these were fitted a number of horizontal timbers, on which the beams of the roadway rested (Rigold 1975, 51, 53–4)

Tusk-tenons A type of mortice and tenon joint, in which the tenon protrudes through the rear of the mortice slot, so it can be secured with a peg

Uncia An ancient measurement (pl *unciae*) equivalent to 24.6mm. Twelve *unciae* make one foot (*pes monetalis*). It is the root word for the modern inch and ounce; it is probably derived from the width of the thumb, used as an alternative to *digitus*

BIBLIOGRAPHY

Manuscript sources

British Library (BL)

Arundel MS 133

Harley MS 6016

Harley MS 6205

Royal MS 16, F.11 (illuminated MS known as the volume of poems by Charles, Duke of Orleans, c 1480)

Stowe MS 942, cartulary of the Hospital of St Thomas the Martyr in Southwark (late 15th century with later additions)

Corporation of London Record Office (CoLRO)

The Bridge House (BH) records now form part of the CoL archive (for details see Harding and Wright 1995, xxiv–vii)

BH accounts (earliest surviving accounts Michaelmas 1381 to Michaelmas 1382)

BHCJ: Bridge House committee journals (volumes of committee minutes relating to construction of the new London Bridge and the demolition of the medieval one)

BHR: Bridge House rentals (series of volumes starting 1404 and containing a full list of weekly and annual receipts)

BHWP: Bridge House weekly payments (series of volumes starting 1404, listing wages, regular expenses, purchases, payment of quit-rents and various other quarterly payments)

CD: comptroller's deeds (a collection of accessioned deeds and plans relating to property owned by the City of London)

CH: Court of Hustings rolls

CP: Corporation of London collection of plans and maps

JCCC: Journal of the Court of Common Council of the City of London

LBC: London Bridge Improvement Committee papers, miscellaneous manuscripts and plans 37.37 (1756–67) (this committee was set up under the act of 1756 which authorised the modification of the medieval bridge)

REP: Repertory (volumes of minutes relating to the Court of Aldermen of the City of London)

SBHR: 'small' Bridge House register

Greater London Record Office (GLRO) (now known as the London Metropolitan Archives)

GH: Guy's Hospital, Southwark, Ledger Book G (1813–42) HA9/GY/D1/7

Guildhall Library, City of London (GL)

St Barts: St Bartholomew's (by the Exchange, City of London) MS 4383/1 (17th-century churchwardens' accounts) and MS 4384/1 (vestry minutes)

Guildhall Museum, City of London (GM)

Files (all numbered) (now housed in Museum of London Archaeological Archive, see below)

Museum of London (MoL) Archaeological Archive
Contains copies of all unpublished MoL and MoLAS archive reports, plus all Guildhall Museum (GM) records

Sir John Soane's Museum, London (Soane)
Soane's collection of drawings and engravings (catalogued)

Legislation: Acts of Parliament (Act)
(For details see HMSO 1999)
Act, 1736 Building Westminster Bridge, 10 Geo 2, ch 16
Act, 1756 Modification of London Bridge, 20 Geo 3, ch 4
Act, 1823 Rebuilding of London Bridge, 4 Geo 4, ch 50

Printed works

Abulafia, D, Franklin, M, and Rubin, M (eds), 1992 *Church and city 1000–1500*, Cambridge

Adalbjarnarson, B (ed), 1945 *Snorri Sturluson: Heimskringla II*, Hid Ìslenzka Fornritafélag, Reykjavík

Addison, J, 1712 *The Spectator*, 20 May, 1

Admiralty, 1977 *Admiralty tide tables: European waters including the Mediterranean Sea*: 1, London

Alberti Alberti, L B, *Ten books on architecture* (trans J Leoni), 1955, London

Alexander, J, and Binski, P (eds), 1987 *Age of chivalry: art in Plantagenet England 1200–1400*, Roy Acad Arts exhibition catalogue, London

Allen, T, and Welsh, K, 1996 Eton rowing lake, *Curr Archaeol* 148, 124–7

Anglo, S, 1962 The imperial alliance and the entry of the Emperor Charles V into London: June 1522, *Guildhall Misc* 2 (1960–8), 131–51

Anglo, S, 1963 The London pageants for the reception of Katherine of Aragon: November 1501, *J Warburg Courtauld Inst* 26, 53–89

Anglo, S, 1969 *Spectacle, pageantry and early Tudor policy*, Oxford

Anglo-Saxon Chron *The Anglo-Saxon Chronicle* (ed and trans M Swanton), 1996, London

Anon, 1767 An account of late proceedings of the committee for letting the city lands, so far as those proceedings relate to London Bridge and the waterworks under the bridge, *Gentleman's Mag* 37(1), 337–9, and 37(2), 407–8

Anon, 1833a Note in antiquarian researches on the library and antiquities of Mr William Knight FSA, *Gentleman's Mag* 103(1), 67–8

Anon, 1833b Metropolitan antiquities, *Gentleman's Mag* 103 (1), 69

Anon, 1847 Ancient figure of god the Father found in the Thames, *Archaeologia* 32, 409–10

Anon, 1994 *Whitney-on-Wye toll bridge: a concise history*, privately pub

Anon, n d All Saints Church, Steep, Hampshire [guide], privately pub

Appleby, J S, 1970 Appendix 1, Dendrochronological evidence, in Huggins, 144

Archer, I, Barron, C M, and Harding V (eds), 1988 *Hugh Alley's caveat: the markets of London in 1598: Folger MS Va 318*, London Topogr Soc Pub 137, London

Armitage, P L, 1982 Studies on the remains of domestic livestock from Roman, medieval and early modern London; objectives and methods, in Hall and Kenwood, 94–106

Armstrong, C A J (trans), 1936 *The usurpation of Richard the third – Dominicus Mancinus ad angelum catonem de occupatione regni Anglie per Richardrum Tercium libellus*, Oxford

Astill, G, and Grant, A (eds), 1988 *The countryside of medieval England*, Oxford

Aston, M, 1988 *England's iconoclasts*, London

Ayre, J, and Wroe-Brown, R, with Malt, R, 1996 Aethelred's Hythe to Queenhithe: the origin of a London dock, *Medieval Life* 5, 14–25

Ayre, J, and Wroe-Brown, R, in prep *Queenhithe: excavations at Thames Court, City of London, 1989–1997*, MoLAS Monogr Ser

Baart, J M, 1994 Dutch redwares, *Medieval Ceram* 18, 19–27

Baillie, M G L, 1982 *Tree-ring dating and archaeology*, London

Bang, N E, 1922 *Tabeller over skibsfart og varentransport gennem Øresund 1497–1660*, Copenhagen

Barker, P, and Higham, R, 1982 *Hen Domen, Montgomery, a timber castle on the English-Welsh border: Vol 1*, Roy Archaeol Inst, London

Barron, M B, 1970 The government of London and its relations with the Crown, 1400–1450, unpub PhD thesis, Univ London

Bass, W M, 1987 *Human osteology, a laboratory and field manual*, 3 edn, Missouri Archaeol Soc Spec Pub 2, Columbia Mo

Bateman, N, 1986 Bridgehead revisited, *London Archaeol* 5, 233–41

Bateman, N, and Locker, A, 1982 The sauce of the Thames, *London Archaeol* 4, 204–7

Beard, G, 1978 *The work of Robert Adam*, London

Becker, M J, 1930 *Rochester Bridge: 1387–1856*, London

Beier, A L, and Cannadine, D (eds), 1989 *The first modern society: essays in English history in honour of Lawrence Stone*, London

Bell, W G, 1923 *The Great Fire of London in 1666*, London

Bentham, G, and Hooker, J D, 1947 *Handbook of British flora*, 7 rev edn, Kent

Bentley, J, 1985 *Restless bones – the story of relics*, London

Benton, J F, 1968 *Town origins: the evidence from medieval England*, Lexington, Mass

Beresford, G, 1987 *Goltho: the development of an early medieval manor c 850–1150*, Engl Heritage Archaeol Rep 4, London

Bergeron, D M, 1971 *English civic pageantry, 1558–1642*, London

Bergeron, D M (ed), 1986 *Pageantry in the Shakespearean theatre*, London

Betts, I M, 1987 Ceramic building material: recent work in London, *Archaeol Today* 9, 26–8

Betts, I M, 1990 Building materials, in Schofield et al, 220–9

Betts, I M, 1994 Appendix, Medieval floor tiles in London churches, in Schofield 133–40

Betts, I M, 1995 Procuratorial tile stamps from London, *Britannia* 26, 207–29

Betts, I M, 1996 Report on building material: north-west

Roman Southwark, unpub MoLAS archive rep
Betts, I M, 1997 Report on ceramic building material and geological samples from Fennings Wharf, Southwark (FW84), unpub MoLAS archive rep
Betts, I M, and Foot, R, 1994 A newly identified late Roman tile group from southern England, *Britannia* 25, 21–34
Biddle, M, and Hudson, D M, with Heighway, C M, 1973 *The future of London's past*, Rescue Pub 4, Worcester
Bidwell, P T, and Holbrook, N, 1989 *Hadrian's Wall bridges*, London
Bird, J, and Graham, A H, 1978 Gazetteer of Roman sites in Southwark, in Sheldon et al, 517–26
Bird, J, Chapman, H, and Clark, J (eds), 1978 *Collectanea Londiniensia: studies in London archaeology and history presented to Ralph Merrifield*, London Middlesex Archaeol Soc Spec Pap 2, London
Black, M, 1985 *Food and cooking in medieval Britain: history and recipes*, Engl Heritage, London
Blackmore, L, 1994 Pottery, the port and the populace: the imported pottery of London 1300–1600 (part 1), *Medieval Ceram* 18, 29–44
Blackmore, L, and Vince, A G, 1994 Medieval pottery from south-east England found in the Bryggen excavations 1955–68, *Bryggen Pap Suppl Ser* 5, 9–160
Blagg, T C, and King, A C (eds), 1984 *Military and civilian in Roman Britain: cultural relationships in a frontier province*, BAR Brit Ser 136, Oxford
Blair, J, and Ramsay, N (eds), 1991 *English medieval industries, craftsmen, techniques, products*, London
Bliss, W H, and Jesse, A T (eds), 1895 *Calendar of entries in the papal registers relating to Great Britain and Ireland: papal letters (1198–1492): Vol 2*, London
Boe, G, De, and Verhaeghe, F (eds), 1997 *Urbanism in medieval Europe: papers of the 'Medieval Europe Brugge 1997' conference: Vol 1*, IAP Rapporten 1, Zellik
Boland, D, Breen, C, and O'Sullivan, A, 1996 *Clonmacnoise bridge project: report of the pre-disturbance survey*, Irish Underwater Archaeol Res Team rep
Bond, F, 1914 *Dedications and patron saints of English churches*, London
Booker, J, 1974 *Essex and the Industrial Revolution*, Essex Rec Off Pub 66, Chelmsford
Boulton, J, 1996 Wage labour in seventeenth-century London, *Econ Hist Rev* 49, 268–90
Boyer, M N, 1966 Rebuilding the bridge at Albi, 1408–1410, *Technol Culture* 7, 24–37
Boyer, M N, 1976 *Medieval French bridges, a history*, Med Acad Amer Pub 84, Cambridge Mass
Brett, D W, 1978 Medieval and recent elms in London, in Fletcher, 195–9
Brett, M, 1992 The annals of Bermondsey, Southwark and Merton, in Abulafia et al, 279–310
Brigham, T, 1990a A reassessment of the second basilica in London, AD 100–400: excavations at Leadenhall Court, 1984–86, *Britannia* 21, 53–97
Brigham, T, 1990b The late Roman waterfront in London, *Britannia* 21, 99–183
Brigham, T, 1992a Civic centre redevelopment, in Milne 1992a, 81–95
Brigham, T, 1992b Reused house timbers from the Billingsgate site, 1982–83, in Milne 1992b, 86–105
Brigham, T, 1998 The port of Roman London, in Watson 1998b, 23–34
Brigham, T, in prep, *Excavations at Regis House 1994–6: Part 1*, MoLAS Monogr Ser
Brigham, T, and Watson, B, 1995 Suffolk House, 5 Laurence Pountney Hill and 154–156 Upper Thames Street, London EC4, unpub MoLAS archive rep
Brigham, T, Goodburn, D, Milne, G, and Tyers, I, 1992 Terminology and dating, in Milne 1992b, 14–22
Brigham, T, Goodburn, D, and Tyers, I, with Dillon, J, 1996a A Roman timber building on the Southwark waterfront, London, *Archaeol J* 152, 1–72
Brigham, T, Watson, B, and Tyers, I, with Bartkowiak, R, 1996b Current archaeological work at Regis House in the City of London (part 1), *London Archaeol* 8, 31–7
Brit J, 1730 *British Journal*, 25 July
Britnell, R H, 1994 Rochester Bridge, 1381–1530, in Yates and Gibson, 43–106
Britton, F, 1987 *London Delftware*, London
Brooke, C N L, assisted by Keir, G, 1975 *London 800–1216: the shaping of a city*, London
Brooks, N P, 1971 The development of military obligations in eighth and ninth century England, in Clemoes and Hughes, 69–84
Brooks, N P, 1994 Rochester Bridge AD 43–1381, in Yates and Gibson, 3–40
Brooks, N P, 1995 Medieval bridges: a window onto changing concepts of state power, *J Haskins Soc* 7, 11–29
Brown, S, 1991 Excavations on the medieval Exe bridge, St Edmund's Church and Frog Street tenements, Exeter, 1975–9, unpub Exeter Museums Archaeol Field Unit rep 91.52
Bruce, J C, 1885 The three bridges over the Tyne at Newcastle, *Archaeologia Aeliana* n ser 10, 1–11
The Builder, 1921 [various anon short articles on London Bridge], *The Builder* 121
The Builder, 1922 [various anon short articles on London Bridge], *The Builder* 122–3
Butler, H E, 1934 A description of London by William Fitz Stephen, in Stenton, 47–67
Caesar, Julius *De Bello Gallico*, in *Caesar: the Gallic war* (trans H J Edwards), 1986, London
Cal Pat R *Calendar of Patent Rolls preserved in the Public Record Office*, 65 vols (1291–1509, 1547–63), 1893–1948, London
Cal L Bk F *Calendar of Letter Book F* (ed R R Sharpe), 1904, London
Cal L Bk G *Calendar of Letter Book G* (ed R R Sharpe), 1905, London
Cal L Bk K *Calendar of Letter Book K* (ed R R Sharpe), 1911, London
Campbell, B M S, Galloway, J, Keene, D, and Murphy, M, 1993 *A medieval capital and its grain supply. Agrarian production and distribution in the London region c 1300*, Hist Geogr Res Ser 30, London
Carlin, M, 1983 The urban development of Southwark, c 1200 to 1550, unpub PhD thesis, Univ Toronto

Carlin, M, 1996 *Medieval Southwark*, London

Carlin, M, in prep The historical evidence, in Seeley

Carlin, M, and Belcher, V, 1989 Gazetteer, in *British atlas of historic towns: Vol 3, The City of London* (ed M D Lobel), 63–99, Oxford

Caruana, I, 1983 Carlisle, *Curr Archaeol* 86, 77–81

Caspall, J, 1987 *Making fire and light in the home pre-1820*, Antique Collectors Club, London

Cassius, Dio *Roman history by Dio Cassius* (trans E Cary), 9 vols, 1914–27, London

Certosa di Firenze, 1982 *Iconografia di San Benedetto nella pittura della Toscana*, exhibition catalogue, Florence

Chandler, M J, 1952 London Bridge before the Great Fire, *Guildhall Misc* 1 (1952–9), 19–21

Chapelot, J, and Fossier, R, 1980 *The village and house in the middle ages* (trans H Cleere), France

Chapman, H, 1980 Roman bone objects, in Jones, 93–4

Charles, F W B, and Charles, M, 1995 *Conservation of timber buildings*, London

Chaucer, G, *The Canterbury tales*, in Robinson, F N (ed), 2 edn 1957 repr 1979, *The works of Geoffrey Chaucer*, 17–265, Oxford

Cherry, J, 1991 Pottery and tile, in Blair and Ramsay, 189–209

Christianson, C P, 1987 *Memorials of the book trade in medieval London: the archives of old London Bridge*, London

Clark, J (ed), 1995 *The medieval horse and its equipment*, HMSO Medieval Finds from Excavations in London Ser 5, London

Clemoes, P, and Hughes, K (eds), 1971 *England before the Conquest: studies in primary sources presented to Dorothy Whitelock*, London

Cobb, H S (ed), 1990 *The overseas trade of London: Exchequer customs accounts 1480–1*, London Rec Soc 27, London

Colvin, H, and Foister, S, 1996 *The panorama of London c 1544 by Anthonis van den Wyngaerde*, London Topogr Soc Pub 151, London

Cooke, E W, 1970 *A selection of drawings of old and new London Bridge*, London Topogr Soc Pub 113, London

Cooper, A R, 1998 Obligation and jurisdiction: roads and bridges in medieval England (*c* 700–1300), unpub PhD thesis, Harvard Univ, Cambridge Mass

Cooper, L, and Ripper, S, 1994 The medieval Trent bridges at Hemington Fields, Castle Donnington, *Trans Leicestershire Archaeol Hist Soc* 77, 153–61

Cooper, L, and Ripper, S, in prep *The Hemington bridge: the excavation of three medieval bridges at Hemington Quarry, Castle Donington, Leicestershire*

Cooper, L, Ripper, S, and Clay, P, 1994 The Hemington bridges, *Curr Archaeol* 140, 316–21

Corner, G R, 1860 Observations on the remains of an Anglo-Norman building in the parish of St Olave, Southwark, *Archaeologia* 38, 37–53

Cowan, C, 1992 A possible mansio in Roman Southwark: excavations at 15–23 Southwark Street, 1980–86, *Trans London Middlesex Archaeol Soc* 43, 3–192

Cowan, C, in prep *Roman Southwark: urban development in the north-west quarter*, MoLAS Monogr Ser

Cowie, R, and Whytehead, R, 1989 Lundenwic: the archaeological evidence for middle Saxon London, *Antquity* 63, 706–18

Crumlin-Pedersen, O, Schou Jørgensen, M, and Edgren, T, 1992 Ships and travel, in Roesdahl and Wilson, 42–51

Crummy, N, 1983 *The Roman small finds from excavations in Colchester 1971–9*, Colchester Archaeol Rep 2, Colchester

Cuming, H, S, 1887 Traders' signs on Old London Bridge, *J Brit Archaeol Ass* 43, 166–7

Cüppers, H, 1969 *Die Trier Römerbrücken*, Rheinisches Landesmuseum Trier, Mainz

Curle, J, 1911 *A Roman frontier post and its people: the fort of Newstead in the parish of Melrose*, Glasgow

Currie, I, 1996 *Frosts, freezes and fairs*, London

Daily Post Extraordinary, 1734 [A list of the persons who have polled for Mr William Selwin], no. 4545, 9 April

Dalton, O M, 1912 *Catalogue of the finger rings: Early Christian, Byzantine, Teutonic, medieval and later ... in the [British] Museum*, London

Dam, van, J D, 1991 *Netherlanse Tegel*, 2 edn, Amsterdam/ Antwerp

Dance, G, 1799 Appendix B1, in Second report of the House of Commons Select Committee upon the improvement of the Port of London, printed in *Reports from Committees of the House of Commons 1793–1802: Vol 14*, 1803, 461–542, London (copy in CoLRO)

Davis, A, 1993 An assessment of the plant remains from Fennings Wharf, unpub MoLAS archive rep

Davis, A, 1995 An assessment of the plant remains from the Courage Brewery site, Southwark, unpub MoLAS archive rep

Davis, B, Richardson, B, and Tomber, R S, 1994 *The archaeology of Roman London: 5, A dated corpus of early Roman pottery from the City of London*, CBA Res Rep 98, London

Davis, A, Locker, A, Rackham, J, Tyers, I, West, B, and Winder, J, 1995 Environment and economy in Saxon London, unpub MoLAS rep

Dawson, G J, 1969 Roman London bridge, *London Archaeol* 1, 114–17

Dawson, G J, 1970 Roman London bridge: Part 2, Its location, *London Archaeol* 1, 156–60

Dawson, G J, 1976 Montague Close excavations 1969–73, *Surrey Archaeol Soc Res Vol* 3, 37–58

Dawson, G J, 1977 Roads, bridges and the origin of Roman London, *Surrey Archaeol Collect* 71, 43–55

Deadman, H, and Scudder, E, 1994 *An introductory guide to the Corporation of London Records Office*, London

Dearne, S, 1745 *The state of London Bridge waterworks*, London

Devoy, R J, 1979 Flandrian sea level changes and vegetational history of the lower Thames estuary, *Phil Trans Roy Soc London*, B 285, 355–406

Devoy, R J, 1980 Post-glacial environmental change and man in the Thames estuary: a synopsis, in Thompson, 134–48

DGLA, 1989 Department of Greater London Archaeology, London post-excavation review, unpub rep

Dierick, A L, 1970 *The mystic lamb*, Ghent

Dobson, R B, 1970 *The peasant's revolt of 1381*, London

Dodd, R, 1820 *Report of Mr Ralph Dodd civil engineer and architect on the present decayed and dangerous state of London Bridge* (copy in CoL Guildhall Lib, fo pam 554)

Driesch, A von den, 1976 *A guide to the measurement of animal bones from archaeological sites*, Peabody Mus Bull 1, Cambridge Mass

Driesch, A von den, and Boessneck, J A, 1974 Kritische Anmerkungen zur Widerristhöhenberechnung aus Langenmassen vor- und frühgeschichtlicher Tierknochen, *Säugetierkundliche Mitteilungen* 22, 325–48

Drummond-Murray, J, and Thompson, P, 1998 Did Boudica burn Southwark? The story of the Jubilee Line excavation, *Curr Archaeol* 158, 48–9

Drummond-Murray, J, Saxby, D, and Watson, B, 1994 Recent archaeological work in the Bermondsey district of Southwark, *London Archaeol* 7, 251–7

Dubler, E, 1966 *Das Bild der Heilige Benedikt*, Munich

Duffy, E, 1992 *The stripping of the altars: traditional religion in England 1400–1580*, London

Dymond, D P, 1961 Roman bridges on Dere Street, County Durham, with a general appendix on the evidence in Roman Britain, *Archaeol J* 118, 136–64

Dyson, T, 1975 The 'pre-Norman' bridge of London reconsidered, *London Archaeol* 2, 326–28

Dyson, T, 1978 Two Saxon land grants for Queenhithe, in Bird et al, 200–15

Dyson, T, 1980 Documentary survey of the London riverside wall in the medieval period, in Hill et al, 7–10

Dyson, A G, 1981 The terms 'quay' and 'wharf' and the early medieval London waterfront, in Milne and Hobley, 37–8

Dyson, T, 1990 King Alfred and the restoration of London, *London J* 15(2), 99–110

Eames, E S, 1980 *Catalogue of medieval lead-glazed earthenware tiles in the Department of Medieval and Later Antiquities, British Museum* (2 vols), London

Eames, E S, 1992 *English tilers*, London

Eaton, R, 1996 From medieval times to the eighteenth century, in Murray and Stevens, 36–82

Edwards, R, 1974 Documentary sources in relation to the excavations, in Sheldon, 3–8

Egan, G, 1998 *The medieval household: daily living c 1150–c 1450*, HMSO Medieval Finds from Excavations in London Ser 6, London

Egan, G, and Pritchard, F, 1991 *Dress accessories c 1150–c 1450*, HMSO Medieval Finds from Excavations in London Ser 3, London

Ekwall, E, 1928 *English river names*, Oxford

Ekwall, E, 1954 *Street-names of the City of London*, London

Ellis, B M A, 1995 Spurs and spur fittings, in Clark, 124–56

Elrington, E R (ed), 1987 *The Victoria history of the county of Cheshire: Vol 1*, Oxford

English Heritage, 1998 *Dendrochronology – guidelines on producing and interpreting tree-ring dendrochronological data*, London

Environment Agency, 1998 *Fish found in the tidal Thames*, London

Evans, P, 1973 Excavations at New Hibernia Wharf, *London Archaeol* 2, 99–103

Everard, C E, 1980 On sea-level changes, in Thompson, 1–23

Farmer, D H, 1992 *The Oxford dictionary of saints*, 3 edn, Oxford

Fisher, J (ed), 1981 *A collection of early maps of London 1553–1667*, Kent

Fitzpatrick, A P, and Scott, P R, 1999 The Roman bridge at Piercebridge, North Yorkshire-County Durham, *Britannia* 30, 111–32

Fletcher, J M (ed), 1978 *Dendrochronology in Europe*, BAR Int Ser 51, Oxford

Fletcher, A, and MacCulloch, D, 1997 *Tudor rebellions*, 4 edn, London

Flight, C, 1997 *The earliest recorded bridge at Rochester*, BAR Brit Ser 252, Oxford

Fowler, R C, 1907 Religious houses, in Page and Round, 84–202

Fox, C (ed), 1992 *London – world city 1800–1840*, New Haven and London

Freestone, I, Johns, C, and Potter, T (eds), 1982 *Current research in ceramics: thin-section studies*, Brit Mus Occas Pap 32, London

Frere, S S, and Tomlin, S O (eds), 1993 *The Roman inscriptions of Britain: Vol 2(5), 'Instrumentum Domesticum'*, Oxford

Gage, J, 1831 Letter ... remains of the Prior of Lewes' hostelry in the parish of St Olave, Southwark, *Archaeologia* 23, 299–308

Gardner, A, 1951 *English medieval sculpture*, Cambridge

Garner, F H, and Archer, M, 1972 *English Delftware*, London

Gibbs, M (ed), 1939 *The early charters of the cathedral church of St Paul's, London*, Camden Soc 3 ser, 58

Giles, F, 1843 Lecture on London Bridge, *Minutes Proc Inst Civil Engineers* 2 (1842–3), 87–9

Giorgi, J, 1993 Interim archive report on the botanical remains from the Fleet Valley, City of London, unpub MoLAS archive rep

Giorgi, J, 1997 The plant remains from Fennings Wharf, Southwark, unpub MoLAS archive rep

Giorgi, J, in prep a The plant remains, in Seeley in prep

Giorgi, J, in prep b The plant remains, in Grainger et al in prep

Glasbergen, W, and Groenman-van Waateringe, W, 1974 *The pre-Flavian garrisons of Valkenburg ZH*, Amsterdam and London

Glob, P V, 1971 *The bog people*, London

Good, G L, Jones, R H, and Ponsford, M W (eds), 1991 *Waterfront archaeology: proceedings of the 3rd international conference, Bristol, 1988*, CBA Res Rep 74, London

Goodburn, D, 1991 A Roman timber-framed building tradition, *Archaeol J* 148, 182–204

Goodburn, D, 1993 Period 4, Norman 1066–1154: significant timbers, in McCann, 52–4

Goodburn, D, 1995 Beyond the posthole: notes on the stratigraphy and timber buildings from a London perspective, in Shepherd, 43–2

Goodburn, D, 1997 Aspects of the structural woodwork recorded during the Fennings Wharf Excavations, Southwark (FW84), unpub MoLAS archive rep

Graham, A H, 1978 The geology of north Southwark and its topographical development in the post-Pleistocene period, in Sheldon et al, 501–17

Graham, A H, 1988 District heating scheme, in Hinton, 27–54

Graham, A H, and Hinton, P, 1988 The Roman roads in Southwark, in Hinton, 19–26

Grainger, I, 1996 Three Quays House, Lower Thames Street,

London EC3, unpub MoLAS archive rep

Grainger, I, Falcini, P, and Philpotts, C, in prep Excavations at the Royal Navy Victualling Yard, East Smithfield, London, MoLAS Monogr Ser

Grant, A, 1988 Animal resources, in Astill and Grant, 149–87

Gray-Rees, L, 1996 Assessment report on the plant remains from Borough High Street, Southwark, unpub MoLAS archive rep

Greensmith, J T, and Tucker, E V, 1969 The origin of Holocene shell deposits in the Chenier Plain Facies of Essex, England, *Marine Geol* 7, 403–25

Greensmith, J T, and Tucker, E V, 1973 Holocene transgressions and regressions on the Essex coast and outer Thames estuary, *Geol en Mijnb* 52, 193–202

Greig, J, 1988 Plant resources, in Astill and Grant, 108–27

Greig, J, 1991 The British Isles, in Zeist et al, 299–334

Grew, F (ed), 1984 The finds, in Thompson et al, 33–148

Groos, G W (ed), 1981 *The diary of Baron Waldstein, a traveller in Elizabethan England*, London

Groves, C, and Hillam, J, 1988 The potential of non-oak species for tree-ring dating in Britain, in Slater and Tate, 567–79

Groves, C, and Hillam, J, 1997 Tree-ring analysis and dating of timbers, in Hurst, 121–6

Gwilt, C E, 1834 Crypt discovered in making the approach to new London Bridge ... in the parish of St Olave's Southwark, *Archaeologia* 25, 604–6

Hall, A R, and Kenwood, H K (eds), 1982 *Environmental archaeology in the urban context*, CBA Res Rep 43, London

Hammerson, M, 1988 Roman coins from Southwark, in Hinton, 417–26

Hammond, P W, 1995 *Food and feast in medieval England*, London

Harding, V A, 1990 The population of London, 1550–1700: a review of the published evidence, *London J* 15, 111–28

Harding, V A, 1995 Supplying London Bridge, 1380–1450, *Franco-Brit Stud* 20, 111–24

Harding, V, and Wright, L (eds), 1995 *London Bridge: selected accounts and rentals, 1381–1538*, London Rec Soc 31, London

Harrison, D F, 1992 Bridges and economic development 1300–1800, *Econ Hist Rev 2 ser* 45, 240–61

Harrison, D F, 1996 Bridges and communications in pre-industrial England, unpub DPhil thesis, Univ Oxford

Harvey, I M W, 1991 *Jack Cade's rebellion of 1450*, Oxford

Harvey, J H, 1944 *Henry Yevele c 1320–1400, the life of an English architect*, London

Hassall, J M, and Hill, D, 1970 Pont de l'Arche: Frankish influence on the west Saxon burh?, *Archaeol J* 127, 188–95

Hassall, M W C, and Tomlin, R S O, 1996 Roman Britain in 1995: II, Inscriptions, *Britannia* 27, 439–57

Haughey, F, 1999 The archaeology of the Thames: prehistory within a dynamic landscape, *London Archaeol* 9, 16–21

Hawksmoor, N, 1736 *A short historical account of London Bridge, with a proposal for a new stone bridge at Westminster*, London

Heal, A, 1931 Old London Bridge tradesmen's cards and tokens, in Home, 308–31

Hill, P V, 1967 Appendix II, Barbarous radiates in Britain, in Askew, G, *The coinage of Roman Britain*, London

Hill, D, 1976 London Bridge: a reasonable doubt, *Trans London Middlesex Archaeol Soc* 27, 303–4

Hill, C, Millett, M, and Blagg, T, 1980 *The Roman riverside wall and monumental arch in London, excavations at Baynard's Castle, Upper Thames Street, London, 1974–76* (ed T Dyson), London Middlesex Archaeol Soc Spec Pap 3, London

Hillam, J, 1992 Appendix 2, Tree-ring analysis of oak timbers, in Steedman et al, 143–73

Hillam, J, and Tyers, I, 1995 Reliability and repeatability in tree-ring dendrochronological analysis: tests using the Fletcher archive of panel-painting data, *Archaeometry* 37(2), 395–405

Hillam, J, Morgan, R A, and Tyers, I, 1987 Sapwood estimates and the dating of short ring sequences, in Ward, 165–85

Hinton, P (ed), 1988 *Excavations in Southwark 1973–76, Lambeth 1973–79*, London Middlesex Archaeol Soc and Southwark Archaeol Soc Joint Pub 3, London

Hinton, P, and Graham, A H, 1988 Roman roads in Southwark, in Hinton, 19–26

Hinton, P, and Thomas, R, 1997 The Greater London publication programme, *Archaeol J* 154, 196–213

HMSO, 1965 *Magna Carta* (trans and facs), London

HMSO, 1999 *Chronological tables of the statutes: Part 1, 1235–1962*, London

Hodgson, P, and Liegey, G M (trans), 1966 *The orchard of Syon*, Early Engl Text Soc 243, London

Holinshed, *Chronicles* Holinshed, R, 1586–7 *Chronicles of England, Scotland and Ireland: Vol 3, Chronicles from William the Norman to Queen Elizabeth*, London

Hollaender, A E J, and Kellaway, W (eds), 1969 *Studies in London history*, London

Home, G, 1931 *Old London Bridge*, London

Honeybourne, M B, 1969 The pre-Norman bridge in London, in Hollaender and Kellaway, 17–39

Hope, W St J, 1913 *Windsor Castle: Vol II*, Windsor

Hopkins, H J, 1970 *A span of bridges*, Newton Abbot

Horsman, V, Milne, C, and Milne, G, 1988 *Aspects of Saxo-Norman London: Vol I, Buildings and street development*, London Middlesex Archaeol Soc Spec Pap 11, London

Houghton, F T S, 1919 Parochial chapels of Worcestershire, *Birmingham Archaeol Soc Trans* 45, 23–114

Howard, F E, and Crossley, F H, 1917 *English church woodwork*, London

Howard, R E, Laxton, R R, and Litton, C D, 1997 Tree-ring analysis of timbers from Ware Priory, High Street, Ware, Hertfordshire, unpub Ancient Monuments Lab rep 84/97, Engl Heritage

Howe, G W, 1964 The dockyard at Northfleet, *Port London Authority Monthly* 39, no. 463, 152–5

Huff, C, 1997 Falconers, fowlers and nobles: the sport of falconry in Anglo-Saxon England, *Medieval Life* 7, 7–9

Huggins, P J, 1970 Excavation of a medieval bridge at Waltham Abbey, Essex, in 1968, *Medieval Archaeol* 14, 126–47

Hulme, E W, and Jenkins, R, 1895 Notes on London Bridge waterworks: Parts 1 and 2, *The Antiquary* 31(1), 261–5, and

31(2), 243–6

Hume, I N, 1977 *Early English Delftware from London and Virginia*, Colonial Williamsburg Occas Pap Archaeol 2, Charlottesville Va

Hunter, J (ed), 1833 *Magnus Rotulus Scaccarii de anno 31 Henrici I*, Rec Comm, London

Hurst, J G, 1961 The kitchen area of North Holt Manor, Middlesex, *Medieval Archaeol* 5, 211–99

Hurst, D (ed), 1997 *A multi-period salt production site at Droitwich: excavations at Upwich*, CBA Res Rep 107, York

Hurst, J G, Neal, D S, and van Beuningen, H J E, 1986 *Pottery produced and traded in north-west Europe 1350–1650*, Rotterdam Pap 6, Rotterdam

Hutton, C, 1772 *Principles of bridges*, Cambridge

Hyde, R, 1976 *A large and accurate map of the City of London by John Ogilby*, Kent

Hyde, R, 1981 *The A to Z of Georgian London*, Kent

Hyde, R, 1994 *A prospect of Britain – the town panoramas of Samuel and Nathaniel Buck*, London

Impey, E, and Keevill, G, 1998 The Tower of London moat: a summary history of the Tower moat, based on historical and archaeological research undertaken for the Tower Environs Scheme, unpub Historic Royal Palaces and Oxford Archaeol Unit rep

Ireland, S, 1996 *Roman Britain: a source book*, 2 edn, London

Jackson, C J, 1911 *History of English plate: Vol 2*, London

Jackson, P, 1971 *London Bridge*, London

Jackson, D A, and Ambrose, T M, 1976 A Roman timber bridge at Aldwincle, Northants, *Britannia* 7, 39–72

Jacob, E F, 1961 *The fifteenth century 1399–1485*, Oxford Hist England VI, Oxford

James, M R, 1933 *St George's Chapel Windsor – the woodwork of of the choir*, Windsor

Jannson, I, with Nosov, E N, 1992 The way to the east, in Roesdahl and Wilson, 74–83

Jervoise, E, 1930 *The ancient bridges of the south of England*, London

Johnsen, O A (ed), 1922 *Olafs saga hins helga*, Norske Historiske Kildeskriftkommission, Olso

Johnsen, O A, and Helgason, J (eds), 1941 *Den store saga om Olav den helloge*, Norsk Historisk Kjeldeskrift-Institutt, Olso

Johnson, D, 1969 *Southwark and the City*, London

Johnson, C, and Cronne, H A (eds), 1956 *Regesta Regum Anglo-Normannorum*, London

Jones, A K G, 1978 The fish remains, in Sheldon et al, 414–16

Jones, D M, 1980 *Excavations at Billingsgate Buildings 'Triangle', Lower Thames Street, 1974*, London Middlesex Archaeol Soc Spec Pap 4, London

Jones, G, Straker, V, and Davis, A, 1991 Early medieval plants use and ecology, in Vince 1991a, 347–85

Jónsson, F (ed), 1902–3 *Fagrskinna: Nóregs kononga tal*, Udg for Samfund til udgivelse af gammel Nordisk litteratur XXX, Copenhagen

Jónsson, F (ed), 1912 *Den norsk-islandske Skjaldedigting*, Copenhagen

Jørgensen, M S, 1993 And the bridge is falling down – but when?, *Newsl Wetland Archaeol Res Proj* 13, 8–10

Jørgensen, M S, 1997a The Viking age bridge at Ravning Enge – new studies, *Annu Proc Nat Mus Denmark*, 74–87

Jørgensen, M S, 1997b Borbyggere omkring år 1000, *Årbok Norsk Vegmuseum [Fåvang, Norway]* 1997, 52–65

Juddery, J Z, 1991 The Bridge Wardens' accounts: Appendix 1, in Brown, 19–25

Judge, S M, 1988 Sir Edward Banks – the legend and the facts, *Bygone Kent* 9, 515–20 (part 1), and 10, 601–7 (part 2)

Judson, S, and Kauffman, M E, 1990 *Physical geology*, New Jersey

Keene, D, and Harding, V, 1985 *A survey of sources for property holding in London before the Great Fire*, London Rec Soc 22, London

Keene, D, and Harding, V, 1987 *Historical gazetteer of London before the Great Fire: Vol 1, Cheapside* (microfiche), London

Kenward, H, and Williams, D (eds), 1979 *Biological evidence from the Roman warehouse in Coney Street*, Archaeol York 14 fasc 2, York

King, A C, 1984 Animal bones and the dietary identity of military and civilian groups in Roman Britain, Germany Gaul, in Blagg and King, 187–218

Kingsford, C L (ed), 1905 *Chronicles of London*, London

Kipling, G, 1982 The London pageants for Margaret of Anjou, *Medieval English theatre*, London

Kipling, G, 1998 *Enter the king: theatre, liturgy and ritual in the medieval civic triumph*, London

Kitching, C (ed), 1980 *London and Middlesex chantry certificates 1548*, London Rec Soc 16, London

Knight, W, 1831 Observations on the mode of construction of the present London Bridge, *Archaeologia* 23, 117–19

Knight, W, 1832 [letter], *Gentleman's Mag* 102(2), 98

Knight, W, 1834 Account of some antiquities discovered in excavating for the foundations of London Bridge; and of the ancient northern embankment of the Thames in its neighbourhood, *Archaeologia* 25, 600–2

Knight, W, et al (not named), 1832 A memoir of old London Bridge, with observations made during its demolition by W Knight, *Gentleman's Mag* 102(1), 201–6

Krogstad, M, and Schia, E, 1993 *Guide to Gamlebyen and medieval Olso*, Olso

Laing, S (trans), 1964 *Heimskringla: 1, The Olaf sagas by Snorri Sturluson*, London

Lakin, D, 1997 Summerton Way, Thamesmead SE28, London Borough of Bexley, unpub MoLAS archive rep

Lakin, D, 1999 A Romano-British site at Summerton Way, Thamesmead, London Borough of Bexley, *Archaeol Cantiana* 119, 311–41

Landsberg, S, 1995 *The medieval garden*, London

Langland, W, *Piers the ploughman* *Langland – Piers the ploughman* (ed J F Goodridge), 1959, London

Latham, R, and Matthews, W (eds), 1970a *The diary of Samuel Pepys: Vol 1 (1660)*, London

Latham, R, and Matthews, W (eds), 1970b *The diary of Samuel Pepys: Vol 2 (1661)*, London

Latham, R, and Matthews, W (eds), 1970c *The diary of Samuel Pepys: Vol 3 (1662)*, London

Latham R, and Matthews, W (eds), 1971 *The diary of Samuel Pepys: Vol 4 (1663)*, London

Latham, R, and Matthews, W (eds), 1972 *The diary of Samuel Pepys:*

Vol 7 (1666), London
Lawson, M K, 1993 *Cnut – the Danes in England in the early eleventh century*, London
Lepper, F, and Frere, S, 1988 *Trajan's column*, New Hampshire and Gloucester
Libois, R M, Hallet-Libois, C, and Roseaux, R, 1987 *Eléments pour l'identification des restes crâniens des poissons dulcaquicoles de Belgique et du nord de la France: 1, Anguilliformes, gasterosteiformes, cyprinodontiformes et perciformes*, Fiches d'Ostéologie Animale pour l'Archéologie Sér A Poissons 3, APDCA Juan-les-Pins, France
Lindley, P, 1995 *Gothic to Renaissance*, Stamford
Lipski, L L, and Archer, M, 1984 *Dated English Delftware*, London
Lloyd, R, 1995 Pre-Reformation music in the chapel of St Thomas the Martyr, London Bridge, unpub MMus dissertation, Royal Holloway Univ London
Lockyer, R, 1964 *Tudor and Stuart Britain 1471–1714*, London
Loengard, J S (ed), 1989 *London viewers and their certificates 1508–58*, London Rec Soc 26, London
Loyn, H R, 1962 *Anglo-Saxon England and the Norman Conquest*, London
Luard, H R (ed), 1866 *Annales Monasterii de Bermundeseia AD 1042–1432*, Rolls Ser 36(iii), London
Lyle, H M, 1950 *The rebellion of Jack Cade*, Hist Ass Publ G16, London
Mabey, R, 1972 *Food for free: a guide to the edible wild plants of Britain*, London
McCann, B (ed), 1993 Fleet Valley interim report, unpub MoLAS archive rep
McCormack, J (compiler), 1999 Provins, in *Vernacular Architecture Group Champagne Conference Handbook 1999*, 108–16, London
McDonnell, K, 1978 *Medieval suburbs of London*, London
MacGregor, A, 1975 *Bone, antler, horn and ivory: the technology of skeletal materials*, London
MacGregor, A, 1976 Bone skates: a review of the evidence, *Archaeol J* 133, 57–74
Maclagan, E, and Oman, C C, 1936 An English gold rosary of about 1500, *Archaeologia* 186, 1–22
McLynn, F, 1998 *1066 the year of three battles*, London
Madan, M, 1760, *The book of martyrs: containing an account of the suffering and death of Protestants in the reign of Queen Mary the First* (by J Foxe, revised M Madan), London
Malcolm, G, and Bowsher, D, with Cowie, R, in prep *Lundenwic: excavations at the Royal Opera House 1989–1998*, MoLAS Monogr Ser
Manley, L (ed), 1986 *London in the age of Shakespeare, an anthology*, London
Margary, I D, 1955 *Roman Roads in Britain: Vol 1*, London
Marsden, P, 1965 A boat of the Roman period discovered on the site of New Guy's House, Bermondsey, *Trans London Middlesex Archaeol Soc* 21(2), 118–31
Marsden, P, 1971 Archaeological finds in the City of London 1967–70, *Trans London Middlesex Archaeol Soc* 23(1), 1–14
Marsden, P, 1975 The excavation of a Roman palace site in London, 1961–1972, *Trans London Middlesex Archaeol Soc* 26, 1–102
Marsden, P, 1987 *The Roman forum site in London*, London
Marsden, P, 1994 *Ships of the port of London – 1st to 11th centuries AD*, Engl Heritage Archaeol Rep 3, London
Marsden, P, 1996 *Ships of the port of London – 12th to 17th centuries AD*, Engl Heritage Archaeol Rep 5, London
Massie, W M H, 1849–55 On a wooden bridge, found buried fourteen feet deep under the silt at Birkenhead, *J Architect Archaeol Hist Soc County, City Neighbourhood Chester* 1, 55–60 and 68–76
Meadow, I, 1996 Wollaston: the Nene Valley, a British Moselle?, *Current Archaeol* 150, 212–15
Meddens, F M, 1996 Sites from the Thames Estuary wetlands, England, and their Bronze Age use, *Antiquity* 70, 325–34
Merrifield, R, 1965 *The Roman city of London*, London
Merrifield, R, 1970 Roman London bridge: further observations on its site, *London Archaeol* 1, 186–7
Merrifield, R, 1983 *London, city of the Romans*, London
Merrifield, R, 1987 *The archaeology of ritual and magic*, London
Merrifield, R, and Sheldon, H, 1974 Roman London bridge: a view from both banks, *London Archaeol* 2, 183–91
Miller, L, 1982 Miles Lane, the early London waterfront, *London Archaeol* 4, 143–7
Miller, L, Schofield, J, and Rhodes, M, 1986 *The Roman quay at St Magnus House*, London Middlesex Archaeol Soc Spec Pap 8, London
Mills, P, 1996 The battle of London 1066, *London Archaeol* 8, 59–62
Milne, G, 1982 Further evidence for Roman London bridge?, *Britannia* 13, 271–6
Milne, G, 1985 *The port of Roman London*, London
Milne, G, 1988 Billingsgate study area: excavation summaries, in Horsman et al, 12–21
Milne, G (ed), 1992a *From Roman basilica to medieval market*, London
Milne, G (ed), 1992b *Timber building techniques in London c 900–1400*, London Middlesex Archaeol Soc Spec Pap 15, London
Milne, G, and Hobley, B (eds), 1981 *Waterfront archaeology in Britain and Europe*, CBA Res Rep 41, London
Milne, C, and Milne, G, 1982 *Medieval waterfront development at Trig Lane, London*, London Middlesex Archaeol Soc Spec Pap 5, London
Milne, G, and Milne, C, 1992 Catalogue of waterfront installations, in Milne 1992b, 23–77
Milne, G, Bateman, N, and Milne, C, 1984 Bank deposits with interest, *London Archaeol* 4, 395–400
Mitchell, R J, and Leys, D R, 1958 *A history of London life*, London
Moher, J G, 1987 John Torr Foulds (1742–1815): millwright and engineer, *Newcomen Soc Trans* 58, 59–73
Mommsen, T, and Meyer, P H (eds), 1905 *Theodosiani libri XVI cum constitutionibus Sirmondianis et leges novellae ad Theodosianum pertinentes: Vol I(i)*, Berlin
Mook, W G, 1986 Business meeting: recommendations/resolutions adopted by the 12th international radiocarbon conference, *Radiocarbon* 28, 799
Moorhouse, S, 1972 Medieval distilling apparatus of glass and pottery, *Medieval Archaeol* 16, 79–121

Morgan, A R, and Schofield, J, 1978 Tree-rings and the archaeology of the Thames waterfront in the City of London, in Fletcher, 223–38

Murray, P, 1996 Thames water habitable bridge competition – introduction, in Murray and Stevens, 135–53

Murray, P, and Stevens, M A (eds), 1996 *Living bridges*, Roy Acad exhibition catalogue, London

Myrvoll, S, 1991 Vågen and Bergen: the changing waterfront and the structure of the medieval town, in Good et al, 150–61

Newell, R W, 1994 Thumbed and sagging bases on English medieval jugs: a potter's view, *Medieval Ceram* 18, 51–8

Nunn, P D, 1983 The development of the Thames in central London during the Flandrian, *Trans Inst British Geogr* 8, 187–213

Nuttal, N, 2000 Royalty's lamprey returns to Thames, *The Times*, no. 66,886, 22 July, 8

O'Connor, C, 1993 *Roman bridges*, Cambridge

O'Kelly, M J, 1961 A wooden bridge on the Cashen river, co Kerry, *J Roy Soc Antiq Ir* 91, 135–52

Oleson, J P, 1988 The technology of Roman harbours, *Int J Naut Archaeol Underwater Explor* 17(2), 147–57

Oman, C C, 1930 *Catalogue of rings*, London

Opie, I, and Opie, P, 1955 *The Oxford nursery rhyme book*, Oxford

Opolovnikov, A, and Opolovnikova, Y, 1989 *The wooden architecture of Russia: houses, fortifications, churches*, New York

Orton, C R, 1988 Post-Roman pottery, in Hinton, 295–364

Orton, C R, and Pearce, J E, 1984 The pottery, in Thompson et al, 34–68

Orton, C, Orton, J, and Evans, P, 1974 Medieval and Tudor pottery, in Sheldon, 64–87

Oxley, J E, 1978 The medieval church dedications of the City of London, *Trans London Middlesex Archaeol Soc* 29, 117–25

Page, W, and Round, J H (eds), 1907 *The Victoria history of the county of Essex: Vol 2*, London

Parry, J, 1994 The Roman quay at Thames Exchange, London, *London Archaeol* 7, 263–67

Parsloe, G, 1928 Appendix 5, Notes on the site of the Roman London bridge at London, in RCHM, 192–4

Peacock, D P S, and Williams, D F, 1986 *Amphorae and the Roman economy*, London

Pearce, J E, 1992 *Post-medieval pottery in London, 1500–1700: Vol 1, Border wares*, London

Pearce, J E, 1993 The medieval and post-medieval pottery from excavations at Boston House, 90–94 Broad Street, London, unpub MoLAS archive rep

Pearce, J E, 1997 Medieval and later pottery from excavations at Fennings Wharf, Southwark, unpub MoLAS archive rep

Pearce, J E, in prep Medieval and post-medieval pottery from excavations at Bull Wharf, Upper Thames Street, London, in Ayre and Wroe-Brown

Pearce, J E, and Vince, A G, 1988 *A dated type series of London medieval pottery: Part 4, Surrey whitewares*, London Middlesex Archaeol Soc Spec Pap 10, London

Pearce, J E, Vince, A G, and Jenner, M A, 1985 *A dated type series of medieval pottery: Part 2, London type wares*, London Middlesex Archaeol Soc Spec Pap 6, London

Pearson, E, and Giorgi, J, 1992 Plant remains, in Cowan, 165–70

Pearson, G W, and Stuiver, M, 1986 High-precision calibration of the radiocarbon time scale, 500–2500 BC, *Radiocarbon* 28, 839–62

Pearson, G W, Pilcher, J R, Baillie, M G L, Corbett, D M, and Qua, F, 1986 High-precision C14 measurements of Irish oaks to show the natural C14 variations from AD 1840–5210 BC, *Radiocarbon* 28, 911–34

Pennant, T, 1813 *Some account of London*, 5 edn, London

Perring, D, and Roskams, S, with Allen, P, 1991 *The archaeology of Roman London: Vol 2, Early development of Roman London west of the Walbrook*, CBA Res Rep 70, London

Petrikovitz, H, von, 1952 Die Ausgrabungen in der Colonia Traiana, *Bonner Jahrb* 152, 138–57

Pevsner, N, 1978 *The buildings of England: Derbyshire*, 2 edn (rev L Williamson), Harmonsdworth

Pevsner, N, and Harris, J, 1989 *The buildings of England: Lincolnshire*, 2 edn (rev N Antram), Harmonsdworth

Pluis, J, 1997 *The Dutch tile, designs and names 1570–1930*, Nederlands Tegelmuseum Leiden

Porter, S, 1996 *The Great Fire of London*, Stroud

Potter, G, 1991 The medieval bridge and waterfront at Kingston-upon-Thames, in Good et al, 137–49

Pritchard, F, 1991 Small finds, in Vince 1991a, 120–278

Pryor, S, and Blockley, K, 1978 A 17th-century kiln site at Woolwich, *Post-Medieval Archaeol* 12, 30–85

Public Advertiser, 1761 [note], *The Public Advertiser* 26 Dec 1761

Quarrell, W H, and Mare, M (eds), 1934 *London in 1710 from the travels of Zacarias Conrad Von Uffenbach*, London

Rackham, D J, 1986 Assessing the relative frequency of species by the application of a stochastic model to a zooarchaeological database, in Wijngaarden-Bakker, 185–92

Rackham, O, 1986 *The history of the countryside*, London

Ray, Revd J, 1737 *Compleat collection of English proverbs*, London

RCHM (Roy Comm Hist Monuments), 1924 *Westminster Abbey: Vol 1*, London

RCHM (Roy Comm Hist Monuments), 1928 *An inventory of the historical monuments in London: Vol 3, Roman London*, London

RCHM (Roy Comm Hist Monuments), 1972 *An inventory of the historical monuments in York: Vol 2, The defences*, London

RCHME (Roy Comm Hist Monuments Engl), 1996 Beaumont Quay, Beaumont-Cum-Moze, Essez, NMR number TM 12 SE 34, request survey April 1996, RCHME unpub rep

Read, H, and Tonnochy, A B, 1928 *Catalogue of silver plate*, London

Rees, A, 1819 *The cyclopaedia or universal directory: Vol 4*, London

Renfrew, J, 1985 *Food and cooking in Roman Britain: history and recipes*, London

Report, 1796 Report from the Committee appointed to enquire into the best mode of providing sufficient accommodation for the increased trade and shipping of the Port of London, printed in *Reports from Committee of the House of Commons: Vol 14, 1793–1802*, 1803, 267–443, London

Report, 1801 Report from the Select Committee upon the

improvement of the Port of London 3 June 1801, printed in *Reports from Committee of the House of Commons: Vol 14, 1793–1802*, 1803, 604–35, London

Rhodes, M, 1991 The Roman coinage from London Bridge and the development of the City and Southwark, *Britannia* 19, 179–90

Richmond, I, 1982 *Trajan's army on Trajan's column*, London

Rielly, K, 1998 The animal bones from Fennings Wharf, Southwark, unpub MoLAS archive rep

Rielly, K, in prep a The animal bones, in Grainger, I, Excavations at Battlebridge Lane in 1995: medieval and early post-medieval development along Tooley Street, Southwark, *Surrey Archaeol Collect*

Rielly, K, in prep b The Roman period animal bones from Winchester, in Yule

Rigold, S E, 1975 Structural aspects of medieval timber bridges, *Medieval Archaeol* 19, 48–91

Riley, H T (ed and trans), 1861 *Liber albus: the white book of the City of London ... translated from the original Latin and Anglo-Norman*, London

Riley, H T, 1863 *Chronicles of the mayors and sheriffs*, London

Rites of Durham *Rites of Durham* (ed J T Fowler), 1903, Surtees Soc 107, Durham

Rivet, A L F, and Smith, C, 1979 *The place-names of Roman Britain*, London

Rixson, D, 1974 Animal bones, in Sheldon, 108–11

Roberts, M, 1994 *Durham*, London

Robertson, A, 1925 *The laws of the kings of England from Edmund to Henry I*, London

Roesdahl, E, and Wilson, D M (eds), 1992 *From Viking to Crusader: the Scandinavians and Europe 800–1200*, 22nd Council of Europe exhibition catalogue, Sweden

Rowlands, M L J, 1994 *Monnow bridge and gate*, Stroud

Rowsome, P, 1999 The Huggin Hill baths and bathing in London: barometer of the town's changing circumstances?, in *Roman baths and bathing* (eds J DeLaine and D Johnston), J Roman Archaeol Suppl Ser 37, 262–77

Ruddock, T, 1979 *Arch bridges and their builders 1735–1835*, Cambridge

Rushforth, G M N, 1936 *Medieval Christian imagery*, Oxford

Ryan, P, 1996 *Brick in Essex*, London

SAEC (Southwark Archaeological Excavation Committee), 1973 Excavations at New Hibernia Wharf, *London Archaeol* 2, 99–103

St Augustine's Ramsgate [the Benedictine monks of St Augustine's Ramsgate], 1989 *A book of saints*, London

Salzman, L F, 1952 *Building in England down to 1450: a documentary history*, Oxford

Sankey, D, 1998 Cathedrals, granaries, and urban vitality in late Roman London, in Watson 1998b, 78–82

Savill, P, and Craven, A, 1999 The bridge chapel at St Ives, Cambridgeshire, *Ecclesiology Today* 16, 8–11

Sawyer, P H, 1968 *Anglo-Saxon charters: an annotated list and bibliography*, Roy Hist Soc Guides Handbooks 8, London

Sawyer, P H, 1987 Cheshire Domesday – translation of text, in Elrington, 342–70

Schofield, J, 1984 *The building of London*, London

Schofield, J, 1994 Saxon and medieval parish churches in the City of London, *Trans London Middlesex Archaeol Soc* 45, 231–45

Schofield, J (ed), with Maloney, C, 1998 *Archaeology in the City of London 1907–91: a guide to the records of excavations by the Museum of London*, MoL Archaeol Gazetteer Ser 1, London

Schofield, J, and Dyson, T, 1980 *Archaeology of the City of London*, London

Schofield, J, Allen, P, and Taylor, C, 1990 Medieval buildings and property development in the area of Cheapside, *Trans London Middlesex Archaeol Soc* 41, 39–237

Schweingruber, F H, 1982 *Microscopic wood anatomy*, 2 edn, Teufen, Switzerland

Schweingruber, F H, 1988 *Tree rings*, Dordrecht

Schweingruber, F H, 1990 *Anatomy of European woodlands*, Berne

Schweingruber, F H, 1993 *Trees and wood in tree-ring dendrochronology*, Berlin

Schweizerisches Landesmuseum, 1980–1 *Der heiligen Benedikt* 480–1980, exhibition catalogue, Zurich

Seeley, D, with Carlin, M, and Phillpotts, C, in prep *Medieval mansions of Southwark: Winchester Palace*, MoLAS Monogr Ser

Seip, D A (ed), 1929 *Den legendariske Olavssaga og fagrskinna*, Norske Videnskaps-Akademi 1 Oslo, II Hist-Filos Klasse 2, Oslo

Seward, D, 1997 *Richard III: England's black legend*, 2 rev edn, London

Seymour, R, 1734 *A survey of London and Westminster, borough of Southwark and parts adjacent: Vol 1*, London

Sharpe, R, 1890 *Calendar of wills proved and enrolled in the Court of Hustings, London AD 1258–1688*, 2 vols, London

Sheldon, H, 1974 Excavations at Toppings and Sun Wharves, Southwark, 1970–72, *Trans London Middlesex Archaeol Soc* 25, 1–116

Sheldon, H L, 1978 The 1972–74 excavations: their contribution to Southwark's history, in Sheldon et al, 11–49

Sheldon, H L, Bird, J, Graham, A, and Townend, P (eds), 1978 *Southwark excavations 1972–74*, 2 vols, London Middlesex Archaeol Soc and Surrey Archaeol Soc Joint Pub 1, London

Shepherd, L (ed), 1995 *Interpreting stratigraphy 5: proceedings of a conference held at Norwich Castle 16 June 1994*, Norwich

Shepherd, L (ed), 1998 *Norfolk Archaeol Unit Annu Rev 1997–8*, Norwich

Sidell, J, Cotton, J, Rayner, L, and Wheeler, L, in prep The topography and prehistory of north Southwark and Lambeth, MoLAS Monogr Ser

Simco, A, and McKeague, P, 1997 *Bridges of Bedfordshire*, Bedfordshire Archaeol Occas Monogr 2, Bedfordshire

Slater, E A, and Tate, J O (eds), 1988 *Science and archaeology, Glasgow 1987*, BAR Brit Ser 196, Oxford

Sloane, B, Swain, H, and Thomas, C, 1995 The Roman road and the river regime: archaeological investigations in Westminster and Lambeth, *London Archaeol* 7, 359–70

Smeaton, J, 1763 Report on improving and enlarging London Bridge, reprinted as Appendix 5B, in Second Report of the House of Commons Select Committee upon the improvement of the port of London (1799), printed in *Reports from Committee of the House of Commons: Vol 14, 1793–1802*,

1803, 461–552, London

Smith, C R, 1840 On some Roman bronzes discovered in the bed of the Thames in January 1837, *Archaeologia* 28, 38–46

Smith, C R, 1841 On the Roman coins discovered in the bed of the Thames, near London Bridge from 1834 to 1841, *Numis Chron* 4, 147–68, 187–94

Smith, C R, 1846 Roman London, *Archaeol J* 1, 108–17

Smith, C R, 1854 *Catalogue of the Museum of London antiquities*, London

Smuts, R M, 1989 Public ceremony and royal charisma: the English royal entry in London, 1485–1642, in Beier and Cannadine, 65–93

Spencer, B W, 1974 The lead ampulla, in Sheldon, 113–15

Spencer, B, 1978 King Henry of Windsor and the London pilgrim, in Bird et al, 235–64

Spencer, B, 1996a Appendix 5, Expenditure on ship buildings and repair by London Bridge, 1382–98, in Marsden 1996, 209–12

Spencer, B, 1996b Appendix 6, Expenditure on cargoes of stone by London Bridge, in Marsden 1996, 212–13

Spencer, B, 1998 *Pilgrim souvenirs and secular badges*, HMSO Medieval Finds from Excavations in London Ser 7, London

Stace, C A, 1991 *New flora of the British Isles*, London

Steedman, K, Dyson, T, and Schofield, J, 1992 *Aspects of Saxo-Norman London: Vol 3, The bridgehead and Billingsgate to 1200*, London Middlesex Archaeol Soc Spec Pap 14, London

Steinman, D B, and Watson, S R, 1941 *Roman bridgebuilders and their bridges*, New York

Stenton, F M, 1947 *Anglo-Saxon England*, 2 edn, Oxford

Stenton, F M, with Butler, H E, 1934 *Norman London: an essay and William Fitz Stephen's description*, Hist Ass Leaflets 93–4, London

Stott, P, 1991 Saxon and Norman coins from London, in Vince 1991a, 279–325

Stow, J, 1603 *A survey of London* (ed C L Kingsford), 2 vols, 1908 repr 1971, Oxford

Straker, V, 1984 First and second century carbonised grain from Roman London, in Zeist and Casparie, 323–9

Stroud, D, 1971 *George Dance architect 1741–1825*, London

Stuiver, M, and Reimer, P J, 1986 A computer program for radiocarbon age calculation, *Radiocarbon* 28, 1022–30

Stuiver, M, Pearson, G W, and Braziunas, T, 1986 Radiocarbon age calibration of marine samples back to 9000 cal yr BP, *Radiocarbon* 28, 980–1021

Sunter, N J, 1976 The bridge: a discussion, in Jackson and Ambrose, 47–54

Symonds, R P, and Groves, J, 1997 The Roman pottery from Fennings Wharf, Southwark, unpub MoLAS archive rep

Symonds, R P, and Tomber, R S, 1991 Late Roman London: an assessment of the ceramic evidence from the City of London, *Trans London Middlesex Archaeol Soc* 42, 59–99

Tatton-Brown, T, 1974 Excavations at the Custom House site, City of London, 1973, *Trans London Middlesex Archaeol Soc* 25, 117–219

Tatton-Brown, T, 1991 Medieval building stone at the Tower of London, *London Archaeol* 6, 361–6

Taylor, W, 1833 *Annals of St Mary Overy*, London

Taylor, K, 1998 A pious undertaking: the retrieval of a medieval bridge chantry, *Ecclesiology Today* 16, 8–11

Taylor, F, and Roskell, J S (eds), 1975 *Gesta Henrici Quinti – the deeds of Henry V*, Oxford

Taylor-Wilson, M, 1990 Preliminary report on the excavations carried out in advance of the Guy's Hospital phase III development SE1, unpub MoL archive rep

Telford, T, 1823 *Report on the effects which will be produced on the river Thames by the rebuilding of London Bridge*, copy in CoLRO, PD96.7

Thomas, A H, and Thornley, I, 1983 (1938), *The great chronicle of London*, repr, London

Thomas, C, and Rackham, J (ed), 1996 Bramcote Green, Bermondsey: a Bronze Age trackway and palaeo-environmental sequence, *Proc Prehist Soc* 62, 221–53

Thompson, F H (ed), 1980 *Archaeology and coastal change*, Soc Antiq London Occas Pap 1, London

Thompson, A, Grew, F, and Schofield, J, 1984 Excavations at Aldgate 1974, *Post-Medieval Archaeol* 18, 1–148

Thompson, A, Westman, A, and Dyson, T (eds), 1998 *Archaeology in Greater London 1965–90: a guide to the records of excavations by the Museum of London*, MoL Archaeol Gazetteer Ser 2, London

Thomson, R, 1827 *Chronicles of London Bridge*, London

Tyers, I, 1988a The prehistoric peat layers (Tilbury IV), in Hinton, 5–12

Tyers, I, 1988b Environmental evidence from Southwark and Lambeth, in Hinton, 443–77

Tyers, I, 1991 Interim report on timbers from the Horsefair site, unpub MoLAS archive rep (dendro 6/91)

Tyers, I, 1993 Project 53a: medieval London Bridge dendrochronology research archive and updated project design, unpub MoLAS archive rep (dendro 7/93)

Tyers, I, 1994 Tree-ring dendrochronology of Roman and early medieval ships, in Marsden, 201–9

Tyers, I, 1995 Report on the timbers from Regis House (KWS94): 1, Timbers from the perimeter trenches, unpub MoLAS archive rep (dendro 3/95)

Tyers, I, 1997a Updated dendrochronology archive for MoLAS project 53a: medieval London Bridge, unpub ARCUS (Archaeol Res Consultancy Univ Sheffield, Res Sch Archaeol) rep 257

Tyers, I, 1997b Tree-ring analysis of timbers from the excavations of the Tower of London moat 1996–7, unpub ARCUS (Archaeol Res Consultancy Univ Sheffield, Res Sch Archaeol) rep 293

Tyers, I, in prep The dendrochronology of beech and elm from medieval excavations in England and Wales

Vince, A G, 1982 Medieval and post-medieval Spanish pottery from the City of London, in Freestone et al, 135–44

Vince, A G, 1985 Saxon and medieval pottery in London: a review, *Medieval Archaeol* 29, 25–93

Vince, A, 1990 *Saxon London: an archaeological investigation*, London

Vince, A (ed), 1991a *Aspects of Saxo-Norman London: 2, Finds and environmental evidence*, London Middlesex Archaeol Soc Spec Pap 12, London

Vince, A G, 1991b Early medieval London: refining the

chronology, *London Archaeol* 6, 263–71

Vince, A, and Jenner, A, 1991 The Saxon and early medieval pottery of London, in Vince 1991a, 19–119

Vitruvius *Vitruvius: the ten books on architecture* (trans M H Morgan), 1914 repr 1960, New York

Walker, R J B, 1979 *Old Westminster Bridge – the bridge of fools*, Newton Abbot

Ward, R G W (ed), 1987 *Applications of tree-ring studies: current research in dendrochronology and related areas*, BAR Int Ser 333, Oxford

Ward Perkins, J B, 1940 *London Museum medieval catalogue*, London

Wardle, A, 1996 The accessioned finds from Fennings Wharf, unpub MoLAS archive rep

Watkin, W T, 1886 *Roman Cheshire, or a description of the Roman remains in the county of Chester*, Liverpool

Watson, B, 1997a Old London Bridge in Surrey today, *Surrey Archaeol Soc Bull* 315, 7–9

Watson, B, with Bartkowiak, R, and Dyson, T, 1997b Old London Bridge in Kent today, *Dartford Antiq Hist Soc Newsl* 34, 13–17

Watson, B, with Dyson, T, 1997c London Bridge is broken down, in Boe and Verhaeghe, 311–27

Watson, B, 1998a Dark earth and urban decline in late Roman London, in Watson 1998b, 100–6

Watson, B (ed), 1998b *Roman London, recent archaeological work*, *J Roman Archaeol* Monogr Suppl Ser 24, Portsmouth, Rhode Island

Watson, B, and Tyers, I, 1995 Wood and water or rebuilding London Bridge, in Shepherd, 61–6

Weatherill, J, 1963 The eighteenth-century Rievaulx bridge and its medieval predecessor, *Yorkshire Archaeol J* 161, 71–81

Webster, G, 1979 *The Roman imperial army of the first and second centuries*, London

Webster Smith, B (ed), 1956 *The wonderful story of London*, 3 edn, London

Welch, C, 1894 *History of Tower Bridge*, London

West, B, 1982 A note on bone skates from London, *Trans London Middlesex Archaeol Soc* 33, 303

West, B, 1995 The case of the missing victuals, *Historical Archaeol* 29, 20–42

Westfehling, U (ed), 1982 *Die Messe Gregors des Grossen*, Cologne

Wheeler, R E M, 1927 *London and the Vikings*, London Mus catalogue 1, London

Wheeler, A, 1979 *The tidal Thames*, London

White, W, in prep The cremations, in Sidell et al

Wijngaarden-Bakker, van L H (ed), 1986 *Database management and zooarchaeology*, *PACT (J European Study Grp Physical, Chemical, BiologicalMathematical Techniques applied to Archaeology)* Res Vol 40, Amsterdam

Wilkinson, K, 1997 An investigation into the geoarchaeology of the foreshore deposits at Bull Wharf, unpub MoLAS archive rep

Wilkinson, K, in prep The geoarchaeology of the Thames foreshore at Bull Wharf, in Ayre and Wroe-Brown

Wilkinson, T J, and Murphy, P, 1986 Archaeological survey of an intertidal zone: the submerged landscape of the Essex coast, England, *J Fld Archaeol* 13, 177–93

Willcox, G, 1977 Exotic plants from Roman waterlogged sites in London, *J Archaeol Sci* 4, 269–82

Williams, G A, 1963, *Medieval London from commune to capital*, Univ London Hist Stud 11, London

Williams, D, 1979 The plant remains, in Kenward and Williams, 52–62

Wilson, C A, 1973 *Food and drink in Britain: from the Stone Age to recent times*, London

Wilson, D M, 1976 Craft and industry, and the Scandinavians in England, in Wilson, D M (ed), *The archaeology of Anglo-Saxon England*, 253–81, 393–403, Cambridge

Wilson, K, and White, D J B, 1986 *The anatomy of wood: its variability and diversity*, London

Wright, L, 1996 *Sources of London English: medieval Thames vocabulary*, London

Yates, N, and Gibson, J M (eds), 1994 *Traffic and politics: the construction and management of Rochester Bridge*, Kent Hist Ser 1, Rochester Bridge Trust and Woodbridge

Young, S, Clark, J, and Barry, T B (eds), 1984 Fennings Wharf, Southwark, in Medieval Britain and Ireland in 1983, *Medieval Archaeol* 28, 203–65

Yule, B, 1988 Natural topography of north Southwark, in Hinton, 13–17

Yule, B, 1989 Excavations at Winchester Palace, Southwark, *London Archaeol* 6, 31–9

Yule, B, in prep Roman Southwark: riverside development in the north-west quarter, MoLAS Monogr Ser

Yule, B, and Rankov, N B, 1998 Legionary soldiers in 3rd-century Southwark, in Watson 1998b, 67–77

Zeist, W van, and Casparie, W A (eds), 1984 *Plants and ancient man*, Rotterdam

Zeist, W van, Wasylikowa, K, and Behre, K (eds), 1991 *Progress in old world palaeoethnobotany: a retrospective view on the occasion of 20 years of the international work group for palaeoethnobotany*, Rotterdam

INDEX

Compiled by Susanne Atkin

Page numbers in **bold** refer to illustrations